D1802617

Particles and Fields–1979
(APS/DPF Montreal)

AIP Conference Proceedings
Series Editor: Hugh C. Wolfe
Number 59
Particles and Fields Subseries, No. 19

Particles and Fields–1979
(APS/DPF Montreal)

Editors
B. Margolis and D.G. Stairs
McGill University

American Institute of Physics
New York 1980

L.C. Catalog Card No. 80-66631
ISBN 0–88318–158–4
DOE CONF- 7910116

WELCOMING ADDRESS BY PRINCIPAL D. L. JOHNSTON
TO THE ANNUAL MEETING OF THE AMERICAN PHYSICAL SOCIETY,
DIVISION OF PARTICLES AND FIELDS,
POLLACK CONCERT HALL, MCGILL UNIVERSITY

It is indeed an honour that you have chosen McGill to host the Annual Meeting of the Division of Particles and Fields of the American Physical Society this year. A good fraction of our Physics Department is in fact engaged in research in Elementary Particle Physics and our university has a serious dedicated commitment to this exciting and important forefront area of science.

We at McGill are not alone in supporting this meeting. We have in fact been joined by the Department of Education of the Province of Quebec, The Canadian Association of Physicists, and the National Sciences and Engineering Research Council, the Granting Agency of our Federal Government. We are all indebted and proud to have this important meeting in Canada for the first time.

I understand that there is much excitement in your field of science at present and that impressive new results from the Laboratories and Theory Groups in both North American and Europe are to be presented here. We at McGill wish you much success in the interpretation of these results with a view to moving science ahead to improve our understanding of the natural world.

Welcome to all of you and we do hope you enjoy your stay with us.

FOREWORD

The Annual Meeting of The Division of Particles and Fields of The American Physical Society was held October 25, 26 and 27, 1979 at McGill University, Montreal. This was the first time the Division met in Canada and there were approximately 300 physicists in attendance. The theme of the program was recent developments in experimental and theoretical high energy physics. The organizers selected some twenty speakers to present papers which either surveyed an interesting area or described some new, important piece of research. The talks were uniformly outstanding.

Along with our colleagues on the organizing committee, Y. Afek, S. Conetti, R. Henzi, C. Leroy, T. F. Morris, D. Ryan and P. Valin, we wish to acknowledge the honor we feel at having been selected to host this conference. Our secretaries, especially Betty Desjardins and Charlotte Benabdallah from Special Events have been indispensible. We thank them as well as our other colleagues and research associates and graduate students who helped with the various aspects of the meeting.

We were pleased to see our colleagues from so many centers of high energy physics in attendance. The success of the meeting was to a great extent the result of their interest and participation.

Bernard Margolis
D. G. Stairs

Montreal
February 20, 1980

TABLE OF CONTENTS

Chapter 4. Theoretical Papers

NUCLEON STRUCTURE FUNCTIONS

J. Steinberger
CERN, Geneva, Switzerland

INTRODUCTION

This talk is addressed to non-experts in the field. I will try to give an overview of the types of structure functions which can be measured at present, how they may be measured, some of the latest experimental results, as well as the relevance of these results to the quark parton model (QPM) and quantum chromodynamics (QCD).

The nucleon structure functions are defined by the following expression for lepton (charged and neutral) inelastic scattering on nuclei which follow from the Feynman diagram below:

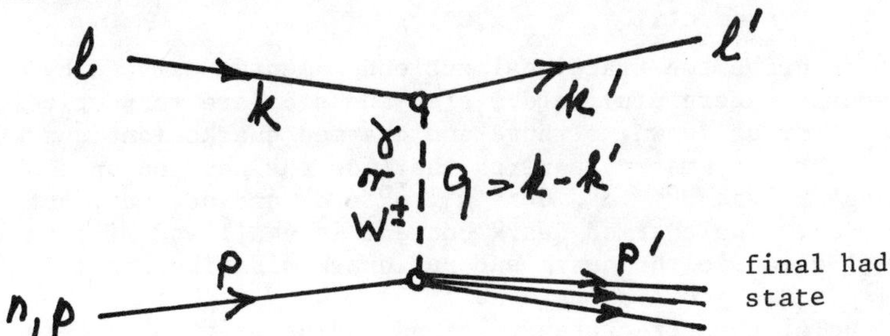

final hadronic state

For charged lepton scattering the intermediate particle is a photon. For neutrino scattering it may be an intermediate boson; in any case it is assumed that the weak current is of the V-A form, and that the mass of the $W^{\pm}$ is so large that $Q^2/M_W^2 = -q^2/M_W^2$ can be neglected. The corresponding cross sections, if polarizations are averaged or summed over, can be written:

$$\frac{d^2\sigma}{dxdy} = \left.\begin{matrix} \frac{4\pi\alpha^2}{Q^2} \times \\ G^2 \times \end{matrix}\right\} \frac{EM}{\pi} \left\{ \left[1 + (1-y)^2 - \frac{xyM}{E} \right] F_2(x,Q^2) \right.$$
$$+ y^2 \left[2 \times F_1(x,Q^2) - F_2(x,Q^2) \right]$$
$$\left. \begin{matrix} 0 \\ + \\ - \end{matrix} \left[1 - (1-y)^2 \right] xF_3(x,Q^2) \right\} \qquad (1)$$

Here the x^2 and G^2 coupling constant refer to charged leptons and neutrinos respectively, and the term in xF_3 is absent for charged leptons, and has opposite sign for neutrinos and antineutrinos. Also:

$$Q^2 \equiv -q^2$$
$$x \equiv \frac{Q^2}{2MyE} \qquad 0 \le x \le 1$$
$$y \equiv \frac{\nu}{E} \qquad 0 \le y \le 1$$
$$\nu = \frac{p \cdot q}{M} = E_h - M$$
$$E = \text{Lab. energy of incident lepton}$$

SSN:0094-243X/80/590001-24$1.50

E_h = Lab. energy of outgoing hadron system

M = mass of the proton.

Charged lepton scattering is described by the two structure functions xF_1 and F_2, while neutrino and antineutrino scattering requires the three functions xF_1, F_2 and xF_3. If we respect the fact that charged lepton, neutrino and antineutrino structure functions are not in general the same, and neither are those for neutrons and protons, then there are altogether 16 structure functions.

In the QPM these may be described in terms of six quark distributions:

$$\begin{aligned} &u(x,Q^2) \quad \bar{u}(x,Q^2) \\ &d(x,Q^2) \quad \bar{d}(x,Q^2) \\ &s(x,Q^2) = \bar{s}(x,Q^2) \\ \text{and } &c(x,Q^2) = \bar{c}(x,Q^2). \end{aligned}$$

These functions describe the fractional nucleon momentum carried by the different quarks. Here $u(\bar{u})$, $d(\bar{d})$, $s(\bar{s})$ and $c(\bar{c})$ are respectively the distributions for up, down, strange and charmed quarks (antiquarks) in the proton. Charge symmetry requires that for the neutron up and down quarks change roles: $u^p = d^u$, $d^p = u^u$, $\bar{u}^p = \bar{d}^u$ and $\bar{d}^p = \bar{u}^u$, but $s^p = s^u$ and $c^p = c^u$. The charmed quark content is small and will be ignored. We will also use the quark and antiquark distributions: $q = u + d + s + c$ and $\bar{q} = \bar{u} + \bar{d} + \bar{s} + \bar{c}$.

For I = 0 nuclei the structure functions can be written in terms of the quark distributions as follows:

$$F_2^{\bar{\nu}} = F_2^{\nu} = \frac{18}{5} F_2^{\ell\pm} = 2xF_1 = q + \bar{q} \tag{2}$$

$$xF_3^{\nu} = q - \bar{q} + 2s\,, \quad xF_3^{\bar{\nu}} = q - \bar{q} - 2s \tag{3}$$

Relation (2) contains the Callan-Gross relation[1)] $2xF_1 = F_2$, which for $E >> M$ is equivalent to the absence of longitudinal currents for spin ½ quarks.

For I = 0 nuclei the cross sections can then be written:

$$\frac{d^2\sigma^{\ell\pm}}{dxdy} = \frac{4\pi\alpha^2}{Q^2}\frac{EM}{2\pi}\left[1 + (1-y)^2\right](q+\bar{q}) \tag{1'a}$$

$$\frac{d^2\sigma^{\nu}}{dxdy} = \frac{G^2EM}{2\pi}\left[(q+s) + (1-y)^2\ (\bar{q}-\bar{s})\right] \tag{1'b}$$

$$\frac{d^2\sigma^{\bar{\nu}}}{dxdy} = \frac{G^2EM}{2\pi}\left[(\bar{q}+\bar{s}) + (1-y)^2\ (q-s)\right] \tag{1'c}$$

It can be seen that the total quark distribution $F_2 = q+q$ is measured in charged lepton interactions, as well as for the <u>sum</u> of neutrino and antineutrino scattering.

The <u>valence</u> structure function $xF_3 = q-\bar{q}$ is measured by taking the <u>difference</u> of neutrino and antineutrino cross section. Furthermore the antiquark <u>sea</u> $q + \bar{s}$ is measured directly in antineutrino interactions at large y. The strange sea structure function $s(x) = \bar{s}(x)$ is directly measured in like-sign dimuon production by antineutrinos.

In this process the extra muon is the result of charm decay, and the charmed quark, according to the GIM mechanism[2] (which has substantial experimental verification), is produced dominantly on strange antiquarks.

In the QPM the fact that the nucleon has three valence quarks is reflected in the sum rule[3]

$$\int_0^1 xF_3 \frac{dx}{x} = \int_0^1 (q-\bar{q}) \frac{dx}{x} = 3. \qquad (4)$$

We will see that the data give strong support to the QPM.

The naive QPM predicts that the structure functions are independent of Q^2 for Q^2 large. This however is modified in the strong interaction theory QCD of coloured quarks and gluons; logarithmic scaling violations are predicted. The observation of such violations in deep inelastic scattering experiments has been a big boost to QCD, and it is now a key problem, both experimentally and theoretically, to achieve a critical, quantitative confrontation of theory and experiment.

In the following I have chosen the experimental examples whereever I thought best; no attempt has been made to be complete.

LONGITUDINAL CURRENTS

How large are the longitudinal currents, or how much is the Callan-Gross relation violated?

$$\frac{\sigma_L}{\sigma_T} \equiv R = \frac{F_2(1 + \frac{2xM}{yE}) - 2xF_1}{2xF_1} \xrightarrow[E \gg M]{} \frac{F_2 - 2xF_1}{2xF_1}.$$

The experimental determinations of σ_L/σ_T rest on measuring the y distributions of the charged lepton cross section or of the sum of neutrino and antineutrino cross sections, and finding the amount of a y^2 term superposed on the dominant $1 + (1-y)^2$ term, after correction for a small y^2 term due to the strange sea (Eq. 1). Some published results are given in Table I.

TABLE I. Experimental determinations of R

Experiment	Reference	Beam	Energy	R
SLAC	4,5	$e^{\pm}$	∿ 20 GeV	0.21 ± 0.10
CHIF	6	$\mu^{\pm}$	∿200 GeV	0.52 ± 0.16
GGM	7	ν	∿100 GeV	0.32 ± 0.15
BEBC	7	ν	∿100 GeV	0.11 ± 0.14
CDHS	8	ν	∿100 GeV	-0.03 ± 0.05
CDHS	9	ν	∿100 GeV	0.03 ± 0.11

It can be seen that there is considerable variation in the reported values. In part this may be due to the difficulty of the experimental problem, in part it may be because the data are obtained in different domains of x and Q^2. Some of the results have not been radiatively corrected, some have not been corrected for the strange sea. All the neutrino results, except the last entry in Table I, suffer from the fact that the y variation is coupled to Q^2 variation:

that is low values of y have low Q^2 and high y have high Q^2, so that the small scaling violations may be mixed up with the y^2 term. The last CDHS analysis avoids this difficulty, but at the expense of increasing the uncertainty. The CDHS y distributions which underlie the analysis are shown in Fig. 1. The CHIF results for R are shown in more detail in Figs. 2 and 3, where, subject to the large uncertainties, it would seem that R varies rapidly with x and Q^2, being large at very small x and Q^2, and dropping to quite small values at larger x and Q^2. This is qualitatively expected in QCD. Better data and comparison with QCD are both possible and desirable.

A ZOO OF STRUCTURE FUNCTIONS

The following figures show various structure functions. In each case, these are averaged over some domain of incident energy; Q^2 variations are ignored. In general, the values at smaller x are obtained at smaller Q^2 and vice versa. We are interested here in comparing the x distributions of different structure functions. In Figs. 4-7 the results for the total quark distribution $F_2(x)$, the valence quark distribution $xF_3(x)$, the $\bar{q}(x) + \bar{s}(x)$ sea distribution, and the strange sea distribution $\bar{s}(x)$ are given as obtained by the CDHS collaboration[8-10] on an iron target. In Fig. 8 the SLAC results[5] for protons and neutrons for $F_2(x)$ and their difference, are shown. In Fig. 9 deuterium bubble chamber neutrino results[11] on F_2 for protons and neutrons, and finally, in Fig. 10 a recent result[12] on the pion structure function obtained by applying the Drell-Yan mechanism to muon pair production by pions, are given.

The various structure functions are quite different. $\bar{q}(x)$ and $\bar{s}(x)$ can be approximately fitted with $(1-x)^n$, with $n \simeq 6$-10. Recent results[10] indicate that $s(x)$ is approximately 15% narrower than $\bar{q}(x)$ for the same Q^2. xF_3 can be fitted reasonably with $\sqrt{x}(1-x)^n$ with $n \simeq 3$-4. Also, $xF_3(x)$ is consistent with approaching zero as x goes to zero, in agreement with the expectations of the QPM. $F_2{}^p(x)$ and $F_2{}^n(x)$ are different. Notice that $F_2{}^{p,\ell\pm} \neq F_2{}^{p,\nu}$ and $F_2{}^{n,\ell\pm} \neq F_2{}^{n,\nu}$ (with obvious notation, I hope); in the QPM:

$$F_2{}^{p,\ell\pm} = \frac{4}{9}(u+\bar{u}) + \frac{1}{9}(d+\bar{d})$$
$$F_2{}^{p,\nu} = d + \bar{u}$$
$$F_2{}^{u,\ell\pm} = \frac{1}{q}(u+\bar{u}) + \frac{4}{9}(d+\bar{d})$$
$$F_2{}^{u,\nu} = u + \bar{d}.$$

Nevertheless, it can be seen that the neutrino and charged lepton results for $F_2{}^p$ and $F_2{}^u$ are in good agreement, since the curves fitting the neutrino data of Fig. 9 marked FF are based on the fits obtained by Field and Feynman[13]: $u(x) \propto \sqrt{x}(1-x)^3$ and $d(x) \propto \sqrt{x}(1-x)^4$, to the SLAC data. The pion structure functions extend appreciably more to large x than those of the nucleon.

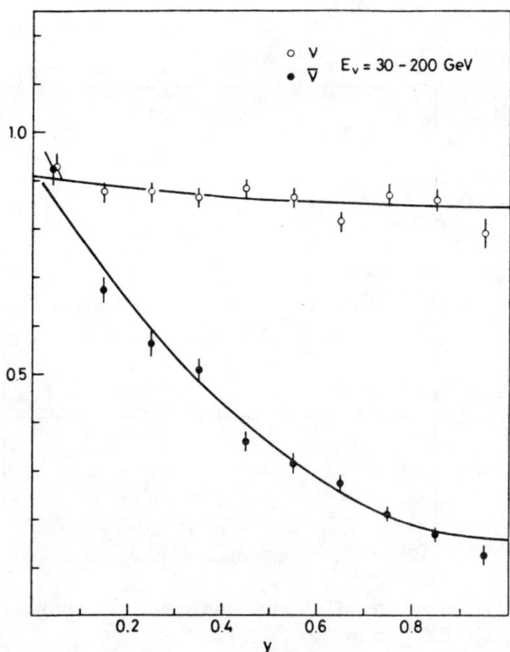

Fig. 1 The experimental ν and $\bar{\nu}$ y distributions obtained by CDHS (Ref. 8) and which underlie the R determination of that group.

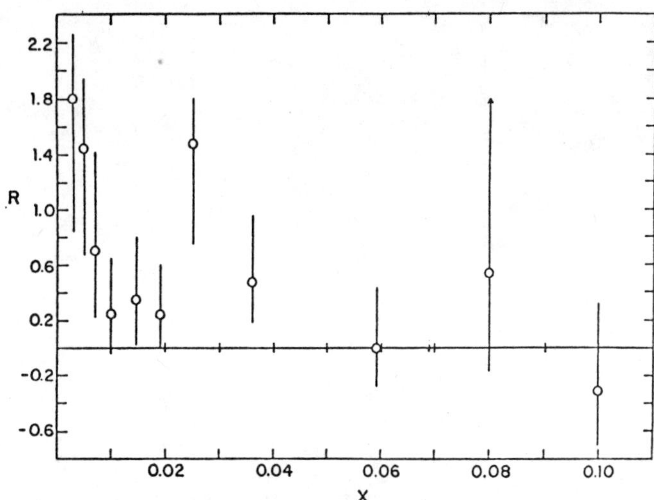

Fig. 2 R as function of x as observed in muon scattering by the CHIFR group (Ref. 6).

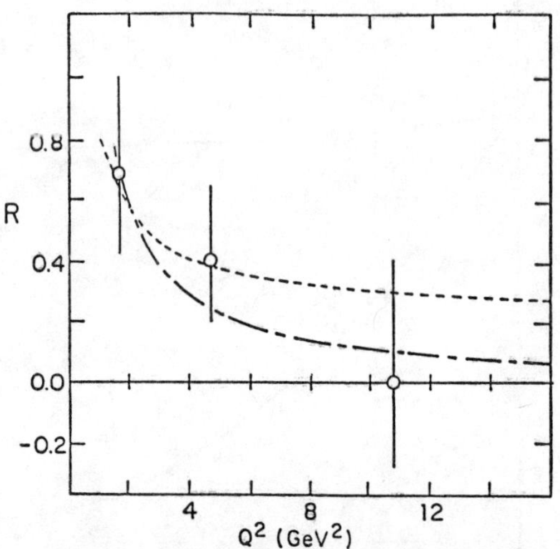

Fig. 3 R as function of Q^2 as observed in muon scattering by the CHIFR group (Ref. 6).

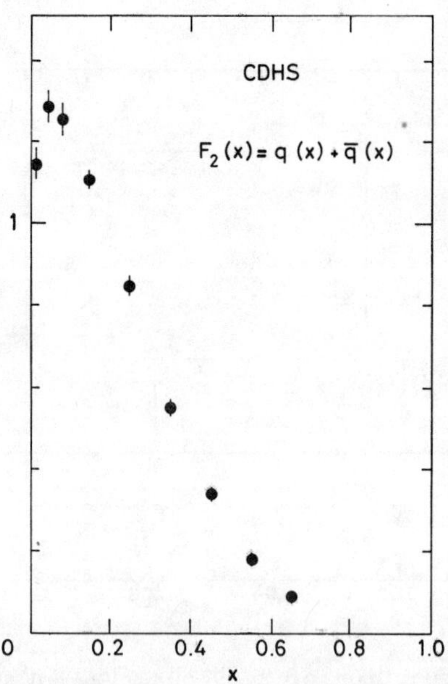

Fig. 4 CDHS determination of $F_2(x)$ for $30 < E_\nu < 200$ GeV (Ref. 8).

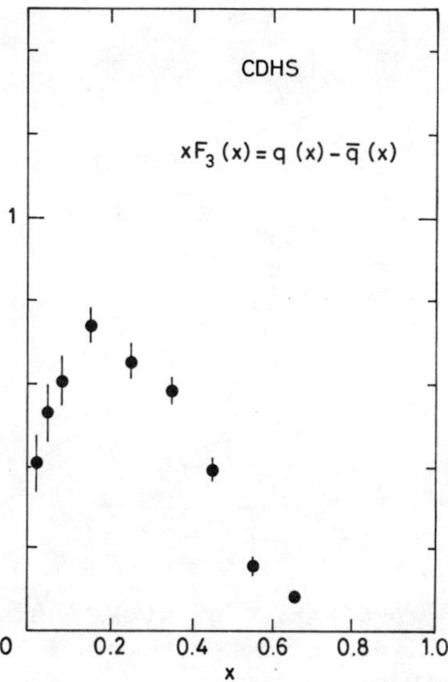

Fig. 5 CDHS determination of $xF_3(x)$ for $30 < E_\nu < 200$ GeV (Ref. 8).

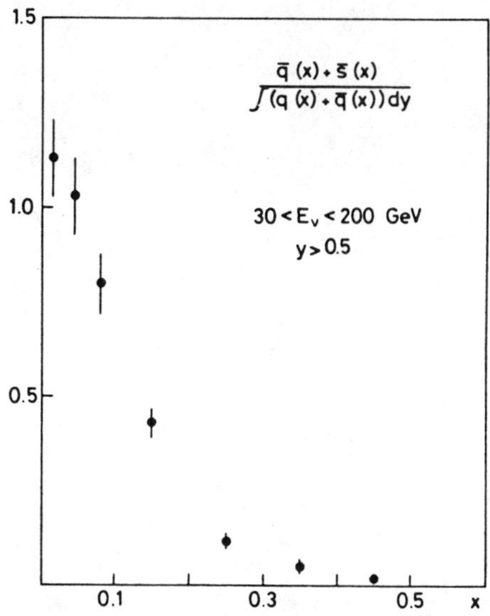

Fig. 6 CDHS determination of $\bar{q}+\bar{s}$ for $30 < E_\nu < 200$ GeV (Ref. 8).

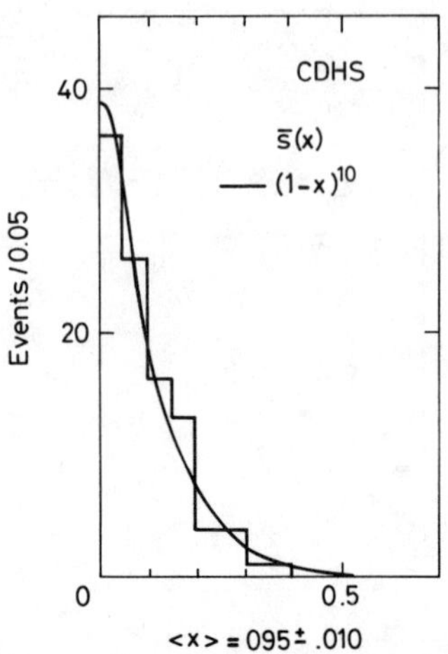

Fig. 7 CDHS determination of $\bar{s}(x)$. This structure $\bar{s}(x)$ is measured in the process $\bar{\nu} + Fe \to \mu^+ + \mu^- + X$ (Ref. 10).

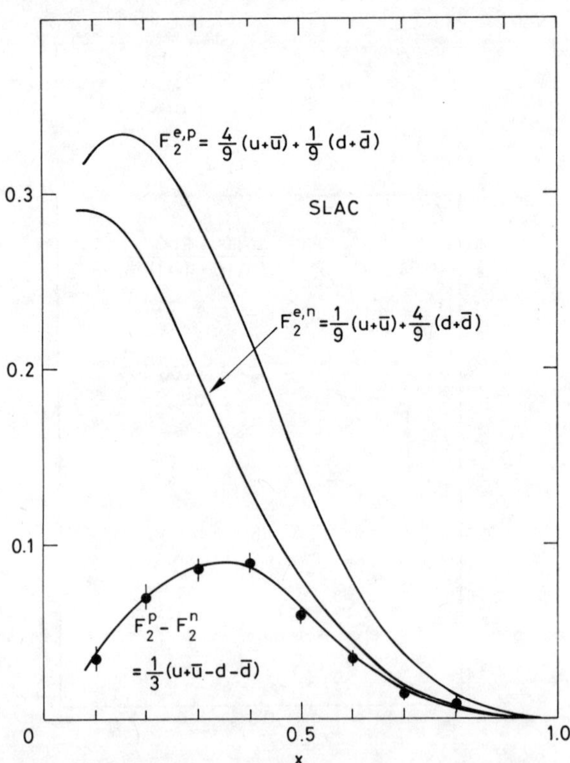

Fig. 8 $F_2{}^p(x)$, $F_2{}^n(x)$ observed at SLAC (Ref. 5) as well as the difference $F_2{}^p - F_2{}^n$.

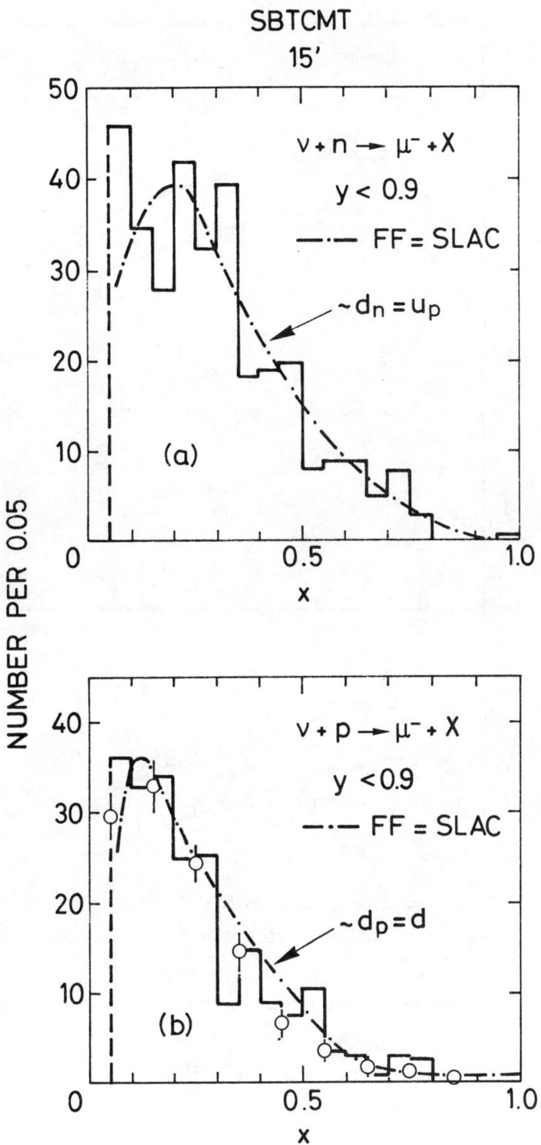

Fig. 9 $F_2{}^p(x)$ and $F_2{}^n(x)$ as observed in neutrino scattering (Ref. 11).

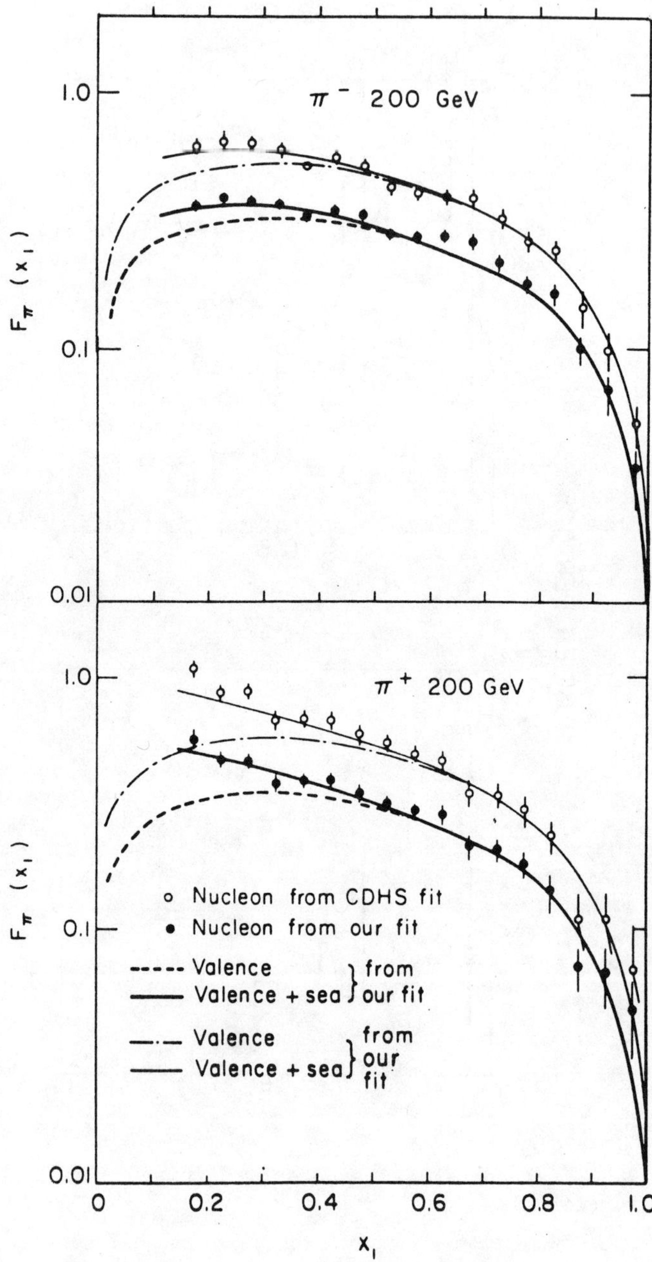

Fig. 10 π^+ and π^- structure functions determined from muon pair production by π^+ and π^- mesons, assuming the Drell-Yan mechanism (Ref. 12).

The "fractional nucleon momentum" carried by the various components are roughly the following, in the CDHS energy range:

a) valence quarks = $\int(q-\bar{q})dx$ $\quad$ 0.36 ± 0.02

b) sea quarks = $2\int\bar{q}(x)\,dx$ = 0.14 ± 0.01

c) strange quarks = $2\int\bar{q}(x)dx$ = 0.013 ± 0.003*)

d) all quarks = $\int(q(x) + \bar{q}(x))dx$ = 0.50 ± 0.02

In QCD, the fractional nucleon momentum not carried by quarks must be carried by gluons. This gluon fraction is then also one half, 0.50 ± 0.02. It may be noted that the s and $\bar{q}$ integrals are not in the ratio of 1:3 as SU_3 would predict, but the strange sea is appreciably smaller, presumably reflecting the heavier mass of the strange quark.

COMPARISON OF CHARGED LEPTON AND NEUTRINO NUCLEON STRUCTURE FUNCTIONS, AND QPM

At the Batavia Lepton-Photon conference, three experimental groups reported first results for $F_2(x,Q^2)$ obtained by scattering high energy muons in iron, and for which the x and Q^2 domain overlaps that of the CDHS neutrino experiment for the most part. What is more, the x bin sizes have been chosen the same by all groups, so that comparison is easy. These results are shown in Figs. 11-14[14-16]. The Berkeley - Fermilab - Princeton results are compared with CDHS in Fig. 15, while the European Muon Collaboration (EMC) results and the Bologna-CERN-Dubna-Munich-Saclay collaboration (NA4) results show the neutrino results superposed. In these comparisons the neutrino results were multiplied by 5/18 according to the expectations of the QPM, relation (2). The QPM predictions, in x dependence, Q^2 dependence and absolute magnitude are confirmed by the experiment to within the experimental uncertainties. This is a most important verification of the QPM. Another quantitative prediction which is confirmed is the valence quark number sum rule (4). Experimentally it is found that $\int xF_3(x)\frac{dx}{x}$ = 3.2 ± 0.5, in good agreement with the prediction of (3).

There is however one conceivable prediction which does not seem to be confirmed experimentally. It is often assumed that the $\bar{u}$ and $\bar{d}$ seas should be the same. Then in the QPM it would be expected for charged leptons that

$$\int (F_2{}^p - F_2{}^u)\frac{dx}{x} = \frac{1}{3}\int\left[(u-\bar{u}) - (d-\bar{d})\right]\frac{dx}{x} = \frac{1}{3}.$$

The experimental data of Fig. 16, if I didn't make an error, lead to the result of ≃ 1/6 rather than 1/3. If true, the $\bar{u}$ and $\bar{d}$ seas are different. On the whole however, the QPM is brilliantly confirmed in these experiments. It is difficult to doubt the quark's existence, even though it has not been observed in the free state.

---

*) This result is based on the assumption that charm particle production is dominated by D and D^* mesons.

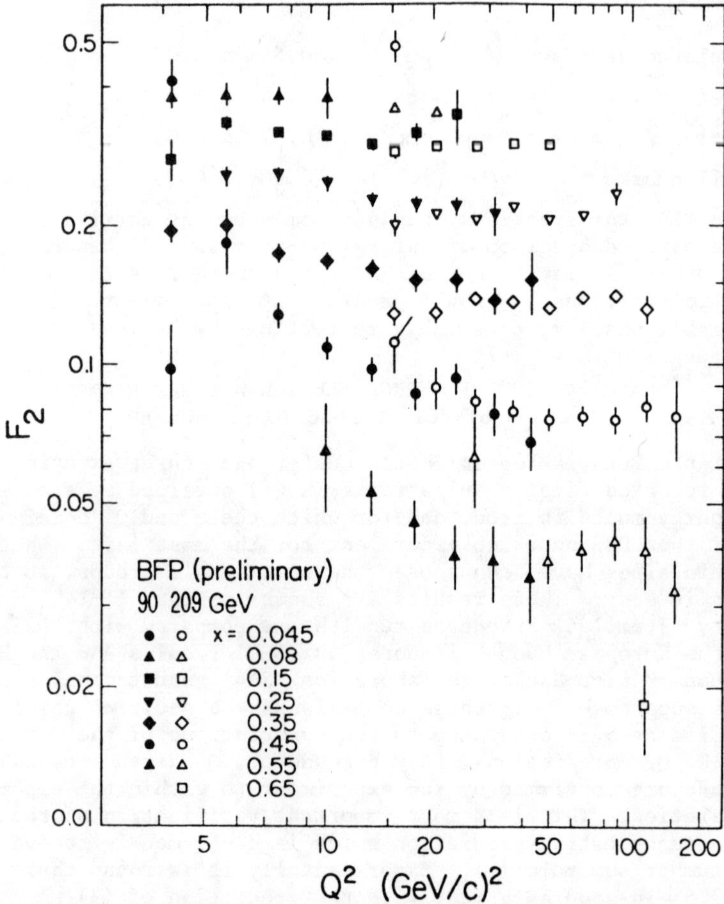

Fig. 11 Muon scattering result for $F_2(x,Q^2)$ obtained by the Berkeley-Fermilab-Princeton group (Ref. 14).

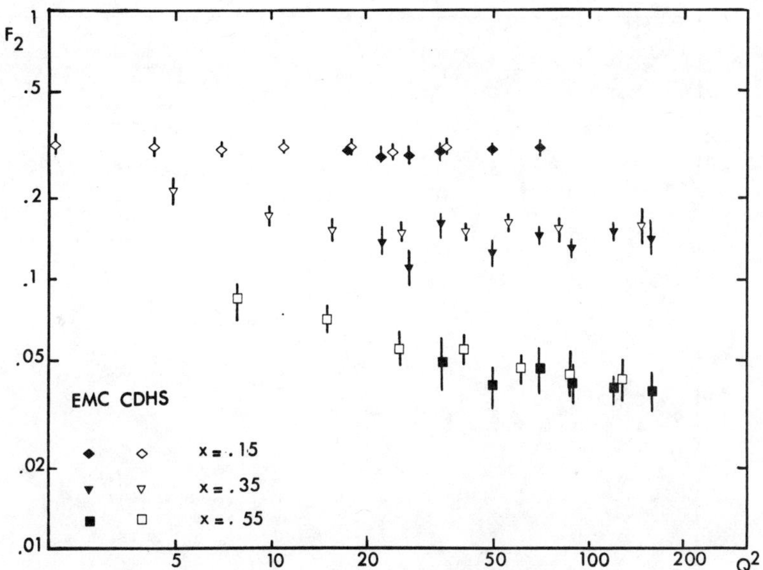

Fig. 12 Muon scattering results for $F_2(x,Q^2)$ obtained by the European Muon Collaboration (Ref. 15) as well as the CDHS results (multiplied by 5/18).

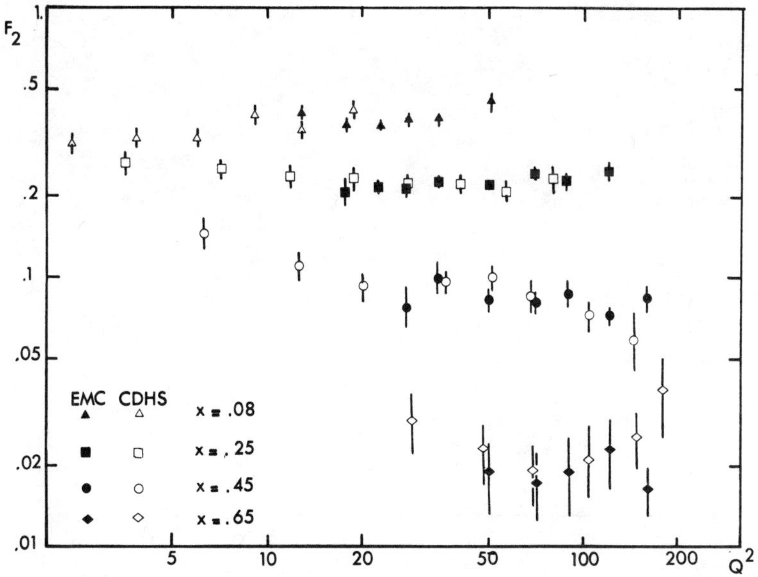

Fig. 13 Muon scattering results for $F_2(x,Q^2)$ obtained by the European Muon Collaboration (Ref. 15) as well as the CDHS results (multiplied by 5/18).

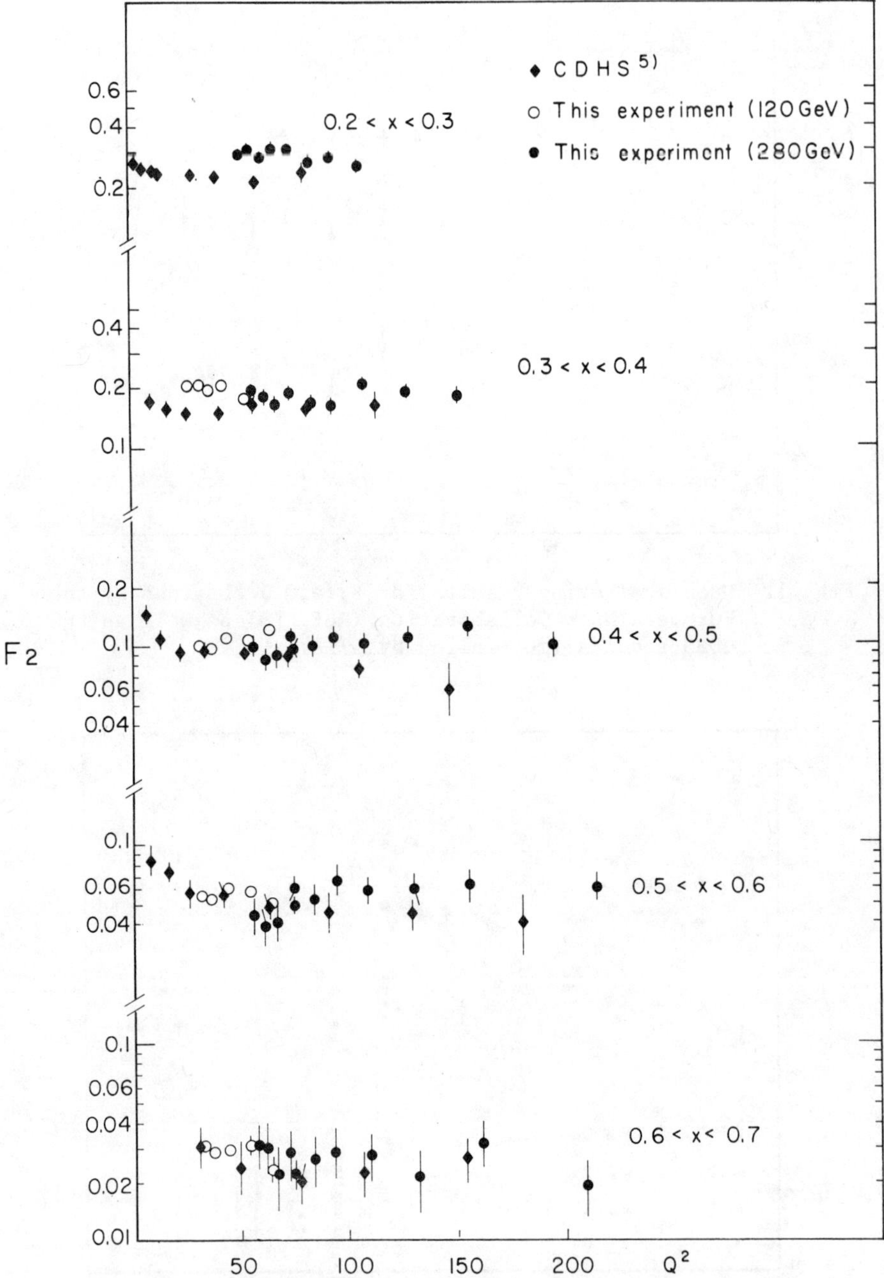

Fig. 14 Muon scattering results for $F_2(x,Q^2)$ obtained by the Bologna-CERN-Dubna-Munich-Saclay collaboration (Ref. 16) compared with the CDHS neutrino results after multiplying by 5/18.

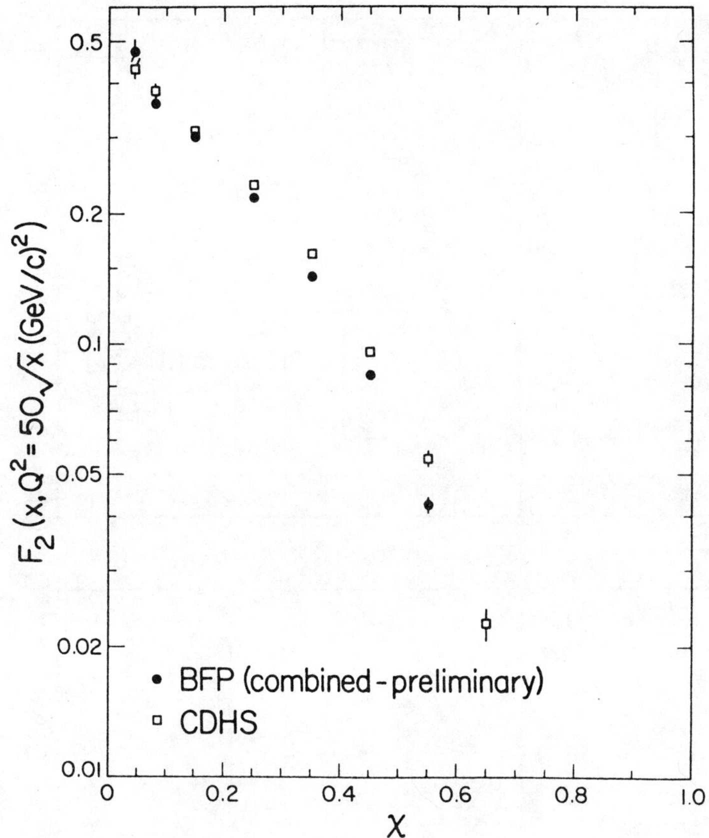

Fig. 15 BFP muon results for F_2 (Ref. 14) compared with the CDHS results.

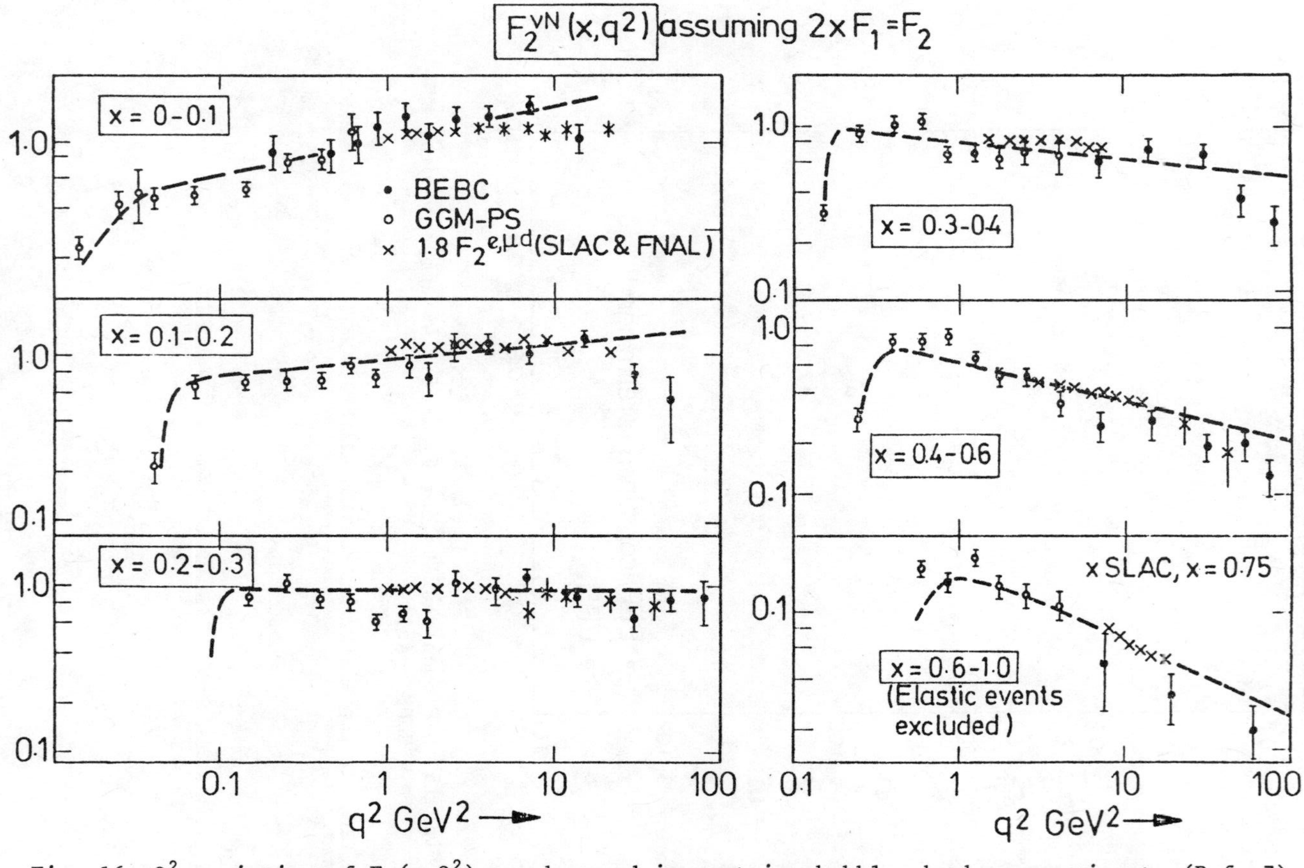

Fig. 16 Q^2 variation of $F_2(x,Q^2)$ as observed in neutrino bubble chamber experiments (Ref. 7). The results of electron scattering obtained at SLAC (Refs. 4, 5) are also shown.

Q^2 EVOLUTION OF STRUCTURE FUNCTIONS AND QCD

The Q^2 evolution of $F_2(x,Q^2)$ can be seen in Figs. 11-14. The variations are small, but visible. The evolutions amount to a shrinking of F_2 in x while maintaining its integral: at small x the function increases with Q^2 and at large x it decreases. At smaller Q^2 the effect is more pronounced, as can be seen from the BEBC and Gargamelle results[7)] shown in Fig. 16, where also the SLAC results are shown.

The CDHS results for xF_3 are given in Fig. 17, and some extremely preliminary CDHS results for the $\bar{q} + \bar{s}$ sea in Fig. 18. The question is: can these Q^2 variations be quantitatively understood in QCD and do they provide a critical check of the theory? The question is not yet definitively answered, partly because the experimental data are still rather primitive, and partly because the theoretical predictions are not as direct and free of ambiguity as one could wish. The theory does not predict the structure functions yet, and not even the rate of Q^2 evolution (there is a parameter, Λ, which is still inaccessible to calculation). Furthermore, where the effects are large, at low Q^2, higher order corrections are important, and at high Q^2, where these may be considered negligible, the expected effects are very small and therefore difficult to measure.

The most vivid comparison of experiment and theory is in terms of the moments of xF_3. The first order QCD equation is:

$$xF_3{}^n(Q^2) = xF_3{}^n(Q_0^2)e^{-d_n s}, \qquad (5)$$

where d_n are numbers (the anomalous dimensions) given by QCD, $s \equiv \ln\left[\ln(Q^2/\Lambda^2)/\ln(Q_0^2/\Lambda^2)\right]$, Λ is a constant to be found experimentally, and $F^n(Q^2) \equiv \int x^{n-2}\, F(x,Q^2)dx$. (It is better at low Q^2, to use somewhat more complicated moments, the so-called Nachtmann moments[18)], which take account of some finite mass effects, and reduce to the above definition for $Q^2 >> M^2$.)

It follows from (5) that when the logarithms of two different moments of xF_3 are plotted against each other, first order QCD requires the points at different Q^2 to lie on a straight line of slope d_n/d_m:

$$\ln(xF_3{}^n) = \text{const.} + \frac{d_n}{d_m}\ln(xF_3{}^m).$$

The results of BEBC and Gargamelle, as well as CDHS, are shown in Fig. 19. The data do fall on straight lines, even for the points which correspond to rather low Q^2 where the higher order corrections may be expected to be serious. The slopes are summarized in Table II. The data are in good agreement with the theoretical prediction. It should be noted however that the different orders of the moments are far from independent.

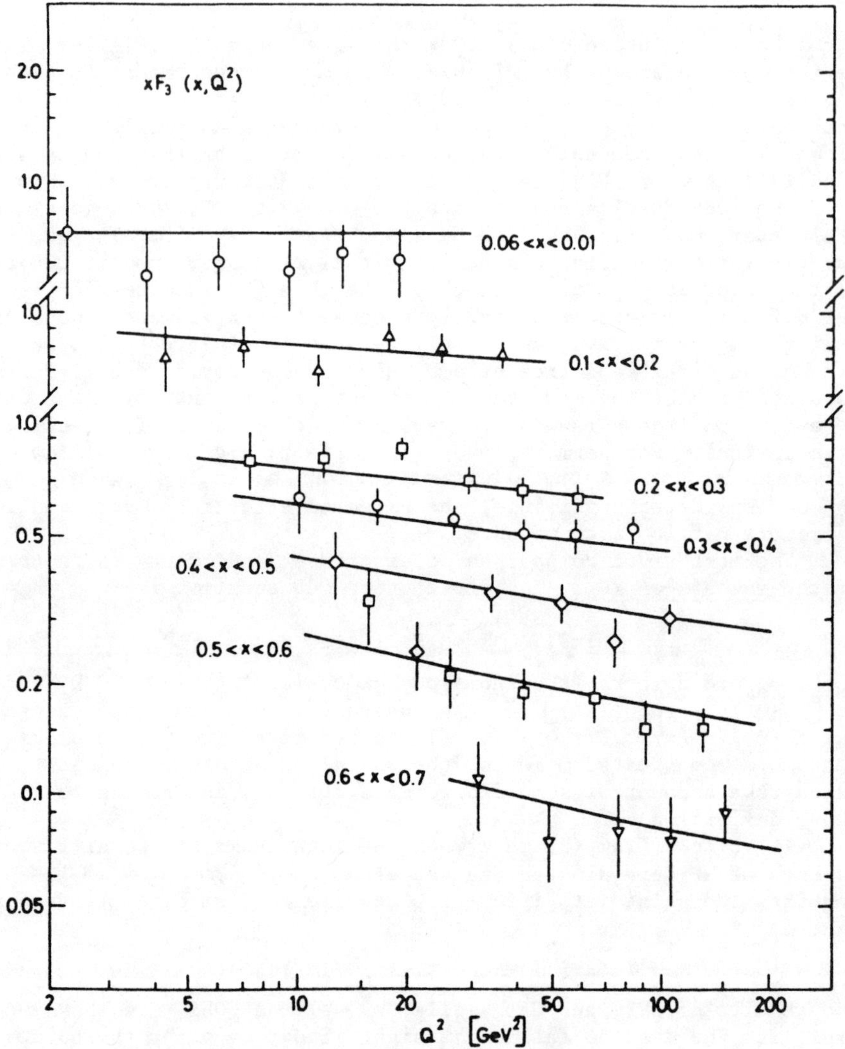

Fig. 17 CDHS results for the Q^2 variation of $xF_3(x,Q^2)$ (Ref. 8). The solid lines are a QCD fit based on the Buras-Gaemers parametrization.

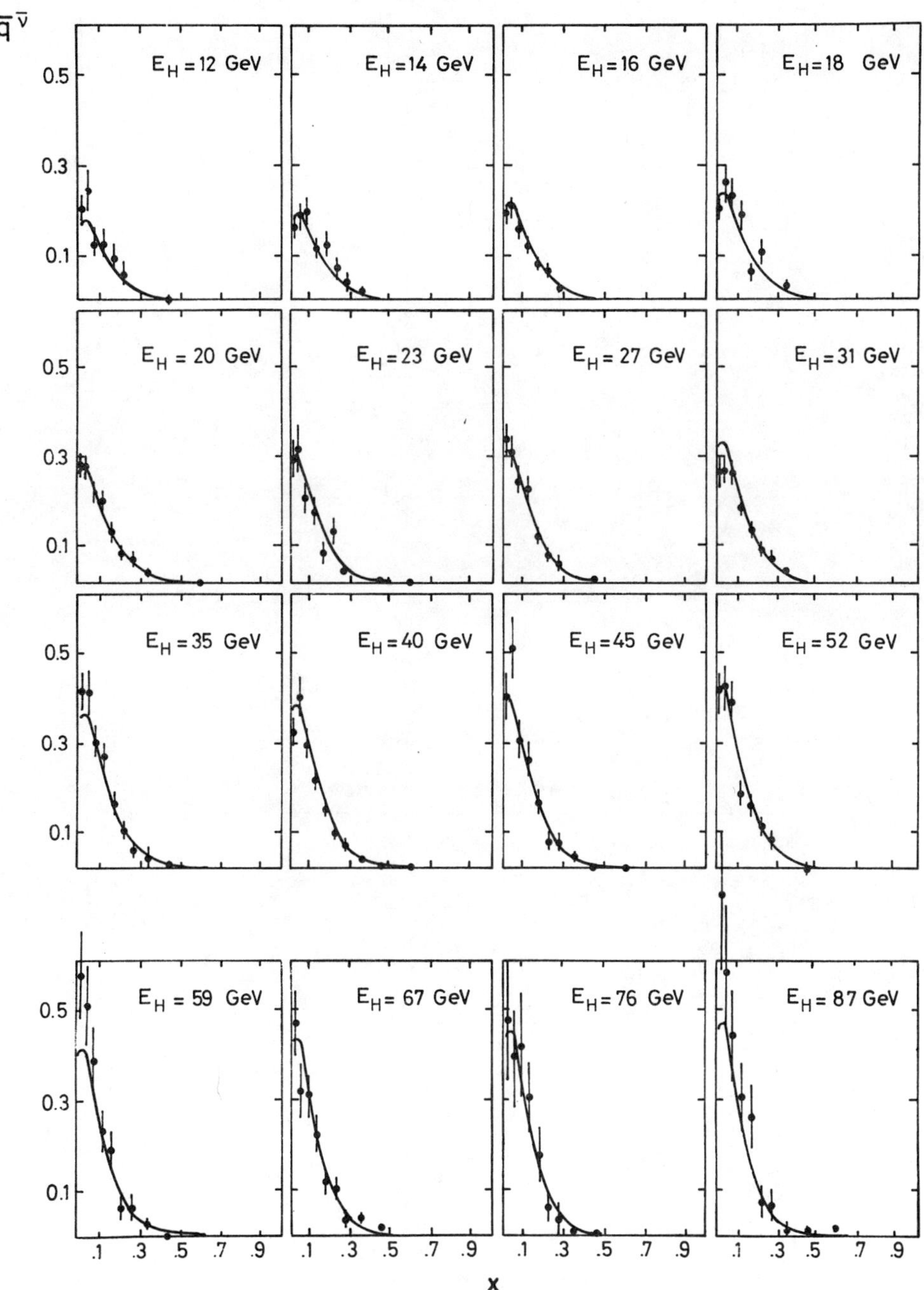

Fig. 18 Very preliminary results of the E_h variation of the $\bar{q}+\bar{s}$ sea (unpublished).

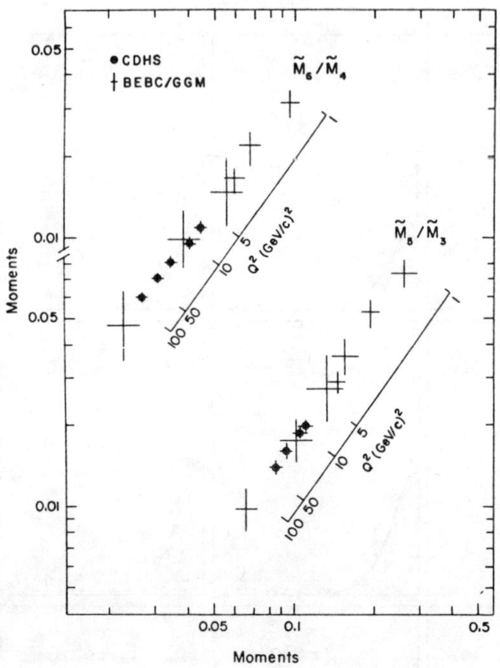

Fig. 19 ln $xF_3{}^m(Q^2)$ vs ln $xF_3{}^n(Q^2)$ for BEBC-GGM (Ref. 7) and CDHS (Ref. 18).

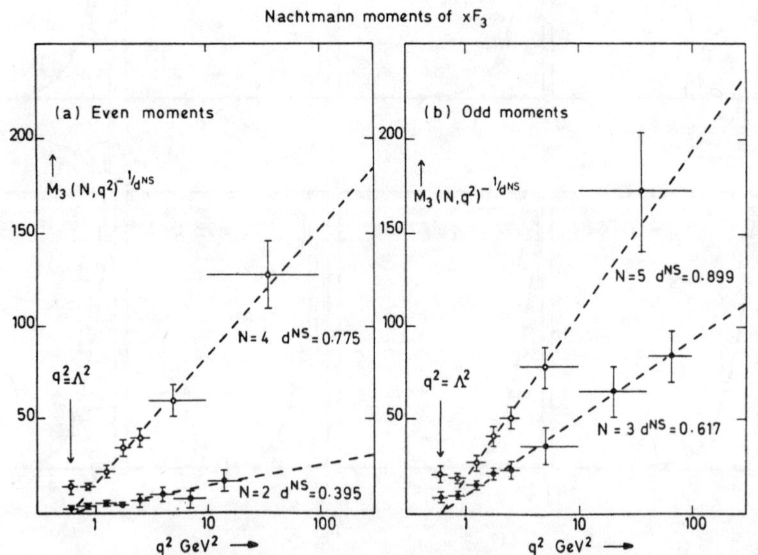

Fig. 20 $\left[xF_3{}^n(Q^2)\right]^{-1/dn}$ vs ln Q^2. BEBC-GGM result (Ref. 7).

TABLE II. Slopes for the logs of Nachtman moments plotted against each other, and expectation of first order QCD

Order of moments	BEBC/GGM Ref. 7	CDHS Ref. 18	QCD d_n/d_m
3/5	1.50 ± 0.08	1.34 ± 0.12	1.46
4/6	1.29 ± 0.06	1.18 ± 0.09	1.29
3/6		1.38 ± 0.15	1.62

A further simple prediction of first order QCD for the moments $xF_3{}^n(Q^2)$, and which follows from (5), is that the quantities $(xF_3{}^n)^{-1/d_n}$ plotted against $\ln Q^2$ should lie on straight lines, intersecting the Q^2 axis at Λ^2. Again this is experimentally confirmed, both by the BEBC-GGM (Fig. 20) and the CDHS result, however the values of Λ^2 so obtained are substantially different: $\Lambda_{BEBC/GGM} = 0.75 \pm 0.05$ and $\Lambda_{CDHS} = 0.33 \pm 0.07$. The difference is probably in part due to differences in the experimental data, and in part due to the different Q^2 range (for the bubble chamber result the low Q^2 GGM data are decisive in the Λ determination) and consequently the effect of higher order corrections is different. The two sets of data have been re-examined with the inclusion of higher order QCD corrections by Para and Sachrajda[19]. The definition of Λ is then n dependent. The bubble chamber and counter data now give Λ values which are closer together, (0.45-0.75), see Fig. 21.

Another method of comparison of QCD with experiment has been suggested by Buras and Gaemers[20]. In this method the structure functions are taken to be simple analytic functions, whose parameters are assumed to be linear in s. The parametrization adopted by the authors is:

$$F_2(x,Q^2) = Ax^{\alpha}(1-x)^{\beta}$$
$$\bar{q}(x,Q^2) = B(1-x)^{\gamma}$$
$$G(x,Q^2) = C(1-x)^{\delta}.$$

Here $G(x,Q^2)$ is the gluon structure function. The method has the advantage that not only xF_3 but also F_2 can be compared with theory, and G can be determined. It has the disadvantage that the parametrized functions do not satisfy the QCD equations exactly, and although it has been shown to be quite adequate for xF_3, it may be, depending on x and Q^2, somewhat poor for $\bar{q}$ and G. A Buras-Gaemers type fit[21] to the CDHS data for $F_2(x,Q^2)$ and $xF_3(x,Q^2)$ is shown in Figs. 17 and 22.

Finally, in Fig. 23 results are shown for the gluon structure function G(x) which follow from the CDHS data with the help of an analysis of the type proposed by Beaulieu and Kounas[21], based on the inversion of the Altarelli-Parisi equations[22].

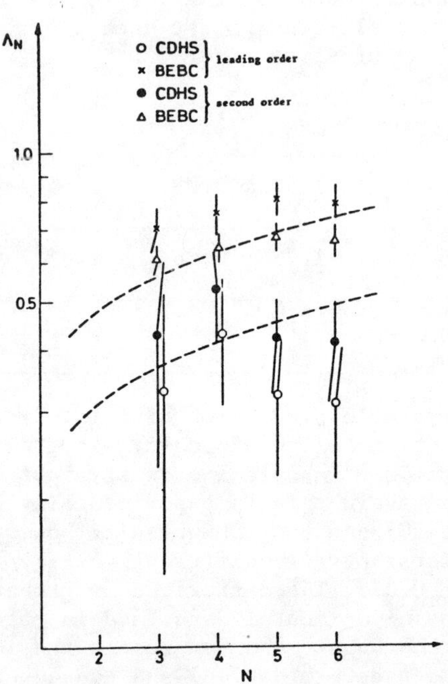

Fig. 21 Λ_n obtained from higher order QCD analyses of BEBC-GGM and CDHS data (Ref. 19).

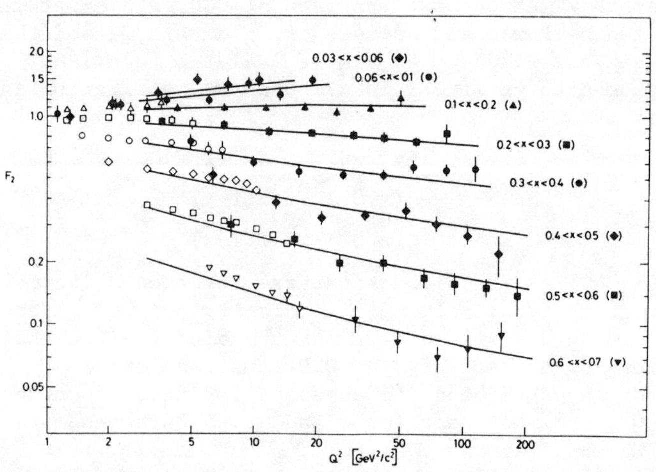

Fig. 22 CDHS (Ref 9) and SLAC (Ref. 4,5) results for $F_2(x,Q^2)$ together with QCD fit.

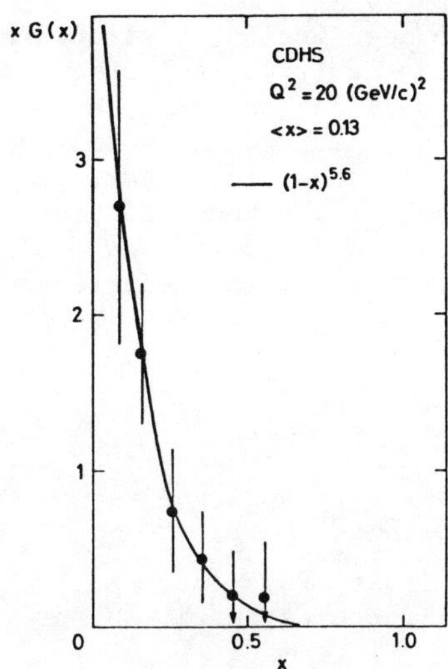

Fig. 23 Gluon structure function extracted from CDHS data (Ref. 8) using the method of Beaulieu and Kounas (Ref. 21).

SOME FINAL REMARKS

Measurements of the deep inelastic structure functions obtained both in charged lepton and neutrino inelastic scattering give a great deal of support to the quark parton model, as well as to its modificati by quantum chromodynamics. Considerable improvement in the experimental results might be expected in the next years, and this will be interesting in particular for the understanding of the longitudinal current terms $\sigma_L(x,Q^2)/\sigma_T(x,Q^2)$, as well as a more critical confrontation of QCD, although the latter may require also improvement in the theoretical methods.

REFERENCES

1. G. Callan and D.J. Gross, Phys. Rev. Lett. 22, 156 (1969).
2. S.L. Glashow, J. Illiopoulos and L. Maiani, Phys. Rev. D2, 1285 (1970).
3. D.J. Gross and C.H. Llewellyn Smith, Nucl. Phys. B14, 337 (1969).
4. M.E. Mestayer (Thesis), SLAC Report 214 (1978).
5. A. Bodek et al., SLAC-Pub-2248 (Jan. 1979)
6. Chicago-Harvard-Illinois-Oxford collaboration, H.L. Anderson et al., Fermilab Pub - 79/30.
7. P.C. Bosetti et al., Nucl. Phys. B142, 1 (1978).
8. J.G.H. de Groot et al., Z. Physik C, Particles and Fields 1, 143 (1979).
9. CDHS collaboration, A. Savoy-Navarro, ν'79 conference, Bergen, Norway (1979).
10. CDHS collaboration, H.J. Willutzki, ν'79 conference, Bergen, Norway (1979).
11. T. Kafka et al., High Energy Charged Current Neutrino Interactions in Deuterium, presented at the Lepton and Photon Conf., Batavia, U.S.A. (1979).
12. J. Badier et al., EPS Int. Conf. on High Energy Phys., Geneva, Switzerland, June 1979. CERN/EP Preprint 79-67.
13. R. Field and R.P. Feynman, Phys. Rev. D15, 2590 (1976).
14. Berkeley-Fermilab-Princeton collaboration, reported by M. Strovink at the Lepton and Photon Conf., Batavia, U.S.A. (1979), LBL Preprint 9912.
15. European Muon collaboration, reported by H.E. Stier at the Lepton and Photon Conf., Batavia, U.S.A. (1979).
16. Bologna-CERN-Dubna-Munich-Saclay Muon collaboration, reported by A. Benvenuti at the Lepton and Photon Conf., Batavia, U.S.A. (1979)
17. O. Nachtmann, Nucl. Phys. B63, 337 (1973); B78, 455 (1978).
18. J.G.H. de Groot et al., Phys. Rev. Lett. 82B, 292 (1979).
19. A. Para and G.T. Sachrajda, Phys. Lett. 86B, 331 (1978).
20. A.J. Buras and K.J.F. Gaemers, Nucl. Phys. B132, 249 (1978).
21. L. Beaulieu and C. Kounas, Preprint, Ecole Normale Superieure et Université Paris Sud, LPTENS 78/27.
22. L. Beaulieu and C. Kounas, Nucl. Phys. B141, 423, (1978).

REVIEW OF NEUTRAL CURRENT WEAK INTERACTIONS

C. Baltay
Columbia University, New York, NY 10027

I. INTRODUCTION

An impressive amount of experimental data on the neutral current weak interactions has been accumulated since the discovery of neutral currents in 1973-74. At the time of the XIXth International Conference on High Energy Physics in Tokyo last summer, it was apparent that the data were in good agreement[1] with the predictions of the Weinberg-Salam model[2] of the weak and electromagnetic interactions, and that a "model independent" analysis of the structure of the neutral currents yielded a unique solution that corresponded to the couplings of the Weinberg-Salam model. At that time some assumptions had to be made in the so-called "model independent" analysis i.e. that the same Z^o mediated all of the neutral current processes, and that the neutrinos had the standard two-component coupling strength.

In the year since the Tokyo conference, some new experimental data have become available, but the situation has not changed significantly. The improvements in the electron-hadron experiments (atomic parity violations and measurement of the y dependence of the parity violating effects in polarized electron-deuteron scattering) now allow some checks in the assumptions that had to be made previously in the "model independent" analysis.[3]

II. REVIEW OF THE EXPERIMENTAL SITUATION

The neutral current processes that have been experimentally studied are summarized briefly in Table I, with some comment on the difficulty of the theoretical interpretation of each process. The bulk of the experimental data have been discussed in previous reviews;[4] we will confine the discussion here to new data obtained this past year.

The results on the purely leptonic process, $\nu_\mu + e^- \rightarrow \nu_\mu + e^-$, are summarized in Table II. The results from the Gargamelle and the Aachen-Padova experiments at the CERN PS and the Columbia-BNL experiment at Fermilab have not changed. There is a change in the cross section measured in the Gargamelle experiment[5] at the CERN SPS. Their new result is in agreement with the Weinberg-Salam model. There are also two new measurements, by the CHARM Colla-

ISSN:0094-243X/80/590025-29$1.50

boration[6] at the CERN SPS and a spark chamber experiment[7] at Fermilab, which are also in good agreement with the W-S model. The weighted average of all of the experiments is:

$$\sigma(\nu_\mu + e^- \rightarrow \nu_\mu + e^-) = (1.6 \pm 0.3) \times 10^{-42}\ E_\nu\ cm^2 \quad ,$$

which is in excellent agreement with the $1.5 \times 10^{-42}\ E_\nu cm^2$ predicted by the W-S model for $\sin^2\theta = 0.23$.

There have been no new results on $\bar{\nu}_\mu + e^- \rightarrow \bar{\nu}_\mu + e^-$ scattering; the previous results are summarized in Table III.

The results on the elastic scattering processes, $\nu_\mu + p \rightarrow \nu_\mu + p$ and $\bar{\nu}_\mu + p \rightarrow \bar{\nu}_\mu + p$, and the single pion production processes are summarized in Tables IV and V, respectively. There has been no change in these results this past year.

The results on the inclusive neutral current to charged current ratios are summarized in Table VI. The measurements from the CHARM collaboration at the CERN SPS are new,[8] and the new ratios obtained by the CDHS experiment,[9] also at the CERN SPS, are impressively precise. The improvements in these data and the agreement with the W-S model, both with and without QCD predictions, are shown in Fig. 1.

There are some new data on the x and y distributions of the inclusive neutral current ν interactions. The y distributions obtained by the CHARM collaboration[8] at the CERN SPS are shown in Fig. 2. The agreement with the W-S model predictions with $\sin^2\theta = 1/4$, shown by the curves on Fig. 2, is excellent.

There are two new results on the x distribution of the inclusive neutral current ν scattering process, from a bubble chamber experiment[10] at the Brookhaven AGS using a narrow band neutrino beam and the 7-ft chamber filled with heavy neon (Fig. 3) and from the CHARM collaboration[8] at the CERN SPS (Fig. 4). These are the first measurement of the x distributions in neutral current processes since these measurements used both an incident narrow band beam and a detector that is sufficiently fine-grained to allow a measurement of the direction as well as the energy of the hadrons in the final state, so that the energy and direction of the outgoing neutrino can be obtained by subtraction. These distributions indicate that the x distribution measured by the neutral currents is similar to those measured by the charged current neutrino interactions.

There has been some progress in the measurements of atomic parity violating effects, which are summarized in Table VII. All of the experiments now observe a non-zero effect.

The SLAC experiment that has observed the parity violating asymmetry in the scattering of polarized electrons in deuterium, previously at a fixed value of y, the fraction energy transfer from the electron to the hadrons, has now more extensive data[11] at a larger range of y. These results are shown in Fig. 5. The observed y dependence of the asymmetry is consistent with what is predicted by the W-S model.

A summary of the experimental results on neutral current processes is given in Table VIII. The second column shows the weighted average of the results from all of the experiments on each of the processes listed in the first column. The third column gives the predictions of the W-S model for each process for $\sin^2\theta_\omega$ = 0.23. The agreement between the experimental results and the predictions of the model is excellent. The last column on Table VIII gives the value of $\sin^2\theta$ determined from the average of all of the experiments on each process. All of the processes give a value of $\sin^2\theta$ consistent with the average value of $\sin^2\theta = 0.23 \pm 0.015$ as is shown graphically on Fig. 6.

III. "MODEL INDEPENDENT" DETERMINATION OF THE NEUTRAL CURRENT COUPLINGS

In the previous section, the experimental data were compared with the predictions of the Weinberg-Salam model. We saw that the data are consistent with these predictions (see Table VIII). In this section, we will discuss a "model independent" analysis in which no specific model is assumed at the outset. The experimental data will be used to determine the structure of the weak neutral currents in a general way. By structure of the weak neutral current, we mean the space-time structure: vector (V), axial vector (A), scalar (S), pseudoscalar (P) or tensor (T); and the isotopic spin structure: isoscalar (I=0) or isovector (I=1). In the usual analyses carried out in the past, however, two assumptions have been made. The first is that the same Z^0 mediates all of the neutral current processes considered, and the second is that the coupling strength of the neutrino is that given by the standard two component neutrino theory. In section A that follows, we will describe the analysis with these assumptions. In section B, we will discuss the experimental checks of these assumptions that are made possible by recent improvements in the data.

A. Analysis with the Usual Assumptions

1. We start out by considering the general space-time structure.

(a) Any pure interaction (V,A,S,P,or T) must give the same neutrino and antineutrino cross sections for any given process. A difference in the ν and $\bar{\nu}$ cross section can only be caused by an interference between two different interactions with different C (like a VA interference term). Experimentally, the results of the HPD experiment[12] for the elastic scattering process is

$$\frac{\sigma(\bar{\nu}_\mu+p \rightarrow \bar{\nu}_\mu+p)}{\sigma(\nu_\mu+p \rightarrow \nu_\mu+p)} = 0.53 \pm 0.17$$

and for the inelastic inclusive interactions the CDHS experiment[13] finds

$$\frac{\sigma(\bar{\nu}_\mu+N \rightarrow \bar{\nu}_\mu+ \ldots)}{\sigma(\nu_\mu+N \rightarrow \nu_\mu+ \ldots)} = 0.58 \pm 0.05 \quad .$$

These ratios are significantly different from unity so we can conclude that the neutral currents are not pure V,A,S,P,or T).

(b) The distribution in $y = (E_\nu \text{ in} - E_\nu \text{ out})/E_\nu \text{ in}$ for the inclusive neutral current process is sensitive to the space time structure of the interaction. For example,

$$\frac{d\sigma}{dy} \approx g(V\text{-}A)+g(V\text{+}A)(1-y)^2+g(S,P)y^2$$

for ν scattering on real quarks. The y distributions for various interactions are shown in Fig. 7. Several experiments have looked at the y distribution of the inclusive neutral current reaction, using narrow band neutrino beams where the incident ν energy is known to some extent: the CITF experiment[14] at Fermilab, the CDHS experiment[15] at the CERN SPS, and the BEBC experiment[16] with a heavy neon fill at the CERN SPS. The conclusions are that:

i) S and/or P can be ruled out

$$\frac{g(S,P)}{g(V\text{-}A)+g(V\text{+}A)} \begin{array}{ll} = 0 \pm 0.03 & \text{BEBC} \\ = 0.02 \pm 0.07 & \text{CDHS} \end{array}$$

ii) Pure V, pure A, and pure (V+A) can be ruled out and pure (V-A) is unlikely (see Fig. 8).

(c) The parity violating asymmetry observed in polarized electron scattering at SLAC is due to an interference between the neutral currents and the electromagnetic interaction, which is a vector interaction. Thus the weak neutral current must have some V or A part to it.

2. In view of the above conclusions it seems reasonable to proceed with the analysis in terms of an arbitrary mixture of V and A interactions.

The various neutral current couplings between the leptons and the quarks are illustrated in Fig. 9. The purely leptonic processes, $\nu+e \rightarrow \nu+e$ depend only on the V and A couplings of the electron, g_V and g_A. In neutrino hadron scattering, isotopic spin is also relevant, so we have four couplings, V and A with I = 0 and I = 1 each, which we write as $G_V{}^1$, $G_A{}^1$, $G_V{}^o$, and $G_A{}^o$, which are the same as α, β, γ, δ, respectively, introduced by Sakurai.[17] Sehgal introduced an alternate four coupling constants U_L, d_L, U_R, and d_R. The subscript L and R are for (V-A) and (V+A) combinations, and the u and d are the isospin combination appropriate for the u and d quarks. The two sets of coupling constants are obviously just linear combinations of each other:

$$u_L = 1/4(\alpha+\beta+\gamma+\delta)$$
$$d_L = 1/4(-\alpha-\beta+\gamma+\delta)$$
$$u_R = 1/4(\alpha-\beta+\gamma-\delta)$$
$$d_R = 1/4(-\alpha+\beta+\gamma-\delta) \quad .$$

The parity violating effects in atomic bismuth or polarized electron scattering depend on six coupling constants: g_A, g_V for the electrons, and u_L, d_L, u_R, and d_R for the quarks. At this point some theoretical assumptions have crept in. The fact that g_A, g_V for the last case are the same as g_A, g_V for $\nu+e \rightarrow \nu+e$ scattering depends on the assumption that the same Z^o mediates both processes and that the neutrino couples to this Z^o with the normal V-A coupling.

3. Determination of the ν-quark couplings. The analysis follows the work of Sehgal,[18] Hung and Sakurai,[19] Abbott and Barnett,[20] Sidhu and Langacker,[21] Paschos,[22] Claudson, Paschos and Sulak,[23] and Langacker, Kim, Levine, Williams, and Sidhu.[24]

(a) The neutral to charged current ratios for the inclusive processes for neutrinos and antineutrinos are used to determine the overall strengths $(u_L{}^2+d_L{}^2)$ and $(u_R{}^2+d_R{}^2)$. These are the circular bands on the u_L vs d_L and the u_R vs d_R planes, shown in Fig. 10.

(b) In the original analysis by Sehgal,[18] the inclusive pion production data from Gargamelle[25] at the CERN PS

$$R_\nu = \frac{\nu+N \rightarrow \nu+\pi^+ + \ldots}{\nu+N \rightarrow \nu+\pi^- + \ldots} = 0.77 \pm 0.14$$

$$R_{\bar\nu} = \frac{\bar\nu+N \rightarrow \bar\nu+\pi^+ + \ldots}{\bar\nu+N \rightarrow \bar\nu+\pi^- + \ldots} = 1.64 \pm 0.36$$

were used to select the four allowed solutions A,B,C,D shown in Fig. 10. More recent high energy data[26] of $R_\nu = 1.07\pm0.17$ and $R_{\bar\nu} = 1.54\pm0.45$ from BEBC at the CERN SPS and $R_{\bar\nu} = 1.27^{+.36}_{-.27}$ from the FIIM Collaboration[27] using the 15-ft chamber at Fermilab have confirmed the low energy results from Gargamelle.

There are some recent results on the inclusive neutral current cross sections on protons and neutrons separately that provide information about the I spin structure:

$$\frac{\nu+p \to \nu+\ldots}{\nu+p \to \mu^-+\ldots} = 0.52 \pm 0.06 \quad \text{FIIM}^{28}$$
$$= 0.48 \pm 0.17 \quad \text{FNAL-LBL-Hawaii-Michigan}^{29}$$

$$\frac{\bar{\nu}+p \to \bar{\nu}+\ldots}{\bar{\nu}+p \to \mu^++\ldots} = 0.42 \pm 0.13 \quad \text{Argonne-Purdue Carnegie-Mellon}^{30}$$

$$\frac{\nu+n \to \nu+\ldots}{\nu+p \to \nu+\ldots} = 1.22 \pm 0.35 \quad \text{Marriner et al}^{31}$$

$$\frac{\bar{\nu}+n \to \bar{\nu}+\ldots}{\bar{\nu}+p \to \bar{\nu}+\ldots} = 0.64 \pm 0.18 \quad \text{FIIM}.^{28}$$

These data, together with the elastic scattering data summarized in Table IV, uniquely select solution A shown on Fig. 10. The best fits to the hadronic couplings using all of the latest available data obtained in a recent analysis by Langacker, Kim, Levine, Williams, and Sidhu,[24] are shown in Fig. 11, and the resulting couplings are listed in Table IX.

4. Determination of the ν-electron couplings.

(a) The intersections of the regions allowed in the g_V-g_A plane by the cross sections for the processes, $\nu_\mu + e^- \to \nu_\mu + e^-$ and $\bar{\nu}_e + e^- \to \bar{\nu}_e + e^-$ produces two solutions, one with $g_V \sim 0$, $g_A \sim -1/2$, the other with $g_V \sim -1/2$, $g_A \sim 0$, as shown in Fig. 12. The region allowed by the cross section for $\bar{\nu}_\mu + e^- \to \bar{\nu}_\mu + e^-$ is consistent with these two solutions but does not distinguish between them.

(b) The region in the g_V-g_A plane allowed by the parity violating asymmetry in polarized electron scattering observed at SLAC is also shown on Fig. 12. This region overlaps the solution near $g_V \sim 0$, $g_A \sim -1/2$, but not the other solution and thus resolves the ambiguity. The values of g_V and g_A with errors for this solution are also listed in Table IX.

B. Experimental Checks of the Assumptions Made in the Previous Analysis

Sakurai[32] among others has pointed out that the recent improvements in the polarized e^-d scattering data and the experiments on parity violations in atomic physics allow some checks of the assumptions made in the analysis described in section A above.

1. Factorization. In combining the polarized e^-d scattering data with the $\nu+e \rightarrow \nu+e$ data in Fig. 12, the assumption had to be made that the couplings g_V and g_A in ν-e processes and the couplings $G_V^{0,1}$ and $G_A^{0,1}$ in ν-quark processes are the same as the couplings that enter in the electron-quark (e-q) processes. This assumption, called factorization by Sakurai,[32] is correct if there is only one Z^o that mediates both processes, as can be seen from the diagrams of Fig. 13. This assumption can be checked experimentally by using the polarized e^-d scattering data combined with the atomic parity violation data to determine the products $g_A G_V^{0,1}$ for the e-q process, and comparing it with the product of g_A and $G_V^{0,1}$ determined from the νe and νq data, respectively.

(a) From the recent SLAC polarized e^-d scattering the y dependence of the parity violating asymmetry has been extracted:[11]

$$A/q^2 = a_1 + a_2 \left[\frac{1-(1-y)^2}{1+(1-y)^2}\right]$$

with

$$a_1 = \frac{G}{\sqrt{2}\, e^2} \frac{18}{5} g_A (G_V^{\,1} + 1/3 G_V^{\,o})$$

$$a_2 = \frac{G}{\sqrt{2}\, e^2} \frac{18}{5} g_V (G_A^{\,1} + 1/3 G_A^{\,o})$$

From the experimental results of

$$a_1 = (-9.7 \pm 2.6) \times 10^{-5}$$

$$a_2 = (4.9 \pm 8.1) \times 10^{-5}$$

we obtain

$$g_A G_V^{\,1} + 1/3 g_A G_V^{\,o} = -0.30 \pm 0.08$$

$$g_V G_A^{\,1} + 1/3 g_V G_A^{\,o} = 0.16 \pm 0.25 \quad .$$

The parity violating rotation of the plane of polarization of laser light in Bi^{83} vapor is sensitive to the quantity

$$Q(Bi) = g_A (86\, G_V^{\,1} - 1254\, G_V^{\,o}) \quad .$$

From the experimental value[33] of $Q(Bi) = -127 \pm 17$, we obtain

$$g_A G_V^{\,1} - 14.6\, g_A G_V^{\,o} = -1.48 \pm 0.20 \quad .$$

Combining this with the above results from ed scattering, we obtain

$$g_A G_V^{1} = -0.33 \pm 0.08$$
$$g_A G_V^{0} = +0.08 \pm 0.02 \quad \} \text{ from eq}$$

There is no unique solution for $g_V g_A^{0,1}$ since the Bi^{83} experiments are not sensitive to the $g_V G_A$ terms.

To check factorization we compare these results to the product of g_A, g_V from ν-e scattering with $G_V^{0,1}$ and $G_A^{0,1}$ from the ν-hadron processes. The values of u_L, u_R, d_L, d_R shown in Table IX correspond to the values

$$G_V^{1} = 0.597 \pm 0.055$$
$$G_A^{1} = 0.935 \pm 0.050$$
$$G_V^{0} = -0.254 \pm 0.082$$
$$G_A^{0} = 0.127 \pm 0.083$$

which combined with g_A, g_V from νe scattering gives

$$g_A G_V^{1} = -0.33 \pm 0.05$$
$$g_A G_V^{0} = 0.13 \pm 0.05 \quad \} \text{ from } \nu e \text{ and } \nu q \text{ combined}$$

which are in very good agreement with the values obtained from e-q processes discussed above. This agreement is shown in Fig. 14a. The check for the $g_V G_A^{0,1}$ terms, shown in Fig. 14b, is weaker.

2. The $\nu\bar{\nu}Z^o$ coupling. The usual assumption is that the $\nu\bar{\nu}Z^o$ coupling is 1, as indicated on the upper vertices of the graphs on Fig. 13. This is implicit in writing the Lagrangian for example for ν+e → ν+e scattering as

$$\mathcal{L} = -\frac{G}{\sqrt{2}}[\bar{\nu}\gamma_\mu(1+\gamma_5)\nu][\bar{e}\ \gamma_\mu(g_V+g_A\gamma_5)e] \quad .$$

In the spirit of "model independence" Sakurai[32] suggested to relax this assumption and call the $\nu\bar{\nu}Z^o$ coupling C_ν. The ν term in the Lagrangian then becomes $[\bar{\nu}C_\nu\gamma_\mu(1+\gamma_5)\nu]$. In this case the cross section for the process $\nu+\nu \to \nu+\nu$ would be proportional to C_ν^2.

The possibility that $C_\nu^2 \neq 1$ would affect νe, νq, and eq processes differently. In Fig. 13 the coupling at the upper vertex for νe and νq becomes C_ν, and the constants g_A and g_V for νe scattering becomes $C_\nu g_A$ and $C_\nu g_V$, and $G_A^{0,1}$ and $G_V^{0,1}$ for νq scattering becomes $C_\nu G_A^{0,1}$ and $C_\nu G_V^{0,1}$, but the constants $g_V G_A^{0,1}$ and $g_A G_V^{0,1}$ for eq processes remain unchanged since the $\nu\bar{\nu}Z^o$ vertex does not appear. We can thus set limits on C_ν experimentally by comparing the couplings obtained in νe and ν-hadron scattering with those obtained in the e-hadron processes (assuming a single Z^o this time):

$$C_\nu^2 = \frac{(C_\nu g_A)_{\nu e}(C_\nu G_V^1)_{\nu q}}{(g_A G_V^1)_{eq}} = 0.96 \pm 0.27$$

and

$$C_\nu^2 = \frac{(C_\nu g_A)_{\nu e}(C_\nu G_V^0)_{\nu q}}{(g_A G_V^0)_{eq}} = 1.15 \pm 0.48 \quad .$$

Thus we see that C_ν is consistent with unity within the rather large experimental errors.

3. Relation between the $W^\pm$ and Z^0 masses. In the standard Weinberg-Salam model, we have

$$\mathcal{L} = -\frac{G}{2}[\bar{\nu}\gamma_\mu(1+\gamma_5)\nu][2J_\mu^3 - 2\sin^2\theta J_\mu^{em}]$$

with

$$m_W^2 = m_Z^2 \cos^2\theta \quad .$$

There are however other possibilities within the SU(2) x U(1) gauge models. For example, Glashow suggested the Lagrangian

$$\mathcal{L} = -\frac{G}{\sqrt{2}}\rho[\bar{\nu}\gamma_\mu(1+\gamma_5)\nu][2J_\mu^3 - 2\sin^2\theta J_\mu^{em}]$$

with

$$m_W^2 = \rho\, m_Z^2 \cos^2\theta \quad .$$

The value of ρ can be determined experimentally. The best determination comes from the inclusive neutral current to charge current ratios (see Table VI).

In the standard W-S model, we have

$$(u_L^2 + d_L^2) = 0.299 \pm 0.015$$
$$(u_R^2 + d_R^2) = 0.024 \pm 0.012 \quad .$$

In the Glashow scheme, these relations become

$$\rho^2(u_L^2 + d_L^2) = 0.299 \pm 0.015$$

$$\rho^2(u_R^2 + d_R^2) = 0.024 \pm 0.012 \quad .$$

Now from the SU(2) x U(1) structure of the currents, we have:

$$u_L = 1/2 - 2/3\sin^2\theta$$
$$d_L = -1/2 + 1/3\sin^2\theta$$
$$u_R = -2/3\sin^2\theta$$
$$d_R = 1/3\sin^2\theta \quad .$$

Using these in the above relations, we get

$$\rho^2(1/2-\sin^2\theta+5/9\sin^4\theta) = 0.299 \pm 0.015$$

$$\rho^2(5/9\ \sin^4\theta) = 0.024 \pm 0.012 \quad .$$

We can solve these for ρ and $\sin^2\theta$, and obtain

$$\rho = 0.98 \pm 0.04 \ .$$

In a similar analysis using all of the available neutral current data, Langacker et al[24] obtain

$$\rho = 0.985 \pm 0.023$$

which again is in good agreement with ρ = 1 of the standard W-S model.

IV. CONCLUSIONS

1. The available experimental data is in good agreement with the predictions of the W-S model (see Table VIII).

2. The "Model Independent" analysis, with some assumptions, yields a unique solution. The resulting couplings are just those of the W-S model (see Table IX).

3. There is now enough experimental data to check some of the assumptions made in the "model independent" analysis, and the assumptions seem justified i.e. factorization works, the $\nu\bar{\nu}Z^o$ coupling and the parameter ρ in the intermediate boson mass relations are consistent with unity.

4. All of the experiments are consisent with a single value of $\sin^2\theta$, as shown in Fig. 6. The weighted average value is

$$\sin^2\theta = 0.23 \pm 0.015 \ .$$

This measurement of $\sin^2\theta$ rests mostly on the inclusive neutral current to charge current ratios and the polarized e^-d scattering asymmetry. The error of 0.015 includes the statistical error of 0.010 and an estimated error of 0.005 on the theoretical (mostly quark parton model) uncertainties involved. The corresponding predictions for the masses of the $W^\pm$ and the Z^o are shown in Fig. 15.

This research was supported in part by the National Science Foundation.

Table I

Experimentally Studied N.C. Processes

Process	Comparison with Theory
1. Purely Leptonic $\nu_\mu + e^- \rightarrow \nu_\mu + e^-$ $\bar{\nu}_\mu + e^- \rightarrow \bar{\nu}_\mu + e^-$ $\bar{\nu}_e + e^- \rightarrow \bar{\nu}_e + e^-$	Very Clean (no hadrons involved)
2. Elastic Scattering $\nu_\mu + p \rightarrow \nu_\mu + p$ $\bar{\nu}_\mu + p \rightarrow \bar{\nu}_\mu + p$	Relatively Straightforward Some uncertainty due to proton form factors (M_A)
3. Single Pion Prod. $\nu_\mu + N \rightarrow \nu_\mu + N' + \pi$ $\bar{\nu}_\mu + N \rightarrow \bar{\nu}_\mu + N' + \pi$	Model dependent due to hadronic vertex Also nuclear physics corr.
4. Inclusive $\nu_\mu + N \rightarrow \nu_\mu + \ldots$ $\bar{\nu}_\mu + N \rightarrow \bar{\nu}_\mu + \ldots$	Quark-Parton model dependent
5. Atomic Physics $e^- + Bi \rightarrow e^- + Bi$	Large uncertainties due to atomic physics calculations
6. Electron Scattering $\vec{e}^- + d \rightarrow e^- + \ldots$	Quark-Parton model dependent

Table II

$\nu_\mu + e^- \rightarrow \nu_\mu + e^-$

Experiment	Events Observed	Back-Ground	Cross Section $10^{-42}\ E_\nu\ cm^2$
Gargamelle CERN PS	≤ 1	0.3 ± 0.1	≤ 3
Aachen-Padova CERN PS Spark Chamber	32	21	1.1 ± 0.6
Columbia-BNL Fermilab 15-ft B.C.	11	0.7 ± 0.7	1.8 ± 0.8
Gargamelle CERN SPS	9	0.4 ± 0.4	$2.3^{+1.1}_{-0.8}$
CHARM Collab. CERN SPS Spark Chamber	11	$4\frac{1}{2} \pm 1\frac{1}{2}$	2.6 ± 1.4
E-253 (Mo-Abashian) Fermilab Spark Chamber	46	12	1.4 ± 0.3 *
Average			1.6 ± 0.3

* Statistical error only, upped by $\sim \sqrt{2}$ in average.

Table III

$\bar{\nu}_\mu + e^- \rightarrow \bar{\nu}_\mu + e^-$

Experiment	Total Sample of $\bar{\nu}_\mu + N \rightarrow \mu^+ + ..$	Events Obs.	Back-Ground	Cross Section $10^{-42} E_\nu \; cm^2$
Gargamelle CERN PS		3	0.4 ± 0.1	$1.0 {}^{+2.1}_{-0.9}$
Aachen-Padova CERN PS Spark Ch.		17	7.4 ± 1.0	2.2 ± 1.0
BEBC Wide Band Neon CERN SPS	7500	≤ 1	0.4 ± 0.2	≤ 3.5
FNAL-Mich-IHEP-ITEP Fermilab 15-ft Neon	6300	0		≤ 2.9
Gargamelle CERN SPS	4000	0		≤ 3.3

Table IV

Elastic Scattering on Protons

Experiment	$\nu_\mu+p \to \nu_\mu+p$			$\bar{\nu}_\mu+p \to \bar{\nu}_\mu+p$		
	Events Obs.	Back-Ground	$\frac{\nu_\mu+p\to\nu_\mu+p}{\nu_\mu+n\to\mu^-+p}$	Events Obs.	Back-Ground	$\frac{\bar{\nu}_\mu+p\to\bar{\nu}_\mu+p}{\bar{\nu}_\mu+p\to\mu^++n}$
Harvard-Penn-BNL BNL Counter Detector	255	88	0.11 ± 0.02	69	28	0.19 ± 0.05
Columbia-Ill-Rock BNL Spark Ch.	71	30	0.20 ± 0.06			
Aachen-Padova CERN PS Spark Ch.	155	110	0.10 ± 0.03			
Gargamelle CERN PS	100	62	0.12 ± 0.06			
Weighted Average *+			0.11 ± 0.02			0.19 ± 0.08

* Errors increased by $\sqrt{2}$ since quoted errors statistical only.

\+ Caution - The experiments have slightly different Q^2 cuts.

Table V

Single Pion Production

Experiments	Ratio Measured	Experimental Result	W-S $\sin^2\theta \sim 1/4$
CIR Aachen-Padova Gargamelle Complex Nuclei	$\frac{(\nu+x+\pi^o)}{2(\mu^-+x+\pi^o)}$	0.21 ± 0.07 *	0.24 ±
	$\frac{(\bar{\nu}+x+\pi^o)}{2(\mu^++x+\pi^o)}$	0.46 ± 0.07 *	0.30 ±
Argonne 12 ft. B.C. Deuterium	$\frac{\nu+n+\pi^+}{\mu^-+p+\pi^+}$	0.13 ± 0.06	0.07 ±
	$\frac{\nu+p+\pi^o}{\mu^-+p+\pi^+}$	0.40 ± 0.22	0.17 ±
	$\frac{\nu+p+\pi^-}{\mu^-+p+\pi^+}$	0.12 ± 0.04	0.07 ±
Gargamelle CERN PS Propane	$\frac{(\nu p\pi^o)+(\nu n\pi^o)}{2(\mu^- p\pi^o)}$	0.45 ± 0.08	0.42 ±
	$\frac{(\bar{\nu} p\pi^o)+(\bar{\nu} n\pi^o)}{2(\mu^+ n\pi^o)}$	0.57 ± 0.11	0.60 ±
	$\frac{\nu+p+\pi^o}{\mu^-+p+\pi^o}$	0.56 ± 0.10	0.42 ± 0.13
	$\frac{\nu+n+\pi^o}{\mu^-+p+\pi^o}$	0.34 ± 0.09	0.42 ± 0.13
	$\frac{\nu+p+\pi^-}{\mu^-+p+\pi^o}$	0.45 ± 0.13	0.28 ± 0.08
	$\frac{\nu+n+\pi^+}{\mu^-+p+\pi^o}$	0.34 ± 0.07	0.28 ± 0.08

* Weighted average of three experiments. Agreement between experiments not very good. Errors scaled by $\sqrt{\chi^2/(n-1)}$. Average may not be valid.

Table VI

Inclusive Neutral Current Ratios on ~Isoscalar Targets

$$R_\nu = \frac{\nu_\mu + N \to \nu_\mu + \ldots}{\nu_\mu + N \to \mu^- + \ldots} \qquad R_{\bar{\nu}} = \frac{\bar{\nu}_\mu + N \to \bar{\nu}_\mu + \ldots}{\bar{\nu}_\mu + N \to \mu^+ + \ldots}$$

Experiment	E_ν GeV	E_H Cut GeV	R_ν *	$R_{\bar{\nu}}$ *
Gargamelle CERN PS	1- 10	1	0.26 ±0.04	0.39 ±0.06
HPWF Fermilab	30-200	4	0.30 ±0.04	0.33 ±0.09
CITF Fermilab	30-200	12	0.27 ±0.02	0.40 ±0.08
ABCLOS CERN SPS BEBC Neon	30-200	15	0.31 ±0.04	0.37 ±0.08
CDHS CERN SPS	30-200	12	0.307±0.008	0.373±0.025
CHARM CERN SPS	30-200	2-17	0.30 ±0.02	0.39 ±0.02
Weighted Average			0.301±0.007	0.38 ±0.014

* Rates Corrected for effects of E_{hadron} cuts.

Table VII

Parity Violations in Heavy Atoms

Experiment	Atom	Transition Used	Exptl.Result / W-S Prediction
Seattle Fortson et al	Bismuth 83	8757 Å	0.2 to 0.3
Oxford Sanders et al	"	6480 Å	0.72 ± 0.13
Novosibirsk Barkov et al	"	"	1.07 ± 0.14
Berkeley Commins et al	Thallium 81	2927 Å	$2.3 \begin{array}{l} +3.1 \\ -1.4 \end{array}$

Table VIII

Summary of Neutral Currents - Comparison with Weinberg-Salam

Process	Experimental Result	W-S Prediction with $\sin^2\theta=0.23$	$\sin^2\theta$
1. Purely Leptonic			
$\bar{\nu}_e+e^- \rightarrow \bar{\nu}_e+e^-$	$(5.7\pm1.2) \times 10^{-42}\ E_\nu\ cm^2$	5.0	0.29 ± 0.05
$\nu_\mu+e^- \rightarrow \nu_\mu+e^-$	$(1.6\pm0.3) \times 10^{-42}\ E_\nu\ cm^2$	1.5	0.23 ± 0.05
$\bar{\nu}_\mu+e^- \rightarrow \bar{\nu}_\mu+e^-$	$(1.8\pm0.9) \times 10^{-42}\ E_\nu\ cm^2$	1.3	$0.30^{+0.10}_{-0.30}$
2. Elastic Scattering			
$\nu_\mu+p \rightarrow \nu_\mu+p$	$(0.11\pm0.02)\times\sigma(\nu_\mu+n\rightarrow\mu^-+p)$	0.12	0.26 ± 0.06
$\bar{\nu}_\mu+p \rightarrow \bar{\nu}_\mu+p$	$(0.19\pm0.08)\times\sigma(\bar{\nu}_\mu+p\rightarrow\mu^++n)$	0.11	≤ 0.5
3. Single Pion Prod.			
$\nu_\mu+N \rightarrow \nu_\mu+N+\pi^o$	$(0.45\pm0.08)\times\sigma(\nu_\mu+N\rightarrow\mu^-N\pi^o)$	0.42	0.22 ± 0.09
$\bar{\nu}_\mu+N \rightarrow \bar{\nu}_\mu+N+\pi^o$	$(0.57\pm0.11)\times\sigma(\bar{\nu}_\mu+N\rightarrow\mu^+N\pi^o)$	0.60	0.15-0.52
4. Inclusive			
$\nu_\mu+N \rightarrow \nu_\mu+ \ldots$	$(0.301\pm0.007)\times\sigma(\nu_\mu+N\rightarrow\mu^-+ \ldots)$	0.30	0.23 ± 0.015
$\bar{\nu}_\mu+N \rightarrow \bar{\nu}_\mu+ \ldots$	$(0.38\ \pm0.014)\times\sigma(\bar{\nu}_\mu+N\rightarrow\mu^++ \ldots)$	0.38	$0.3\ \pm0.1$
5. Polarized e^-d $e^-+d \rightarrow e^-+ \ldots$	$\frac{\sigma(\vec{e})-\sigma(\overleftarrow{e})}{\sigma(\vec{e})+\sigma(\overleftarrow{e})} = (-9.5\pm1.6)q^2\times10^{-5}$	-8.3	0.22 ± 0.02

Table IX

Summary of the Weak Neutral Current Couplings

Coupling Constant	Value in Unique Solution	Weinberg-Salam $\sin^2\theta=0.23$	Weinberg-Salam As Function of $\sin^2\theta$
g_V	0.04 ± 0.06	-0.04	$-\frac{1}{2}+2\sin^2\theta$
g_A	-0.54 ± 0.05	-0.50	$-\frac{1}{2}$
u_L	0.35 ± 0.03	0.35	$\frac{1}{2}-\frac{2}{3}\sin^2\theta$
d_L	-0.42 ± 0.03	-0.42	$-\frac{1}{2}+\frac{1}{3}\sin^2\theta$
u_R	-0.18 ± 0.03	-0.15	$-\frac{2}{3}\sin^2\theta$
d_R	-0.01 ± 0.05	0.08	$\frac{1}{3}\sin^2\theta$

References

1 J.J. Sakurai, Proc. of the Conf. on Neutrino Physics, Oxford, p. 338 (1978).
C. Baltay, Proc. of the 19th Int. Conf. on HEP, Tokyo, p. 882 (1978).

2 S. Weinberg, Phys. Rev. Lett. 19, 1264 (1967); A. Salam, in Elementary Particle Theory, edited by N. Svartholm (Stockholm 1968), p. 367.

3 J.J. Sakurai, talk presented at the Neutrino '79 Conf. Bergen, Norway; and P.Q. Hung, J.J. Sakurai, preprint UCLA/79/TEP/9.

4 A complete set of references for the older experiments can be found in Ref. 1 and will not be repeated here.

5 N. Armenise et al, submitted to PRL, April 1979.

6 M. Jonker et al, paper presented at Neutrino '79.

7 L. Mo et al, VPI Preprint.

8 See Ref. 6.

9 C. Geweniger et al, paper presented at Neutrino '79.

10 C. Baltay et al, to be published in Phys. Rev.

11 C.Y. Prescott et al, Phys. Lett. 77B, 347 (1978) and SLAC PUB 2319 (1979).

12 H.H. Williams et al, Proc. of the Int. Conf. on HEP, Tokyo, (1978), p. 325.

13 M. Holder et al, Phys. Lett. 71B, 222 (1977); Phys. Lett. 72B, 254 (1977).

14 F.S. Merritt et al, Preprint CALT-68-600,601 (1977).

15 M. Holder et al, Phys. Lett. 72B, 254 (1977).

16 Aachen-Bonn-CERN-London-Oxford-Saclay Collab., paper #965, Tokyo Conference.

17 J.J. Sakurai, Invited paper, Neutrino '77, Elbrus, USSR , June (1977).

18 L.M. Sehgal, Phys. Lett. 71B, 99 (1977).

19 P.Q. Hung, J.J. Sakurai, Phys. Lett. 72B, 208 (1977).

20 L.F. Abbott, R.M. Barnett, Phys. Rev. Lett. 40, 1303 (1978); SLAC Preprint SLAC-PUB 2136 (1978).

21 P. Langacker, D.P. Sidhu, Phys. Lett. 74B, 233 (1978); Phys. Rev. Lett. 41, 732 (1978).
22 E.A. Paschos, Brookhaven National Laboratory preprint, BNL-24619 (1978).
23 M. Claudson, E.A. Paschos, L.R. Sulak, paper #1158 Tokyo Conference.
24 P. Langacker, et al, Univ. of Pennsylvania preprint C00-3071-243.
25 H. Kluttig et al, Phys. Lett. 71B, 446 (1977).
26 H. Deden, Neutrino '79.
27 B. Roe, Neutrino '79.
28 See Ref. 27.
29 F.A. Harris et al, Phys. Rev. Lett 39, 437 (1977).
30 M. Derrick et al, Phys. Rev. D18, 7 (1978).
31 J. Marriner, Lawrence Berkeley Laboratory Report LBL 6438 (1977), U. of Cal. Ph.D. Thesis.
32 See Ref. 3.
33 L.M. Barkov et al, Pisma Zh. Eskp. Fiz. (JETP Lett. 26, 379 (1978)) and private communication of latest results.

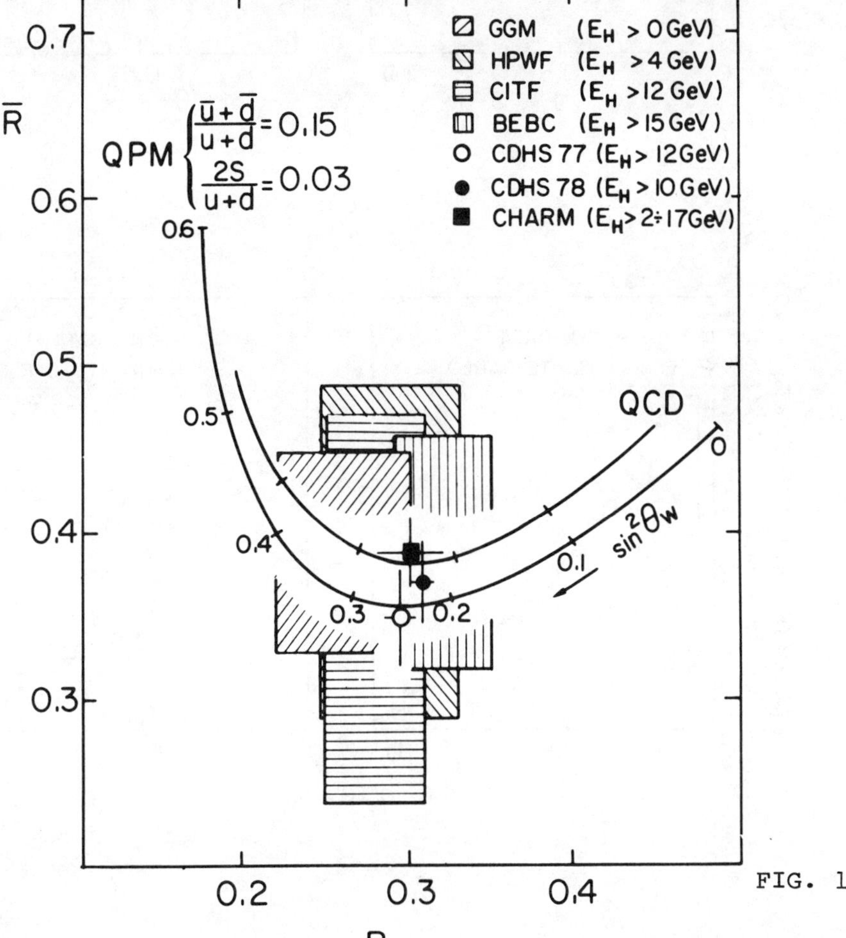

FIG. 1

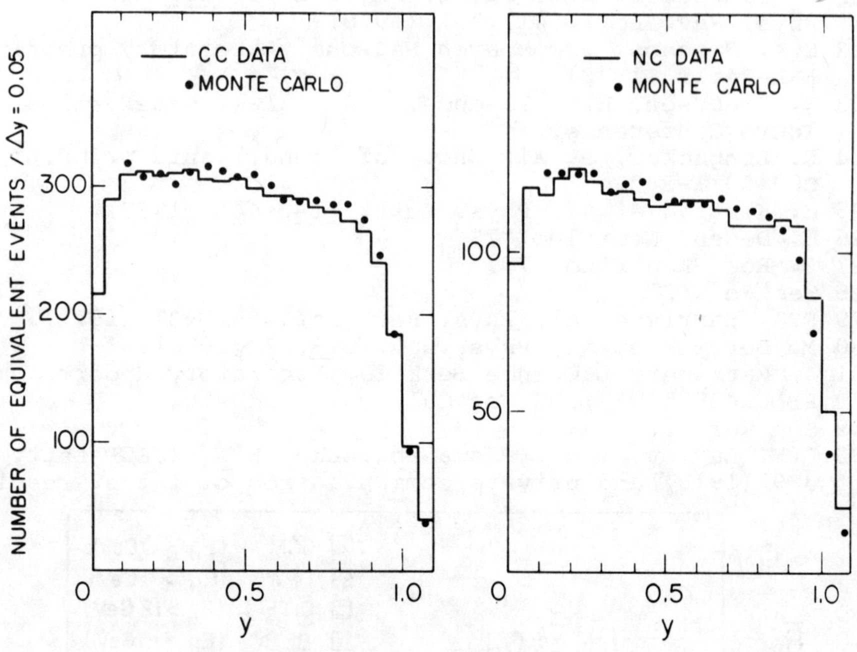
CC DATA
• MONTE CARLO
NC DATA
• MONTE CARLO
NUMBER OF EQUIVALENT EVENTS / Δy = 0.05
300
200
100
50
0
0.5
1.0
y

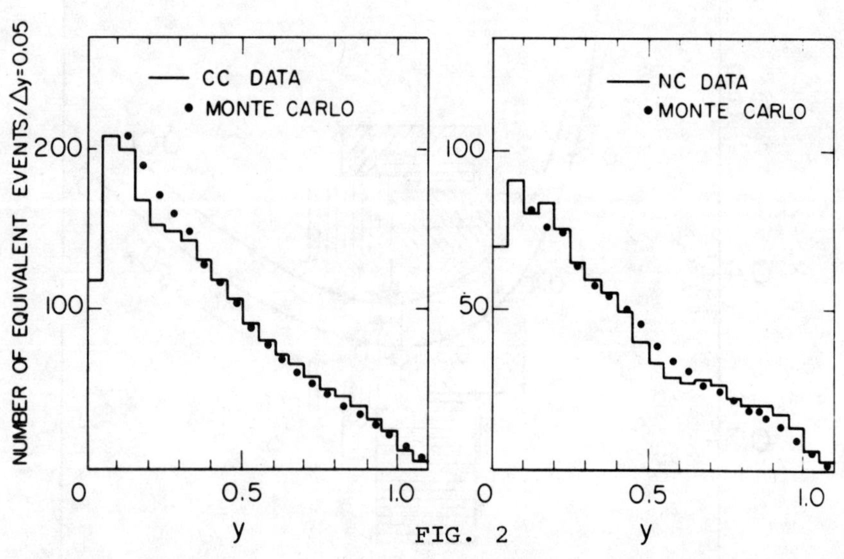
CC DATA
• MONTE CARLO
NC DATA
• MONTE CARLO
NUMBER OF EQUIVALENT EVENTS/Δy=0.05
200
100
50
0
0.5
1.0
y

FIG. 2

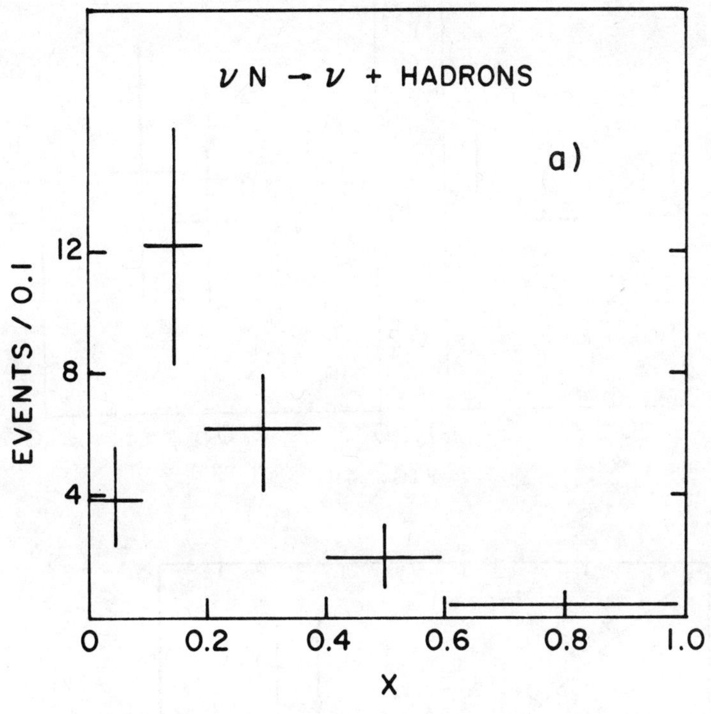

ν N → ν + HADRONS
a)
EVENTS / 0.1
12
8
4
0
0.2
0.4
0.6
0.8
1.0
X

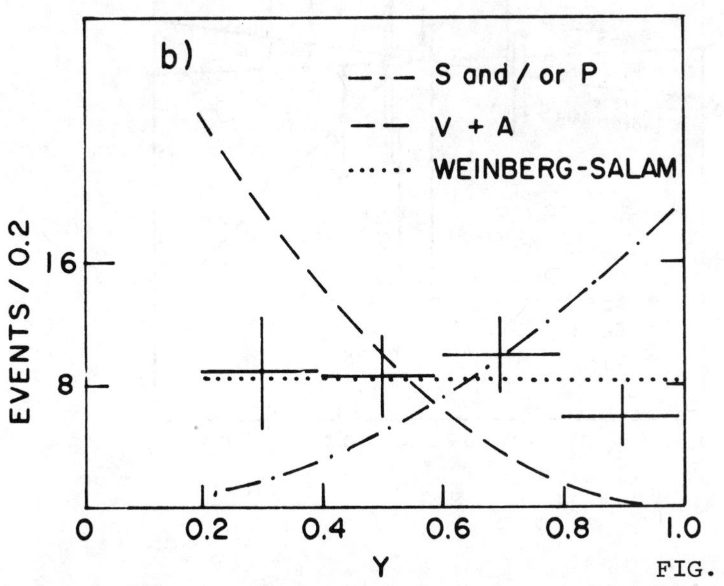

b)
S and / or P
V + A
WEINBERG-SALAM
EVENTS / 0.2
16
8
0
0.2
0.4
0.6
0.8
1.0
Y

FIG. 3

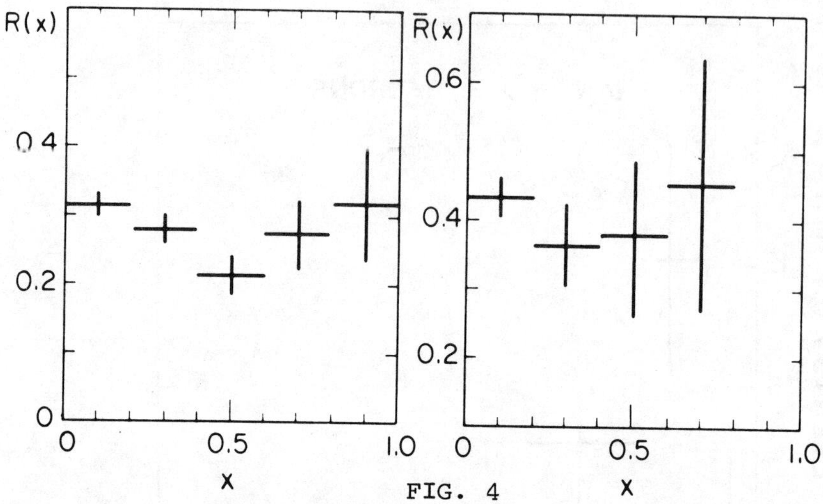
R(x)
0.4
0.2
0
0
0.5
1.0
x
R̄(x)
0.6
0.4
0.2
0
0.5
1.0
x

FIG. 4

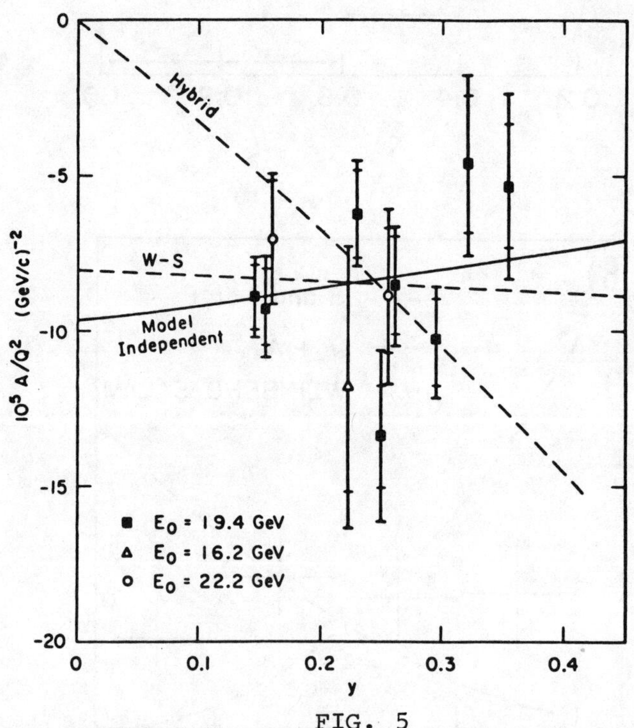
Hybrid
W-S
Model Independent
10^5 A/Q^2 (GeV/c)$^{-2}$
0
-5
-10
-15
-20
0
0.1
0.2
0.3
0.4
y
E_0 = 19.4 GeV
E_0 = 16.2 GeV
E_0 = 22.2 GeV

FIG. 5

SUMMARY OF MEASUREMENTS OF $\sin^2\theta_W$

Weighted Average $\sin^2\theta = 0.23 \pm 0.01$
(statistical error only)

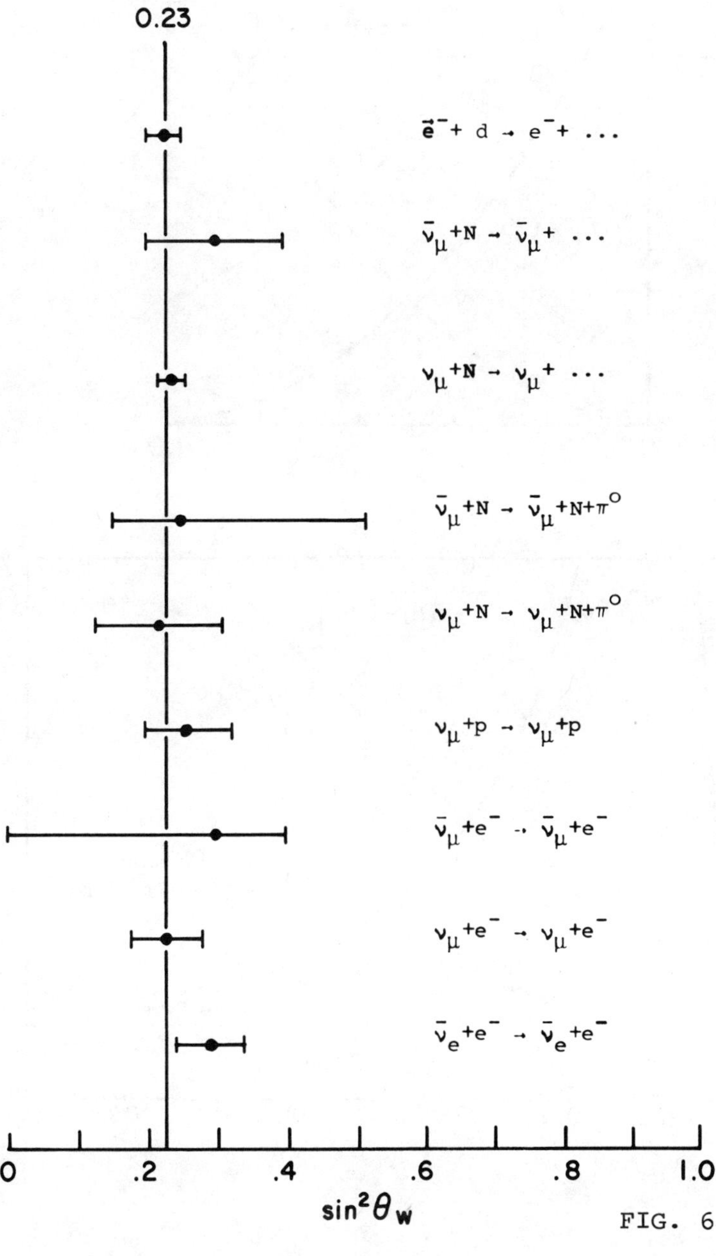

FIG. 6

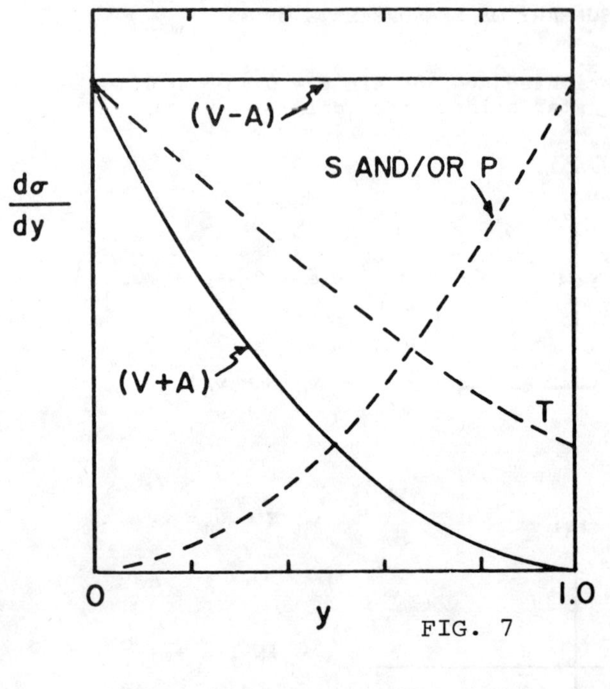
(V-A)
S AND/OR P
dσ/dy
(V+A)
T
0
1.0
y

FIG. 7

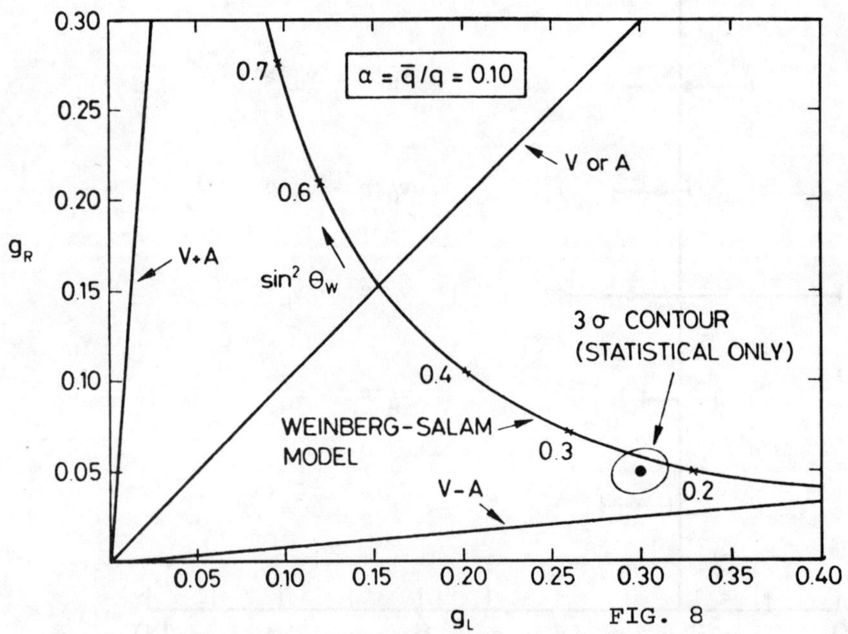
α = q̄/q = 0.10
0.7
0.6
V or A
V+A
sin² θW
gR
3σ CONTOUR
(STATISTICAL ONLY)
0.4
WEINBERG-SALAM
MODEL
0.3
0.2
V-A
0.30
0.25
0.20
0.15
0.10
0.05
0.05
0.10
0.15
0.20
0.25
0.30
0.35
0.40
gL

FIG. 8

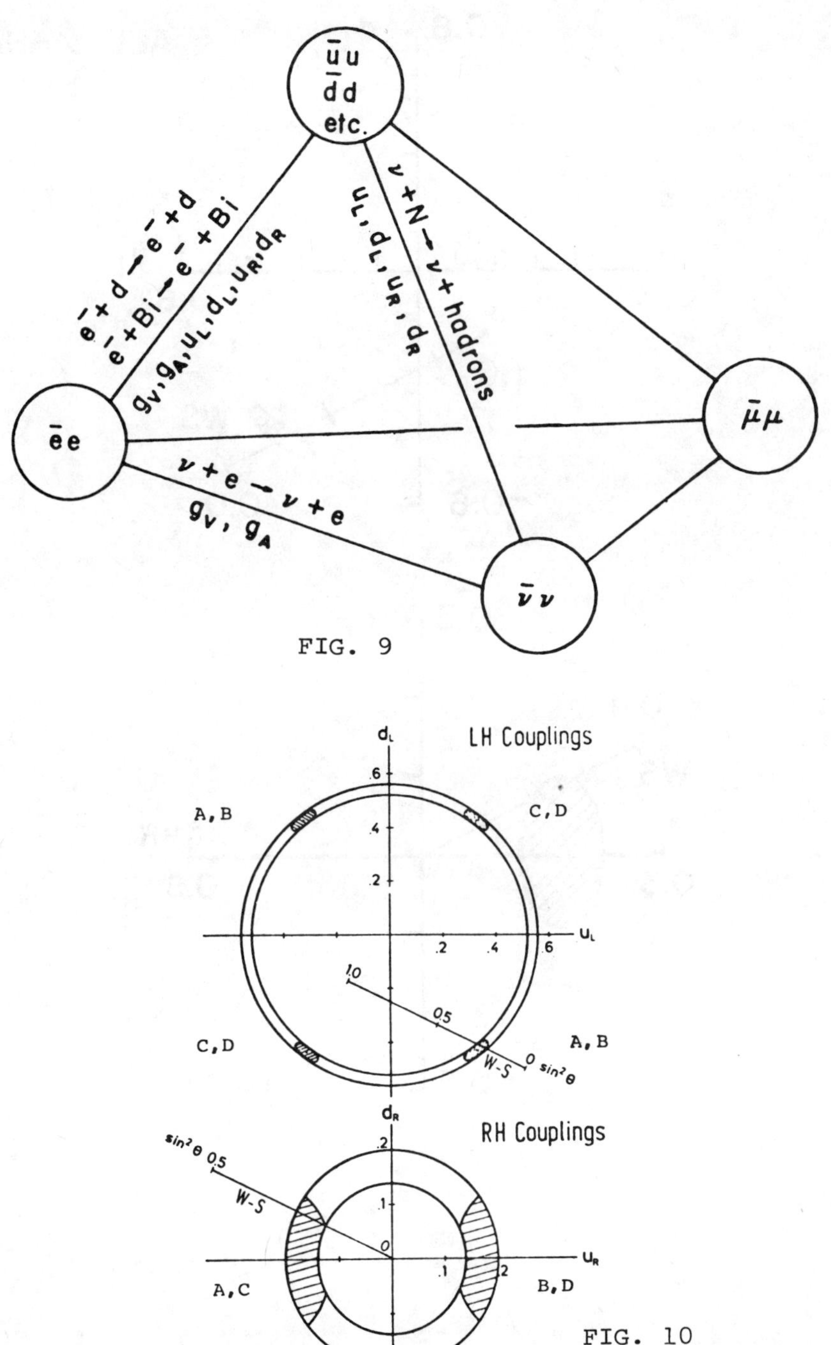
$\bar{u}u$
$\bar{d}d$
etc.
$e^- + d \to e^- + d$
$e^- + Bi \to e^- + Bi$
$g_V, g_A, u_L, d_L, u_R, d_R$
$\nu + N \to \nu +$ hadrons
u_L, d_L, u_R, d_R
$\bar{e}e$
$\bar{\mu}\mu$
$\nu + e \to \nu + e$
g_V, g_A
$\bar{\nu}\nu$

FIG. 9

d_L
LH Couplings
A, B
C, D
C, D
A, B
u_L
W-S
$\sin^2\theta$
d_R
RH Couplings
$\sin^2\theta$ 0.5
W-S
A, C
B, D
u_R

FIG. 10

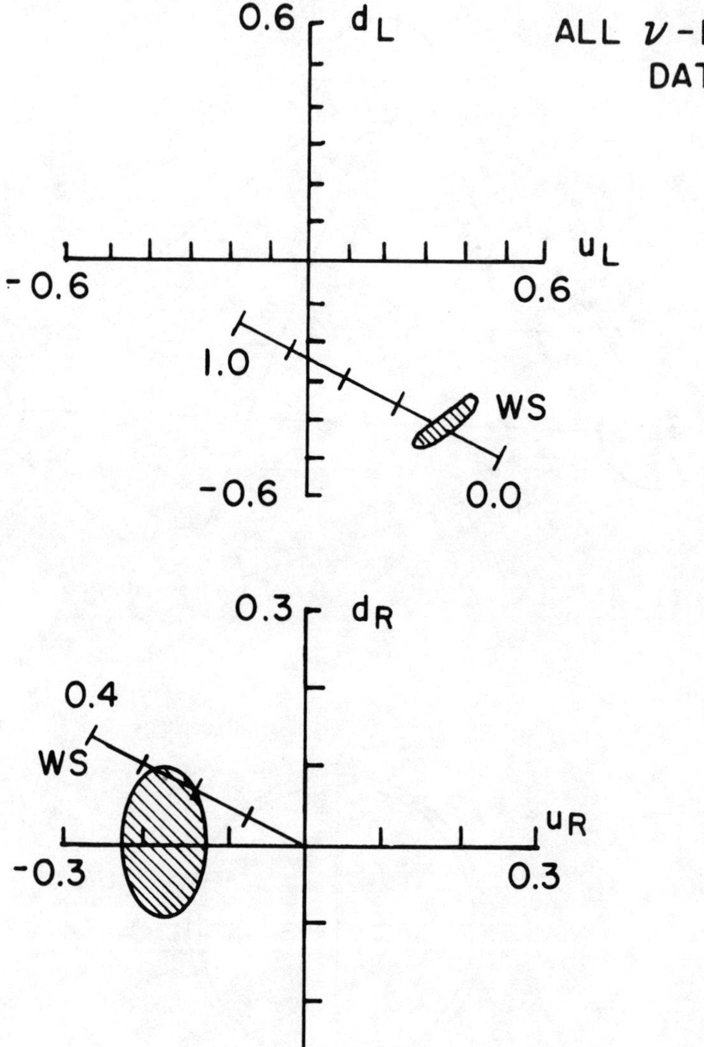

FIG. 11

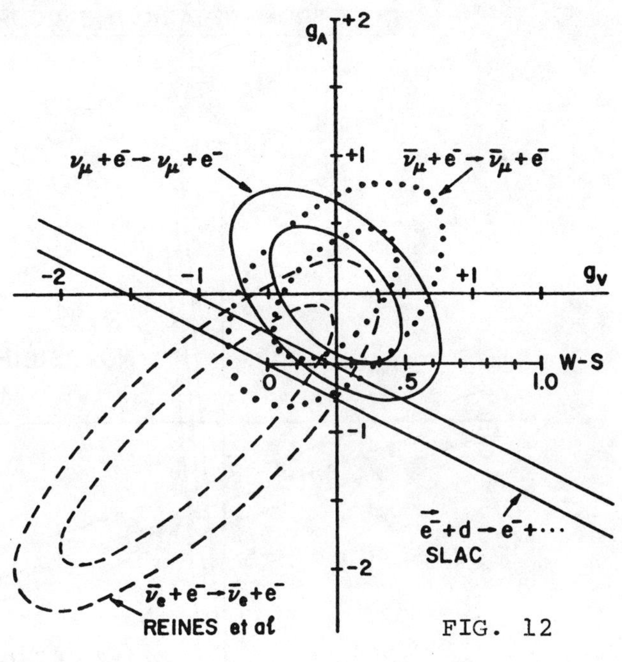

g_A
+2
+1
-2
-1
+1
g_V
0
.5
1.0
W-S
-1
-2
$\nu_\mu + e^- \to \nu_\mu + e^-$
$\bar{\nu}_\mu + e^- \to \bar{\nu}_\mu + e^-$
$\vec{e}^- + d \to e^- + \cdots$
SLAC
$\bar{\nu}_e + e^- \to \bar{\nu}_e + e^-$
REINES et al

FIG. 12

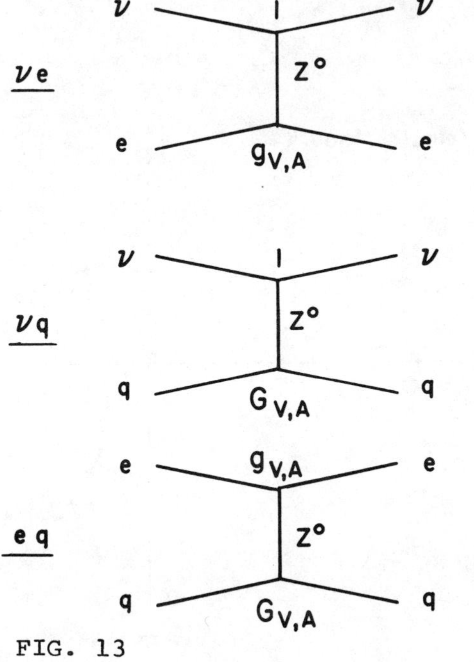

ν
ν
νe
Z°
e
e
$g_{V,A}$
ν
ν
νq
Z°
q
q
$G_{V,A}$
e
$g_{V,A}$
e
eq
Z°
q
q
$G_{V,A}$

FIG. 13

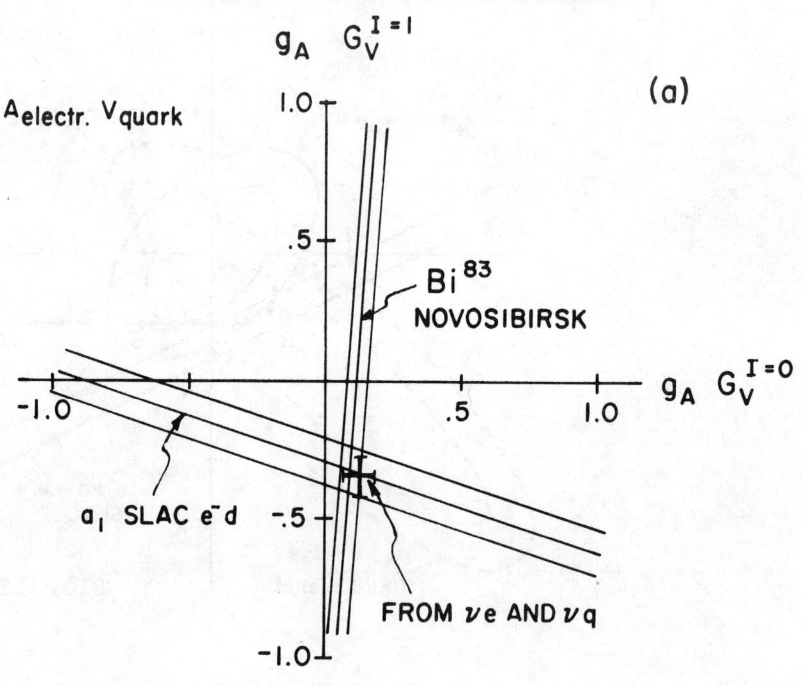
ELECTRON - QUARK N.C. COUPLINGS
$g_A \; G_V^{I=1}$
(a)
$A_{electr.} \; V_{quark}$
1.0
.5
-.5
-1.0
Bi^{83}
NOVOSIBIRSK
-1.0
.5
1.0
$g_A \; G_V^{I=0}$
a_1 SLAC e^-d
FROM νe AND νq

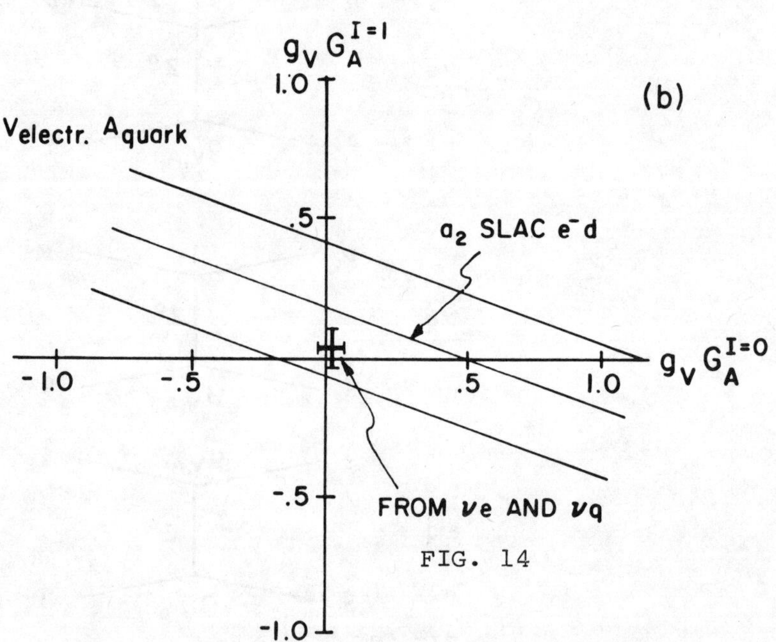
$g_V \; G_A^{I=1}$
1.0
(b)
$V_{electr.} \; A_{quark}$
.5
a_2 SLAC e^-d
-1.0
-.5
.5
1.0
$g_V \; G_A^{I=0}$
-.5
FROM νe AND νq
-1.0

FIG. 14

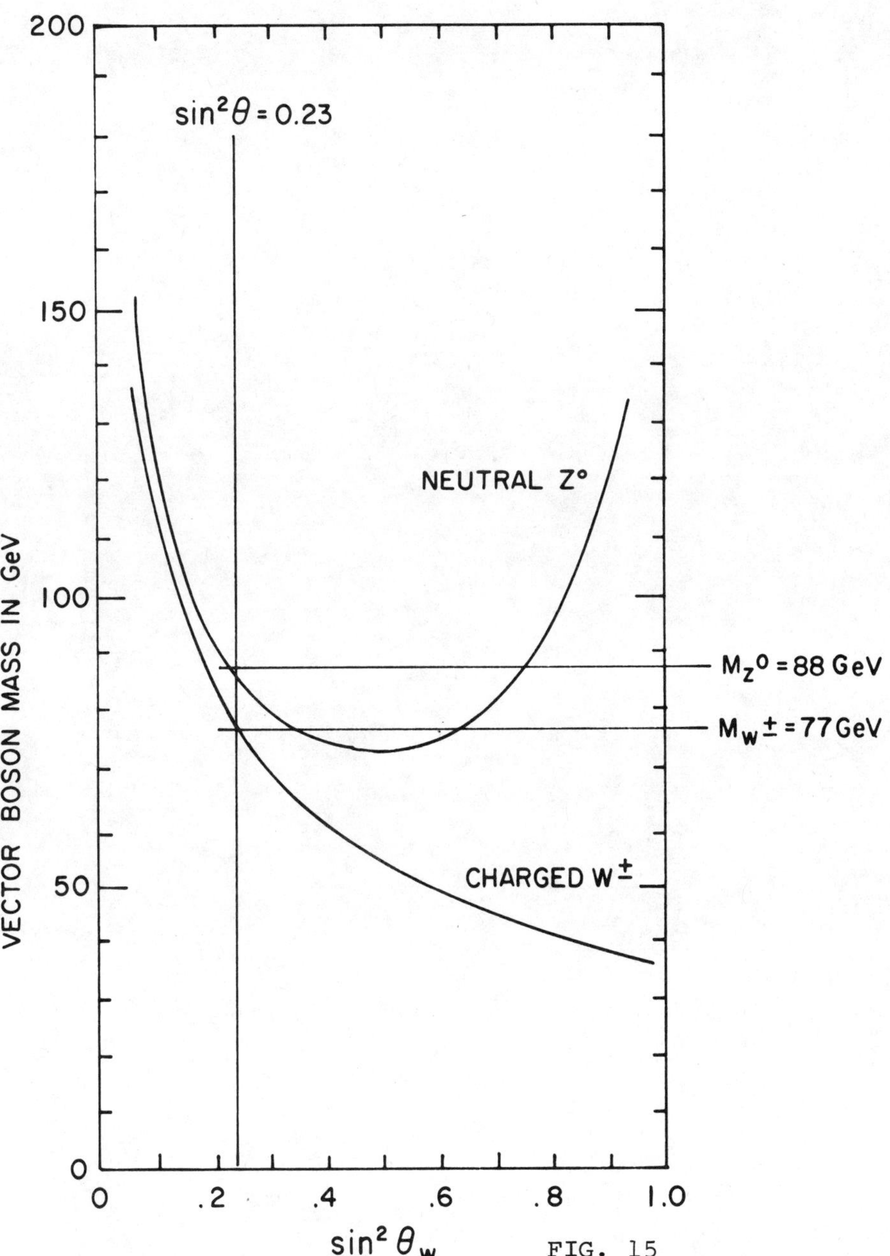
200
150
100
50
0
$\sin^2\theta = 0.23$
NEUTRAL Z°
CHARGED W±
M_{Z^0} = 88 GeV
$M_{W^\pm}$ = 77 GeV
VECTOR BOSON MASS IN GeV
0
.2
.4
.6
.8
1.0
$\sin^2\theta_W$

FIG. 15

MEASUREMENT OF THE CROSS SECTION
For $\nu_\mu + e^- \quad \nu_\mu + e-$

A. Abashian
National Science Foundation, Washington, D.C. 20550

ABSTRACT

A total of ~ 249,000 neutrino interactions were observed at Fermilab in a high angular resolution electromagnetic shower detector, with 0.947×10^{19} protons of 350 GeV energy incident upon the production target. The neutrino beam resulted from single-horn focussing of the charged beam and was wide-band. Based on a data sample of 0.71×10^{19} protons, 46 electrons were observed with $\theta_e < 10$ mr. Of these 46 events, 34 are attributed to the process $\nu_\mu + e^- \rightarrow \nu_\mu + e^-$, and 12 are attributed to background processes. This leads to the following results: $\sigma = (1.40 \pm 0.30) \times 10^{-42}$ Ecm^2; and $\sin^2\theta_W = 0.25$, in agreement with other measurements of $\sin^2\theta_W$ and the Weinberg-Salam theory.

INTRODUCTION

The subject and status of neutral weak interactions has been thoroughly discussed by Professor Baltay in the previous paper. (1.) I will confine my remarks on neutral currents largely to an experiment on $\nu_\mu e$ elastic scattering performed by a collaboration of physicists from VPI, Maryland, Oxford, Peking, and NSF. Determinations of the Weinberg angle, and the vector coupling constants g_a, and g_v will be given and comparisons made with other accurate determinations of the angle.

THEORY AND HISTORY

Prior to the discovery of the tau lepton, the weak neutral current couplings could be shown diagrammatically as shown in Figure 1.

Couplings between the various lepton states such as $e\nu$, and μ involve only point like objects, are particularly simple and provide unique tests of various models of the weak interactions. In the cases where the couplings involve u and d quarks, structure functions of the hadrons need to be known to extract the information about the weak processes. Experimentally, large amounts of data on neutrino-quark scattering are obtained from the deep inelastic scattering of neutrinos from nucleons. Extraction of $\sin^2\theta_W$ involves many theoretical calculations and assumptions about valence and sea quarks, flavors, colors etc. It is probably accurate to say that the uncertainty in $\sin^2\theta_W$ is due not only to statistics but in addition due to theoretical uncertanties.

In the case of electron-quark couplings as determined from the scattering of longitudinally polarized electrons from deuterium, both experimental and theoretical difficulties arise, the former because very small asymmetries of the order of 10 parts per million need to be measured.

ISSN:0094-243X/80/590055-21$1.50

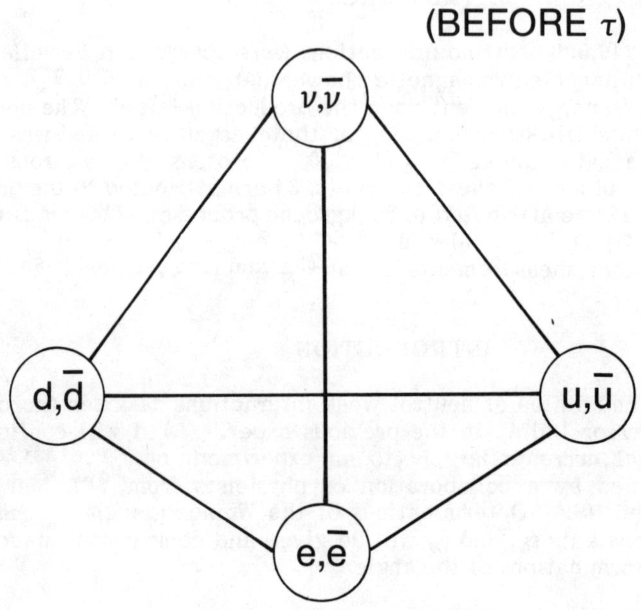

Figure 1. Schematic diagram of weak couplings.

In the case of neutrino electron scattering, difficulties arise from the smallness of the cross section; namely, of the order of $10^{-42}cm^2$. Theoretically, the situation is rather straight forward. Prior to the discovery of the weak neutral currents, the interaction was assumed to be vector, charged current-current and expressable as given in equation (1.).

$$H_{WI} = \frac{Gv}{2} j^{+}_{\lambda} j_{\lambda} \qquad (1)$$

$$\text{With } j_{\lambda}^{lepton} = \bar{\mu} \quad \lambda \ (1 + \quad 5) \ \nu_{\mu} + \bar{e} \quad \lambda \ (1 + \quad 5) \ \nu e \qquad (2)$$

For ν_{μ} scattering, the cross section is predicted to be identically zero; with the observation of the reaction, neutral currents were proven to exist, ν_{μ}. (2)

The diagram ν_{μ} e scattering is shown in Figure 2.

As given by 'tHooft, (3.), the differential cross section is given by

$$\frac{d\sigma = G_V^2 \ m_e \ E_{\nu}}{d_y \quad |2\P} \ \{(g_v + g_a) + (g_v - g_a)^2 \ (1-y)^2 \} \qquad (3)$$

$$\text{where } y = \frac{Ee}{E\nu} = \frac{\text{electron energy}}{\text{incident neutrino energy}} \qquad (4)$$

g_v and g_a = vector and axial vector lepton coupling constants

G_v= Fermi vector coupling constant

M_e= electron mass

The interaction is now given by

$$H_{WI} = \frac{G_V}{|\ 2} \quad \{ \nu_{\mu} \ \Upsilon_{\lambda} \ (1 + \quad 5)\} \times \{e \quad \lambda \quad (g_V + g_a \quad 5)e\} \qquad (5)$$

In the simplest Weinberg-Salam model, (3), the axial and vector coupling constants are given by:

$$g_a = \tfrac{1}{2} \qquad (6)$$

$$g_v = 2 \sin^{2\theta} w - \tfrac{1}{2} \qquad (7)$$

Before discussing the experimental details, it is important to recognize a few essential features of $\nu_{\mu}e$ scattering. Firstly, there is only a single visible particle in the final state. Secondly, because the involved particles are essentially massless, the electrons are emitted into a very small angular core

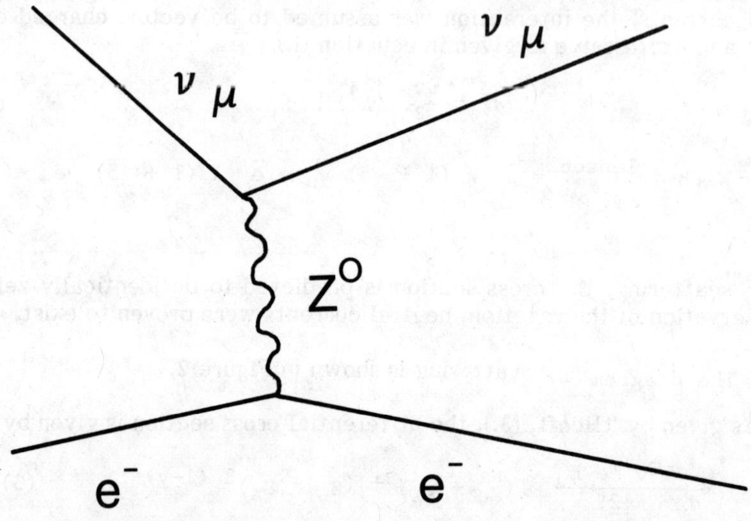

Figure 2. Feynman diagram for $\nu_\mu e$ scattering.

of the order of a few milliradians in the laboratory. Typically, events are contained within an angle given by

$$E_{\theta}^{2} = 2M_e \qquad (8)$$

Finally, although background processes such as neutral current interactions of ν_μ and charged current processes of ν_e have cross sections up to several thousand times as large as the desired reaction, their scattering kinematics is not typically confined to such a narrow forward core.

EXPERIMENTAL DETAILS

The members of the collaboration of this experiment are given below:

R. Heisterberg	VPI& St. U.
L. Mo	"
T. Nunnamaker	"
K. Lefler	Maryland
A. Skuja	"
A. Abashian	NSF
N. Booth	Oxford
C. C. Chang	HEPI Peking
C. Li	"
C. Wang	"

The experiment was conducted from November, 1978 through March, 1979 in the single-horn focussed, wide-band ν beam, at the Fermi National Laboratory. The average neutrino energy was about 20 GeV but energies extended to over 100 GeV. The beam contamination was estimated to be 11% for $\bar{\nu}_\mu$, and less than 0.5% for both νe and $\bar{\nu} e$. Approximately $0.947 \times 10^{+19}$ protons (1.5 Coulomb) at 350 GeV were delivered onto the neutrino production target.

The apparatus, shown in Figure 3, was located approximately 500 m from the end of the pion decay pipe. It consisted primarily of 49 basic modules of detectors. Each module, with details shown in Figure 4, was made of one 9.27 cm thick Aluminum plate ($\sim$1 r. ℓ.), one 1 m x 1 m multiple-wire proportional chamber (MWPC), and one layer of plastic scintillation counters. The cathode planes of each MWPC had delay-lines, of spacing 1.5 mm, along orthogonal directions. Both the x and y-coordinates of the edges of showers were measured in each chamber with five taps equidistantly spaced across the edge of each chamber. The signal from each of the taps was fed to a time-to-digital converter (TDC) to record the distance from a tap as a time interval. Anode wires of each chamber were grouped togther with their common pulses fed to an analogue-to-digital converter (ADC) to record energy deposited in the chamber. The plastic scintillation counters in each module were also used to record the energy deposited in that layer. Hodoscope counters (A) before the apparatus were used to reject muons while, behind the aparatus, two banks of hodoscope counters (B and C), separated by 4' of steel identified charged current interactions events by the muon penetration.

The trigger for an event was given by $\bar{A}$*(CH)*(S), where $\bar{A}$ was the front charged particle veto, CH was the sum of any six consecutive chambers, and S

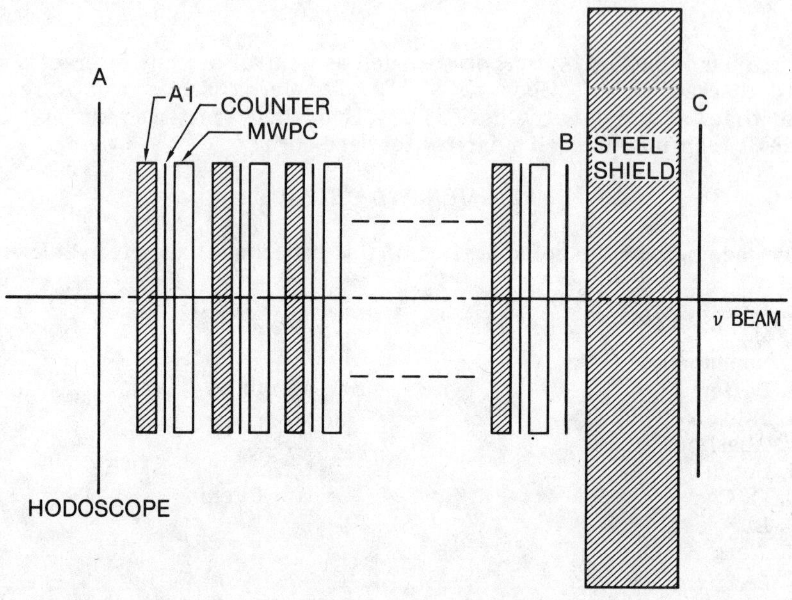

Figure 3. Schematic of experimental apparatus .

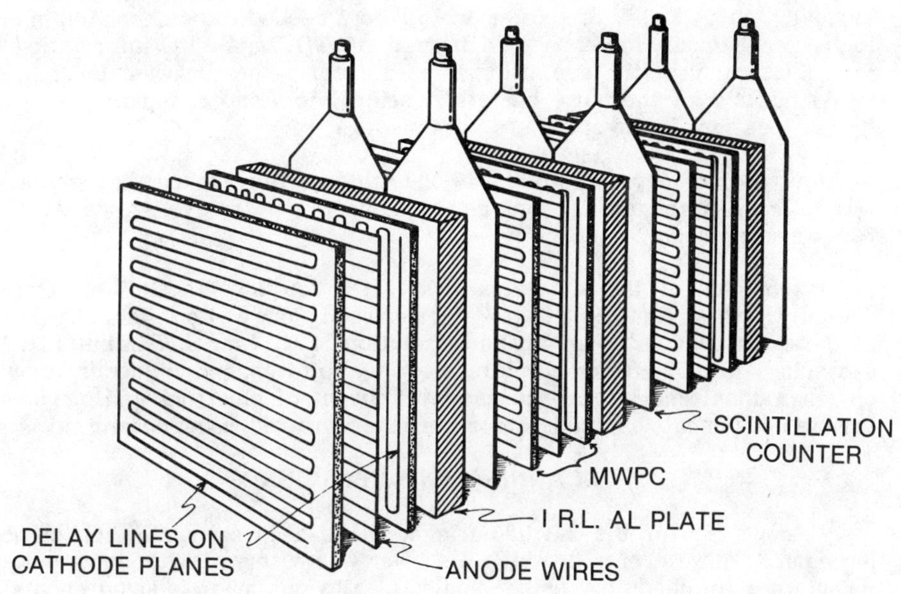

Figure 4. Schematic showing the details of a basic detector modules .

was the sum of any four consecutive layers of scintillation counters. Typically, about 2 GeV of shower energy were needed to be deposited in order to trigger. Upon occurence of a trigger, all TDC and ADC information was read out onto magnetic tape, using an on-line computer. Between beam pulses, visual displays of the pulse heights in scintillators and chambers, as well as shower locations in the chambers were made.

A total of approximately 249,000 triggers were taken during the run. Due to low discriminator setting and high beam rates, an average dead-time of 33% was observed.

A portion of the apparatus was tested for 2.5 - 9.5 GeV electrons at Cornell University and 5 - 30 GeV electrons and pions at Fermilab. The results of these tests indicated an angular resolution of $\pm$ 4 mr (FWHM) at 4 Gev, as shown in Figure 5, gave an absolute energy calibration, and yielded information on longitudinal and lateral shower development of electrons and pions. The various components of the detector were standardized, using cosmic rays.

SCANNING AND MEASURING

About 70% of the 249,000 triggers have been scanned, each one by a physicist. Those events in which a shower was observed without an outgoing muon were retained for further analysis. Obvious muon-induced events and those events obviously developing from outside of the detector assembly were rejected. Approximately 80,000 events passed the initial scan. Of these, only about 1,000 events were observed visually to have a relatively small angle relation to the neutrino beam axis.

For each of the 80,000 events which passed the initial scan, the two "outger edges" of the shower in each chamber were determined, using the redundancy of the TDC information. The midpoint between the edges was then assumed to be the centroid of the shower in that chamber. A straight-line fit to the centroids of the first five chambers for both x and y coordinates was then made. Because most of these events are due to neutral current interactions and involve more than one electromagnetic shower, only about one third of the events could be fitted to the hypothesis of a signal shower. For those events for which straight line fits could be made, only those events with at least one projected angle less than 50 mr were retained.

Each of these events was re-examined carefully by a physicist to see that the event looked like a single electromagnetic shower and that a back-scatter was not involved. This reduced the sample to about 1,000 events. At this point, each event was hand-fitted by a physicists thereby further reducing the sample to about 480 events. Each event was then subjected to a cut in deE/dx for the first 4 chambers which attempted to further enhance the electron signal relative to the hadron background. This cut required that the fraction of the total energy deposited in the first three and four radiation lengths was greater than 12.5% and 25%, respectively. In tests with 10 GeV electrons, it was found that 92% of electrons pass this cut, compared to only 25% for neutral current interaction events.

The final sample of 313 events was then examined for biases. Figures 6a and 6b show longitudinal and transverse vertex distribution for those events

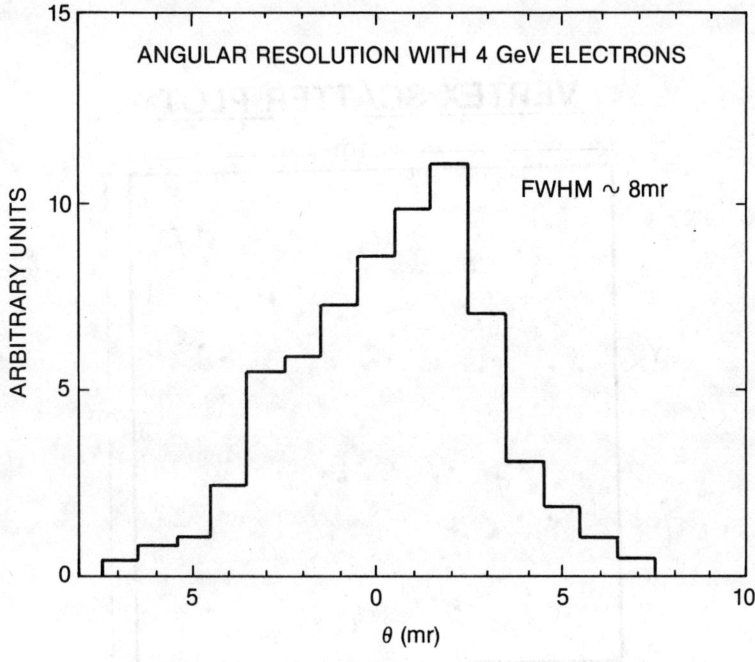

Figure 5. Measured electron angular resolution at 4 GeV.

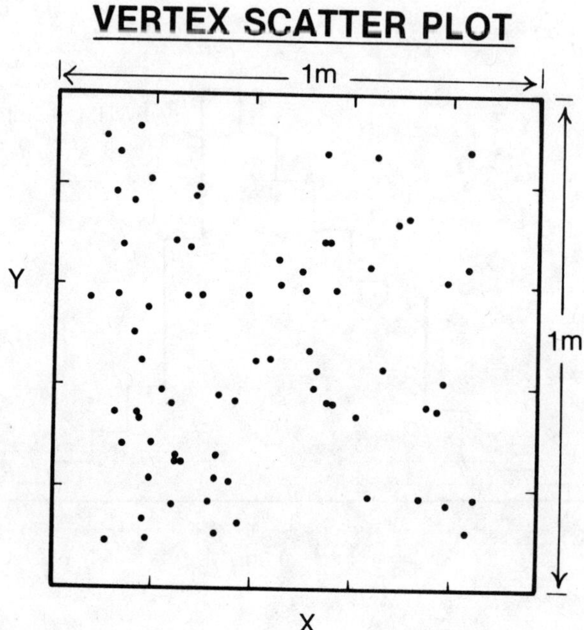

Figure 6a Vertex distribution along the neutrino beam direction for events with $\theta_{-} \leq 20$ mr.

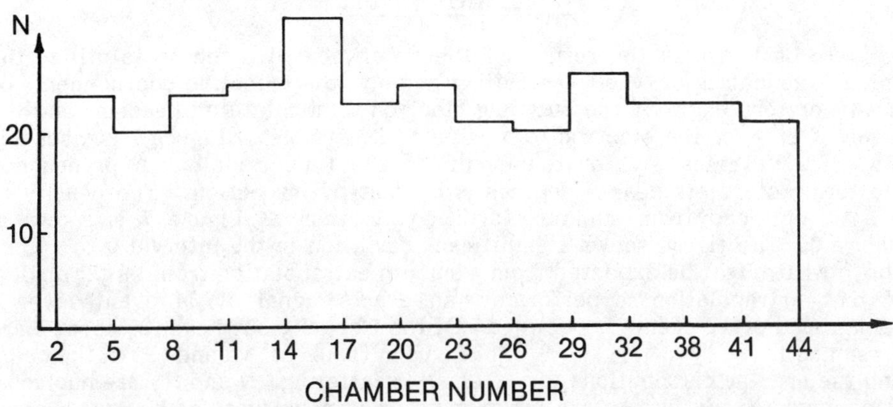

Figure 6b Vertex distribution transverse to the neutrino beam direction for events with $\theta \leq 20$ mr.

with $\theta_e > 20$ mr. Isotropy in all coordinates is apparent indicating spatial biases are absent. A fiducial area of 90 cm x cm transverse to the beam direction is also evident. In order to clearly identify the events, and account for energy leakage, only 40 detector modules were used in the final selection.

DATA AND RESULTS

To best display the results of the experiment, we choose to utilize the unique kinematics of $\nu_\mu e$ scattering; namely, the effective confinements of the laboratory angle of the outgoing electron to about 10 milliradians and the value of $E\,\theta^2$ for the electron to a value of $2m_e$=1 MeV. Figure 7 is a scatter plot of θx versus θy for the events in the final sample. A pronounced clustering of points near 0 degrees is evident by inspection. The density of events, obtained from summing over equal areas of Figure 7 is shown in Figure 8. The figure shows a significant deviation in the interval 0 to 10 mr. above what might be expected from a smooth extrapolation from larger angles. If that extrapolation is performed, an excess signal of 34 events over a background of 12 events is obtained. Of the 12 background events, about 6 are presumed to be ν_e, $\nu_e + N \rightarrow \nu_e, \bar{\nu}_e + N^1$ base upon fluxes of ν_e and $\bar{\nu}_e$ in the beam and the angular distributions for ν_p elastic scattering. N and N^1 are nucleons. The remaining six events are assumed to be various types of neutral current events

Figure 9 shows the $E_\theta{}^2$ versus θ^2 scatter plot for those events with laboratory angles less than 10 mr. Twenty six of the events have an $E\,\theta^2$ less than 1 MeV, a value which is consistent with what we expect based upon our measured angular resolution for electrons in this energy range. To extract the signal we plot in Figures 10a and 10b the observed intergrated $E\theta^2$ distributions for the two angular ranges $0 < \theta \leq 10$ mr and $10 < \theta \leq 20$ mr. Assuming the mean value of the latter plot to be the background level for the former plot, we again arrive at a signal of 34 events with a backgound of 12 events.

To translate this signal for $\nu_\mu e^- \rightarrow \nu_\mu e^-$ into a cross section and eventually into a determination of $\sin^2\theta_W$, it is necessary to normalize the results. As our primary approach, we have chosen to determine the shape of the incident neutrino spectrum based upon the observation of charged current events within a restricted fiducial volume of 0.2m X 0.2m X 40 chambers. This volume is sufficiently small and distant from the edges of the detector to contain the entire hadronic shower and for the hadronic energy to be measured both with the scintillators and the anode planes of the proportional chambers. Because a magnet was not available for measuring the energy of the outgoing muons, we needed to rely strictly upon the hadronic energy to deduce the neutrino spectrum. Above neutrino energies of 30 GeV, there is wide agreement that the neutrino spectrum is well known and given by the calculations of Stefanski-White. Below neutrino energies of 30 GeV, bubble chamber experiments have reported yields in excess of the calculations. We have used as a variable in the calculations the extent of the enhancement below E_ν=30GeV from the calculated values and adjusted it until the observed hadronic charge current spectrum was fitted by Monte Carlo calculations.

The cross section determined from the data and calculated neutrino spectrum then turns out to

$$\sigma\,(\nu_\mu + e \rightarrow \nu_\mu + e) = (1.4 \pm 0.3) \times 10^{-42}\ \mathrm{Ecm}^2 \qquad (9)$$

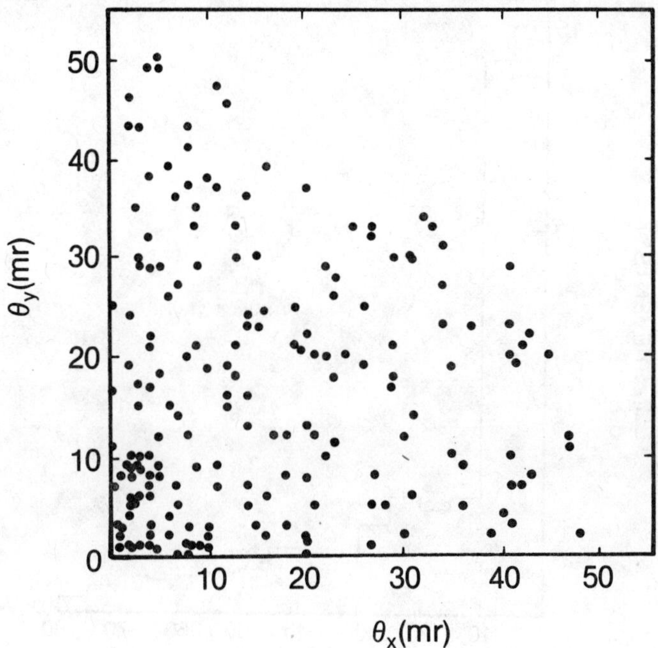

Figure 7. Electron angle scatter plot

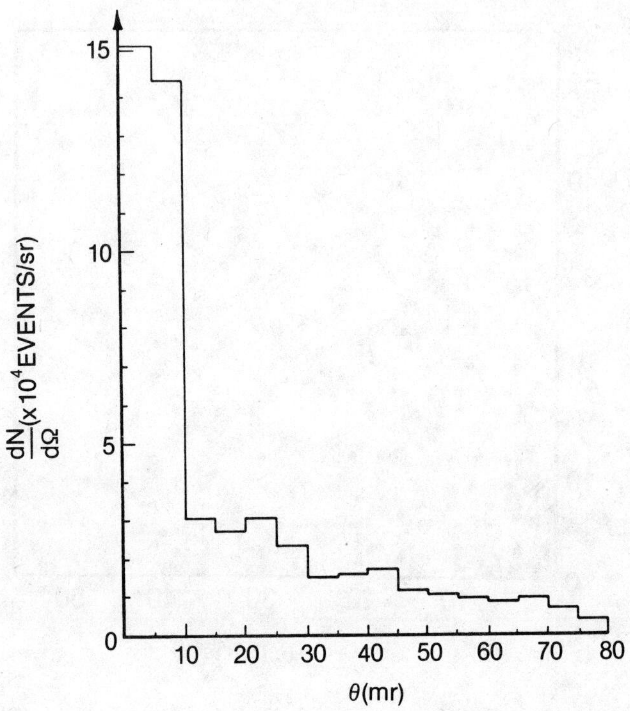

Figure 8. Angular distribution of measured events per unit solid angle which satisfies either θ_x or $\theta y \leq 50$ mr.

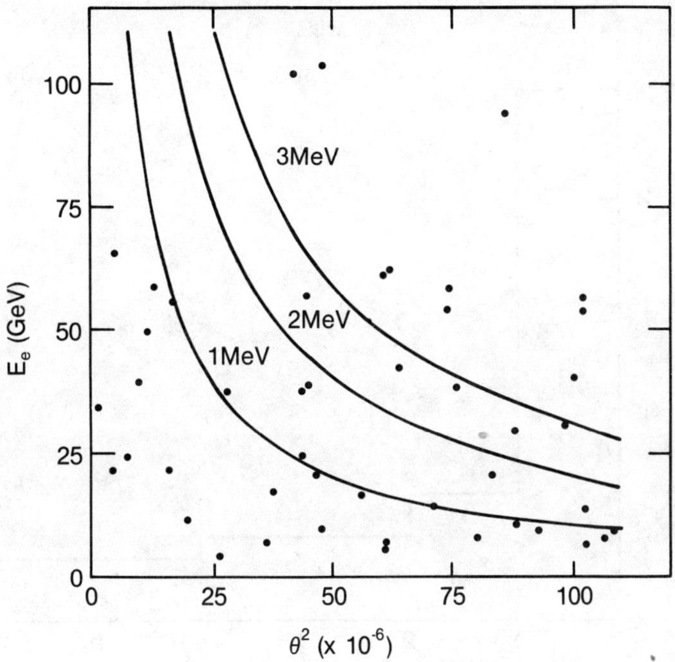

Figure 9. Distribution of observed $\nu_\mu e$ candidates with $\theta \leq$ 10 mr in the E-θ^2 plane.

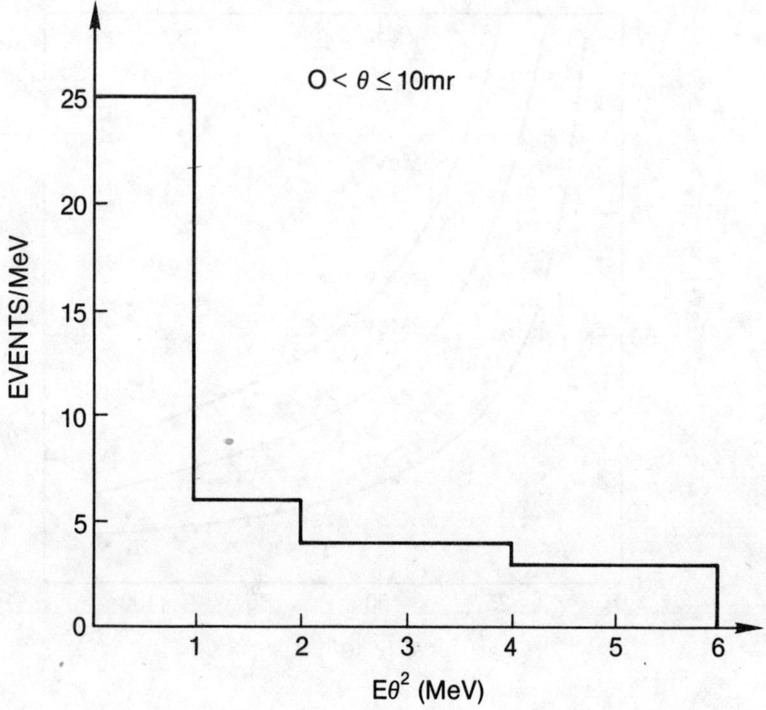

Figure 10a. $E\theta^2$ distribution of observed events in the angular range $\theta \leq 10$ mr.

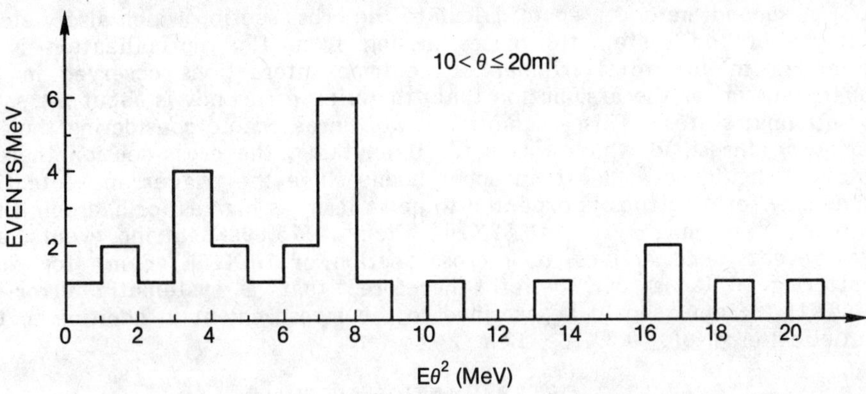

Figure 10b. $E\theta^2$ distribution of observed events in the angular range of $10 < \theta \leq 20$ mr.

In this determination we have assumed the relative fluxes of $\nu_\mu:\bar{\nu}_\mu:\nu_e:\nu_e$ have been taken into account and a value for $\sin^2\theta_W$ of 1/4.

A second method used to calculate the cross section which also yields a measure of the systematic biases arising from flux normalization is to normalize to the total number of neutrino interactions observed in the experiment under the assumption that triggering efficiency is about the same for all final states. This assumption is not unreasonable considering the low triggering threshold which was set. If anything, the cross section thereby obtained should be at least an upper bound since the trigger and detection efficiency for electrons is expected to be at least as high as for hadronic final states. Assuming $\sigma_T(\nu_\mu)=0.82\times10^{-38}$Ecm^2, 175,000 neutrino events, and 34 $\nu_\mu e$ events, one arrives at a cross section of 1.6×10^{-42}Ecm^2 for $\nu_\mu e$ scattering. It is our belief, therefore, that a systematic error of $\pm 0.2\times10^{-42}$Ecm^2 should be ascribed to the measurement in addition to the statistical error of $\pm 0.3\times10$ Ecm^2.

DETERMINATION OF SIN $^2\theta_W$

By integrating equation 3, the production cross section can be expressed in a quadratic form in terms of the two coupling constants g_a and g_V. Figure 11 shows the ellipse obtained by plotting the cross section on a g_a, g_V plot. The errors shown are only the statistical errors. With one measurement, clearly two parameters cannot be deduced. We have, therefore, incorporated the Weinberg-Salam prediction that $g_a = -\frac{1}{2}$ to deduce g_V and thereby from equation 7 to determine $\sin^2\theta_W$. Using the one standard deviation limits, we conclude that $\sin^2\theta_W = 0.25^{+0.07}_{-0.05}$ where we have neglected the solution near 0.5 because of its incompatibility with the deep inelastic scattering results. The systematic errors in cross section do not alter these conclusions markedly.

These results can be compared to those of previous measurements of scattering as given in Table I. (4) Detailed comparisons are inappropriate considering the relatively large errors involved. It is clear that these results are consistent with earlier results. Comparisons with some more accurate determinations of $\sin^2\theta_W$ are given in Table II. The results are in remarkably good agreement with those measurements and indicate quite convincingly that $\sin^2\theta_W$ is in the 0.2 range.

CONCLUSION

This experiment has demonstrated that the measurement of $\nu_\mu e$ scattering cross sections can be accomplished through the use of electronic techniques, providing high angular resolution can be achieved. The leptonic cross sections measured yield a value of $\sin^2\theta_W$ of about $\frac{1}{4}$, in agreement with experiments involving quark states. Together with future measurements of $\bar{\nu}_\mu e$ scattering, the $\nu_\mu e$ results should be capable of determining g_a and g_V to a few percent, independent of particular models.

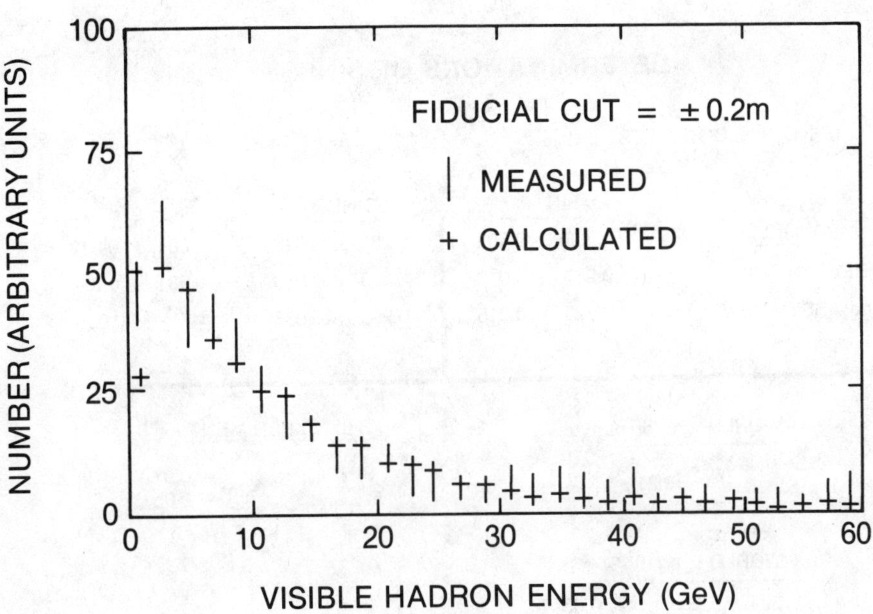

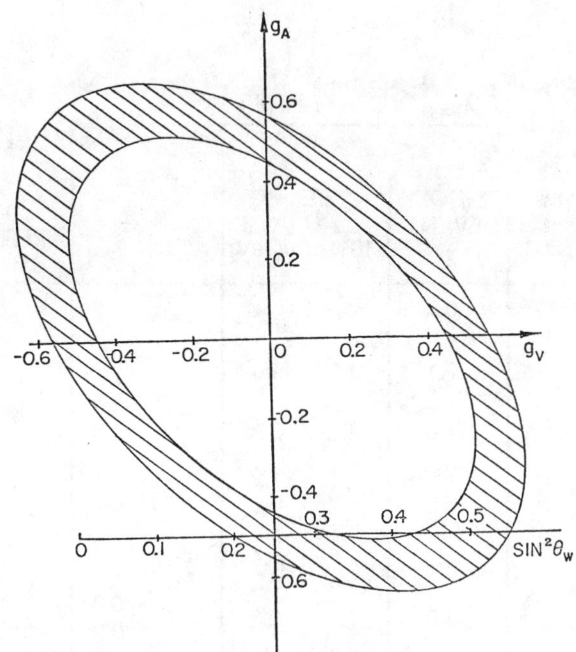

Figure 11. Cross section of $\nu_\mu e$ scattering plotted on g_a, g_V plane using expression od 't hooft.

Table I

DETERMINATIONS OF $SIN^2 \theta_W$

IF NO R.H. DOUBLETS,

	$SIN^2 \theta_W$	$SIN^2 \theta_W$	ϱ
WORLD	0.232 ± 0.009	0.235 ± 0.016	1.004 ± 0.019
DEEP IN. ν	0.233 ± 0.012	0.213 ± 0.038	0.981 ± 0.037
ed ASSYMMETRY	0.223 ± 0.015	$0.293 {+0.033 \atop -0.100}$	1.74 ± 0.036

WORLD ν_μ e (8/78)

$SIN^2 \theta_W = 0.21 {+0.009 \atop -0.06}$

THIS EXPERIMENT

$SIN^2 \theta_W = 0.25 {+0.07 \atop -0.05}$

WORLD ν_μ e (10/79)

$SIN^2 \theta_W = 0.24 {+0.05 \atop -0.05}$

Table II

$\nu_\mu + e^- \rightarrow \nu_\mu + e^-$

EXPERIMENT	GOOD EVENT #	σ/E_ν (10^{-42} cm^2/GeV)	$SIN^2 \theta_W$	REFERENCE
GGM*	0.7	<3.0		4a
AC–PD*	7.1	1.1 ± 0.6		4b
BNL—COLUMBIA	11.0	1.8 ± 0.8		4d
GGM	8.6	$2.4 {+1.2 \atop -0.9}$		4c
CHARM	3.0	2.6 ± 1.4		4e
E-253	34.0	1.4 ± 0.3	$0.25 {+0.07 \atop -0.05}$	

*LOW ENERGY (PS)

REFERENCES

1. C. Baltay, Proceedings of APS Conference, Division of Particles and, Fields, McGill University, ed. B. Margolis, et al, p.

2. G. 'tHooft, Physics Letters 37B, 195 (1971)

3. S. Weinberg, Physical Review Letters 19, 1264 (1967)
   S. Weinberg, Physical Review Letters 27, 1688 (1971)
   A. Salam, Elementary Particle Theory, ed.
   N. Suartholm (Almquist and Wilsells, Stockholm, 1969), p. 367

4. a) J. Blietschau et al., Nucl. Phys. B114, 189 (1976); Phys. Lett. 73B, 232 (1978).
   b) H. Faissner et al., Phys. Rev. Lett. 41, 213 (1978).
   c) P. Alibran et al., Phys. Lett. 74B, 422 (1978); N. Armenise et al., Phys. Lett. B (to be published).
   d) A. M. Cnops et al., Phys. Rev. Lett. 41, 357 (1978).
   e) M. Jonker et al., Neutrino 1979 Conference, Bergen, Norway, June 18-22, 1979.
   f) F. Reines et al., Phys. Rev. Lett. 37, 315 (1976).

5. The complete references and reviews on experiments and theories can be found in (e.g.):
   a) C. Baltay, Proc. 19th Int. Conf. on High Energy Phys., Tokyo, 1978; ed., S. Homma et al. (Physical Society of Japan) p. 882.
   b) L. M. Sehgal, Talk given at "Seminar on Probing Hadrons with, Leptons", Erice, Italy, March 13-21, 1979.
   c) P. Langacker, J. E. Kim, M. Levine, and H. H. Williams; and D. P. Sidhu, Neutrino 1979 Conference, Bergen, Norway, June 18-22, 1979.
   d) I. Liede and M. Roos, Neutrino 1979 Conference, Bergen, Norway, June 18-22, 1979.

MEASUREMENT OF CHARMED PARTICLE LIFETIMES

N.W. Reay
Ohio State University, Columbus, Ohio

ABSTRACT

This talk briefly reviews recent measurements of charmed particle lifetimes, then presents in detail the lifetime measurements of Fermilab Experiment 531.

INTRODUCTION

I was really impressed by the preceding talk of Professor Carnegie, in which results of many experiments were synthesized to give a coherent picture of meson spectroscopy. The reason that this could be done is that all experiments relied on conventional techniques, and therefore could be trusted. Today, I want to talk to you about the charmed particle lifetime business. People in this business have been known to use emulsions and even cosmic rays, so all experiments have a unifying feature -- no one believes them. For example, when Professor Niu of Nagoya[1] discovered charmed particles in a cosmic ray emulsion exposure, very little excitement was created in the high energy physics community. In fact, the field lay dormant until the discovery of the J-Ψ.

Of course, there is some justification for this state of affairs. The GIM coupling of electroweak theory to hadrons made it possible to predict charmed particle lifetimes, and predictions ranged within less than an order of magnitude about 10^{-13} seconds. However, as recently as at the 1978 Tokyo High Energy Conference, experimental results were divided into two camps, one with lifetimes 10^{-13} seconds or longer and another with lifetimes 10^{-14} seconds or shorter. The latter was in flat contradiction to the above mentioned theory.

RECENT EXPERIMENTAL RESULTS

In the past year, results from several CERN and Fermilab experiments have supported longer lifetimes. CERN experiment WA 17 has published five decays, one of which has been mass fit.[2] CERN experiment WA 4 has found a decay generated in a photon interaction.[3] Two bubble chamber experiments at Fermilab recently reported results at Bergen.[4,5] Experiment 546 has four visible decays and experiment 53 has one visible charged decay. These experiments suffer from obvious deficiencies in seeing short lifetimes, but have the distinct advantage that they have a known number of decays. For example, experiment 53 has 250 μ-e di-lepton events, most of which presumably contain the decay of a charmed particle. If charmed lifetimes were long, they would have found 250 visible decays. Professor Baltay has furnished me the graph of expected visible decays versus lifetime shown in Figure 1; his result certainly is consistent with a charged lifetime of roughly 5×10^{-13} seconds.

Now that ballpark numbers exist for charmed particle lifetimes, we must determine lifetimes and branching ratios for each of the

ISSN:0094-243X/80/590077-13$1.50

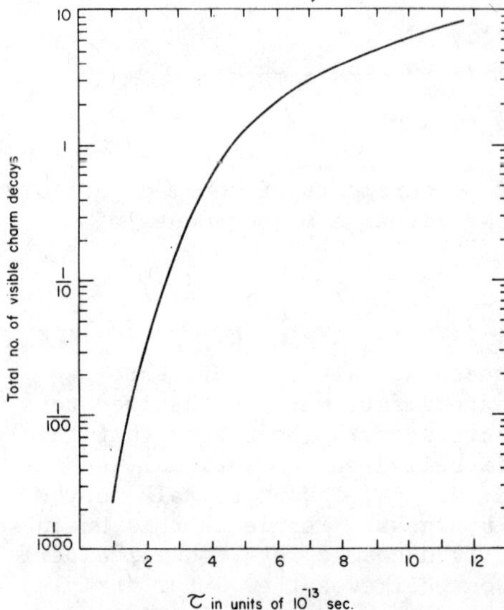

Figure 1: Expected number of visible decays given as a function of lifetime for Fermilab Exp.53.

different types of lightest charmed particles. It is even conceivable that the anomalously short measured lifetimes were the result of an unexpectedly broad spectrum of lifetime values. Three Fermilab experiments designed to answer more detailed questions completed data taking in February, 1979. Though they all exposed hybrid emulsion spectrometers to the same single horn wide band neutrino beam, each was quite different in character.

Experiment 553, shown in Figure 2, used thin emulsion modules, holey spark chambers for event location, a magnetic spectrometer for momentum analysis and a flash tube calorimeter for detection of electromagnetic and hadronic showers.

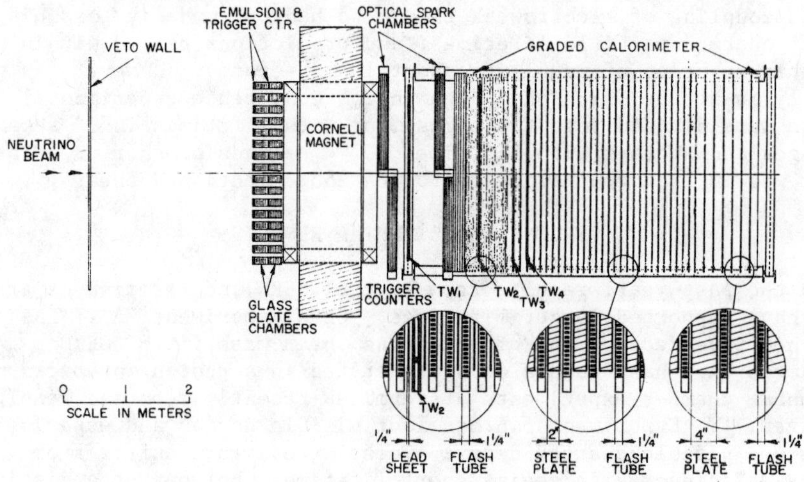

Figure 2: A layout for Fermilab Experiment 553.

The holey spark chamber is conceptually quite clever. The chambers consist of glass plates coated with a metallic layer so thin that sparks burn holes in the coating. Contact prints are then made of fired plates, enabling 50 micron accuracy in location of track coordinates. Professor Hand has informed me that on the basis of 1/3 to 1/2 of their total data sample they have located 25 neutrino interactions and have found one "kink" and one multiprong decay candidate. As neither has been mass fit, proper decay times are not yet available.

Fermilab experiment 564 placed Russian emulsions inside the 15-foot bubble chamber at cold temperature, so that the bubble chamber itself became the downstream detector. Professor Voyvodic has informed me that based on a small fraction of their expected data base, they have located about two dozen neutrino interactions and have discovered two three prong charged charmed particle decays.[6] The proper time and particle identification are not yet available.

FERMILAB EXPERIMENT 531

A. Counter System

Now, I wish to exercise the time honored perogative of rapporteurs everywhere by spending the remainder of the time discussing the experiment on which I work -- Fermilab experiment 531. The apparatus shown in Figure 3 was constructed and operated by a large collabora-

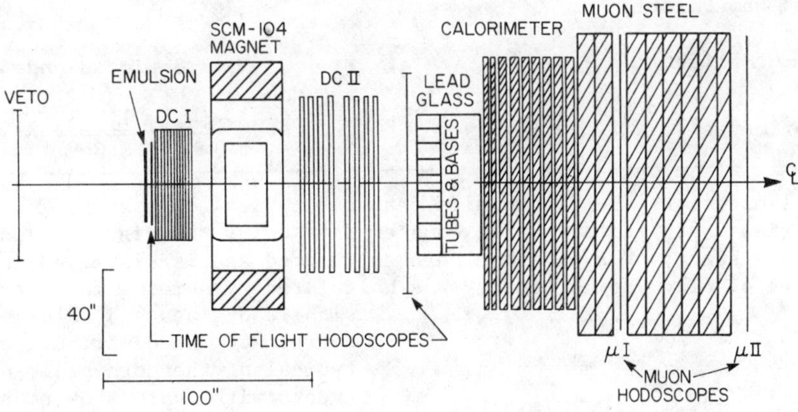

Figure 3: A layout for Fermilab Experiment 553.

tion of Canadian, Japanese, Korean and United States scientists.[7] A simple neutrino trigger required only that no charged particle be incident and that two or more charged particles exit the emulsion and pass through the magnet. Charged and neutral current events were accepted equally. A magnet surrounded by drift chambers provided momentum analysis and aided in finding events within the emulsion. A time-of-flight system separated pions and kaons to 2½ GeV, and protons to 5 GeV. A wall of lead glass detectors was used to identify electromagnetic showers and muons were identified by passage through 1.6 and/or 3 GeV equivalent of steel. Non-electromagnetic hadronic energy was recorded in a conventional iron plate calorimeter.

Information on resolution of spectrometer elements is summarized in Table I.

At this point, let me emphasize the philosophy of our counter system. A hybrid emulsion effort must be viewed as a complete counter effort in addition to the obvious emulsion technology. Many such experiments have suffered grievously because the counters were viewed as a necessary but minimal adjunct. Be assured that within statisti-

Table I

Quantity Measured	Error (σ)
Drift Chamber Position Resolution	120 microns
Charged Particle Momenta for	
Tracks passing through magnet	$\frac{\delta P}{P} = .013 + .005P$
Tracks analyzed only by fringe field	$\frac{\delta P}{P} \approx 0.3P$
Gamma Ray Energy	$\frac{\delta E}{E} = \frac{0.3}{\sqrt{E(GeV)}}$
Gamma Ray Position	5 cm
Time-Of-Flight	120 picoseconds
Calorimeter	$\frac{\delta E}{E} = \frac{1.0}{\sqrt{E(GeV)}}$

cal limitations, E-531 exhibits features typical of most bulk neutrino experiments. Using counter data only, the distribution in visible energy of events has been calculated and is displayed in Figure 4. The visible energy was calculated by summing muon energy with lead glass and calorimeter energies. The distribution is typical of the horn beam, and events with charm show no strong differentiation in shape.

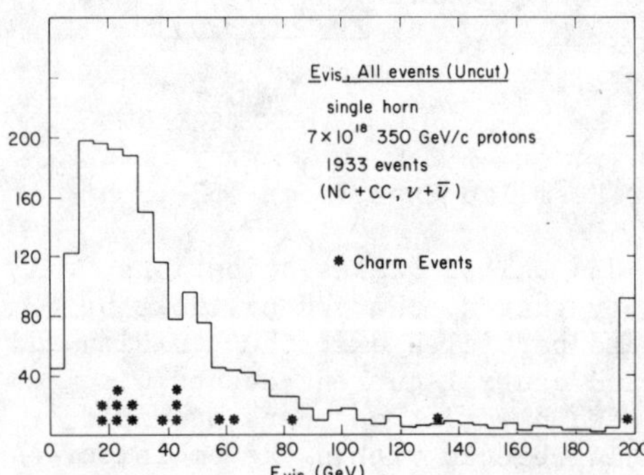

Figure 4: Energy distribution of events for E-531. The asterisks indicate individual events which have charm decays.

A subset selected to have visible energy greater than 10 GeV have been assigned to neutrino or anti-neutrino distributions by the sign of the leading muon. The x and y distributions shown in Figures 5 and 6 exhibit the expected features for charged current interactions. The raw neutrino distributions are flat in y and agree in x with the exhibited $F_2(x)$ shape taken from SLAC inelastic e-D scattering data. The distribution of events containing charm is similar to the overall event distributions. The anti-neutrino distributions as expected are peaked at small x and y, and there is a hint of the expected flat distribution in y of charmed events.

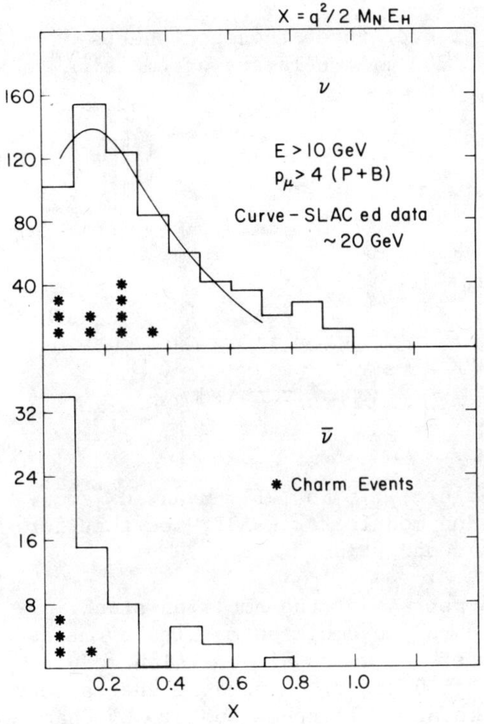

Figure 5: X distribution of charged current events for E-531.

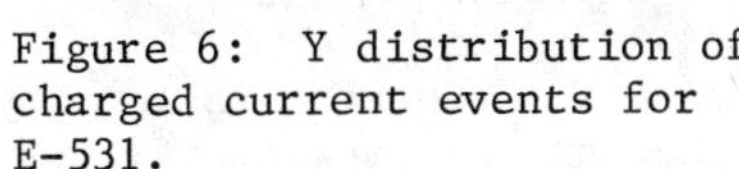

Figure 6: Y distribution of charged current events for E-531.

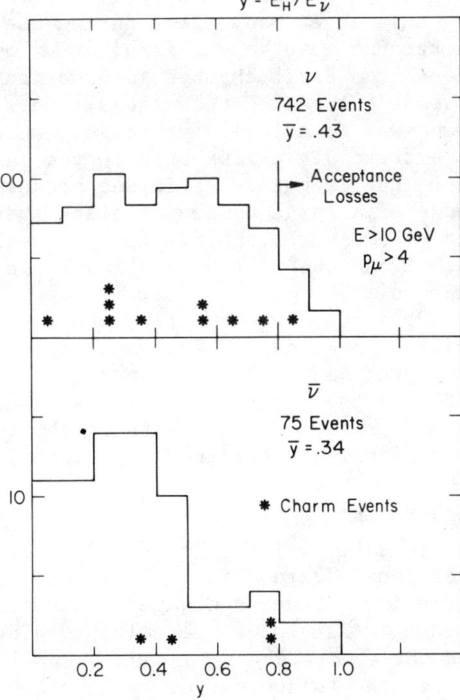

B. Emulsion System

The emulsion target consisted of over three thousand sheets of emulsion, roughly half of which had 300 micron layers of emulsion coated on each side of a 70 micron plastic backing and which were mounted normally to the neutrino beam. One module of this type is exhibited in Figure 7. The remainder were 600 micron pure emulsion pellicles mounted edge-on to the beam. The latter were scanned for events by searching a small volume surrounding the vertex location inferred from intersection of drift chamber tracks. The normally mounted emulsion was scanned using the changeable sheet setup shown in Figure 7. An 800 micron sheet of lucite coated on both sides with a thin layer of emulsion was placed immediately downstream of the emulsion stack. This sheet was located with respect to the emulsion modules by means of X-ray sources residing in the posts on which all emulsion modules were mounted. As there were almost 150 posts, changeable sheets were liberally speckled with precisely located blackened spots. By changing this sheet many times during the experiment, muon and other background were kept sufficiently low that individual tracks could be found from drift chamber information using computer-controlled semi-automatic scanning techniques. Once an individual track from an event was found, highly precise emulsion techniques could be employed to extrapolate tracks back to the interaction vertex. Such a technique is highly efficient because it does not rely on the existence of a large number of black tracks from nuclear breakup as does the volume scan method. Recent developments indicate that the scan-back technique will be applicable as well to emulsions mounted edge on.

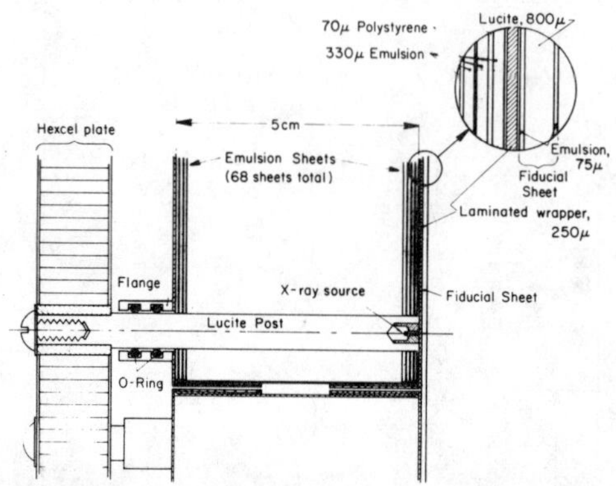

Figure 7: "Normally Mounted" emulsion module for E-531, see text for a description.

Presently, we have 1700 events predicted via computer fits to drift chamber data. By comparing efficiencies of two separate fitting programs, we have determined that a final analysis should locate 2200 events; the final analysis should be available shortly. Scanning of the expected 2200 events is about 40% complete, resulting in the event finding efficiency exhibited in Table II.

C. Analysis

To date, we have located 20 multiprong charm decays, and have an additional dozen "kinks" in which a single track undergoes a large angle instantaneous change in direction without evidence of nuclear breakup. Eleven of the multiprong decays have unique identification, and three more are ambiguous between assignment as D or F mesons.

In the following, decay hypotheses have been shown and particles

Table II

Event Finding Summary

Scanning Method	Computer Predictions	Predicted Events After Fiducial Cuts	Events Searched For	Found
Volume Scan	900	700	460	184
Track Following	850	620	343	300
Total	1750	1320	803	484

Table III

Λ_c^+ Baryon Fits

Decay Length (Microns)	Hypothesis	P (GeV/c)	Mass (MeV)	Decay Time x 10^{-13} Seconds
27	$\Lambda c \to \underline{P}\ \underline{\pi}^+\ \pi^-\ (K^0)$	2.4 4.8	>2070	0.8 0.4
15	$\Lambda c \to K^- \underline{P}\ \underline{\pi}^+ (\pi^0)$	1.9 2.7	>2170	0.6 0.4
40	$\Lambda c \to \underline{\Lambda}^0 \underline{\pi}^+\ \underline{\pi}^-\ \pi^+$	5.3	2220 ±90	0.6

which have been identified as well as momentum analyzed are underlined. Unseen particles added to balance momentum transverse to the direction of the parent particle are indicated in brackets. When all particles are momentum-analyzed a two constraint fit is possible; when a particle is missing the resulting zero constraint fit in general has two solutions. Frequently, one of these solutions may be eliminated, as in cases where the mass solution is at a minimum or when in conjunction with another particle a better higher mass meson (such as D^* or F^*) may be formed. Decay lengths are given in microns, P is the momentum determined for the parent charmed particle and proper decay times are given in units of 10^{-13} seconds. Table III contains the three Λ_c^+ events, Table IV the four D^0 events and Table V the charged D and F events, including the three events which cannot be uniquely assigned.

The preliminary mean life for the Λ_c^+ baryon obtained by averaging the individual proper decay times is

$$< \tau > = 0.6 \times 10^{-13} \text{seconds} \qquad (1)$$

Table IV

D^0 Meson Fits

Decay Length (Microns)	Hypothesis	P (GeV/c)	Mass (MeV)	Decay Time x 10^{-13} Seconds
34	$\overline{D^0} \to \pi^+ \pi^+ \underline{\pi^-} \pi^- (\pi^0)$	10	>1880	0.21
40	$D^0 \to K^- \pi^- \underline{\pi^+} \underline{\pi^+} \underline{\pi^0}$	15	1830 ±40	0.16
74	$D^0 \to \pi^+ \pi^- (K_L^0)$	12.4	1870 ±180	1.9
210	$D^0 \to K^+ \pi^- (\pi^0)$	6.7 8.3		1.9 1.6

For the event with the missing K^0 the lower momentum solution was chosen because the combined mass of the D^0 and another π^+ from the event is closer to the D^* mass. The preliminary mean life for the D^0 obtained by averaging the proper decay times of these events is

$$< \tau > = 0.7 \times 10^{-13} \text{seconds} \qquad (2)$$

In addition to volume scanning downstream of each neutrino interaction for neutral decays, a variation of the track scanback technique was applied to search for decays undiscovered by the first technique. In this variation, tracks found by the drift chambers which had slopes different from those measured in the emulsion at the primary vertex were identified in the changeable sheet and traced back into the emulsion. In 150 events treated in this manner, many nuclear interactions and electron-positron pairs were discovered, but no new charm decays were identified. This fact coupled with the short decay lengths of found events supports a high efficiency for finding D^0 decays.

Table V

F and Charged D Meson Fits

Decay Length (Microns)	Hypotheses	P (GeV/c)	Mass (MeV)	Decay Time x 10^{-13} seconds
	<u>F</u> <u>Mesons</u>			
670	$F^- \rightarrow \underline{\pi}^+\underline{\pi}^-\pi^-\underline{\pi}^0$	12.8	2068 ± 60	3.6
130	$F^+ \rightarrow \underline{K}^+\underline{\pi}^+\underline{\pi}^-(K^0)$	10.4	∿2130	0.8
	<u>Charged</u> <u>D</u> <u>Mesons</u>			
1802	$D^+ \rightarrow K^+K^-\underline{\pi}^+\underline{\pi}^0$	17.8	1851 ± 30	6.4
2145	$D^+ \rightarrow K^-\pi^+\underline{\mu}^+\ (\nu)$	12.8 24.8		10.4 5.4
	<u>Ambiguous</u> <u>Events</u>			
530	$F^+ \rightarrow K^-K^+\pi^+\underline{\pi}^0$ (Favored)	34	2057 ± 60	1.1
	$D^+ \rightarrow K^-\pi^+\pi^+\underline{\pi}^0$	34	1964 ± 60	1.0
2307	$D^- \rightarrow \underline{K}^+\pi^-\underline{e}^-\ (\nu)$ (Favored)	6.7 10.5		21.4 13.7
	$F^- \rightarrow \underline{K}^+K^-\underline{e}^-\ (\nu)$	6.9 7.5		22.7 20.9
457	$D^+ \rightarrow K^-\pi^+\underline{\pi}^+\underline{\pi}^0$	10	1835 ± 50	2.8
	$F^+ \rightarrow K^-K^+\underline{\pi}^+\underline{\pi}^0$	10	2012 ± 49	3.1

If only uniquely determined decays are used, two charged D and two F meson events give

$$\langle \tau \quad \rangle = 7 \times 10^{-13} \text{seconds} \tag{3}$$

$$\langle \tau \quad \rangle = 2 \times 10^{-13} \text{ seconds} \tag{4}$$

After assigning two additional decays to their one sigma favored particle types and weighting the ambiguous event by 0.5 for each of its twofold ambiguous solutions, the following lifetimes may be obtained.

$$\langle \tau \quad \rangle = 9 \times 10^{-13} \text{seconds} \tag{5}$$

$$\langle \tau \quad \rangle = 1.8 \times 10^{-13} \text{seconds} \tag{6}$$

The ratio of D^0 to charged D lifetimes is then

$$\frac{\langle \tau_{D^0} \rangle}{\langle \tau_{D^\pm} \rangle} = 0.08 \pm 0.06 \tag{7}$$

(though with the small statistics involved, a ± error is inferior to values determined by maximum likelihood). Evidence that all four D^0 decays are non-leptonic, whereas the three decays favoring the charged D identification were semi-leptonic or Cabibbo-unfavored non-leptonic also is supportive of the large difference in lifetimes.

We also studying an increasingly interesting amount of dynamical information concerning these decays. All Λ_c^+ baryons were produced slowly in the laboratory, as might be expected if their production is assumed to be target-like. Many, if not most of the D and F mesons appear to come from strong decays of D^* and F^* higher mass states, and about half of the charmed particles appear to contain most of the visible hadronic energy. We should have more to say about these points in the near future.

Scanning should be complete by next summer, obtaining roughly 50 multiprong and an uncertain number of analyzed single track decays. At that time we are approved to run again with improvements both to backgrounds and equipment, and expect to find even more decays. I could bore you with lists of improvements and dreams for the future, but the proper time to discuss this sort of thing is after running and not before. Let me close instead by discussing two "zoo" events which are as yet incompletely understood.

D. Two Unusual Events

We have found the charged two decay track illustrated in Figure 8. Three tracks emerge from a neutrino interaction, two of these are minimum ionizing while the third has been measured by two methods to have an ionization 4.0 ± 0.5 times more dense than nearby minimum ionizing charged tracks plus at least one neutral particle. Multiple scattering measurements in the emulsion have yielded momentum values of $P_{2-1} \gtrsim 3.0$ GeV/c and $P_{2-2} = 0.6 \pm 0.15$ GeV/c. The former track travels about 1 cm in the emulsion before interacting without evidence of nuclear breakup (or decaying) to give two minimum ionizing tracks and an electron of momentum 60 ± 10 KeV/c.

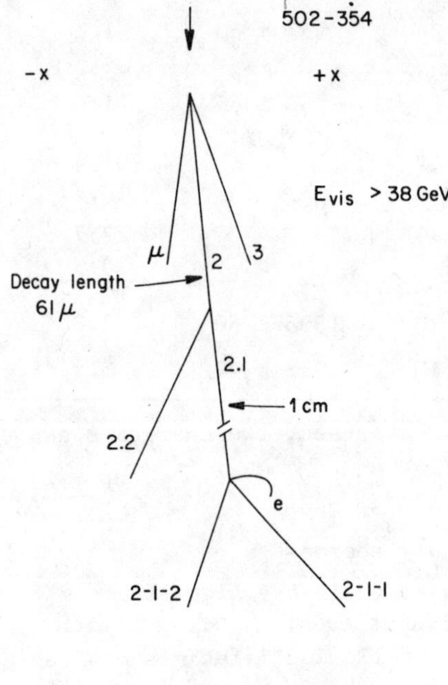

Figure 8: Decay of doubly charged particle, see text for a description.

Unfortunately, this event occurred when the spectrometer magnet was off due to a power supply failure and no momentum measurements could be performed by the magnetic spectrometer. A "vee" materialized downstream of the emulsion and was identified by time-of-flight measurements to be a lambda hyperon. An electron-positron pair also is observed in the emulsion and has been connected to spectrometer tracks which show that it is in time with and therefore associated with this event. We are still working and though we are strongly tempted to speculate on the existence of a weakly decaying charge two baryon, we will make no claims at this time.

The second unusual event, shown in Figure 9 is one which contains three electrons. The table given as an insert in Figure 9 tells the story. Tracks 1 and 2-2 were clearly identified as electrons, as the momentum measured by the magnetic spectrometer equalled the energy measured by the lead glass blocks struck by these tracks. The identification of track 3 is not so clear, as the energy measured by the lead glass is only 70% of that measured by the magnetic spectrometer. This event could be quite exciting, but we are waiting for emulsion measurement on the two tracks found in the emulsion but not picked up by our drift chambers. Visions of many exciting possibilities dance in our heads, but if these two tracks turn out to be electrons the event could be as ordinary as a gamma-induced shower. It is safe to say that we do not yet understand this event, and I am hopeful that after the end of the talk one of you will leap up with an explanation. So without further ado let me close by thanking you for this invitation to talk in the beautiful city of Montreal.

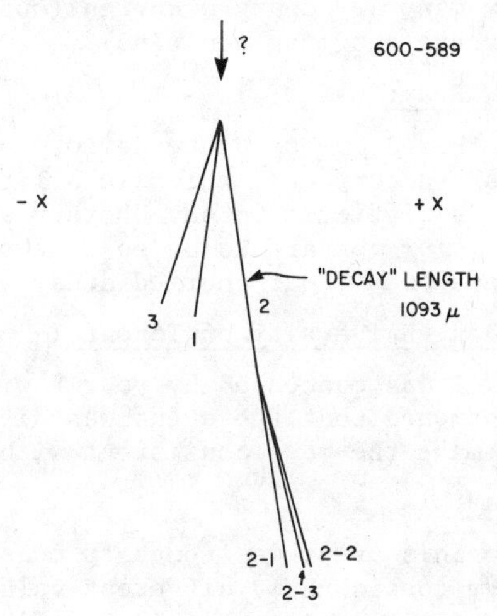

TRACK	$\frac{dx}{dz}$	$\frac{dy}{dz}$	P	I.D.
1	-.059	-.057	+8.0	e+
3	-.157	-.054	-3.7	e- ?
2	.073	.027		
2-1	.075	.027		
2-2	.137	.023	36	e+
2-3	.117	-.091		

Figure 9: Multi-electron event, see text for a description.

We wish to acknowledge the financial support of the U.S.D.O.E., the Canadian N.S.E.R.C., the Japanese Ministry of Education and other private contributors, and the Korean Science Foundation. We thank the many members of the Fermilab staff and the technical personnel in our universities for their contributions to our experiment.

DISCUSSION

A.Abashian, National Science Foundation:

Of the neutrino events in which a charmed particle is produced, how many are charged current(outgoing μ) and how many are neutral current(outgoing neutrino)?

N.W.Reay:

Based on our Monte Carlo, we have a 35-40% ratio of neutral to charged currents, which is a little high. On an event by event basis it is difficult to say whether an event is truly a neutral current event or appears to be so because the muon missed the detector. At least one of our charmed decay events has no tagged muon.

Isgen, University of Toronto:

I was concerned by your F with a mass of around 2130 MeV. You mentioned that the event was difficult; are the problems sufficient to make the mass consistent with the DESY value of about 230 MeV?

Reay:

This event was tough to measure in the emulsion; re-measurements gave considerably different values. I would say that for this event the minimum mass in this OC fit has a large error and could be consistent with 2030 MeV.

REFERENCES

1. K.Niu et al., Prog.Theor.Phys. 46, (1971) 1644.
2. R.Sever, et al., Contributed paper to neutrino 1979 Bergen Conference by the WA 17 Collaboration; Ankara-Brussels-CERN-Dublin-UC London-Open University-Pisa-Roma-Torino.
3. A.Conti et al., Contributed paper to the 1979 International Symposium on Lepton and Photon Interactions at High Energies, Bologna-CERN-Florence-Geneva-Moscow LPI-Paris VI-Santander-Valencia-Emulsion Collaboration and Bonn-CERN-EP Paris-Glasgow-Landcaster-Manchester-Orsay-Paris VI-Parris VII-Rutherfore-Sheffield Omega Collaboration. Also discussed in presentation by R.Richard to this Conference.
4. D.Reeder, Invited talk at the 1979 International Symposium on Lepton and Photon Interactions At High Energies.
5. C.Baltay, private communication.
6. L.Voyvodic, Invited talk at the 1979 International Symposium on Lepton and Photon Interactions At High Energies, Fermilab-Conf-79/80-Exp. 2050.00 .
7. The names and institutions of the physicists who are currently working on this experiment are: N.Ushida, Aichi University of Education, Japan; T.Kondo, Fermi National Accelerator Laboratory; G.Fujioka, H.Fukushima, Y.Homma, O.Minakawa, J.Orimoto, Y. Takayama, S.Tatsumi, Y.Tsuzuki, Kobe University, Japan; S.Y.Bahk, T.G.Choi, C.O.Kim, S.N.Kim, J.N.Park, Korea University; D.C. Bailey, S.Conetti, J.-R.Fisher, J.M.Trischuk, McGill University, Canada; H.Fuchi, K.Hoshino, K.Niu, K.Niwa, H.Shibuya, Y.

Yanagisawa, Nagoya University, Japan; S.M.Errede, M.J.Gutzwiller, S.Kuramata, N.W.Reay, K.Reibel, T.A.Romanowski, R.A.Sidwell, N.R. Stanton, Ohio State University; K.Moriyama, H.Shibata, Okayama University, Japan; T.Hara, O.Kusumoto, Y.Noguchi, Y.Takahashi, N.Teranaka, Osaka City University, Japan; J.-Y.Harnois, C.D.J. Hébert, J.Hébert, B.McLeod, University of Ottawa, Canada; K.Okabe, J.Yokota, Science Education Institute of Osaka Prefecture, Japan; S.Tasaka, University of Tokyo, Japan; P.J.Davis, J.F.Martin, D.Pitman, J.D.Prentice, P.Sinervo, T.-S.Yoon, University of Toronto, Canada; J.Kimura, Y.Maeda, Yokohama National University, Japan.

RECENT RESULTS IN PHOTOPRODUCTION

John P. Cumalat
Fermi National Accelerator Laboratory, Batavia, IL 60510

ABSTRACT

This paper reviews the recent high energy photoproduction data on vector mesons and charmed mesons.

INTRODUCTION

As the title indicates, I have been asked to review the recent results from high energy photon experiments. During the last year there has been three high energy photoproduction experiments which have announced results. The first was performed in the Tagged Photon Facility at Fermilab by a Fermilab - Santa Barbara - Toronto collaboration. Results were obtained on the ρ^o, ω, and ϕ photoproduction cross sections at the same time as a precise determination[1] of the photon total cross section was being made.

The second experiment, performed in the broadband photon beam at Fermilab by a Columbia - Fermilab - Illinois collaboration, has reported results on the high mass dipion mass spectrum[2] and on the charmed meson photoproduction cross section.[3]

The third experiment was located in the tagged photon beam at CERN and was executed by a large British-French-German collaboration using the OMEGA spectrometer. Preliminary results[4] on higher mass vector meson recurrences and on charmed meson photoproduction cross sections have been presented at the 1979 Photon-Lepton Symposium at Fermilab.

The results of these three experiments can be organized into three topics; low mass vector mesons, high mass vector mesons, and charmed mesons.

II. LOW MASS VECTOR MESONS

Most information on higher energy ($E > 20$ GeV) ρ^o, ω, and ϕ photoproduction comes from the tagged photon experiment performed at Fermilab. A schematic diagram of the apparatus is shown in Figure 1. The analysis of ρ^o, ω, and ϕ states was done in two parts; the study of the two charged track decays of the $\rho \to \pi^+ \pi^-$ and $\phi \to K^+K^-$ and the study of the all neutral decay of $\omega \to \pi^o \gamma$.

ISSN:0094-243X/80/590091-25$1.50

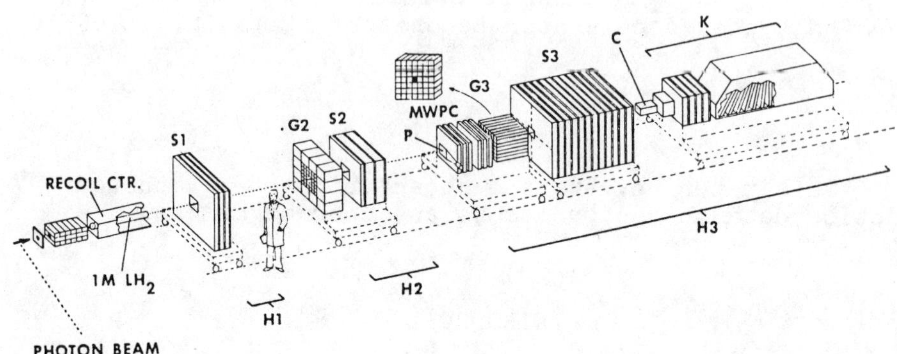

Fig. 1. Apparatus for the Fermilab-Santa Barbara-Toronto experiment, configured for $E_0 = 90$ GeV. Vacuum extended to H3, with helium between theMWPC's and C, the central Pb-scintillator counter. Hadronic detectors: S1(three planes Pb(scintillator), G2(12 X_0 Pb glass), S2(Pb/scintillator/Fe/scintillator/Fe/scintillator), G3(21 X_0 Pb glass), S3 and K(Fe/scintillator calorimeters).

A. RHO AND PHI PHOTOPRODUCTION

Since the apparatus shown in Fig. 1 did not have a magnet or a Cerenkov counter, it was not possible to make a direct mass spectra measurement or to discriminate between pions and kaons for ρ^0 and ϕ meson production. However the full coverage of the forward hemisphere in the γp center of mass frame allowed events to be selected which had exactly two tracks and no extra particles. Under these conditions a plot of Δ (track separation) multiplied by $E\gamma$ provided a good substitute for the actual invariant mass, in that the ρ^0 and ϕ decays show up as two distinct peaks.

For a particle of mass M and energy E > > M decaying into two particles of mass $m \neq 0$, the track separation, Δ , at a distance Z from the target is given by

$$\Delta = \frac{4\, Z \sin\theta \sqrt{\left(\frac{M}{2}\right)^2 - m^2}}{E\left\{\sin^2\theta + \left(\frac{2m}{M}\right)^2 \cos^2\theta\right\}} \qquad (1)$$

whereas nearly all ρ^{0} decays at $M = M_{\rho}(770)$ have $\Delta \geq 1.432\ Z/E$. From equation (1) it can be seen that the distribution of $E\,\Delta$, or more conveniently $R \equiv \Delta / \Delta\phi_{max}$, is independent of E. By plotting a R-spectrum, it is possible to combine data from the entire tagged photon energy region. A typical R-distribution (with target empty rates subtracted) is shown in Fig. 2.

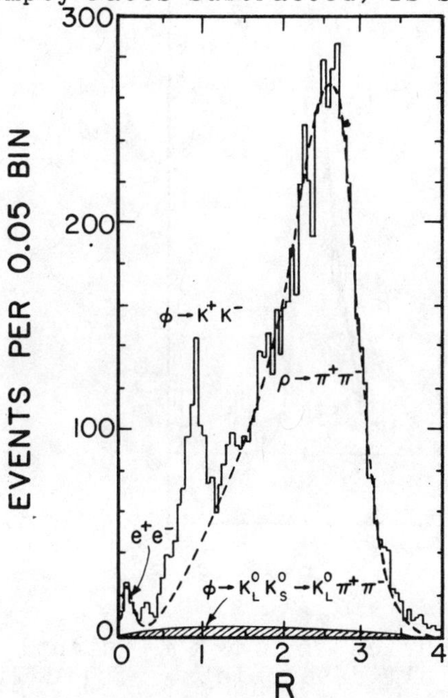

Fig. 2. R distribution for E = 90 GeV ρ^{0} and ϕ candidates with target empty rates subtracted. Dashed lines are fits to the $\rho \rightarrow \pi^{+}\pi^{-}$ and $e^{+}e^{-}$ contributions.

A Monte Carlo program was used to extract the ρ^{0} and ϕ yields from plots like Fig. 2. The Monte Carlo calculation incorporated effects from the ρ^{0} mass shape, geometrical acceptance, target length, beam size, and resolution in the photon energy and in Δ. Cross sections were assumed to vary with t like e^{bt} with $b(\rho) = 8.5\ GeV^{-2}$ and $b(\phi) = 6.5\ GeV^{-2}$. A Söding parameterization[5] with a mass dependent width[6] was used for the ρ^{0} mass spectrum. A contribution due to $e^{+}e^{-}$ pairs was included for cases in which one electron interacted hadronically. A five parameter fit of the R plots involving the Söding slope parameters and the amounts of ρ^o, ϕ, and e^+e^- was used. Also included was a small calculated contribution from $\phi \to K^o_L K^o_S \to K^o_L \pi^+ \pi^-$. Fig. 3 shows the phi data (summed over all data points) with all other processes subtracted.

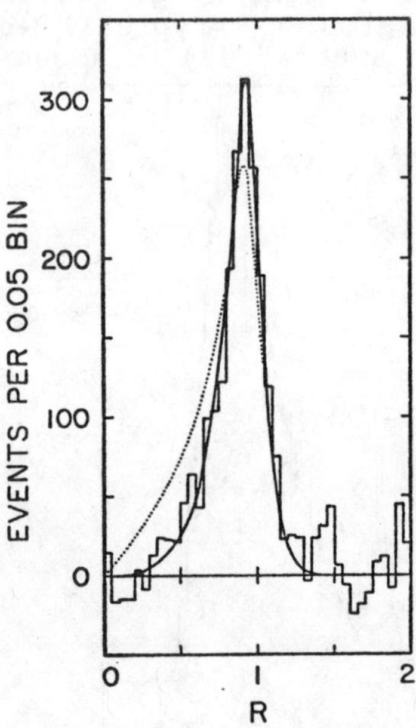

Fig. 3. R distribution for $\phi \to K^+K^-$ signal above background summed over all data points. Solid line is Monte Carlo prediction assuming s-channel helicity conservation; Dotted line is prediction for isotropic decay.

Corrections to the ρ^o and ϕ yields were made for geometric acceptance, decays in flight, branching ratios, inelastic downstream interactions, and inelastic events involving target dissociation. The inelastic events eliminated amounted to 13% for the ρ^o and 18% for the ϕ.

The resulting ρ^o photoproduction cross section is shown in Figure 4 along with results from previous experiments.[7,8,9] For both the ρ^o and ϕ cross sections the total energy independent systematic uncertainty is less than 5%. The ρ has an additional 5% systematic uncertainty due to geometrical acceptance. By using additive quark model relations and vector meson dominance the ρ photoproduction can be expressed in terms of $\pi^{\pm}p$ elastic scattering cross sections at the same s and t as:

$$\frac{d\sigma}{dt}(\gamma p \rightarrow \rho^{o} p) = \frac{e^2}{4\gamma_p^2}\left[\frac{p^*_\pi}{2p^*_\gamma}\left\{\sqrt{\frac{d\sigma}{dt}(\pi^+ p)} + \sqrt{\frac{d\sigma}{dt}(\pi^- p)}\right\}\right]^2 \qquad (2)$$

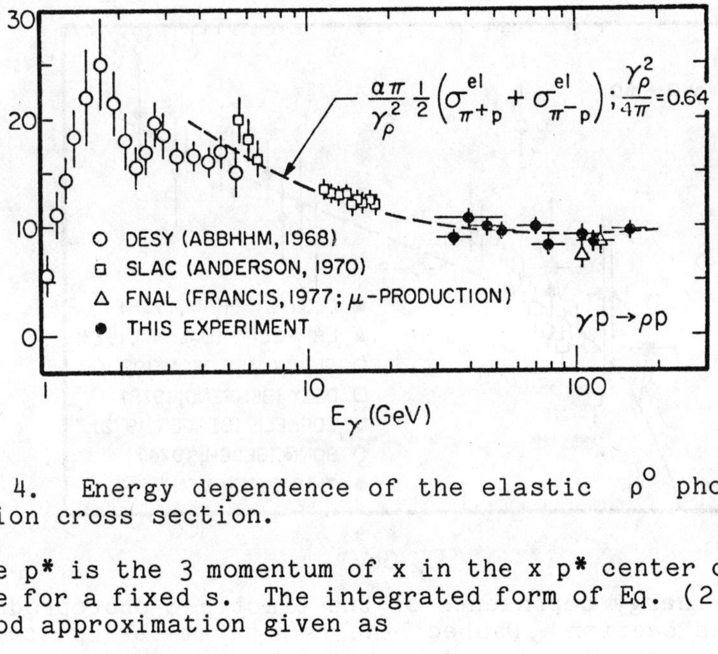

Fig. 4. Energy dependence of the elastic ρ^{o} photoproduction cross section.

where p* is the 3 momentum of x in the x p* center of mass frame for a fixed s. The integrated form of Eq. (2) is to a good approximation given as

$$\sigma(\gamma p \rightarrow \rho^{o} p) = \frac{e^2}{4\gamma_p^2}\,\frac{1}{2}\left[\sigma(\pi^+ p \rightarrow \pi^+ p) + \sigma(\pi^- p \rightarrow \pi^- p)\right] \qquad (3)$$

Formula (3) is shown as a dashed line in Fig. 4 using smoothed πp measurements with $\gamma_\rho^2/4\pi = .64$.

In Fig. 5, the ϕ data points are shown together with low energy measurements.[8-12] The dashed line is a parameterization of the data using the same prescription that was followed for eq. 2, the following prediction is obtained:

$$\frac{d\sigma}{dt}(\gamma p \rightarrow \phi p) = \frac{e^2}{4\gamma_\phi^2} \frac{1}{(p_\gamma^*)^2} \left[p_K^* \left\{ \sqrt{\frac{d\sigma}{dt}(K^+p)} + \sqrt{\frac{d\sigma}{dt}(K^-p)} - p^*_\pi \sqrt{\frac{d\sigma}{dt}(\pi^-p)} \right] ^2 \quad (4)$$

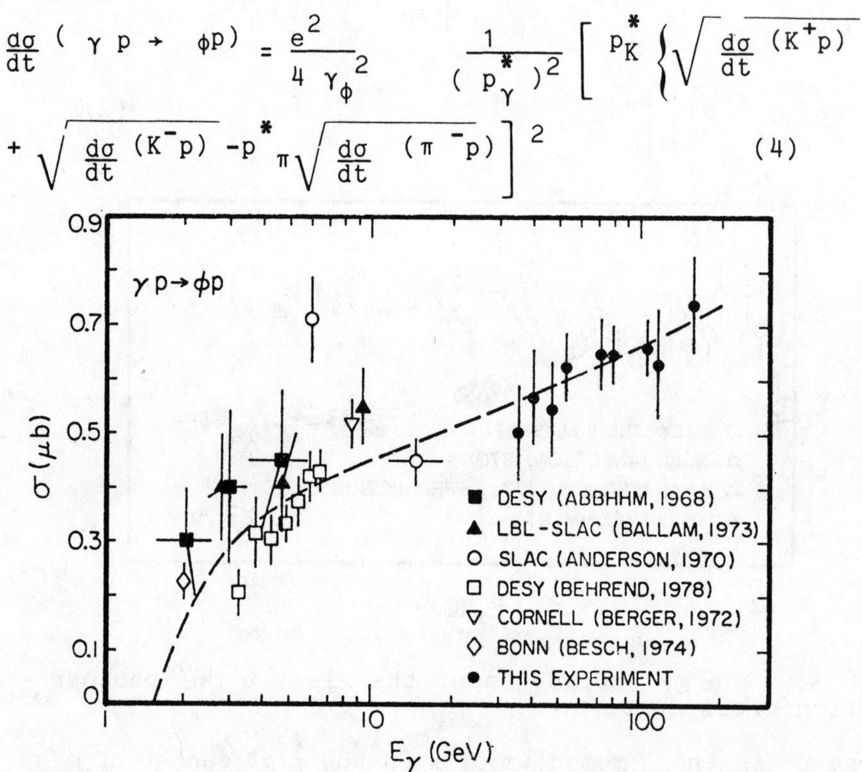

Fig. 5. Energy dependence of the elastic ϕ photoproduction cross section. Dashed line is a parameterization of the data.

Using forward hadronic data[13,14] and assuming a e^{bt} form for the photoproduction cross section with $b(\phi)=4.66 + 0.38 \ln s$,[9] $\sigma(\gamma p \rightarrow \phi p)$ can be evaluated. The resulting predictions, normalized to the photoproduction data, are shown in Fig. 6, along with the curve from Fig. 5. The normalization procedure yields a value of $\gamma_\omega^2/4 = 4.7 \pm 0.3$ as compared to 5.5 ± 2.4 obtained from photoproduction on complex nuclei and $2.83 \pm .2$ from colliding beam measurements.

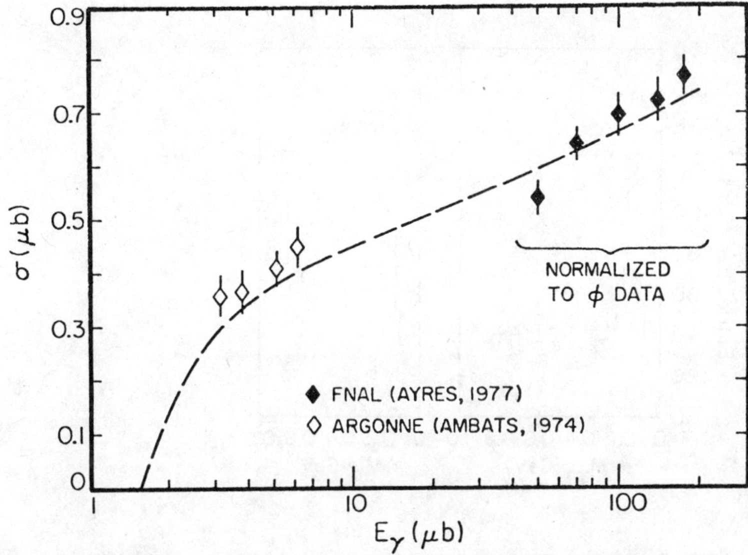

Fig. 6. VMD-quark model predictions for ϕ photoproduction, using data from hadron-beam experiments, and normalized to the phi photoproduction data. The curve is in the same as in Fig. 5.

B. Omega Photoproduction

For ω photoproduction, the $\pi^o \gamma$ decay mode was used with all three γ rays being measured in two lead glass arrays (see Fig. 1). Omega candidates were selected with 3 photons and no other observed particles. The two photon invariant mass spectrum for ω candidates is shown in Fig. 7a. Each event contributes 3 entries to the plot in Fig. 7a. Only the 2γ combination closest to the π^o mass is plotted in Fig. 7b. The three-photon invariant mass spectrum for events which have a two photon mass combination between 80 MeV and 220 MeV is shown in Fig. 8. The decay angle distribution in the omega center of mass for $\omega \to \pi^o \gamma$ is presented in Fig. 9. After corrections are made, a $1 + \cos^2 \phi$ dependence is observed, as expected for s-channel helicity conservation.

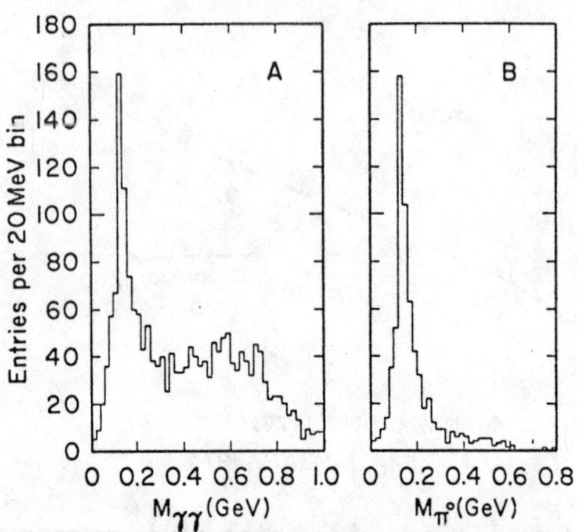

Fig. 7. Two-photon mass spectrum for ω candidates. (a) All three combinations for each event. (b) Combination nearest the π^0 mass.

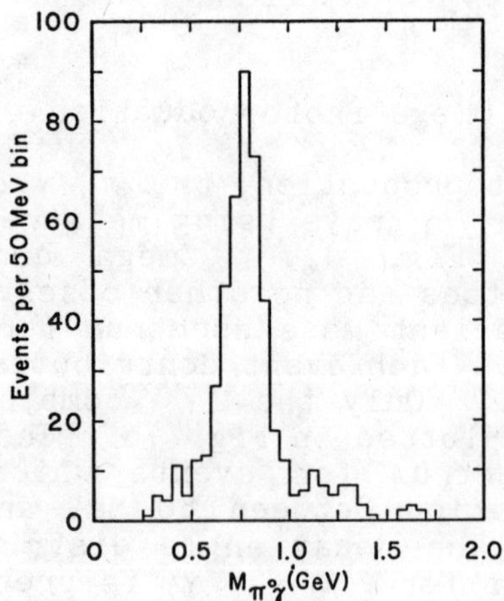

Fig. 8. Three-photon invariant mass spectrum for ω candidates passing all analysis cuts.

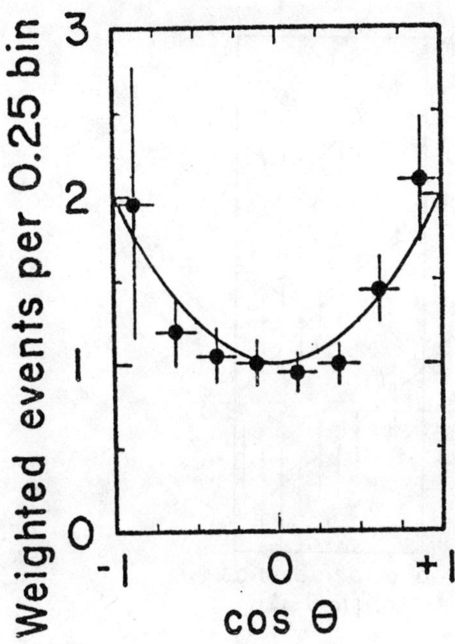

Fig. 9. Decay angle distribution in the helicity frame for $\omega \to \pi^{0} \gamma$. The curve has the form $1 + \cos^{2}\theta$.

Corrections have been applied for geometric acceptance, event reconstruction, branching ratio, and inelastic production. The inelastic event contribution was determined to be 26 ± 4%. A correction factor for $\rho \to \pi^{0}\gamma$ of 0.96 was applied using a $\rho \to \pi^{0}\gamma$ branching ratio of 0.043 ± 0.005%.[15] An additional correction for ρ^{0}-ω interference was made assuming the ρ^{0} and ω are produced in phase utilizing simple Breit-Wigner decay amplitudes.

The differential cross section for ω photoproduction was determined from fits to the data shown in Fig. 10. of the form $\frac{d\sigma}{dt} = A\, e^{bt}$. The average value of b is found to be 8.42 ± 0.74 GeV^{-2}. Results of σ ($\gamma p \to \omega p$) are shown in Fig. 11 along with lower energy data.[7,10,16] The solid curve in Fig. 11 is a calculation using VMD and the additive quark model and is

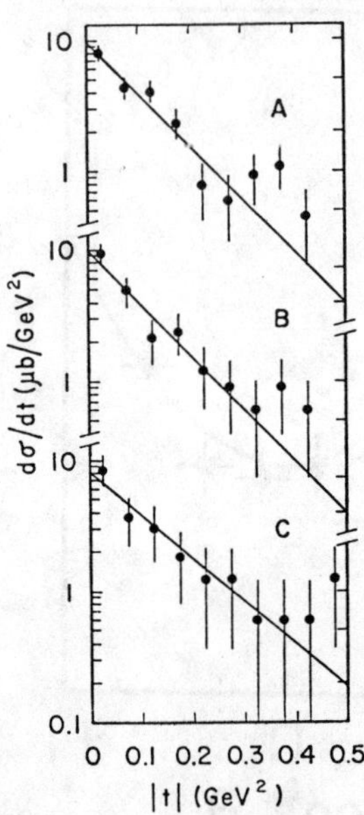

Fig. 10. Differential cross section for elastic ω photoproduction. The lines are fits to $A\ e^{bt}$.

given by

$$\sigma(\gamma p \to \omega p) = \frac{p_\pi^{*2}}{K_\gamma^*} \frac{e^2}{4\gamma_\omega^2} 2 \quad \frac{1}{2}\left[\sigma_{el}(\pi^+ p) + \sigma_{el}(\pi^- p)\right] e^{-b\left|t_{min}\right|} \qquad (5)$$

where p_π^* and K_γ^* are momenta of the π and the photon in the πp and the γp center of mass sytems evaluated at the same s. The curve has been normalized to the high energy data.

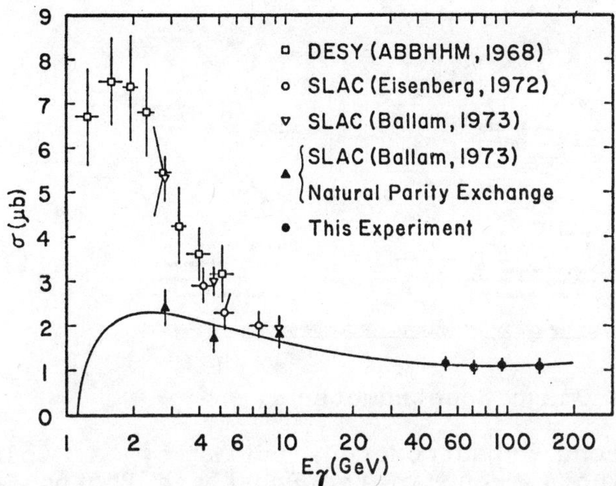

Fig. 11. Energy dependence of the elastic ω photoproduction cross section. The curve is a VMD-quark model prediction normalized to the data above 40 GeV.

The value of the coupling constant resulting from the normalization is $\gamma_\omega^2/4\pi = 6.5 \pm .5$. This compares with 4.6 ± 0.5 from colliding beam measurements and 7.5 ± 1.3 from photoproduction on complex nuclei. The solid curve deviation from the low energy data is attributed to unnatural parity (pion) exchange. The two natural parity points are in good agreement with the VMD-quark model curve.

In summary the ρ and ω cross sections are consistent with being flat with increasing energy while the ϕ cross section is rising rapidly with energy. The energy dependences are all well represented by the VMD-quark model.

III. High Mass Vector Mesons

Two large open geometry multiparticle spectrometers with good particle identification have been used in photon beams to search for high mass vector mesons and for charmed particles. The first is the CERN OMEGA spectrometer shown in Fig. 12 and operated in the tagged photon beam in the West Area of CERN. The spectrometer has a 32 cell Cerenkov counter filled with CO_2 and a large lead glass array. The tagged photon energies range from 20 GeV to 70 GeV.

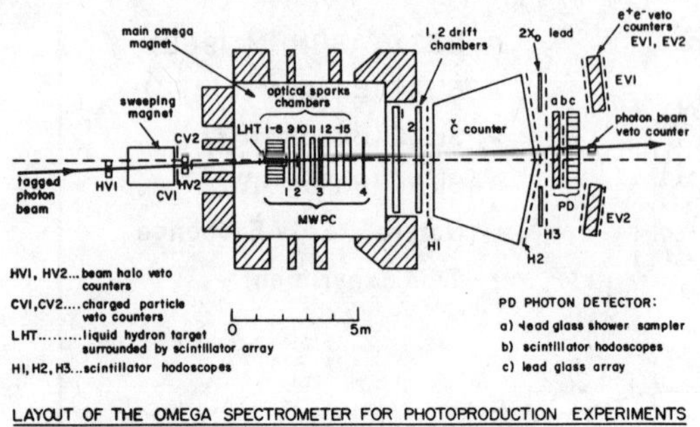

Fig. 12. The Omega Spectrometer

The second spectrometer built by a Columbia-Fermilab-Illinois group in the Broadband Photon Beam at Fermilab is shown schematically in Fig. 13. The spectrometer contains two Cerenkov counters, a 90 block lead glass array, and a good muon identifier. The photon energies which were used extended from 50 GeV to 250 GeV with an average photon energy of 100 GeV.

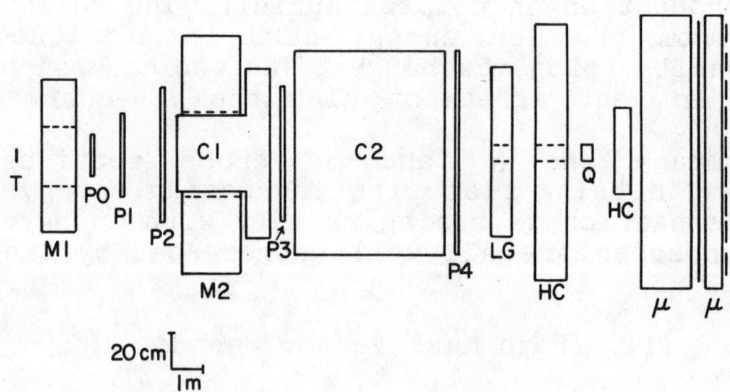

Fig. 13. A schematic view of Columbia-Fermilab-Illinois spectrometer. T is a segmented target consisting of 20 1 mm thick scintillator pieces; M1, M2 are dipole magnets; C1, C2 are Cerenkov counters; P0-P4 are MWPC's; LG is a lead glass array; and HC is a hadron calorimeter.

In the OMEGA spectrometer analysis diffractively produced events are chosen by requiring that the net charge be 0 or 1 and that the energy of the decay products balance the energy of the incident photon to ± 1 GeV. In the analysis of the Fermilab experiment, exclusive charged track events are selected by requiring the following: 1) The total charge is 0. 2) There is less than 15 unused hits in the proportional chambers. 3) All visible tracks must be retraceable to the production vertex. 4) There is no evidence for gamma rays in the lead glass.

Results from the two spectrometer groups are organized by specific decay channels.

A. Charged Pion Final States 2π, 4π, ρ' (1600).

The first attempts to look for the photoproduction of high mass vector mesons were made in dipion final states.[17, 18] Previous experiments[19-23] have observed a broad structure centered at 1600 MeV. With improved spectrometers and improved mass resolution the structure has become more pronounced. The CERN OMEGA experiment has observed a dipion peak at 1.60 GeV with a width of $\Gamma = 0.23 \pm 0.08$ GeV.[4] A histogram of the $\pi^+\pi^-$ mass spectra is presented in Fig. 14. The cross section associated with the peak is calculated to be 130 ± 20 nb. The Broad band photon beam experiment has found a similar structure in the dipion mass spectrum with the mass peak at 1600 MeV with a width of 283 ± 10 MeV.[2] The plot is shown in Fig. 15. The solid curve is a fit to a monotonically decreasing background plus a Breit-Wigner resonance shape. The integrated cross section of the resonance is

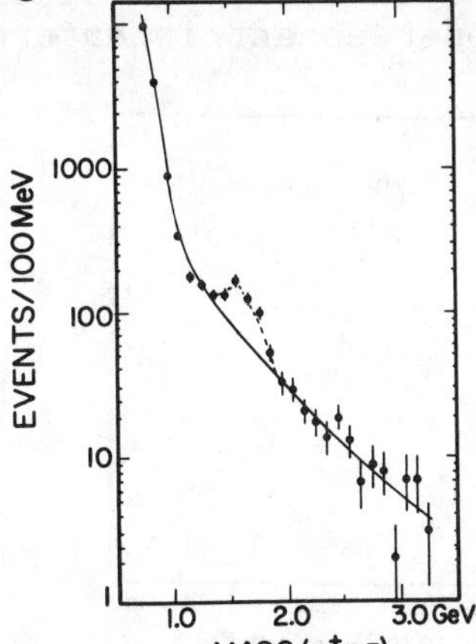

Fig. 14. Dipion mass spectrum of CERN OMEGA spectrometer group.

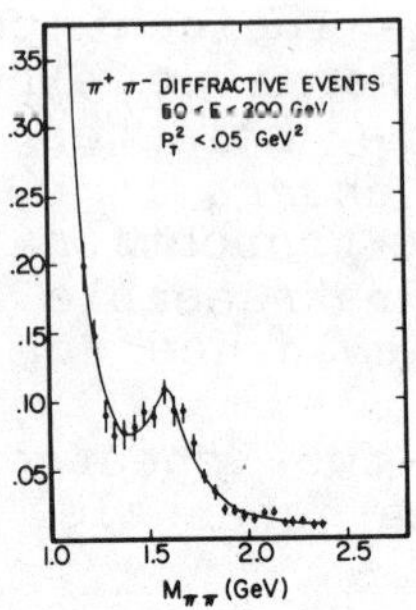

Fig. 15. Two pion mass spectrom from Columbia-Fermilab-Illinois collaboration for those events having $p_T^2 < 0.05$ GeV^2.

The CERN OMEGA spectrometer 4π mass spectrum is shown in Fig. 16. The reported cross section is 0.8 ± 0.3 πb.[4] for the broad resonance at 1600 MeV. A 90% probability of finding a ρ in the four $\pi^+\pi^-$ mass combinations was reported. By comparing the ratio of cross sections for $\sigma(\rho' \to 2\pi)/\sigma(\rho' \to 4\pi)$ a relative branching ratio of 0.16 ± 0.05 is obtained.

In both experiments the dipion mass peak is narrower than the corresponding mass peak in the 4π mass spectrum. In Fig. 17 the 2π data from the broadband photon beam experiment is superimposed on the 4π data. The peak of the distributions occur at different mass values. The difference in the peak values and the width may be explained by non-resonant interference effects.

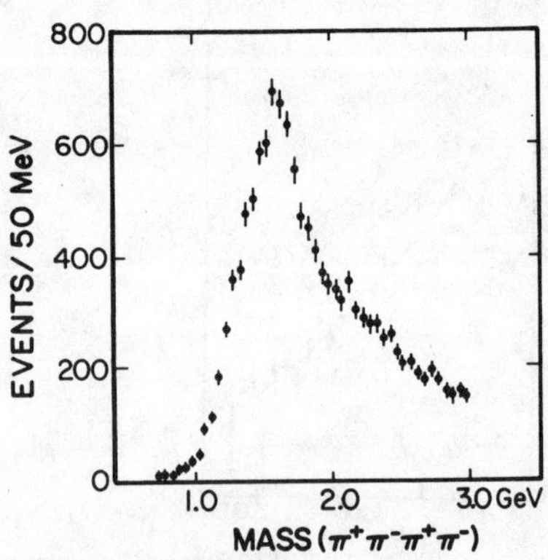

Fig. 16. Four pion mass spectra of the CERN OMEGA spectrometer experiment.

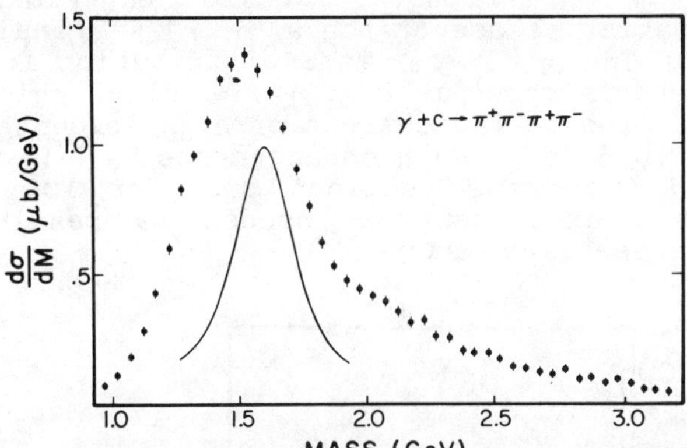

Fig. 17. Four pion mass spectrum for all events. Solid curve is a Breit-Wigner shape with the observed two pion mass of 1600 MeV. and a width of 283 MeV.

B. $\pi^+ \pi^- \pi^0 \pi^0$ $\rho'(1250)$. Previous evidence for a resonance at 1250 MeV decaying into $\omega^0 \pi^0$ comes from data of the SLAC-Berkeley bubble chamber collaboration.[24] That collaboration estimated a cross section for the inferred peak of roughly 2 μb. The collaboration however was not able to positively identify the two π^0's. In the OMEGA experiment with the very large lead glass array, two π^0's could be detected and identified. In Fig. 18 a mass plot of $\pi^+ \pi^- \pi^0 \pi^0$ events is presented. The shaded region requires a $\pi^+ \pi^- \pi^0$ combination to be an ω. The cross section of the 400 MeV wide shaded peak centered at 1.25 GeV is estimated to be $\sigma(\gamma p \rightarrow \omega \pi^0 p) = 1.6 \pm .3$ μb.[4]

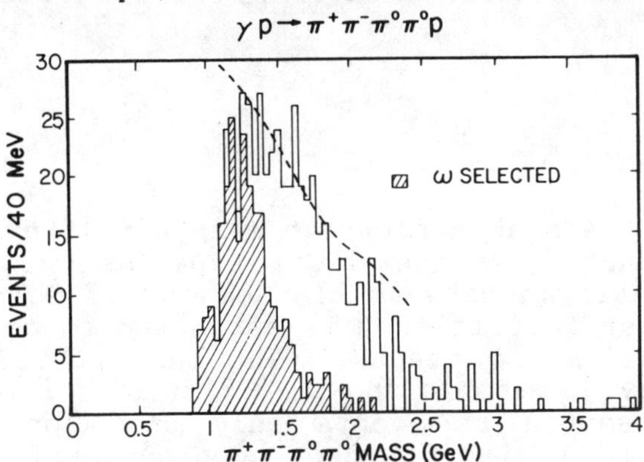

Fig. 18. $\pi^+ \pi^- \pi^0 \pi^0$ mass spectrum of CERN OMEGA experiment.

C. $\pi^+\pi^-\pi^o$. No photoproduction experiment has done a high statistics search for a $\pi^+\pi^-\pi^o$ final state above the ω. The $\pi^+\pi^-\pi^o$ mass distribution from the OMEGA experiment is shown in Fig. 19. A clean phi signal is seen in addition to indications of signals centered at 1.275 GeV and 1675 GeV. Both enhancements have widths of about 100 MeV and cross sections of order 100 nb.[4] Clearly another experiment is needed to resolve the existence of these final states.

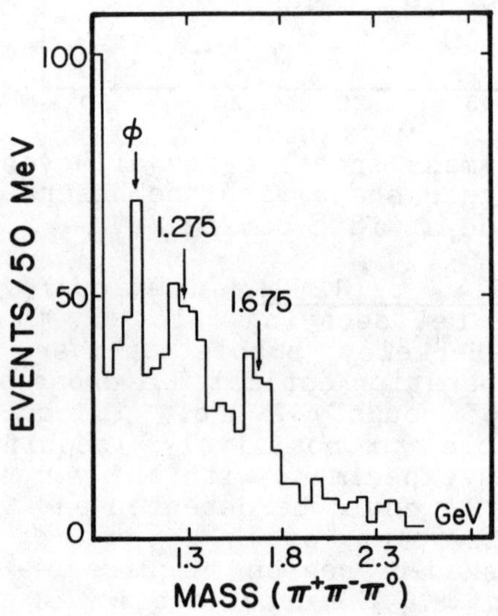

Fig. 19: 3π mass distribution in $\gamma p \rightarrow \pi^+\pi^-\pi^o p$ above the mass.

D. K^+K^-. A search for high mass K^+K^- final states has been conducted by the OMEGA experiment. Charged kaons are identified between 5.5 GeV and 18 GeV. Their Cerenkov counter inefficiency is estimated to be 2% from studying the ρ meson events. The mass plot for K^+K^- pairs above the ϕ mass is shown in Fig. 20. In order to increase the sample of events only one kaon need be identified. Due to the Cerenkov counter inefficiency a peak due to $\rho^o \rightarrow \pi^+\pi^-$ and $\rho' \rightarrow \pi^+\pi^-$ are visible. A new structure at 1.75 GeV is seen. The excess events due to the peak corresponds to a cross section of a few nb.[4]

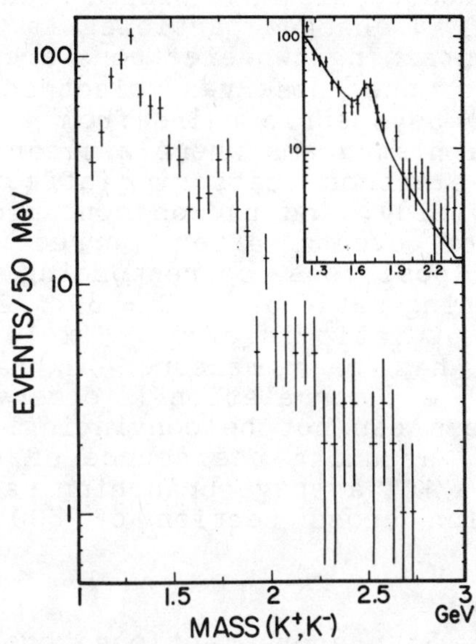

Fig. 20. K^+K^- mass distribution in $\gamma p \to K^+K^-p$ above the ϕ. The data are fitted by a gaussian on top of an exponentially falling background. It gives a mass of 1.76 ± .01 GeV and a width of 120 ± 30 MeV.

IV. Charmed Mesons

The fraction of charm production in photon interactions has been theorized to be approximately 1%.[25] Interpretations of the results of the photon total cross section experiment[1] at Fermilab seem to agree with the predictions. Two experiments, the CERN OMEGA collaboration and the Fermilab Broadband beam group, have undertaken searches for charmed mesons. Preliminary results on evidence for charmed meson production is presented in this section.

A. Prompt Electrons. One method for determining the charm contribution to the total cross section is to measure the direct electron yield. The OMEGA collaboration has measured the direct electron yield using their large lead glass array. Electrons are identified by making E/p (energy in lead glass divided by momentum) histograms. The Cerenkov counter is used to eliminate pions below the pion threshold and to eliminate purely electromagnetic processes a p_T cut plus a requirement that there be at least $\geq$ 4 charged particles is imposed. In addition, events containing two electrons are rejected.

Corrections to the observed electron (position) yields are made for asymmetric pairs from γ conversions (estimated from γe histograms where a clear π^o peak is observed), Compton electron scattering (affecting only the negative electron yield), and hadron contamination. The resultant excess of events after corrections is 200 electrons and 200 positions corresponding to a cross section times branching ratio of $\sigma . B = 80 \pm 20$ nb.[4] The direct electron to pi ratio is $e/\pi = 5 \times 10^{-4}$, a factor of 10 higher than has been measured in hadron beams. Unfortunately, no $K^{\mp} e^{\pm}$ correlation is observed due to $D\bar{D}$ decays, so the excess can not be convincingly attributed to charm. However, assuming the source of electrons is charm and assuming a 10% average branching ratio, a total charm photoproduction cross section of 800 ± 200 nb is obtained.

<u>B. D Mesons</u>. Both above mentioned photoproduction experiments have some evidence for D meson production. In the Broadband experiment a pair production model (i.e. $D\bar{D}$) has been assumed to reduce the data to manageable proportions. Only events with two kaons of opposite charge and no other identified heavy hadrons have been used for the D search. The events where the K^+K^- mass form a ϕ meson (1.01 GeV $> M_{KK} >$ 1.03 GeV) have been eliminated. An additional requirement that the total visible energy be less than 200 GeV is required to suppress the hadron background. An observed enhancement in the $K^{\pm}\ \pi^{\mp}$ decay mode is shown in Fig. 21A. The solid line is a fit of the form of a polynomial background plus a gaussian centered at 1.862 GeV. In Fig. 21b. the same $K^{\mp}\ \pi^{\pm}$ decay mode data is plotted subject to the requirement that the total visible mass lie in the range 3 GeV-4 GeV. In Fig. 22 the corresponding mass plot for $K^{\mp}\ \pi^+ \pi^- \pi^{\pm}$ is presented.

A Monte Carlo program was written based on photons producing a $D^o\bar{D}^o$ at a mass of 4.0 GeV carrying the full energy of the incident photon. The simulation required an average charge multiplicity of 2.2 and a K^+ to $\bar{K}^o$ ratio of 57 to 43. Using a $K^{\pm}\ \pi^{\mp}$ branching ratio of 2.5% yields a cross section for $D^o\bar{D}^o$ production of $\sigma\,(D^o\bar{D}^o)$ = 520 ± 110 nb. For the $K^{\pm}\ \pi^-\pi^+\pi$ with a branching ratio of 4.4%, a $\sigma\,(D^o\bar{D}^o)$ = 540 ± 210 nb is obtained.

In the CERN OMEGA experiment where the average photon energy is only 35 GeV, an associated production

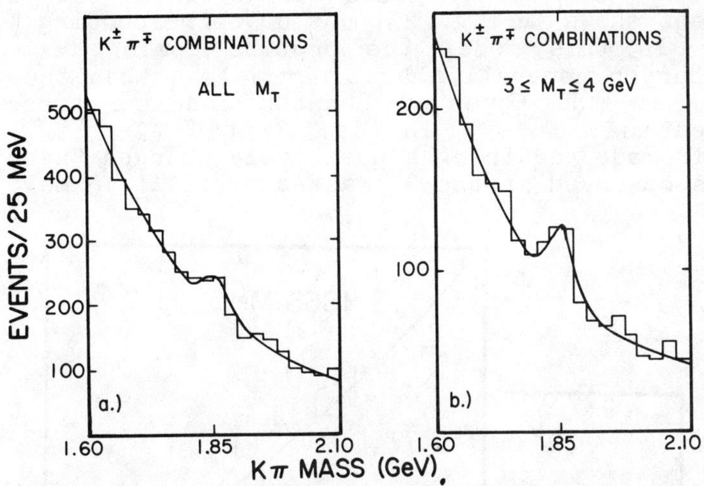

Fig. 21. The $K^{\pm}\pi^{\mp}$ mass distribution of Columbia-Fermilab-Illinois collaboration with (a) no cut on the total visible mass, (b) total mass between 3 GeV-4 GeV. Solid curves are fits to the data described in text.

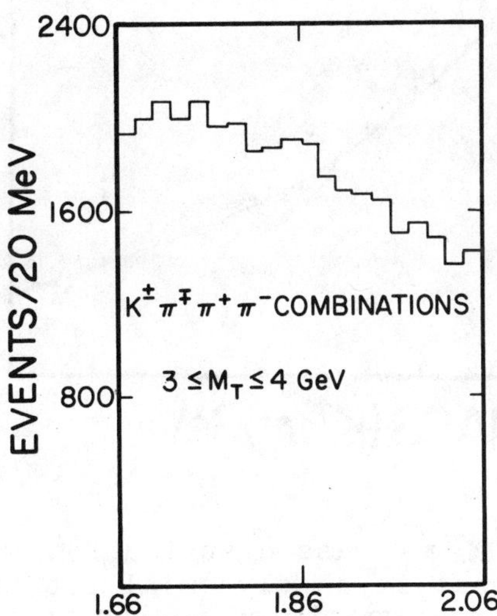

Fig. 22. $K^{\pm}\pi^{\mp}\pi^{+}\pi^{-}$ mass distribution from Columbia-Fermilab-Illinois collaboration.

model of $\Lambda_c \bar{D}^o$ is proposed to explain the $K^+ \pi^-$ enhancement shown in Fig. 23, but no corresponding $K^- \pi^+$ signal. In this model the protons coming from the charmed baryon generally have energy less than the kaon threshold in the Cerenkov counter and are therefore identified as a K^+'s. In Fig. 24 the $K^o_s \pi^+ \pi^-$ mass spectra is made requiring a positively charged "kaon". A signal is observed at the D^o mass of 1862 ± 13 MeV.

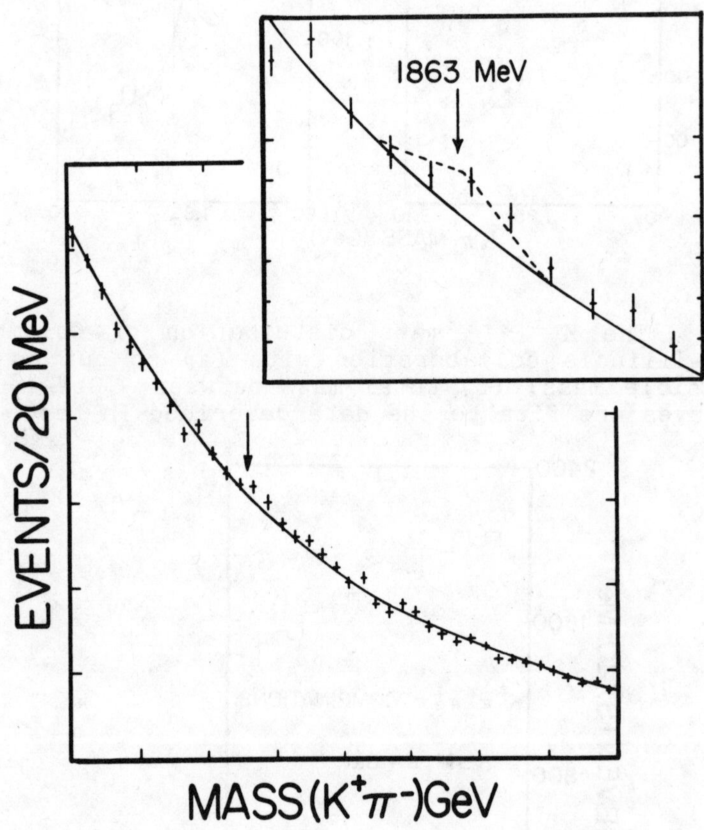

Fig. 23. Inclusive $K^+ \pi^-$ mass distribution of CERN OMEGA experiment. The curve is a polynomial fit to the data. The dotted curve in the magnified region is the expected shape for a $\bar{D}^o$ signal.

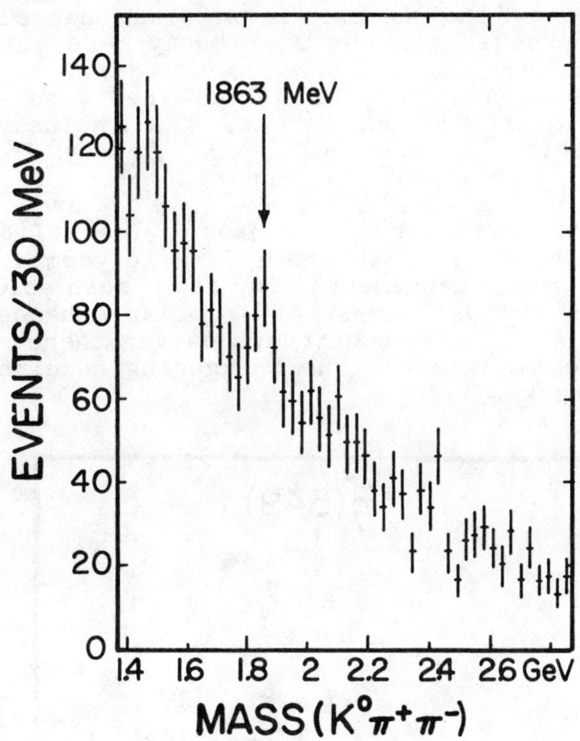

Fig. 24. Inclusive K^o π^+ π^- mass distribution of CERN OMEGA experiment with:

An extra-K^+ (proton), total visible mass above 3.7 GeV, and incident photon energy above 40 GeV.

In both experiments, there is no evidence for charged D mesons. In the Broadband experiment using the pair production model a limit of $\sigma(D^+D^-)$ < 300 nb. is found. In the OMEGA experiment, an inclusive limit of $\sigma(D^{\pm} \rightarrow K^{\mp}\ \pi^{\pm}\ \pi^{\pm})$ 500 nb[4] is obtained.

B. <u>F Mesons</u>. The only evidence which supports the existence of F_o mesons comes from DASP.[26] In that experiment an η $\pi^{\pm}$ mass is plotted against the recoil mass for events with an additional gamma ray with energy less than 140 MeV. Six events are found at an η $\pi^{\pm}$

mass of 2.04 ± .02 GeV. With this evidence in mind the Omega spectrometer group has looked for photoproduced F mesons decaying into an eta meson plus charged pions. The η^o is detected via the 2 γ decay mode in a large lead glass array.

A sample of 14,000 η^o's is obtained with a signal to background ratio of one to one. The inclusive $\gamma\gamma$ mass distribution is shown in Fig. 25. In Fig. 26 a clear resonance of the $\eta'(958)$ is seen in the $\eta\pi^+\pi^-$ mass spectrum. In Fig. 27, the $\eta\pi^+\pi^-\pi^\pm$ and $\eta\pi^+\pi^-\pi^+\pi^-\pi^\pm$ mass plots are presented. The solid curves are fits to the data of the form of a polynomial plus a gaussian. The enhancements in both mass plots are centered near the DASP mass. No similar enhancement is seen in the $\eta\pi^\pm$ mass spectrum, however there may have been a severe bias due to the triggering requirement of at least 4 tracks.

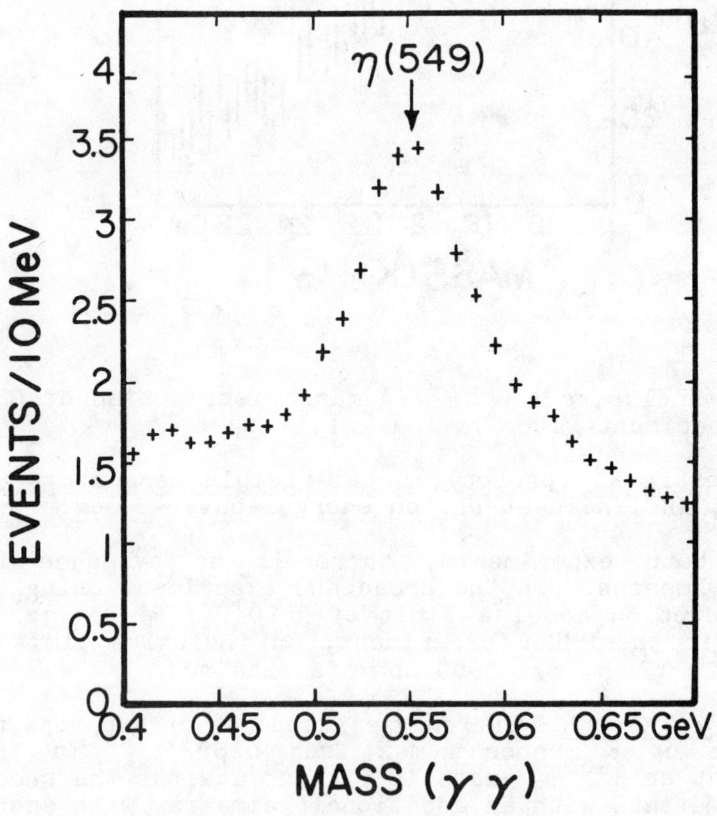

Fig. 25. Inclusive $\gamma\gamma$ invariant mass spectrum.

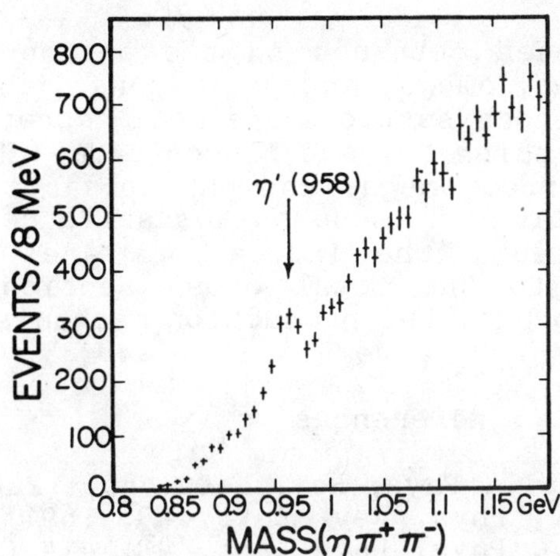

Fig. 26. Inclusive η^{o} π^{-} π^{+} mass distriubution.

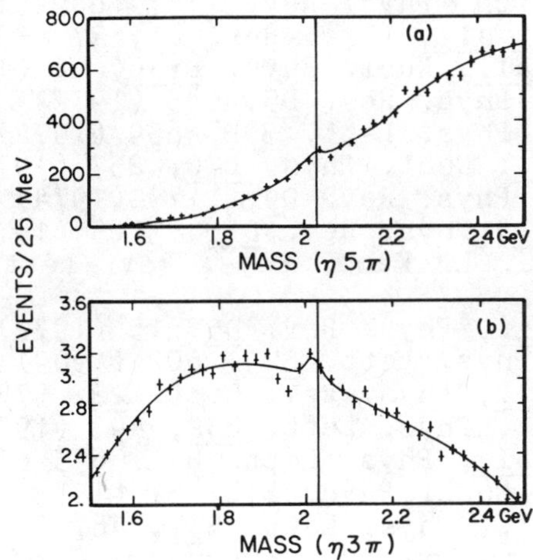

Fig. 27. (a) Inclusive $\eta\pi^{\pm}\pi^{+}\pi^{-}\pi^{+}\pi^{-}$ fitted mass distribution. (b) Inclusive $\eta\pi^{\pm}\pi^{+}\pi^{-}$ fitted mass distribution. The fit uses a polynomial plus a gaussian.

V. Summary

Results have been obtained on the photoproduction cross sections of rho, omega, and phi mesons. The energy dependences of the cross sections are well represented by a VMD - quark model using forward hadron data. High mass recurrences of the rho, omega, and phi appear to exist, but the characteristics of these states still need to be investigated. Finally, the indications are that the charm contribution to the total cross section may be large, of order 1 b but, the production mechanism is not yet understood.

References

[1] D. O. Caldwell et al., Phys. Rev. Lett. 40, 1222 (1978)
[2] M. S. Atiya et al., Phys. Rev. Lett. 43, 1691 (1979)
[3] M. S. Atiya et al., Phys. Rev. Lett. 43, 414 (1979)
[4] F. Richard, invited talk at Photon-Lepton Symposium at Fermilab (1979)
[5] P. Söding, Phys. Lett. 19, (1965); R. Spital and D. R. Yennie, Phys. Rev. D9, 126 (1974)
[6] J. D. Jackson, Nuovo Cimento 34, 1644 (1964)
[7] ABBHHM Collaboration, Phys. Rev. 175, 1669 (1968)
[8] R. L. Anderson, et al., Phys. Rev. D1, 27 (1970)
[9] H. J. Behrend et al., Nucl. Phys. B144, 22 (1978)
[10] J. Ballam et al., Phys. Rev. D5, 545 (1972)
[11] C. Berger et al., Phys. Lett. 39B, 659 (1972)
[12] H. J. Besch et al., Nucl. Phys. B70, 257 (1974)
[13] I. Ambats et al., Phys. Rev. D9, 1179 (1974)
[14] D. S. Ayres et al., Phys. Rev. D15, 3105 (1977)
[15] A. N. Kamal and G. L. Kane, Phys. Rev. Lett. 43, 551 (1979)
[16] Y. Eisenberg et al., Phys. Rev. D5, 15 (1972)
[17] N. Hicks et al., Phys. Lett. B29, 602 (1969)
[18] G. McClellan et al., Phys. Rev. Lett. 22, 374 (1969)
[19] M. Conversi et al., Phys. Lett. B52, 493 (1974)
[20] H. H. Bingham et al., Phys. Lett. B41, 635 (1972)
[21] M. Davier et al., Nucl. Phys. B58, 31 (1973)
[22] G. Alexander et al., Phys. Lett. B57, 487 (1975)
[23] Unpublished results on photoproduction of ' are included in J. Bronstein, thesis, University of Illinois (1977) and J. Knauer, thesis, University of Hawaii, 1975.
[24] J. Ballam et al., Nucl. Phys. B76, 375 (1974)
[25] D. Silvers et al., Phys. Rev. D13, 1234 (1976)
B. Margolis et al., Phys. Rev. D17, 1310 (1978)
[26] R. Brandelik et al., Z. Physik C. Particles and Fields 1, 233 (1979)

Discussion

Q. (Luste, U. of Toronto) Given a 10% branching ratio for the semi-leptonic decay of charmed mesons. What fraction of the total cross section is due to charm from the observed direct electron yield?

A. Approximately 1%.

Q. (Luste, U. of Toronto) Is it significant that no $K^{\mp}$ $-e^{\pm}$ correlation is seen in the CERN OMEGA experiment?

A. No, I don't think it is significant. The average photon energy in that experiment was only 30 GeV, so the D energy would vary between 10-20 GeV depending on the production mechanism. Since the Kaon Cerenkov threshold turns on at 5.5 GeV. I don't think it is surprising that no $K^{\mp}$ $e^{\pm}$ correlation was seen.

RECENT RESULTS FROM PETRA

B.H.Wiik
Deutsches Elektronen-Synchrotron DESY, Hamburg, Germany,
Notkestrasse 85 - 2000 Hamburg 52 - Germany

1. INTRODUCTION

The experimental program at the new DESY electron-positron storage ring PETRA[1] got underway late 1978, more than half a year ahead of schedule. Initially data[2] were collected at c.m. energies of 13 GeV, 17 GeV and 27.4 GeV using three large detectors MARK J, PLUTO and TASSO. In June of 1979 a fourth detector, JADE, was installed and these detectors have since taken data at c.m. energies between 22 GeV and 31.6 GeV, the highest energy available with the present complement of klystrons. The energy region between 29.5 GeV and 31.6 GeV, have been scanned in steps of 20 MeV in a search for narrow 1^{--} states.

Several new results have been obtained during the first year of data taking:

The data on various QED processes agree with the theoretical predictions down to distances of $2 \cdot 10^{-16}$ c.m. and confirm lepton universality at small distances.

The data on multihadron final states give clear evidence for jets and they show that the threshold for $t\bar{t}$ production, where t is a quark with charge 2/3 e, must be above 31.5 GeV.

The outstanding experimental result has been the observation of three-jet events, first seen[3,4] by the TASSO Collaboration and since confirmed[5,6,7] by all the other groups. Such events are evidence for hard gluon bremsstrahlung which is expected in any field theory of strong interactions.

The first data on $e^+e^- \to e^+e^-$ hadrons at high energies have been obtained by the PLUTO Collaboration.

In this talk I'll first describe the status of PETRA and then discuss these experimental results in more detail.

2. STATUS OF PETRA

A schematic layout of the DESY accelerator complex is shown in Fig. 1. PETRA (= Positron Elektron Tandem Ring Anlage) is made of eight 45° bends joined by eight straight sections, four 108 m long and four 68.4 m long. The total circumference is 2.3 km. Electrons are injected from the DESY synchrotron directly into PETRA, the positrons are first accumulated in a small storage ring PIA and then injected into PETRA via DESY.

The energy and luminosity of PETRA has been climbing from 13 GeV and a peak luminosity of 2×10^{29} $cm^{-2}sec^{-1}$ at the beginning of the year to 31.6 GeV and 5×10^{30} $cm^{-2}sec^{-1}$. This luminosity should be compared to a predicted maximum luminosity of 2.2×10^{31} $cm^{-2}sec^{-1}$ for 2 bunches of positrons colliding with 2 bunches of electrons and a free distance of $\pm$ 7.5 m in the interaction region between the quadrupoles. On the average, an integrated luminosity

ISSN:0094-243X/80590117-42$1.50

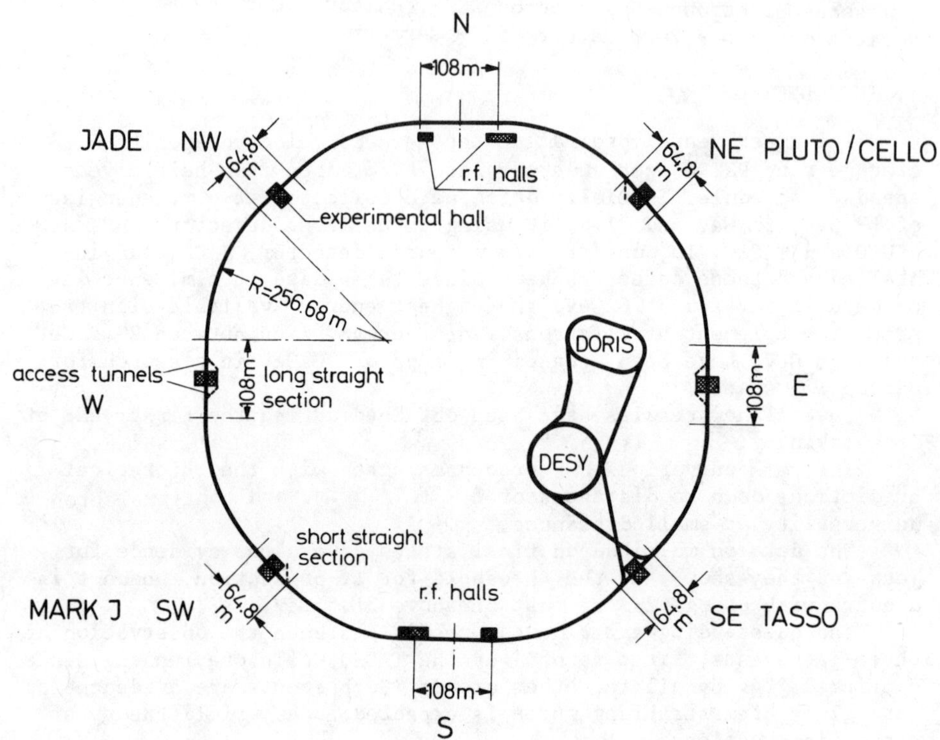

Fig. 1 - Layout of the DESY accelerators

of about 100 nb^{-1} per day is obtained at 30 GeV, this corresponds to some 25 hadron events per day.

The circulating electron beams radiate and this leads to a build up of transverse polarization with a time constant[8] $\tau = 98.8 sec$ $(R\rho^2/E^5)$. R and ρ are the geometrical and the bend radia in meters respectively, and E the beam energy in GeV. From this equation, at a beam energy of 15 GeV, we predict a polarization time τ of about 30 min. Since this time constant is short compared to the storage time we expect, in absence of strong depolarization effects, that the beams are polarized. A measurement of $e^+e^- \to q\bar{q}$ can be used to determine the degree of polarization. The distribution of the jet axis, assuming the quarks to be fermions, is given by

$$\frac{d\sigma}{d\Omega} \sim 1 + \cos^2\theta + P^2 \sin^2\theta \cdot \cos 2(\phi - \Delta) \qquad (1)$$

where P is the degree of transverse polarization and Δ a possible angle in the interaction region between the polarization vector and the transverse direction. θ and ϕ denotes the production and azimuthal angle. The azimuthal distribution of 2-jet events observed by the JADE Collaboration and selected using thrust (see below) is plotted in Fig. 2. The data show a strong azimuthal dependence and a fit to the form given above yields $\Delta = -10^\circ \pm 14^\circ$ and $P = 0.85 \pm 0.14$.

During the long shut down at the end of the year the number of r.f. cavities will be increased from 32 to 64 and the r.f. power will be doubeled. With this r.f.system PETRA will be able to explore energies up to 38 GeV in c.m. Some of the PETRA parameters are listed in table I.

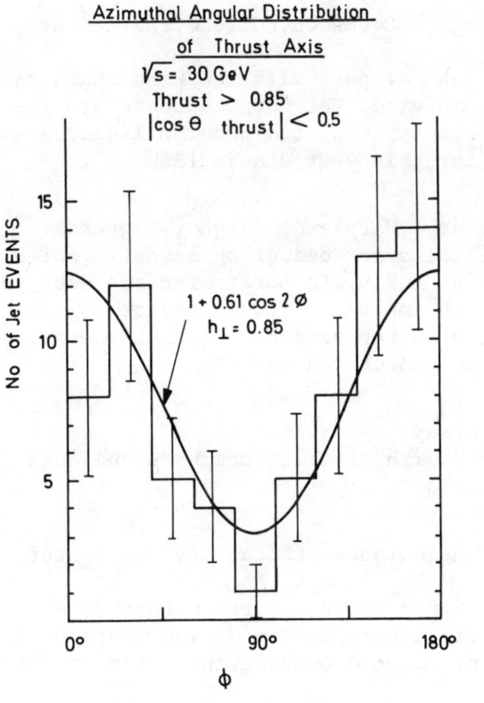

Fig. 2 - Azimuthal distribution of the jet axis in two-jet events as measured by JADE at 30 GeV in c.m.

Table I - PETRA parameters

Maximum c.m. energy:	32 GeV (38 GeV early 1980)
Average circulating current:	$\sim$ 8 mA/Ring
Number of Bunches:	2/Ring (4 possible)
Luminosity after fill:	$5 \times 10^{30} cm^{-2} sec^{-1}$ at 30.0 GeV
Tune shift Q_V^{max}, Q_H^{max}:	0.027 / Interaction Region
Lifetime:	$\sim$ 6.5 hours just after filling $\sim$ 24 hours towards the end of the fill

Table I (continued)

Number of interaction regions:	4 (can be extended to 6)
Length of interaction region:	15 m
Interaction volume:	$\sigma_t \leq 0.07$ cm at 14 GeV $\sigma_L \approx 3$ cm
Vacuum in the interaction regions:	$\sim 10^{-9}$ torr after fill $\sim 10^{-10}$ torr towards the end of the fill

There are four interaction regions equipped with experiments:

SE TASSO Collaboration
Aachen, Bonn, DESY, Hamburg, I.C.London, Oxford, Rutherford, Weizmann and Wisconsin

Large conventional solenoid. Central part with tracking chambers and time of flight counters completed. The muon chambers are installed. The hadron arms are nearly completed and the liquid Argon detector will be installed later this year and in 1980.

NE PLUTO Collaboration
Aachen, Bergen, DESY, Hamburg, Maryland, Siegen, Wuppertal

This is a modified version of the superconducting magnetic solenoid detector which collected data at DORIS. In particular the muon and electron detection has been much improved. This detector is now taking data but will presumably be replaced by the CELLO detector during the long shutdown at the end of the year.

CELLO Collaboration
DESY, Karlsruhe, Munich, Orsay, Paris, Saclay

Superconducting solenoid filled with tracking chambers and surrounded by a liquid Argon detector

NW JADE Collaboration
Daresbury, DESY, Hamburg, Heidelberg, Lancaster, Manchester, Tokyo

Conventional solenoid with a high pressure drift chamber as a tracking detector. Particles are identified by measurements of dE/dx. Large leadglass photon detector mounted outside the solenoid. This detector was installed in June.

SW MARK J Collaboration
Aachen, DESY, MIT, NIKEF Amsterdam, Peking

This detector is a fine grained calorimeter well suited to measure electrons, muons and the energy flow of hadrons.

3. TEST OF QED AND RELATED TOPICS

Electron-positron colliding beams make it possible to test the structure of quantum-electrodynamics at very high momentum transfer in a clean environment with negligible corrections due to strong interactions.

So far the following reactions have been investigated:

1) $e^+e^- \rightarrow e^+e^-$ [7,9,10]

2) $e^+e^- \rightarrow \mu^+\mu^-$ [9]

3) $e^+e^- \to \tau^+\tau^-$[9]

4) $e^+e^- \to \gamma\gamma$[7]

5) $e^+e^- \to e^+e^- \mu^+\mu^-$[9], $e^+e^- \to e^+e^-(\mu^+\mu^- + e^+e^-)$[10]

Reactions 1 - 4 are proportional to α^2 and decreases with energy as 1/s - i.e. $sd\sigma/d\Omega$ is independent of energy. $sd\sigma/d\Omega$ for $e^+e^- \to e^+e^-$ measured[9] by the MARK J group at various energies are plotted as a function of scattering angle in Fig. 3. The results for $e^+e^- \to e^+e^-$ and $e^+e^- \to \gamma\gamma$ measured[7] by the JADE Collaboration are shown in Fig. 4. Both experiments are in agreement with the QED prediction represented by the solid line in Figs. 3 and 4. To quantify the agreement it is assumed that a breakdown of QED in $e^+e^- \to e^+e^-$ can be parametrized by introducing spacelike and timelike form factors $F_S(q^2)$ and $F_T(q^2)$. With $F_S(q^2) = F_T(q^2) =$

$$1 \pm \frac{q^2}{q^2 - \Lambda_\pm^2}$$

the agreement with QED can be expressed as a lower limit on the cut off parameter $\Lambda_\pm$.

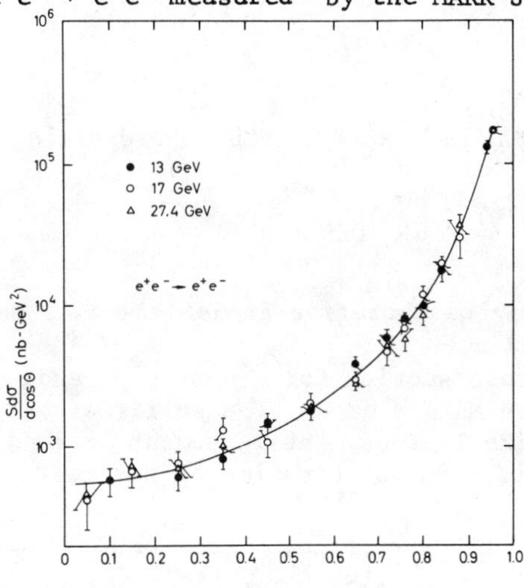

Fig. 3 - The cross section $s \cdot \frac{d\sigma}{d\Omega}$ for $e^+e^- \to e^+e^-$ measured by MARK J between 13 and 27.4 GeV in c.m.

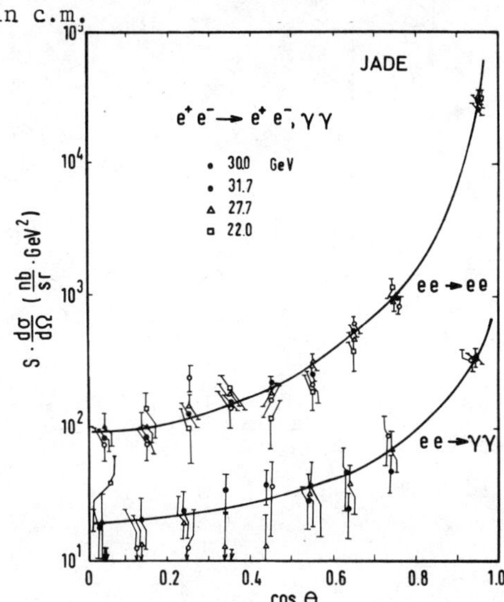

Fig. 4 - The cross section $s \cdot \frac{d\sigma}{d\Omega}$ for $e^+e^- \to e^+e^-$ and $e^+e^- \to \gamma\gamma$ measured by JADE between 22 and 31.7 GeV in c.m.

The results are listed in Table II.

Table II - Lower limits on $\Lambda_\pm$ at 95% confidence in $e^+e^- \to e^+e^-$

	Λ_+(GeV)	Λ_-(GeV)
MARK J[9]	65	64
JADE[7]	89	74
PLUTO[10]	71	67

A possible breakdown of QCD in $e^+e^- \to \gamma\gamma$ can be expressed as

$$d\sigma/d\Omega = (1 \pm \frac{s^2}{2\Lambda_\pm^4} \sin^2\theta) \left(\frac{d\sigma}{d\Omega} \right)_{QED} \quad (2)$$

From a fit to the data the JADE Collaboration finds[17] $\Lambda_+ > 43$ GeV and $\Lambda_- > 31$ GeV with 95% confidence.

Fig. 5 shows the total cross section for $e^+e^- \to \mu^+\mu^-$ and $e^+e^- \to \tau^+\tau^-$ as reported[9] by the MARK J group. The solid(dashed) line the QED prediction for pointlike leptons. The agreement is good and to express the agreement in terms of a radius of the leptons

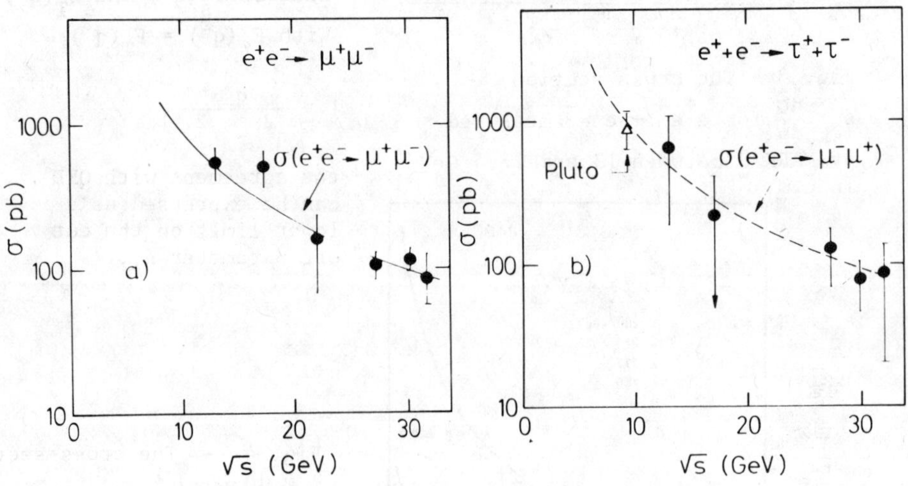

Fig. 5 - The total cross section for $e^+e^- \to \mu^+\mu^-$ and $e^+e^- \to \tau^+\tau^-$ measured by MARK J

the MARK J group assumes a form factor:

$$F_\ell(q^2) = 1 \mp \frac{q^2}{q^2 - \Lambda^2_{\ell\pm}} \tag{3}$$

A fit to the data give:

ℓ	electron	muon	tau
Λ_-	95 GeV	97 GeV	53 GeV
Λ_+	74 GeV	71 GeV	97 GeV

Lepton universality is thus valid down to distances of 10^{-16} cm.

In models incorporating several neutral vector bosons, the lightest vector boson always has a mass below the mass given in the Standard[11] Weinberg-Salam model. Therefore, although they can be made to yield the same prediction as the standard model at low energies, they will differ at high energies. JADE[7] has fitted their data to a particular version[12] of such a model with the mass M, and the width Γ_1 of the lightest neutral vector boson as a parameter. The result is shown in Fig. 6.

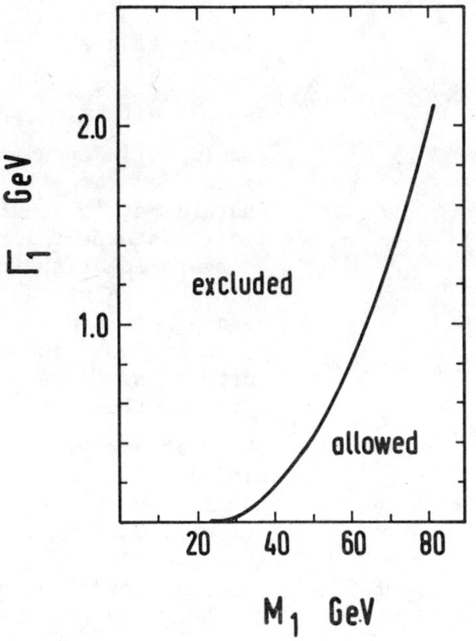

Fig. 6
Limits on the width and mass at the lightest neutral vector boson in a two pole model. The data are from JADE.

3. HADRON PRODUCTION IN e^+e^- ANNIHILATION

It has been conjectured[13-16] that hadron production in e^+e^- annihilation proceeds by quark pairproduction as shown in Fig. 7a, where the electromagnetic current couples directly to the charge of a pointlike quark. (The neutral weak current is expected to contribute on the order of 1% to the total cross section at PETRA energies and is neglected). The total cross section for hadron production is therefore proportional to the cross section for muon pairproduction with the constant of proportionality

$$R = \frac{(e^+e^- \rightarrow \text{hadrons})}{(e^+e^- \rightarrow \mu^+\mu^-)} = 3 \sum_i \left(\frac{e_i}{e} \right)^2 \qquad (4)$$

Here e_i is the charge of the i-th flavour and the sum is over all flavours. The hadrons will appear in two nearly collinear jets of hadrons with small and maybe constant momenta transverse and large and growing momenta parallel to the jet axis. The single particle distribution should scale i.e.

$$s \cdot \frac{d\sigma}{dx} \text{ with } x = E_h/E_{beam}$$

should be independent of energy for large x. The charged particle multiplicity is expected to increase logarithmically with $s = (2E)^2$. The data[17] from SPEAR and DORIS at lower energies support the gross features of this picture.

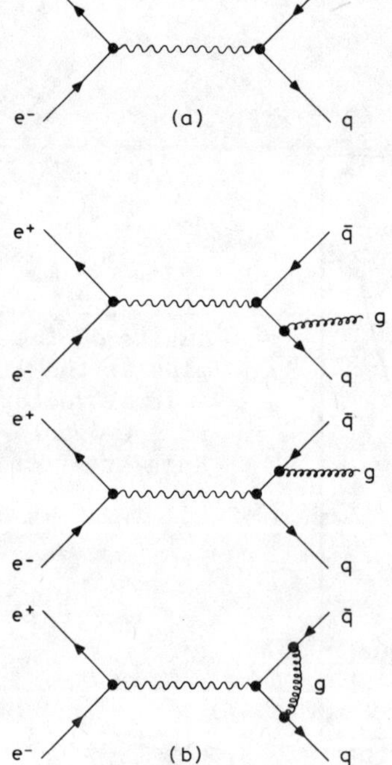

Fig. 7 - Quark pairproduction and gluon bremsstrahlung to first order

This picture will be modified in any field theory[18] of strong interactions. In a field theory e^+e^--annihilation proceeds to lowest order by the Feynman graphs shown in Fig. 7b. The produced quark radiate field quanta (gluons) and the gluons are expected to materialize as hadron jets in the final state.

This has well defined experimental implications[19,20]: The mean transverse momentum of the hadrons with respect to the jet axis will increase with energy. If the quark-gluon coupling constant is small only one of the jets will broaden. The premordial $q\bar{q}g$ state is necessarily planar and the final hadron configuration will retain the planarity. In a small fraction of the events the gluon is radiated at an angle which is large compared to the angular spread of the hadron jet. Such events will be very striking with three visible jets of hadrons defining a plane.

A field theory of the strong interactions will also modify the value for R given above, the multiplicity will grow faster than ln s and the single particle distribution will no longer scale.

At present quantum chromodynamics (QCD)[21] is the leading candidate for a theory of strong interactions. The coupling constant in this theory depends on $q^2 = (2E)^2$ and is given by

$$\alpha_s(q^2) = g^2/4\pi = \frac{12\pi}{(33-2N_f)\ \ln q^2/\Lambda^2} \quad . \qquad (5)$$

Here N_f is the number of flavours with mass below E and Λ is determined[22] in deep inelastic lepton-hadron interactions to be about 500 MeV. The data will be confronted with the QCD predictions. However it is important to bear in mind that most of the general features outlined above will be true in any field theory of strong interactions.

3.1 THE TOTAL CROSS SECTION

In some respects a measurement of the total hadronic e^+e^- annihilation cross section is easier at high than at low energies. The final state hadrons are in general confined in two back to back jets and the angular distribution of the jet axis is proportional to $1 + \cos^2\theta$. The new generation of detectors cover a large solid angle and this together with the high multiplicity result in a high detection efficiency (75-80%) which is subject to small systematic uncertainties only. Events resulting from cosmic radiation or from interactions between the beam and the environment are easily identified and removed in the off line analysis. The background from $e^+e^- \rightarrow \tau\bar{\tau}$ lead to events with low multiplicity (90% of all $\tau\bar{\tau}$ lead to events yield 4 or less prongs) and are removed by a cut on multiplicity. The contribution from two photon processes $e^+e^- \rightarrow e^+e^-$ hadrons are removed by a cut on visible energy.

The values[2,9,23-25] for R at PETRA energies, corrected for radiative effects and with the contribution from $e^+e^- \rightarrow \tau^+\tau^-$ removed, are plotted in Fig. 8 together with data[26] measured at lower energies. In addition to the statistical errors shown, there is a systematic uncertainty on the order of 10%. The different groups collected data using different trigger conditions and they applied different cuts to extract the R values. The good agreement among the various groups demonstrate that systematic effects are well understood.

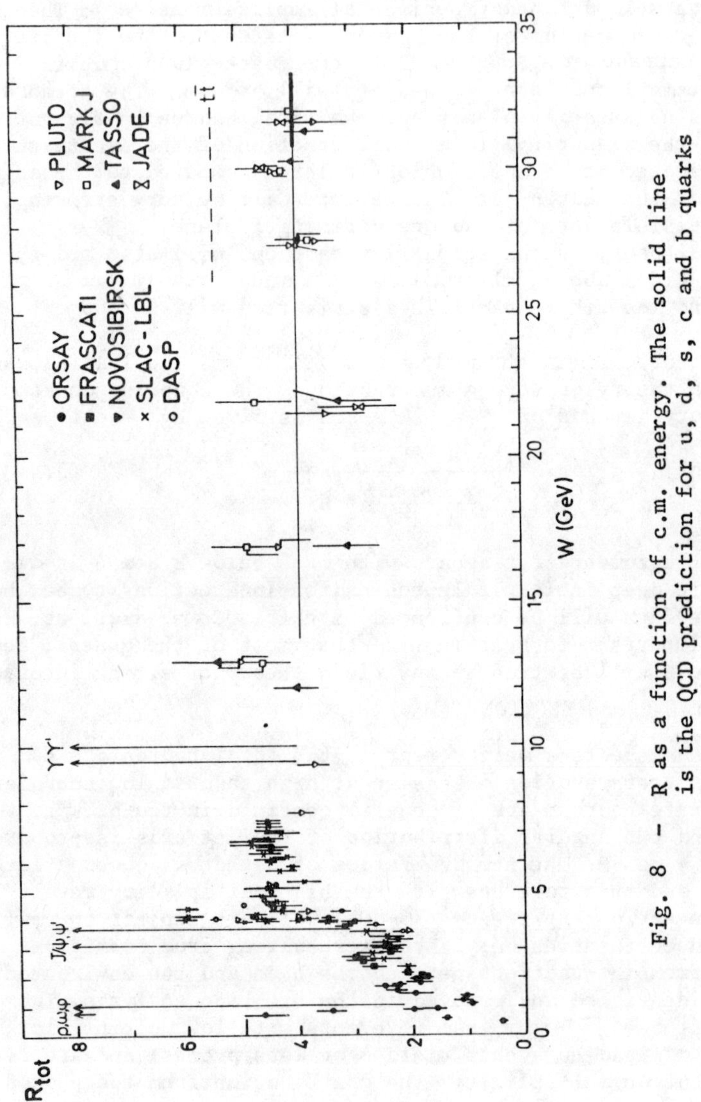

Fig. 8 – R as a function of c.m. energy. The solid line is the QCD prediction for u, d, s, c and b quarks

The new data finds R to be constant above 13 GeV in c.m. In the parton model with u, d, s, c and b quarks R = 3.7. First order QCD predict[27] $R = (1 + \alpha_s(q^2)/\pi)\ 3 \cdot \Sigma(e_i/e)^2$, i.e. a small increase of about 6% only. This prediction, shown as the solid line in Fig. 8, is in good agreement with the data. If we naivly average the R values for all experiments above 27 GeV in c.m. and ignore systematic uncertainties we find <R> = 3.94 compared to the QCD value of 3.92. Of course the data are not yet accurate enough, once systematic errors are included, to discriminate between the quark-parton and the QCD predictions for R.

A charge 2/3 e quark would lead to R $\approx$ 5.4, shown as the dotted line in Fig. 8, in disagreement with the observed R values. However, the data are not yet precise enough to exclude a charge 1/3 e quark. There is also no evidence in the data for pairproduction of new leptons. However, note that new leptons will yield final states with a lower multiplicity than those observed in multihadron states and hence have a lower trigger efficiency.

3.2 CHARGED MULTIPLICITY AND RAPIDITY DISTRIBUTIONS

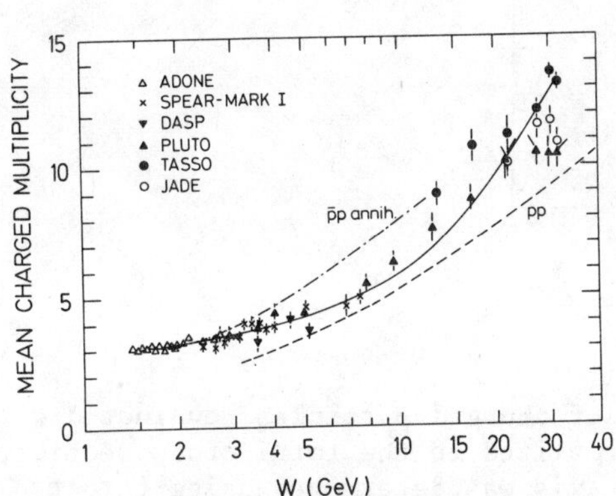

Fig. 9 – Average charged particles multiplicity versus the c.m. energy. The solid line is a combined fit to the low energy data and the TASSO data at high energies

The average charged multiplicity observed[10,24,28] at high energies $\langle n_{ch} \rangle$ is plotted in Fig. 9 together with data[29] at lower energies as a function of W (W = 2E). For comparison the multiplicities observed in pp[30] and $\bar{p}p$[31] collisions are shown as the dashed and the dashed-dotted line in Fig. 10. The data are clearly not proportional to a + b ln s over the whole energy range, as predicted in the naive quark-parton model. Indeed a fit to the data for 1.4 GeV < W < 7 GeV yield a = 2.67 $\pm$ 0.04 and b = 0.48 $\pm$ 0.02. At W = 30 GeV this fit predicts a charged multiplicity of 6 compared to an observed multiplicity of more then 10. Pair production of bottom quarks is expected to increase the multiplicity by 0.2 and cannot account for the strong increase.

The multiplicity in QCD is expected[32] to behave as $\langle n_{ch} \rangle = n_o + a \exp (b \sqrt{\ln s/\Lambda^2})$. A fit to the data[28] for 1.4 GeV W < 31.5 GeV yields:

$$\langle n_{ch} \rangle \;=\; (2.92 \pm 0.04) + (0.0029 \pm 0.005) \exp(2.85 \pm 0.07)(\ln s/\Lambda^2)^{1/2}$$

and is shown as the solid line in Fig. 9. The general trend of the data is reproduced by the fit.

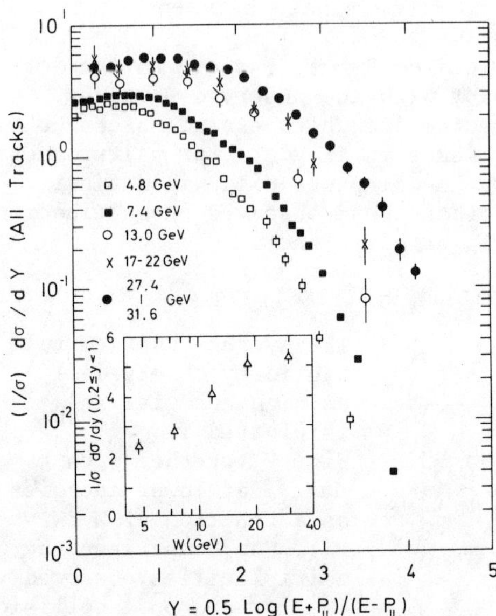

Fig. 10
Rapidity distributions for charged particles assuming they all to be pions. The data at 4.8 GeV and 7.4 GeV are from SLAC-LBL. The high energy data are from TASSO

The rapidity distribution of charged particles, evaluated with respect to the jet axis and normalized to the total cross section, is plotted in Fig. 10. The jet axis was determined using thrust (see below). The rapidity $Y = 0.5 \ln (E+p_{\parallel})/(E-p_{\parallel})$ where E is the energy and $p_{\parallel}$ the momentum of the particle with respect to the jet axis, was evaluated assuming the particles to be pions. The high energy data were obtained by the TASSO Collaboration[28] and they are compared with data from the SLAC-LBL Collaboration[29] at 4.8 and 7.4 GeV. The distribution has a clear plateau at Y = 0 and this plateau becomes longer with increasing energy. The height of the plateau, however, is not constant as expected in quark-parton models, but also increases with energy. This is shown clearly in the insert to Fig. 10, where the normalized cross section $1/\sigma\, d\sigma/dy$ for $0.2 \leq Y < 1.0$ is found to increase linearely with ln W.

To compare the fragmentation regions at various energies the data are replotted in Fig. 11 versus $Y-Y_{max}$ where $Y_{max} \approx 0.5 \ln(s/m^2)$.

The width and the shape of the fragmentation region are nearly independent of energy with the high energy data slightly below the data obtained at lower energies. This is in qualitative agreement with scaling violations expected to occur in QCD. However, it is important to note that the intrinsic resolution is about 1 unit in rapidity caused by the uncertainty in determining the jet direction.

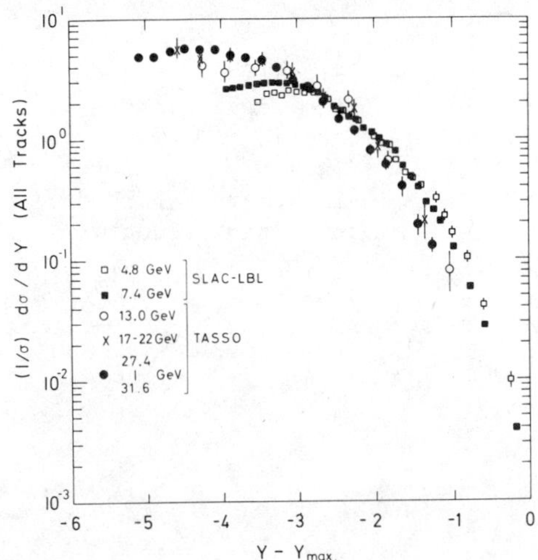

Fig. 11 - The same data as in Fig. 10 plotted versus $Y - Y_{max}$

The rise of the plateau with energy and the near energy independence of the distributions in the fragmentation region shows that the rise in multiplicity is caused by an excess of low energy particles

3.3 SCALING

The cross section $s d\sigma/dx$ with $x = p/p_{beam}$ can be expressed at high energies ($\beta \approx 1$) in terms of two scaling functions $\bar{W}_1$ and $\bar{W}_2$

$$s \frac{d\sigma}{dx} = 4\pi \alpha^2 \cdot x \left(m\bar{W}_1 + \frac{1}{6} s\nu\bar{W}_2 \right) \qquad (6)$$

Here $\nu = \sqrt{s} \cdot E/m$ is the photon energy seen in the rest system of the particle.

Data from DASP[29] at 5 GeV, from SLAC-LBL[33] at 7.4 GeV and from TASSO[28] for energies between 13 and 31.6 GeV are plotted in Fig. 12 The cross sections for $x > 0.2$ scale to within 30% between 5 GeV and 31.6 GeV. For $x < 0.2$ the cross section shows a dramatic rise with energy between 5 GeV and 30 GeV. This rise is related to the strong growth of the multiplicity discussed above.

Gluon emission as indicated in Fig. 7b will lead to a depletion of particles at large x and a corresponding increase in the yield at small x, since the energy is now shared between the quark and the gluon. In QCD, however, these effects are only on the level of 20% since the Q^2 values are large compared to the scale breaking parameter $\Lambda^2 = 0.25\ GeV^2$.

More precisely[34] the 30 GeV data for $x = 0.2$ are predicted to be higher by 10% and for $x = 0.7$ lower by $\sim$ 20% in comparison with the 5 GeV data. The present data are not precise enough to test this prediction.

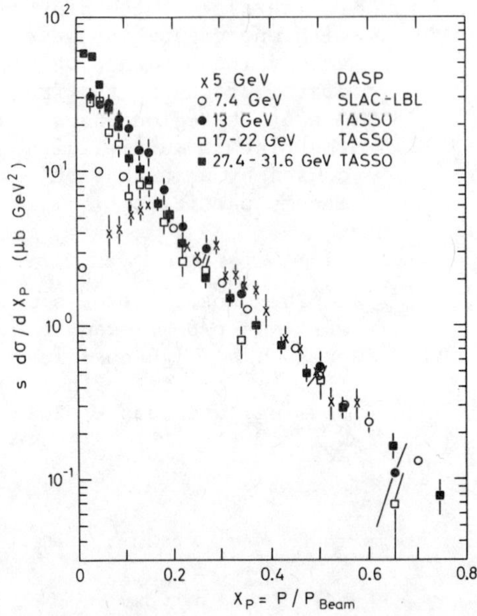

Fig. 12
The scaling cross section s $d\sigma/dx$ ($x = p/p_{beam}$) for inclusive charged particle production

4. EVIDENCE FOR GLUONS AND AGAINST NEW FLAVOURS

The topology of the hadrons in e^+e^- annihilation can be used to identify the production mechanismn:

a) Pairproduction of light quarks manifests itself as two back to back narrow jets of hadrons
b) pairproduction of heavy quarks will, close to threshold, lead to nearly spherical events
c) Gluon bremsstrahlung $e^+e^- \to q\bar{q}g$ leads to planar events with large momenta in the plane and small momenta with respect to the plane. A fraction of the events will have a clear three-jet structure.

All groups have made extensive Monte Carlo computations to confront various production mechanismns with the data. The computations are in general based on the formalismn developed by Feynman and Fields[35]. In their model the various quark flavours are pairproduced proportional to e_f^2. Light quarks pairs are created in vacuum in the ratio: $u\bar{u} : d\bar{d} : s\bar{s} = 2 : 2 : 1$. The quarks fragments according to the distribution function $f(z) = 1-a+3a\ (1-z)^2$ with $a = 0.77$ and $z = E_{meson}/E_{quark}$. A flat distribution function is also used for heavier quarks. The primary mesons are created with a Gaußian distribution $\exp(-p_T^2/2\sigma_q^2)$ around the jet axis. From fits to deep inelastic lepton-hadron data σ_q was found[35] to be about 250 MeV/c. The decay modes for light primary mesons are taken from the particle data tables. Decay mode for heavier mesons were estimated[36] using various models. The fragmentation for the gluon is discussed in the paper by Hoyer et al.[20].

4.1 THRUST AND SPHERICITY DISTRIBUTIONS

Two methods to determine the jet axis, sphericity[15] and thrust[37,38] are in general use:

Sphericity S is defined as $$S = \frac{3}{2} \min \frac{\sum_i (p_T^i)^2}{\sum_i (p^i)^2} \qquad (7)$$

Here p^i is the momentum and p_T^i the transverse momentum of a track with respect to a given axis. The jet axis is defined as the axis which minimizes transverse momentum squared. Sphericity measures the square of δ, the jet cone opening angle. $S = 3/2 \langle\delta^2\rangle$ and is 0 for a perfect jet and 1 for a spherical event.

Thrust T is defined as $$T = \max \frac{\sum_i |p_{/\!/}^i|}{\sum_i |p^i|}$$

Here p^i is the momentum of a track and $p_{/\!/}^i$ its projection along a given axis. The jet axis is defined as the axis which maximizes the directed momentum. Expressed in terms of δ $T \approx (1-\langle\delta^2\rangle)^{1/2}$ and it will approach 1 for a perfect jet event and 1/2 for an isotropic event.

The deviation between the true jet axis and the axis found by either the sphericity or the thrust method was determined by a Monte Carlo computation. The result is plotted in Fig.13 as a function of energy. The jet axis is determined to 5° or better nearly independent of method for c.m. energies above 20 GeV.

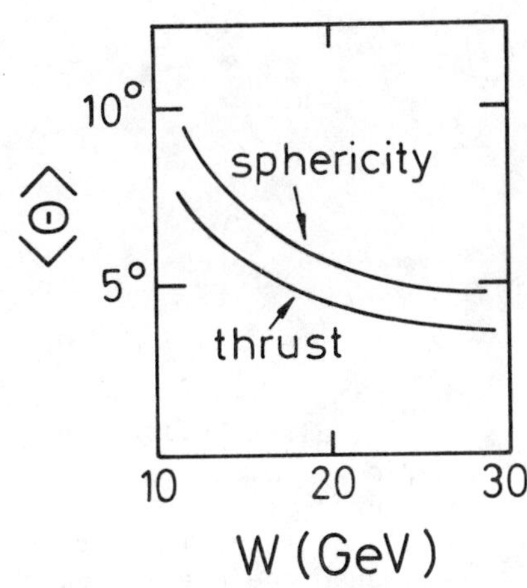

Fig. 13
A Monte Carlo calculation of the angular deviation between the true jet axis and the axis determined using either sphericity or thrust

The sphericity distribution 1/N(dN/dS) as measured by TASSO[25] and the differential thrust distribution as measured by PLUTO[23] are shown in Fig. 14 and Fig. 15 respectively. The jets are expected to become more collimated with increasing energy and this trend is clearly observed in the data. The solid line shows the $q\bar{q}$ prediction including u, d, s, c and b quarks. The TASSO Monte Carlo also includes gluon bremsstrahung.

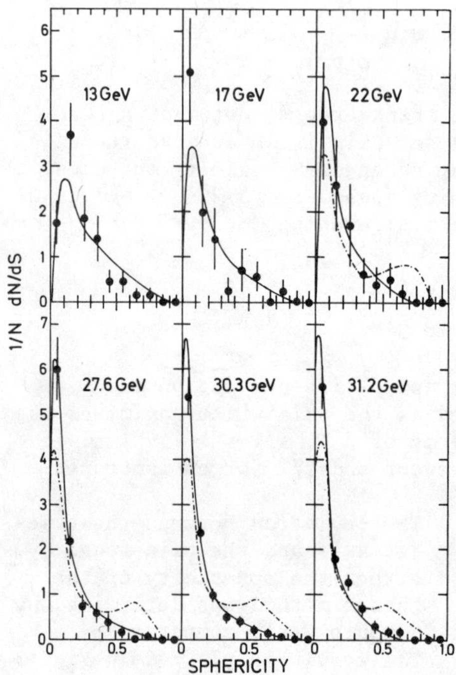

Fig. 14
Sphericity distributions for different c.m. energies measured by the TASSO Collaboration. The curves show the prediction for u, d, s, c and b quarks (solid) plus t quark contribution (dashed-dotted) curves

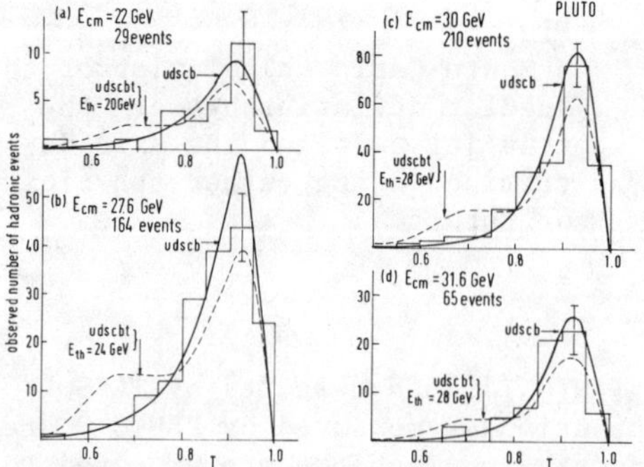

Fig. 15
Thrust distributions for various c.m. energies measured by the PLUTO Collaboration. The curves show the prediction for u, d, s, c and b quarks (solid) plus t quark (dotted)

The energy dependence of the average sphericity <s> and average thrust <T> is plotted in Fig. 16 and Fig. 17 together with data at lower energies. Both <S> and 1-<T> show a smooth decrease with energy without any steps at high energies. The observed energy dependence of <S> is approximately reproduced by

$$\langle S\rangle \simeq 0.8\ (2E)^{-1/2} \qquad (8)$$

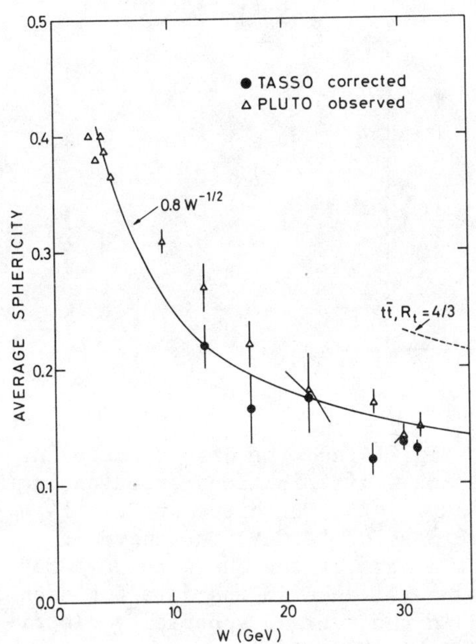

Fig. 16
The average sphericity as a function of c.m. energy. The solid line shows the function $0.8\ (2E)^{-1/2}$

This shrinkage is less than $\langle S\rangle \sim (2E)^{-1}$ expected in the naive quark-parton picture but in agreement with jet broadening due to gluon bremsstrahlung. The QCD corrected prediction for <T> is shown as the solid curve in Fig. 17.

As mentioned above the event shape is very sensitive to contributions from continuum production of $t\bar{t}$. $t\bar{t}$ events are expected to result in a phase space equivalent final state with a large multiplicity. At threshold the t-quarks are produced at rest and will fragment more or less isotropically. As the energy increases the heavy quarks will receive a bost resulting in more jet like events. However, note that a t-quark of mass 10 GeV will reach $\beta = 0.7$ only at 28 GeV in c.m. and hence the effect in the sphericity and thrust spectra will persist well above threshold. The final state particles from $t\bar{t}$ events will therefore be distributed according to phase space near threshold leading to an average sphericity of $\langle S_{PS}\rangle \simeq 0.5$ and a step in thrust $\Delta T = 0.5\ (M/E)^2$.

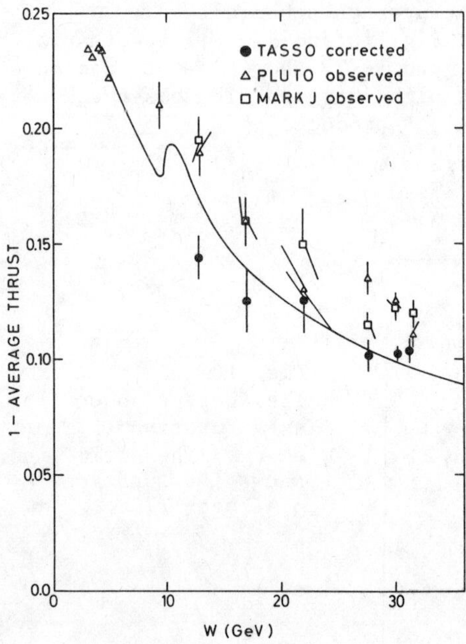

Fig. 17
1-<T> as a function of c.m. energy. The solid line represents QCD prediction, smeared for pionization

The dashed dotted curve in Fig. 14 and the dashed curve in Fig. 15 show the differential S and T distributions expected in the presence of $t\bar{t}$ production with $\Delta R_t = 4/3$. The curve in Fig. 14 was estimated assuming a fixed quark mass of 10 GeV. The curve in Fig. 15 was estimated assuming the mass of the quark to be 1 GeV below the beam energy. There is no evidence in the data for such a contribution. The step expected in the average sphericity distribution is shown as the dotted line in Fig. 16.

The data on sphericity and thrust obtained by the various groups are in agreement. They all give strong evidence against the production of a heavy quark with charge 2/3e.

4.2 EVIDENCE FOR GLUONS

Gluon bremsstrahlung[19] $e^+e^- \rightarrow q\bar{q}g$ has well defined experimental signatures.

A) The transverse momentum with respect to the jet axis will grow with energy.
B) Predominantly only one of the jets in an event will broaden.
C) The events will become increasingly planar with energy - with small and nearly constant momenta normal to the plane and large and growing momenta in the plane.
D) A fraction of the events will have a three-jet structure. The transverse momentum of the hadrons with respect to these axis should be similar to that found in two-jet events at lower energies.

A) GROWTH OF TRANSVERSE MOMENTUM WITH ENERGY

The naive quark-parton model assumes that the transverse momentum of the hadrons with respect to the jet axis remains constant independent of energy. In QCD the transverse momentum squared will grow proportional to Q^2 i.e. effects from QCD will become increasingly prominent with energy.

The normalized transverse momentum distribution $(1/\sigma)d\sigma/dp_T^2$ measured[4] by TASSO and evaluated with respect to the sphericity axis is plotted in Fig. 18 versus p_T^2. The data at 13 GeV and 17 GeV are identical within statistics and are averaged, similarly

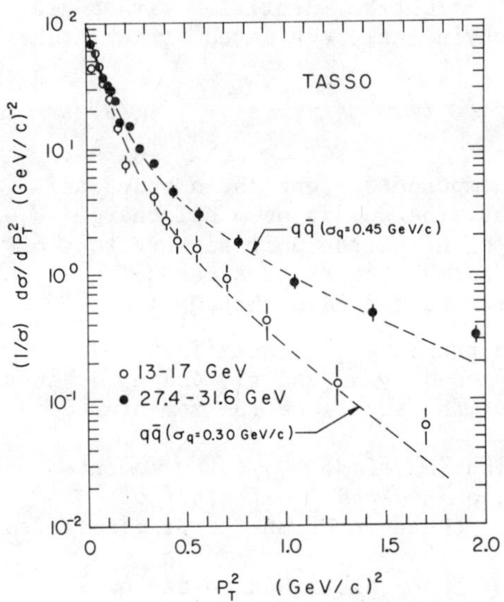

Fig. 18
$1/\sigma\ d\sigma/dp_T^2$ at 13 and 17 GeV combined and for c.m. energies between 27.4 and 31.6 GeV combined as a function of p_T^2. The curves are $q\bar{q}$ fits to the data for $p_T^2 < 1.0\ (GeV/c)^2$ including u, d, s, c and b quarks with σ_q as a free parameter

the data between 27.4 and 31.6 GeV are combined. The data at both energies are in resonable agreement for $p_T^2 < 0.2\ (GeV/c)^2$ but the high energy data are well above the low energy data for large values of p_T^2 - i.e. the average p_T^2 is clearly increasing with energy. The data at 13 - 17 GeV have been fit to $e^+e^- \to q\bar{q}$ for $p_T^2 < 1.0\ (GeV/c)^2$ with $<\sigma_q>$ as a variable. The results, shown as the dashed line in Fig. 18, yield $<\sigma_q> = 300$ MeV/c compared to $<\sigma_T> \sim 250$ MeV/c extracted from deep inelastic lepton hadron data. To fit the higher energy data $<\sigma_q>$ must be increased to 450 MeV/c. A good fit to the high energy data can also be obtained by using $<\sigma_q> = 300$ MeV/c and including gluon bremsstrahlung. Hence the p_T^2 distribution alone cannot be used to discriminate between the $q\bar{q}$ model with an energy dependent value of σ_q and the $q\bar{q}g$ model with a constant value for σ_q.

C) PLANARITY OF THE EVENTS

Regardless of the value of $<p_T>$ hadrons resulting from the fragmentation of a quark must on the average be uniformly distributed in azimuthal angle around the quark axis. Therefore, apart from statistical fluctuations, the two-jet process $e^+e^- \rightarrow q\bar{q}$ will not lead to planar events whereas the radiation of a hard gluon, $e^+e^- \rightarrow q\bar{q}g$, will result in an approximately planar configuration of hadrons with large transverse momenta in the plane and small transverse momenta with respect to the plane. Thus the observation of such planar events at a rate significantly above the rate expected from statistical fluctuations of the $q\bar{q}$ jets shows in a model independent way that there must be a third particle in the final state which might be identified with a gluon.

The shape of an event is conviniently evaluated by constructing the second rank tensor[15,17]

$$M_{\alpha\beta} = \sum_{j=1} p_{j\alpha} \cdot p_{j\beta} \qquad (\alpha,\beta = x,y,z) \qquad (9)$$

where $p_{j\alpha}$ and $p_{j\beta}$ are momentum components along the α and β axis for the jth particle in the event. The sum is over all charged particles in the event. Let $\vec{n}_1$, $\vec{n}_2$ $\vec{n}_3$ be the unit eigenvectors of this tensor associated with the normalized eigenvalues Q_i, $Q_i = \sum_j(\vec{p}_j \cdot \vec{n}_i)^2 / \sum_j p_j^2$, which are ordered such that $Q_1 \leq Q_2 \leq Q_3$. Note that $Q_1 + Q_2 + Q_3 = 1$. The principal jet axis is then $\vec{n}_3$ direction, the event plane is spanned by $\vec{n}_2$ and $\vec{n}_3$; and $\vec{n}_1$ defines the direction in which the sum of the square of the momentum components is minimized.

We first compare the distribution of $<p_T^2>_{out}$, the momentum component normal to the event plane squared, with that of $<p_T^2>_{in}$, the momentum component in the event plane perpendicular to the jet axis.

The data on the $<p_T^2>_{in}$ and $<p_T^2>_{out}$ distribution at low and high energies obtained by TASSO[4] and JADE[7] are plotted in Figs. 22-24. All groups observe that $<p_T^2>_{out}$ changes little with energy in contrast to the distribution of $<p_T^2>_{in}$ which grows rapidly with energy, in particular there is a long tail of events not observed at lower energies. The $<p_T^2>_{out}$ distributions at both low and high energies are described reasonably well with $e^+e^- \rightarrow q\bar{q}$ - i.e. the momenta transverse to the event plane is consistent with the quarks fragmenting into hadrons with a constant transverse momentum independent of energy. The same model also describes the $<p_T^2>_{in}$ distribution at low energies, but it completely fail to reproduce the long tail observed in $<p_T^2>_{in}$ at high energies. This discrepancy cannot be removed by increasing the mean transverse momentum of the jet. Fig. 22 shows a fit assuming σ_q = 450 MeV/c (which gave a good fit to $1/\sigma\, d\sigma/dp_T^2$). The agreement is poor. We therefore conclude that the data include a number of planar events not reproduced by the $q\bar{q}$ model independent of the average p_T assumed. However, as shown in Figs. 23 and 24 the long tail of the $<p_T^2>_{in}$ distribution can be accounted for in the $q\bar{q}g$ model.

In Fig. 19 $\langle p_T^2 \rangle$ measured by PLUTO and TASSO is plotted versus c.m. energy. These data have not been corrected for detector acceptances. Both groups find that $\langle p_T^2 \rangle$ is increasing with energy in agreement with a QCD calculation done by Hoyer et al[20]. A $q\bar{q}$ model with constant σ_q is excluded, however, a good fit can also be obtained in this model if σ_q is allowed to vary. This is not excluded on general grounds.

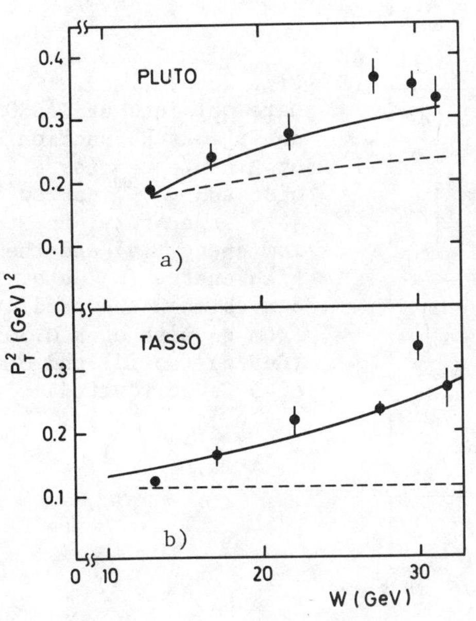

Fig. 19
a) The observed transverse momentum squared of charged particles with respect to the jet axis defined by thrust. The data were obtained and are not corrected for acceptance by PLUTO. The solid line is the prediction for $e^+e^- \to q\bar{q}$ with constant σ_q. The solid line is a QCD prediction by Hoyer et al.[20]
b) Same as in a) except the data were obtained by TASSO

B) ONLY ONE JET BROADN

If hard noncollinear gluon emission is a rare process, as expected in QCD, then there should usually be only one gluon per event. In fact the probability of emitting two gluons in one event compared to single gluon emission is proportional to α_s. Only one of the jets should therefore broadn.

To test this prediction the jets in an event are divided into a narrow and a wide jet. In Fig. 20 and Fig. 21 the data obtained by PLUTO[6] and TASSO[4] are shown. Plotted are $\langle p_T^2 \rangle$ versus $z = p/p_{beam}$ at low and high energies for the wide and the narrow jet separately. A large asymmetrie between the two jets is observed by both experiments at high energy. At low energies the observed distributions are more symmetric and they are well reproduced by $e^+e^- \to q\bar{q}$ using the canonical value for $\langle \sigma_q \rangle$. The observed narrow-wide asymmetry is due to statistical fluctuations. The model with constant σ_q fails to describe the data at high energies. Increasing

$<\sigma_q>$ to 450 MeV for TASSO and to 350 MeV for PLUTO improves the fit. PLUto has also computed the distributions for $e^+e^- \rightarrow q\bar{q}g$. The agreement with the data is good. Also the TASSO data are reproduced by $e^+e^- \rightarrow q\bar{q}g$.

QCD explains naturally the large asymmetry observed in the transverse spread of the two jets in an event. A $q\bar{q}$ model with σ_q increasing with energy results in a worse fit, however such an explanation cannot be completely excluded by the present data.

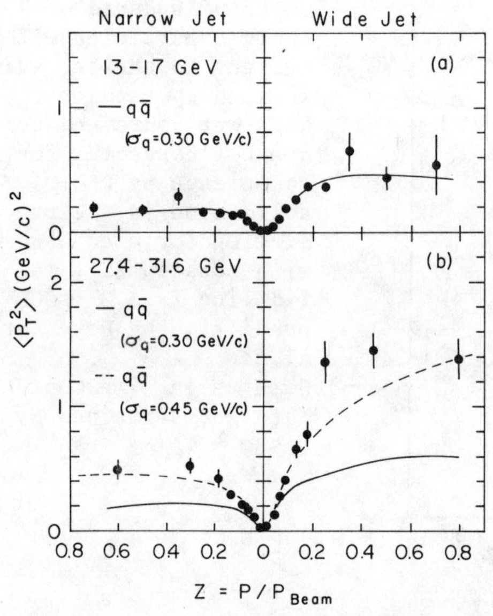

Fig. 20
Data obtained by TASSO on $<p_T^2>$ as a function of $Z = p/p_{beam}$ for wide and narrow jets separately, for the low energy (a) and the high energy (b) data. The curves show the prediction from $q\bar{q}$ with $\sigma_T = 0.30$ (GeV/c) (solid) and $\sigma_T = 0.45$ GeV/c (dotted)

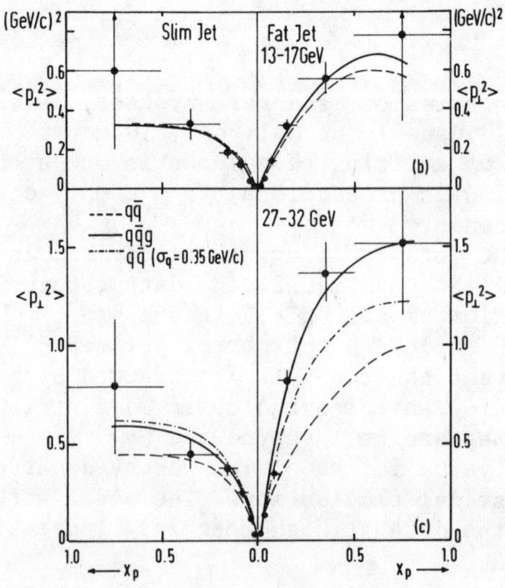

Fig. 21
Data obtained by PLUTO on $<p_T^2>$ as a function of $Z = p/p_{beam}$ for wide and narrow jets. The solid and dashed lines are the $q\bar{q}g$ and $q\bar{q}$ predictions, respectively

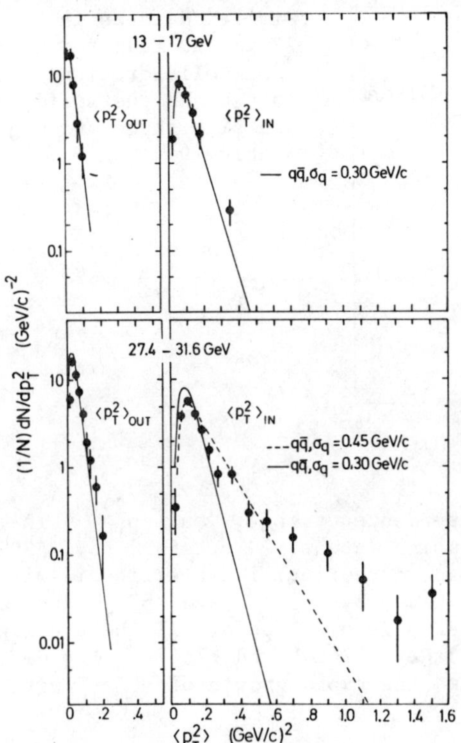

Fig. 22
Distributions of mean transverse momentum squared per event for charged particles, normal to ($\langle p_T^2 \rangle_{out}$) and in ($\langle p_T^2 \rangle_{in}$) the event plane measured by the TASSO Collaboration at low and high energies. The prediction for a $q\bar{q}$ final state with σ_q = 300 MeV/c and σ_q = 450 MeV/c are shown as the solid and the dotted curve respectively

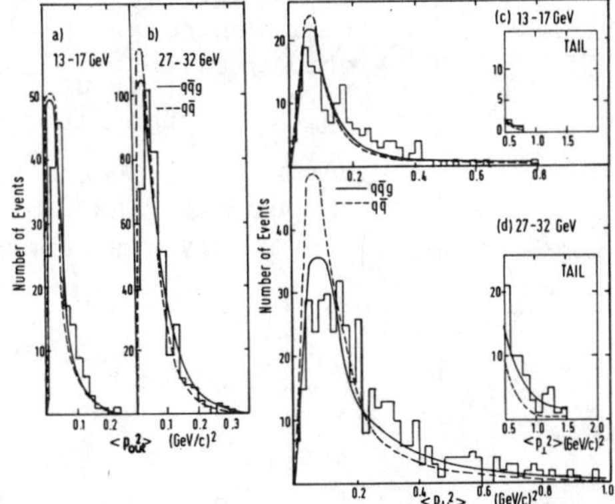

Fig. 23
Similar data as in Fig. 22 obtained by the PLUTO Collaboration. Solid and dashed lines are the $q\bar{q}g$ and $q\bar{q}$ predictions respectively

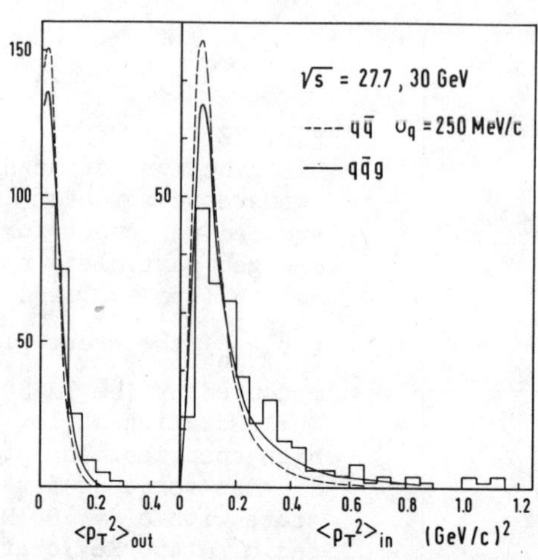

Fig. 24
Similar data as in Figs. 22 and 23 measured by the JADE Collaboration. The dotted and the solid lines are fits to $q\bar{q}$ and $q\bar{q}g$ respectively

The different energy dependence of $<p_T^2>_{in}$ and $<p_T^2>_{out}$ is shown strikingly in Fig. 25 using data[4] obtained by TASSO. Each event, viewed along the $\vec{n}_3$ direction, is an ellipsoid with the small and the large axis given by $<p_T^2>_{out}$ and $<p_T^2>_{in}$ respectively. Fig. 26 shows the computer made overlapp of all these ellipsoids for c.m. energies of 13 GeV, 17 GeV and 27.4 to 31.6 GeV. The slow growth of $<p_T^2>_{out}$ and the rapid growth of $<p_T^2>_{in}$ with energy are seen rather clearly .

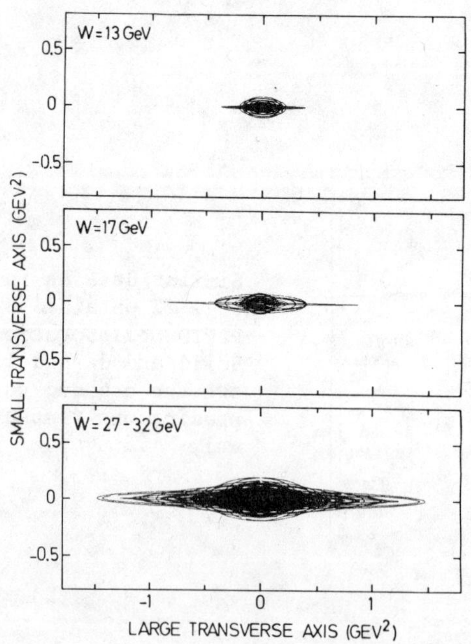

Fig. 25
Events viewed along the jet direction in momentum space. Each event is represented as an ellipsoid with $<p_T^2>_{out}$ and $<p_T^2>_{in}$ as the minor and major axis. Shown are the sum of all TASSO events at various c.m. energies

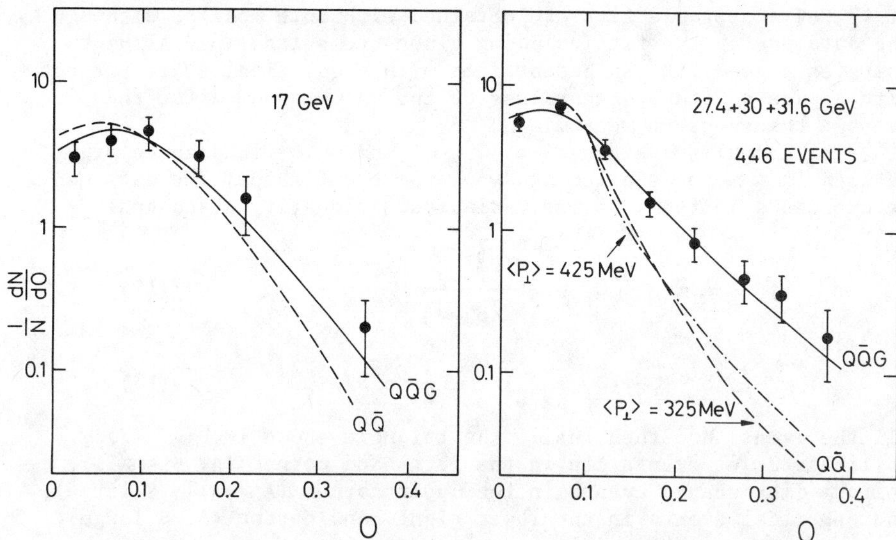

Fig. 26 - The distribution 1/N dN/dO as a function of oblateness for W = 17 GeV and for W between 27.4 GeV and 31.6 GeV. The solid curves are predictions based on $e^+e^- \to q\bar{q}g$, the dotted curve shows the prediction for $e^+e^- \to q\bar{q}$ with a mean transverse momentum of 325 MeV/c. The dashed-dotted curve is the $q\bar{q}$ model prediction with an average transverse momentum of 425 MeV.

The planarity of the events is also observed[9] by the MARK J group using a different technique. They define a coordinate system as follows: the $\vec{e}_1$ axis coincide with the thrust axis which is defined as the direction of maximum energy flow measured in their segmented calorimeter. They next investigate the energy flow in a plane perpendicular to the thrust axis. The direction of maximum energy flow in that plane defines a direction $\vec{e}_2$ with a normalized energy flow

$$\text{major} = \sum_i |\vec{p}^{\,i}\,\vec{e}_2| \,/\, E_{vis} \tag{10}$$

where $E_{vis} = \sum_i \vec{p}^{\,i}$. The third axis $\vec{e}_3$ is orthogonal to both the thrust and the major axis $\vec{e}_3$, and it is very close to the minimum of the momentum projection along any axis i.e.

$$\text{minor} \sim \sum_i |\vec{p}^{\,i}\,\vec{e}_3| \,/\, E_{vis} \quad . \tag{11}$$

They then define the quantity oblatness O = major-minor as a measure of the planarity. This quantity, apart from statistical fluctuations, will be zero for phase space and two jet events and finite for three jet final states. The normalized event distribution is plotted versus oblatness in Fig. 26 for the 17 GeV data and data between 27.4 GeV and 31.6 GeV separately and compared with predictions for $e^+e^- \to q\bar{q}$ (dashed curve) and $e^+e^- \to q\bar{q}g$ (solid line).

At 17 GeV acceptable fits are obtained with both models, although the data prefer the fit including gluon bremsstrahlung. At high energies a good fit can be obtained with a $q\bar{q}g$ final state but not with a pure $q\bar{q}$ state, regardless of the value assumed for the average transverse momentum.

The normalized eigenvalues Q_1, Q_2 and Q_3 defined above might be used in a more detailed study of the event shape. The data can be expressed in terms of two variables, aplanarity A and sphericity S

$$A = \frac{3}{2} Q_1 = \frac{3}{2} \frac{\langle p_T^2 \rangle_{out}}{\langle p^2 \rangle} \tag{12}$$

$$S = \frac{3}{2} (Q_1 + Q_2) = \frac{3}{2} \langle p_T^2 \rangle / \langle p^2 \rangle \tag{13}$$

All the events are then inside the triangle shown in Fig. 27. Collinear 2 jet events lie in the left hand corner (A, S small), uniform disk shaped events in the upper corner (A small, S large), and spherical events in the lower right hand corner (A, S large), while coplanar events will occupy a band along the larger of the two small sides of the triangle in Fig. 27.

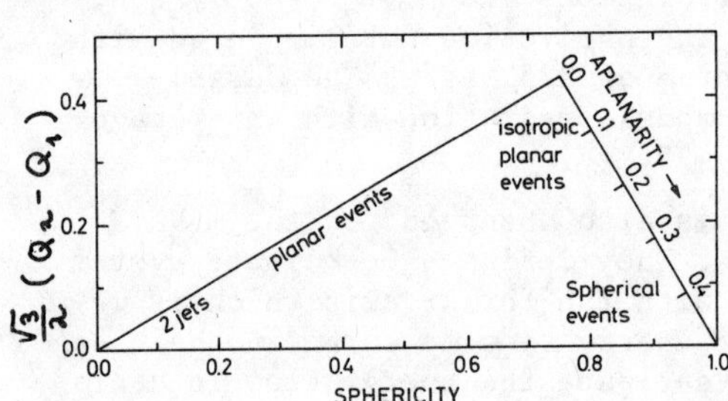

Fig. 27

Distribution of events as a function of aplanarity and sphericity S. Regions in A and S populated by two-jet, planar and spherical events.

The distributions in A and S resulting from:

a) $e^+e^- \to q\bar{q}$ with u, d, s, c and b quarks

b) $e^+e^- \to t\bar{t}$ with m_t = 10 GeV

c) $e^+e^- \to q\bar{q}g$ with q = u, d, s, c and b quarks

are shown in Fig. 28 for a c.m. energy of 30 GeV.

The $q\bar{q}$ model with light quarks indeed populate the region of small A and S with very few events either in the spherical or the planar region.

The $t\bar{t}$ model with m_t = 10 GeV, leads to events with medium values of A and S. Assuming a heavier quark mass will increase values. Note that there are very few planar events in the plot.

The $q\bar{q}g$ model also tends to populate low S and A values but there is a band of planar events resulting from wide angle gluon radiation.

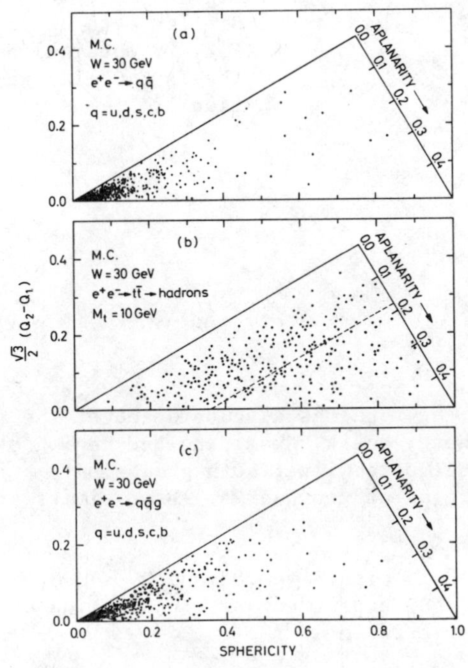

Fig. 28
Monte Carlo created events in aplanarity and sphericity at 30 GeV in c.m. for:

a) $e^+e^- \to q\bar{q}$ with q = u,d,s,c and b quarks

b) $e^+e^- \to t\bar{t}$ with m_t = 10 GeV

c) $e^+e^- \to q\bar{q}g$ with q = u,d,s,c and b quarks

The data obtained by TASSO[25] and JADE[38] are shown in Figs. 29 and 30 respectively.

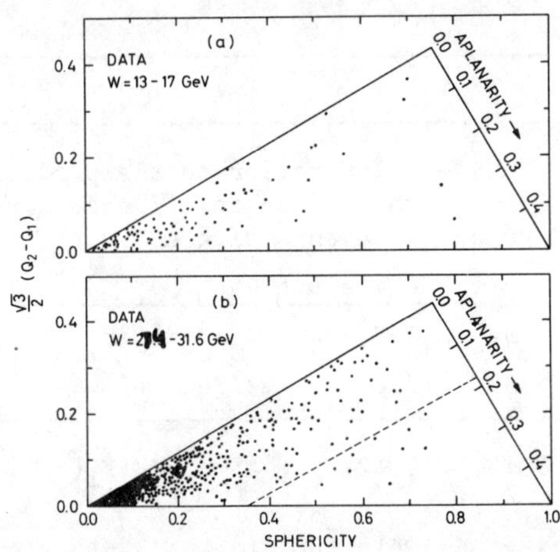

Fig. 29
The event distribution in aplanarity and sphericity observed by the TASSO Collaboration at

a) 13 - 17 GeV

b) at 29.4-31.6 GeV.

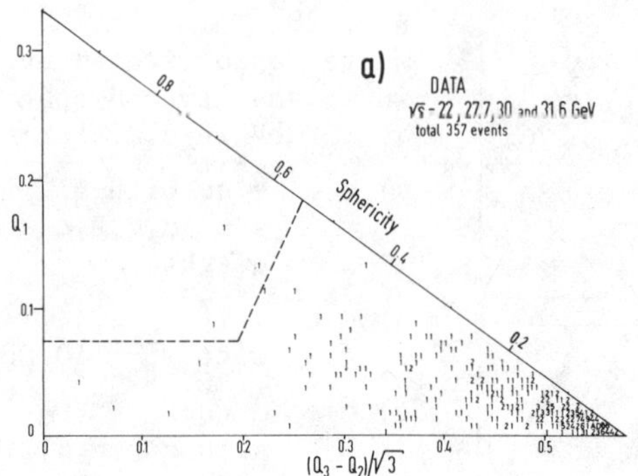

Fig. 30
Similar data as in Fig. 29 obtained by the YADE Collaboration

First, in agreement with the findings discussed above, the data shows no evidence for a heavy quark. Examining the region of large A and S, indicated by the dotted line, both groups put stringent limits on the production of new quarks. These limits are listed in table III and IV.

Table III - The number of events with $S > 0.55$ and $Q_1 < 0.075$ expected in the JADE experiment from production of a quark with charge 2/3 e

Number of events	m_t (GeV)	W (GeV) 22	27.7	30	31.6
expected	8	4	3	2	0
	11		15	22	4
	14			25	5
observed	-	0	1	1	1

Table IV - Limits set by the TASSO-Collaboration on the production of a heavy quark Q with charge 2/3e and 1/3e obtained from the observed number of events with $A > 0.18$

	Number of events					
W(GeV)	ob-served	predicted $q\bar{q}g$	m_Q = 10 GeV 2/3e	1/3e	m_Q = 8 GeV 2/3e	1/3e
29.9-31.6	10	8.2±1.0	80±8	20±2	45±3	11±1
30.9-31.6	2	2.6±0.4	25±1.5	6±0.6	14±1	3.5±0.2

The production of a heavy quark with charge 2/3e in the energy range of PETRA is clearly excluded. The production of a quark with charge 1/3 e seems unlikely.

The production of collinear 2-jet events is seen to dominate at all energies. However, the data at high energy show a band of planar events with small values of A in agreement with a $q\bar{q}g$ final state.

The TASSO data at high energies consist of a total of 949 events (including the data observed during the scan). In a band defined by $A < 0.05$ and $S > 0.25$ there are 62 planar events compared to 49 events predicted by $q\bar{q}g$ and 11 events predicted by $q\bar{q}$.

The JADE Collaboration[38] finds 23 events in band defined by $(Q_3 - Q_2)/\sqrt{3} < 0.35$ and $Q_1 < 0.07$ compared to 22 events predicted for $q\bar{q}g$ and 6 events for $q\bar{q}$.

The PLUTO Collaboration[6] observe 35 events with $S > 0.25$ and $Q_1 < 0.03$. The $q\bar{q}g$ model predict 30 events and the $q\bar{q}$ model 12 events, in the same strip.

The data discussed above proves conclusively that planar events, which cannot result from quark pair production with a Gaußian distribution of transverse momentum around the jet axis are produced in e^+e^- annihilation. Wide angle gluon bremsstrahlung[19] $e^+e^- \to q\bar{q}g$ would naturally result in planar events. The observed rate for such events is consistent with the QCD predictions. Besides this origin, however, there are two ad hoc possibilities ; a flat phase space of unknown origin or that the transverse momentum distribution of the quark fragmentation has a long non Gaußian tail. The first possibility can be excluded by observing events with 3 axes, the second by excluding that the 3 axes are not defined by 2 jets and a single high momentum particle at a large angle with respect to the jet axis.

D) PROPERTIES OF PLANAR EVENTS

The PLUTO Collaboration has searched for three-jet events[6] using a generalization[39] of thrust. In this method the final state hadrons with momenta $\vec{p}_1, \vec{p}_2 \ldots \vec{p}_N$ are grouped into three classes (Fig. 31a) C_1, C_2 and C_3 with momenta $\vec{P}(C_N) = \Sigma|\vec{p}_i|$ where the sum is over all particles assigned to the class i C_1. A new quantity, triplicity T_3, is then defined as

$$T_3 = \frac{1}{\Sigma|p_i|} \cdot \text{Max } |\vec{P}(C_1)| + |\vec{P}(C_2)| + |\vec{P}(C_3)| \qquad (14)$$

T_3 is 1 for a perfect 3-jet event and $3(\sqrt{3/8}) = 0.65$ for a spherical event. The momenta of the three jets are given by $\vec{P}_1 = \vec{P}(C_1)$, $P_2 = P(C_2)$ and $P_3(C_3)$ (Fig.31b) and the angles between these vectors θ_1, θ_2 and θ_3 are the angles between the three jets. The angles are ordered such that $\theta_1 \leq \theta_2 \leq \theta_3$ and $\theta_1 + \theta_2 + \theta_3 = 360$ since by momentum conservation the three vectors $\vec{P}_1$, $\vec{P}_2$ and $\vec{P}_3$ define a plane. All the events can thus be located in the hatched triangle in Fig. 31b. A totally symmetric 3-jet event ($\theta_1 = \theta_2 = \theta_3 = 120$) would be located in the corner A. To select the candidates for a three-jet event they make a scatter plot of T_3 versus T (Fig. 31 d). A three-jet event will have large triplicity and small thrust. In the region $T_3 > 0.9$ and $T < 0.8$ they find 48 events at c.m. between 27.4 GeV and 31.6 GeV. With $\sigma_q = 250$ MeV/c a $q\bar{q}$ model predicts 11 events and a $q\bar{q}g$ model 43 events. The angular Dalitz plot for the high energy data is shown in Fig. 31e. The candidates for three-jet events ($T_3 > 0.9$, $T < 0.8$) are shown as large circles, the others as dots. Events with a clear jet structure will have large values of Θ_1 i.e. for Θ_3 much less than 180°, and Θ_1 large, whereas the black dots tend to have small values for Θ_1. At high energies they find 52 events with $\Theta_1 < 150^o$. The $q\bar{q}g$ model predicts 51 and $q\bar{q}$ predicts 19.

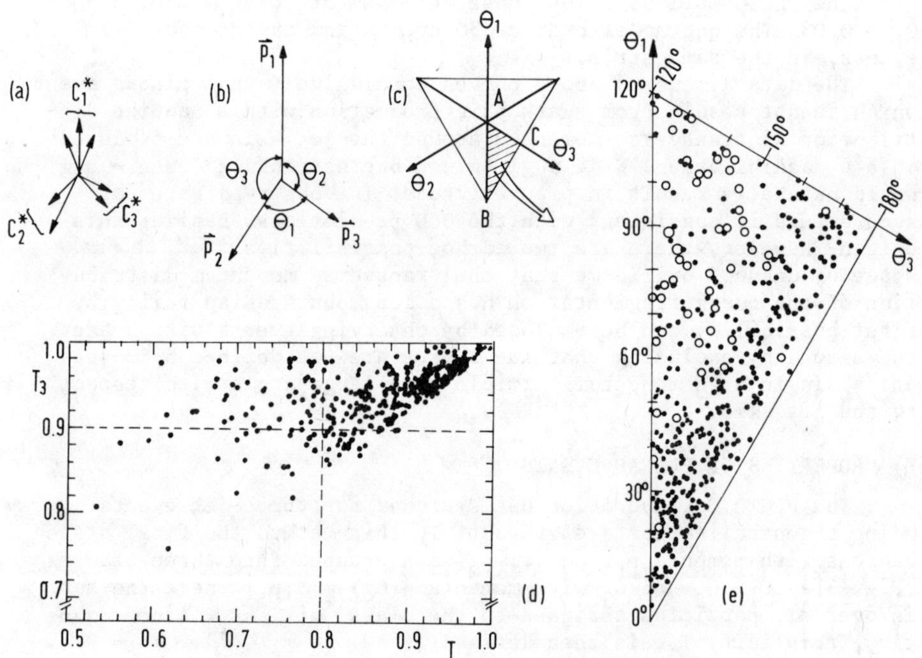

Fig. 31 - a,b) Momentum configuration of hadrons and jets
c) Definition of the angular Dalitz plot with $\Theta_3 > \Theta_2 > \Theta_1$
d) Data obtained by the PLUTO Collaboration at 27.6, 30 and 31.6 GeV shown in a scatter plot of triplicity versus thrust
e) Same data as above in the angular Dalitz plot: events with $T < 0.8$ and $T_3 > 0.9$ are shown as open circles.

The TASSO Collaboration used a generalization of sphericity[40] to define three-jet events. In this method the tracks are projected on to the event plane defined by $\vec{n}_2$ and $\vec{n}_3$ (see above). The projections are divided into three groups and the sphericity for each group S_1, S_2 and S_3 determined. The three axes and the particle assignment to the three groups are defined by minimizing the sum of S_1, S_2 and S_3. This defines the direction of the three jets and assigns the particles to these jet directions.

The distribution of the TASSO events above 27.4 GeV in the angular Dalitz plot is shown in Fig. 32c. The results of a Monte Carlo calculation for $e^+e^- \to q\bar{q}$ (Fig. 32a) and $e^+e^- \to q\bar{q}g$ (Fig. 32b) are also shown. The data clearly favours the $q\bar{q}g$ mechanismn. The TASSO observe 50 events with $\theta_3 < 160^\circ$ compared to 47 events predicted for $q\bar{q}g$ and 20 events for $q\bar{q}$.

In Fig. 33 the TASSO events are plotted versus tri-jettiness J_3. The tri-jettiness is defined as

$$J_3 = \langle p_T^2 \rangle_{in} \, / \, (1/2 \; (300 \text{ MeV}/c)^2) \qquad (15)$$

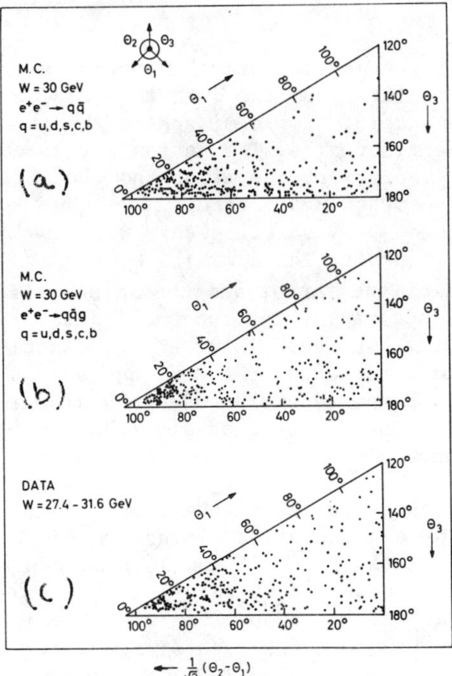

Fig. 32
Events displayed in an angular Dalitz plot with the axis determined by the generalized sphericity method.

a) Distributions expected for $e^+e^- \to q\bar{q}$

b) Distribution expected for $e^+e^- \to q\bar{q}g$

c) Data obtained by the TASSO Collaboration for c.m. energies between 27.4 GeV and 31.6 GeV where $<p_T^2>_{in}$ is computed for all charged tracks with respect to their assigned axis. Thus for three-jet events with a mean transverse momentum of 300 MeV with respect to the jet axis we expect to find the events clustered around $J_3 = 1$, compared with a wide distribution in J_3 in case of a flat phase space distribution. The data agree with the expectations for $e^+e^- \to q\bar{q}g$, shown as the solid line. The fit result in χ^2/degree of freedom of 2.3/5. The data disagree strongly with a phase space calculation[41] shown as the dashed line. This fit has χ^2/degree of freedom of 223/5. Thus the data are not consistent with a phase space distribution.

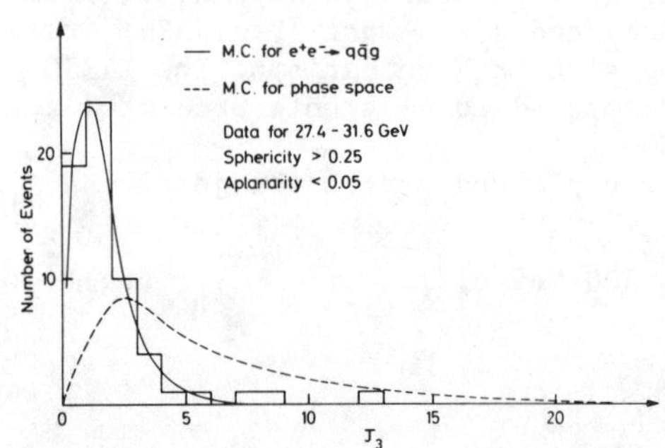

Fig. 33
Planar events ($S \geq 0.25$, $A < 0.05$) measured by the TASSO Collaboration plotted versus the tri-jettiness J_3. The M.C. predictions for $e^+e^- \to q\bar{q}g$ (solid) and for $e^+e^- \to q\bar{q}$ (dotted) are shown

The MARK J group observe[5] a three-jet structure in their energy flow analysis. To extract this structure from the data they divide the energy distribution of each event into two hemisspheres by the plane containing the major and the minor axis. The narrow jet is contained in the forward hemisphere. The major, the minor and the oblateness are calculated separately for each hemisphere. To enhance effects resulting from gluon emission they select events with low thrust $T < 0.8$ and large oblateness $O > 0.1$. The accumulated energy distribution in the plane defined by the thrust and the major axis shown in Fig. 34a has a three-jet structure. The two small jets have been oriented according to size. The calculated energy distribution is in agreement with predictions based on $e^+e^- \to q\bar{q}g$ and is shown as the solid line. In Fig. 34b the accumulated energy distribution in the thrust-minor plane is shown. The flat distribution is consistent with the $q\bar{q}g$ predictions.

The remaining question is then to decide if the third jet is defined by a single particle or by a group of particles. This can be done simply by examining the events. Figs. 35 and 36 show typical candidates for three-jet events observed by the TASSO Collaboration and by the PLUTO Collaboration. For comparison also a two-jet event is shown. The main inserts shows the events viewed along the $\vec{n}_1$ direction, i.e. down onto the event plane. The three axes are

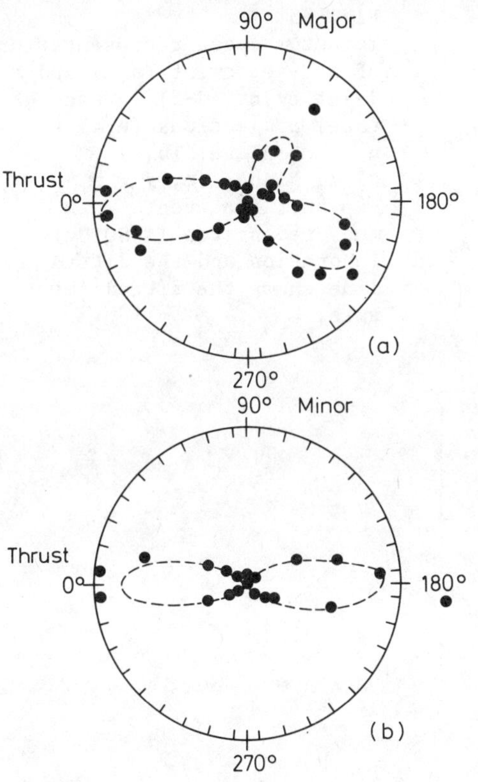

Fig. 34
a) A polar plot of the energy distribution in the plane defined by the thrust and the major axes for all events with thrust < 0.8 and oblateness > 0.1. The measurements were done by the MARK J group at c.m. energies of 27.4, 30 and 31.6 GeV. The energy value is proportional to the radial distance. The superimposed dashed line is the distribution calculated using $q\bar{q}g$ model.
b) The measured and calculated energy distribution in the plane defined by the thrust and the minor axes.

indicated by the dotted lines in the TASSO picture and by the fat bars at the border of the picture in the PLUTO event. The TASSO events contain only charged tracks, the direction of neutral tracks in the PLUTO event is indicated by dotted lines. In both events we see a clear 3-jet structure and each jet contains many tracks. In the two small insertions the events are viewed in the event plane along and transverse to the $\vec{n}_3$ direction. Viewed along the $\vec{n}_3$ direction there is a striking difference between two-jet events and three-jet events.

The TASSO group has also evaluated the transverse momentum of charged particles from three-jet events with respect to the jet axes to which they were assigned by the generalized sphericity method[40]. This distribution, $1/N\ dN/dp_T^2$, is plotted versus p_T^2 in Fig. 37. It is compared to the p_T^2 distribution found with respect to the jet axis in two-jet events at lower energies shown as the solid line. The agreement is very good and demonstrates that $\langle\sigma_q\rangle$ can be taken to be constant independent of energy, when the events at high energy are analyzed as three-jet events.

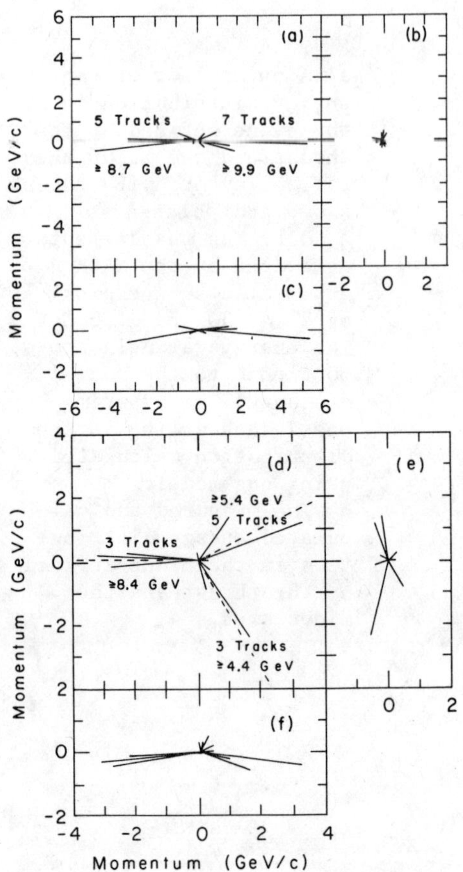

Fig. 35
Momentum space representation of a 2-jet event (a,c) and a 3-jet event (d-f) in each of three projections (a,d) = $\vec{n}_2 - \vec{n}_3$ plane; (b,e) = $\vec{n}_1 - \vec{n}_2$ plane; (c,f) = $\vec{n}_1 - \vec{n}_3$ plane. The events were measured by the TASSO Collaboration and the dotted line shows the fitted jet axis.

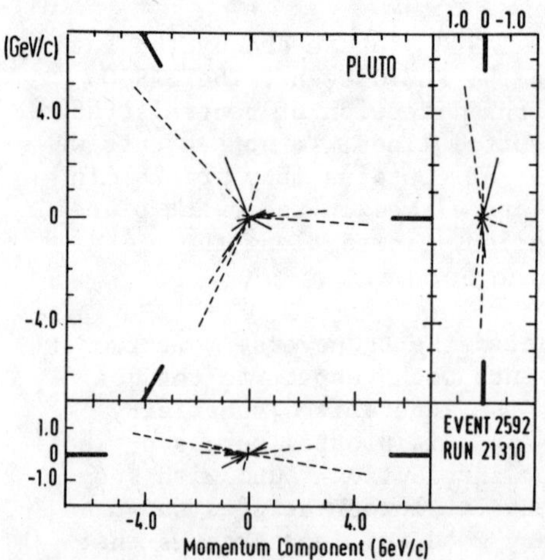

Fig. 36
Momentum vectors of an high triplicity low thrust event, measured by the PLUTO Collaboration at 31.6 GeV in c.m. The event is shown projected onto the triplicity plane (top left), onto a plane normal to the fastest jet (top right) and onto a plane containing the direction of the fastest jet (bottom). Solid and dotted lines correspond to charged and neutral particles, respectively. The directions of the jet axes are shown as the fat bars.

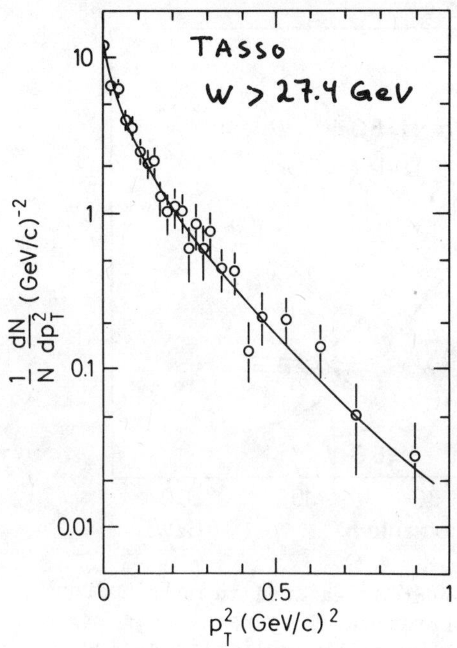

Fig. 37
The transverse momentum distribution $1/N\ dN/dp_T^2$ of the hadrons in the planar events with respect to the three axes found by the generalized sphericity method. The solid line shows the transverse momentum distribution with respect to the jet axis found in 2-jet events at lower energies. The data are from the TASSO Collaboration.

E) CONCLUSION

The naive quark-parton model does not describe the data at high energies. The data show a clear three-jet structure, which result from an initial state of three basic particles. All the properties of these events are consistent with those expected for gluon bremsstrahlung $e^+e^- \rightarrow q\bar{q}g$ in QCD.

5. SEARCH FOR FRACTIONALLY CHARGED OR HEAVY STABLE PARTICLES

$e^+e^- \rightarrow q\bar{q}$ is an ideal place to look for free quarks. The JADE Collaboration[7] has searched for quarks and heavy stable particles using the dE/dx information from the drift chamber. The drift chamber measures dE/dx at 48 points along the track. To reduce the fluctuations from the Landau tail of the ionization loss the highest 20% of the measurements are ignored and the mean of the remaining 80% used. This results in a dE/dx resolution of $\pm$ 6% determined from a measurement of Bhabha scattering.

The dE/dx for each track in multihadron events are plotted in Fig. 38 versus momentum. To be accepted, the ionization must have been measured at least 30 times. Note that the threshold for an ionization measurement is set to 1/7 of the energy loss of a minimum ionizing particle emitted at 90° to the beam.

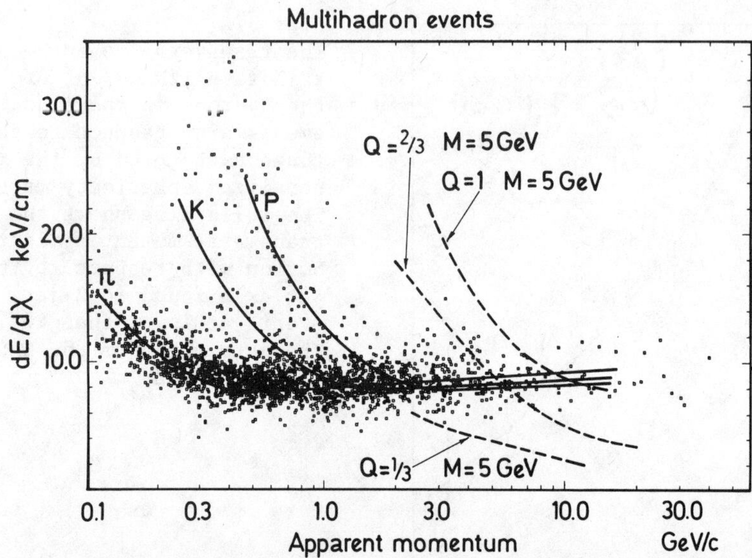

Fig. 38 – dE/dx versus apparent momentum measured in multihadron events by the JADE Collaboration. The energy loss expected for π, K, p are shown as the solid lines. The energy loss expected for fractionally charged or heavy stable particles is shown as the dotted lines.

The energy loss curves predicted for pions, kaons and protons are plotted as solid lines in Fig. 38. The energy loss curves for 1/3 e, 2/3 e particles and a stable particle of mass 5 GeV and unit charge are plotted as dotted lines. The points accumulate along the energy loss curves of the ordinary particles with no accumulation of points along the quarks or the heavy stable particle lines. The upper limits on the production cross section in the corresponding momentum ranges are listed in table V.

Table V - Upper limits ΔR on the production of fractionally charged particles and heavy particles with unit charge

Charge	Mass (GeV)	Range of true momenta GeV/c	upper limit ΔR
1/3	3	$1.5 < p < 2$	0.1
	5	$2 < p < 3$	0.1
	10	$1 < p < 7$	0.1
2/3	3	$p < 1$ $p > 3.5$	0.1
	5	$p < 2$ $p > 4$	0.1
	10	$p < 4.5$ $p > 10$	0.1
1	3	$p < 2$	0.08
	5	$p < 6$	0.08
	10	$p < 10$	0.08

The JADE Collaboration use the absence of heavy particles to set an upper limit on the lifetime τ of the B. Assuming a mass of 5 GeV, a $B\bar{B}$ production cross section of $\Delta R = 1/9$ and a flat momentum distribution of the B mesons they find $\tau < 3 \times 10^{-9}$ sec.

6. TWO PHOTON REACTIONS

Electron-positron collisions are a source of photon-photon collisions[42] as shown in Fig. 39. The mass $(-q^2)$ of the virtual photon is determined by the kinematics of the outgoing lepton. Photon-photon collisions therefore offer a unique opportunity to vary the mass of both the target and projectile over a wide range.

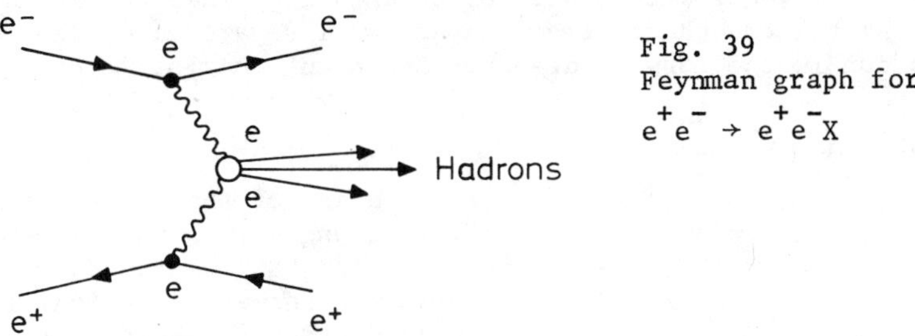

Fig. 39
Feynman graph for
$e^+e^- \to e^+e^-X$

One might investigate the whole kinematic region from collisions of two nearly real photons (hadron-hadron interaction) via deep inelastic electron-photon interactions to the collisions of two heavy photons. Unlike the annihilation process the two photon processes will lead to a final state with two leptons, in general with high energies and at forward angles plus hadrons with a visible energy much less than the c.m. energy. The two processes can therefore readily be separated.

The PLUTO Collaboration has reported the first results on $e^+e^- \to e^+e^-$ hadrons at high energies. The number of observed events, with the beam-gas contribution subtracted, are plotted in Fig. 40 as a function of visible energy. The distribution has two peaks,

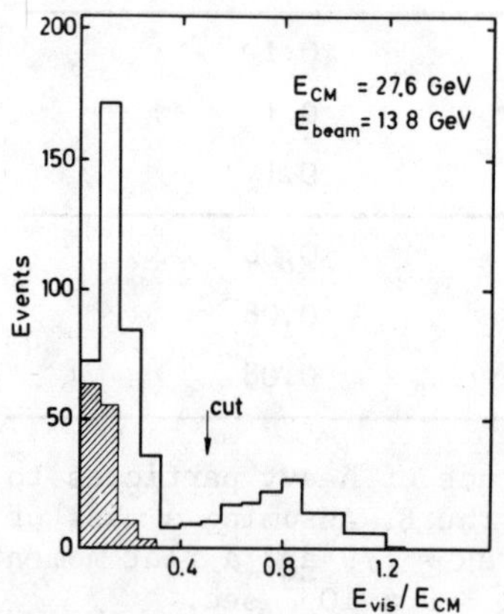

Fig. 40
Distribution of the measured energy per event (charged + neutral) measured by the PLUTO Collaboration at 27.6 GeV. The hatched area shows the energy distribution of events with a single tag in the forward detector.

one corresponding to events with a large visible energy, resulting from annihilation events, the second peak to events with low visible energy. The second peak with a very steep drop towards large visible energies is naturally explained by two photon reaction where the sharp drop is caused by the falling bremsstrahl spectrum of the interacting photons. This interpretation is reinforced by the distribution of events with a lepton in one of the forward spectrometers, which covers angles between 23 and 70 mrad.

This distribution, shown as the hatched area in Fig. 41, only lead to events with small observed energy. Thus the annihilation and the two-photon reaction are well separated by demanding an electron in the forward detector and a cut on visible energy.

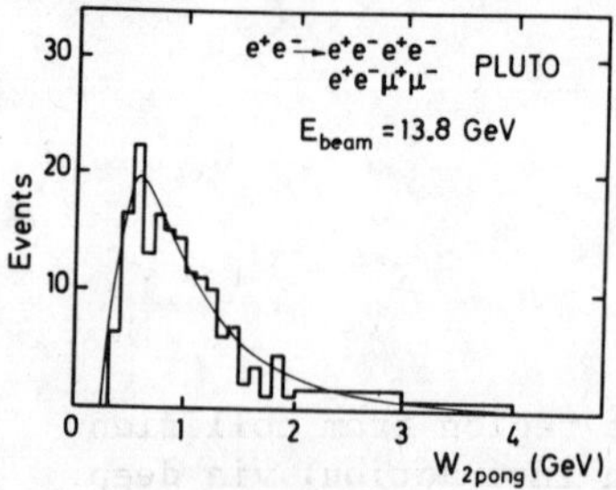

Fig. 41
Distribution of the invariant mass of two prong events in the central part of the PLUTO detector with a tag in the forward detector. The QED prediction is shown by the solid line.

The PLUTO group[10] first selected events of the type $e^+e^- \to e^+e^- + 2$ charged prongs. This class of events will mainly be populated by the two QED reactions $e^+e^- \to e^+e^-e^+e^-$ and $e^+e^- \to e^+e^- \mu^+\mu^-$. The number of two prong events at 27.6 GeV in c.m. are plotted versus the invariant two prong mass in Fig. 41. The distribution peaks at a low invariant mass is in excellent agreement with the QED prediction shown as the solid line.

The hadronic events are also selected from the single tag class by demanding either three or more tracks in the central detector or two tracks in the central detector plus one neutral particle.

In these events the tagged photon had $\langle Q^2 \rangle \simeq 0.1$ GeV^2 for beam energies of 6.5 GeV (8.5 GeV) and $\langle Q^2 \rangle = 0.4$ GeV^2 for a beam energy of about 15 GeV. The virtual photon from the untagged electron will be almost real.

The PLUTO Collaboration therefore analyze their data in terms of electron-photon scattering. The cross section for this process can be written[42] as

$$d\sigma = \Gamma_T(\sigma_T + \varepsilon\, \sigma_L)\, d\Omega'\, dE'\, N(E_\gamma)\, dE_\gamma . \qquad (16)$$

Γ_T and ε are respectively the flux factor and the polarization of the tagged photon. σ_T and σ_L are the total cross section for hadron production by virtual transverse and longitudinal photons. $N(E_\gamma)\, dE_\gamma$ is the spectrum of untagged photons. The values of $\sigma_T + \varepsilon\sigma_L$ extracted from the data are plotted in Fig. 42 as a func-

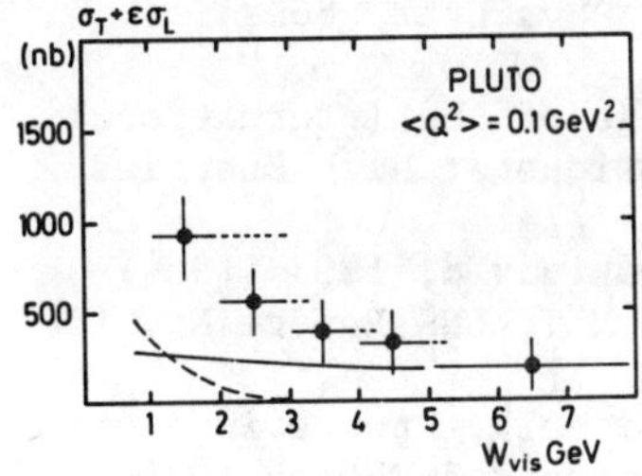

Fig. 42
Total hadronic cross section for two photons initiated events measured by PLUTO for
a) $Q^2 \sim 0.1$ (GeV/c)^2 and
b) $Q^2 \sim 0.4$ (GeV/c)^2

of the visible energy W_{vis} at $\langle Q^2 \rangle = 0.1$(GeV/c)^2 and $\langle Q^2 \rangle = 0.4$(GeV/c)^2 The range of true c.m. energies which contribute to W_{vis} is the indicated by the length of the dashed horizontal bars.

Besides the statistical error there is an overall systematic uncertainty of ± 25%.

The predicted[10] cross section assuming a pure Regge behaviour of the $\gamma\gamma$ scattering extrapolated to low energies using duality arguments and factorization is shown by the solid line in Fig. 42. The estimated contribution from a point-like coupling of the photons to the quark is indicated by the dashed line. Adding the two contributions describes the energy dependence of the cross section qualitatively.

REFERENCES

1. PETRA, updated version of the PETRA Proposal, DESY, Hamburg,1976
2. MARK J.Collaboration, P.Barber et al., Phys.Rev.Lett.42,1113 (1979)
   PLUTO Collaboration, Ch.Berger et al., Phys.Lett.81B, 410 (1979)
   TASSO Collaboration, R.Brandelik et al., Phys.Lett. 83B 261 (1979)
3. B.H.Wiik, Proceedings of the International Neutrino Conference, Bergen, Norway, 18-22 June 1979
   R.Cashmore, Proceedings of the EPS International Conference on High Energy Physics, Geneva, Switzerland, 27 June-4 July 1979
   P.Söding, ibid
   G.Wolf, ibid
4. TASSO Collaboration, R.Brandelik et al., Phys.Lett. 83B, 261 (1979)
5. MARK J Collaboration, P.Barber et al., Phys.Rev.Lett. 43, 830 (1979)
6. PLUTO Collaboration, Ch.Berger et al., Phys.Lett. 86B, 418 (1979)
7. JADE Collaboration, S.Orito, Proceedings of the International Symposium on Lepton and Photon Interactions at High Energies, FNAL 23-29 Aug, 1979 and DESY Report 79/77
8. A.A.Sokolov and I.M.Ternov, Sov.Phys.Doklady 8, 1203 (1964)
9. MARK J Collaboration, P.Barber et al., MIT, LNS Report Nr. 107 Submitted to Nuclear Physics B (1979)
10. PLUTO Collaboration, Ch.Berger, invited talk, Proceedings of the 1979 International Symposium on Lepton and Photon Interactions at High Energies, FNAL, Aug. 23-29, 1979, FITHA 79/29
11. S.Weinberg, Phys.Rev.Lett. 19, 1264 (1967)
    A.Salam, Elementary Particle Physics, ed. N.Svartholm (Almkvist and Wicksell, Stockholm 1968)
12. E.H.de Groote, G.J.Gounaris and D. Schildknecht, Bielefeld Report, BI-TP 79/15
13. S.D.Drell, D.J.Levy, and T.M.Yan, Phys.Rev. 187, 2159 (1969) and Phys.Rev. D1, 1617 (1970)
14. N.Cabibbo, G.Parisi, and M.Testa, Lett.Nuovo Cimento 4, 35 (1970)
15. J.D.Bjorken and S.J.Brodsky, Phys.Rev.D1, 1416 (1970)
16. R.P.Feynman, Photon-Hadron Interactions (Benjamin, Reading, Mass., p. 166, 1972)

17. R.F.Schwitters et al., Phys.Rev.Lett. 35, 1230 (1975)
G.G.Hanson et al., Phys.Rev.Lett. 35, 1609 (1975)
G.G.Hanson, Proceedings of 13th Rencontre de Moriond, edited by J.Tran Thanh Van, Vol. II, p. 15 and SLAC-PUB 2118 (1978)
Ch.Berger et al., Phys.Lett. B78, 176 (1978)

18. J.Kogut and L.Susskind, Phys.Rev. D9, 697, 3391 (1974
A.M.Polyakov, Proceedings of the 1975 International Symposium on Lepton and Photon Interactions at High Energies, Stanford, Aug. 21-27, 1975

19. The first discussion on the experimental implications of gluon bremsstrahlung in e^+e^- annihilation was given by: J.Ellis, M.K.Gaillard and G.G.Ross, Nucl.Phys. B 111, 253 (1976) - erratum B 130, 516 (1977)

20. T.A.DeGrand, Yee Jack Ng, and S.-H.H.Tye, Phys.Rev. D 16, 3251 (1977)
A.de Rujula, J.Ellis, E.G.Floratos and M.K.Gaillard, Nucl.Phys. B 138, 387 (1978)
G.Kramer and G.Schierholz, Phys.Lett. 82B, 102 (1979)
G.Kramer, G.Schierholz and J.Willrodt, Phys.Lett. 79B, 249 (1978)
P.Hoyer, P.Osland, H.G.Sander, T.F.Walsh and P.M.Zerwas, DESY 78/21 - to be published
G.Curci, M.Greco and Y.Srivastava, CERN-Report 2632-1979
G.Kramer, G.Schierholz and J.Willrodt, DESY Report 79/69

21. H.Fritzsch, M.Gell-Mann and H.Leutwyler, Phys.Lett. 47B, 365 (1973)
D.J.Gross and F.Wilczek, Phys.Rev.Lett. 30, 1343 (1973)
H.Politzer, Phys.Rev.Lett. 30, 1346 (1973)
S.Weinberg, Phys.Rev.Lett. 31, 31 (1973)

22. See J.Ellis for a review of the status of QCD in deep inelastic reactions. Proceedings of the International Neutrino Conference, Bergen, Norway 18-22 June, 1979

23. PLUTO Collaboration, Ch.Berger et al., Phys.Lett. 86B, 413 (1979)

24. JADE Collaboration, W.Bartel et al., DESY Report 79/64

25. TASSO Collaboration, R.Brandelik et al., DESY Report 79/74

26. A.Quenzer, thesis, Orsay Report LAL 1299 (1977)
A.Cordier et al., Phys.Lett. 81B, 389 (1979)
V.A.Sidorov, Proceedings of the XVIIIth International Conferenence on High Energy Physics, Tbilisi, USSR, B13 (1976)
R.F.Schwitters, Proceedings of the XVIIIth International Conference on High Energy Physics, Tbilisi, B 34 (1976)
J.Perez-Y-Jorba, Proceedings of the XIXth International Conference on High Energy Physics Tokyo, p. 277 (1978)
G.P.Murtas, Proceedings of the XIXth International Conference on High Energy Physics, Tokyo, p. 269 (1978)
PLUTO Collaboration, J.Burmester et al., Phys.Lett. 66 B, 395 (1977)
DASP Collaboration, R.Brandelik et al., Phys.Lett. 76 B, 361 (1978)

27. T.Appelquist and H.Georgi, Phys.Rev. D8, 4000 (1973)
A.Zee, Phys.Rev. D8, 4038 (1973)
G.'t Hooft, Nucl.Phys. B62, 444 (1973)
M.Dine and J.Sapirstein, Phys.Rev.Lett. 43, 668 (1979)
28. TASSO Collaboration, R.Brandelik et al., DESY Report 79/73
29. C.Bacci et al., Phys.Lett. 86B, 234 (1979)
SLAC-LBL Collaboration, G.G.Hanson, 13th Rencontre de Moriond (1978) ed. by J.Tran Thanh Van, Vol. III (1978)
PLUTO Collaboration, Ch.Berger et al., Phys.Lett. 81 B, 410 (1979) and
V.Blobel private communication
DASP Collaboration, R.Brandelik et al., Nucl.Phys. B 148, 189 (1979)
30. W.Thomé et al., Nucl.Phys. B 129, 365 (1977)
See also review by E.Albini, P.Capiluppi, G.Giacomelli, and A.M.Rossi, Nuovo Cimento 32A, 101 (1976)
31. R.Stenbacka et al., Nuovo Cimento 51A, 63 (1979)
32. W.Furmanski, R.Petronzio and S.Pokorski, Nucl.Phys. B 155, 253 (1979)
33. G.J.Feldman and M.L.Perl, Phys.Reports 33, 285 (1977)
34. R.Baier, J.Engels and B.Peterson, University of Bielefeld Report BI-TP 79/10
W.R.Frazer and J.F.Gunion, Phys.Rev. D20, 147 (1979)
35. R.D.Field and R.P.Feynman, Nuclear Phys. B 136, 1 (1978)
see also
B.Anderson, G.Gustafsen and C.Peterson, Nuclear Physics B 135, 273 (1978)
36. A.Ali, J.G.Körner, J.Willrodt and G.Kramer, Particles and Fields C1, 269 (1979) ibid 2, 33 (1979
37. E.Fahri, Phys.Rev.Lett. 39, 1587 (1977)
38. JADE Collaboration, W.Bartel et al., DESY Report 79/70
39. S.Brandt and H.D.Dahmen, Particles and Fields C1, 61 (1979)
40. S.L.Wu and G.Zobernig, Particles and Fields C2, 107 (1979)
41. The phase space calculation shown here is for a detector without neutral detection, it includes corrections for acceptance and the model is adjusted to reproduce the high multiplicity observed
42. N.Arteago Romero, A.Jaccarini and P.Kessler, C.R.Acad.Sci. B 129, 153 (1969) and
C.R.Acad.Sci. B 269, 1129 (1969)
V.E.Balakin, V.M.Budnev and I.F. Grinzburg, Zh.Eksp.Theor.Fiz.Pis'ma Red. 11, 559 (1970) (JETP Lett. 11, 338 (1970))
S.J.Brodsky, T.Kinoshita and H.Terazawa, Phys.Rev.Lett. 25, 972 (1970)
43. L.N.Hand, Phys.Rev. 129, 1834 (1963)

RECENT RESULTS FROM THE MARK II DETECTOR AT SPEAR*

SLAC-LBL Collaboration[1]

Presented by Jonathan Dorfan
Stanford Linear Accelerator Center
Stanford University, Stanford, California 94305

ABSTRACT

Recent results from the Mark II Detector at SPEAR are presented. These include measurements of the decays $\tau^- \to \rho^- \nu_\tau$ and $\tau^- \to K^{*-}(890)\nu_\tau$, observation of direct photons at the ψ, inclusive proton and Λ production and measurements of charmed baryon and charmed meson decays.

INTRODUCTION

The Mark II detector was installed in the west interaction region at SPEAR in October 1977. All the major detector systems were operational by March 1978 and since then data were obtained in the center-of-mass energy region 3-7.4 GeV. In particular data were recorded at the $\psi(3096)$, $\psi(3684)$ and $\psi(3770)$ resonances and at 3.52 GeV, 3.67 GeV, 4.16 GeV, 4.4 GeV, 5.2 GeV, 6.5 GeV and 7.4 GeV. The region from 3.7-6 GeV was scanned using a step size of 6-20 MeV recording 100-300 μ-pairs per scan point. A broad range of physics measurements has resulted from these data and a subset of the more recent results are presented herein. A notable exception in this presentation is the first observation of the production of a resonance (η') in two photon physics and a measurement of its radiative width.[2] Topics which are forthcoming include a measurement of the total hadronic cross section, limits on rare decay modes of the heavy lepton and measurements of the cascade and hadronic decays of the χ charmonium states. The Mark II magnetic detector is described below as a preface to the following physics topics:

(1) The decays $\tau^- \to \rho^- \nu_\tau$ and $\tau^- \to K^{*-}(890)\nu_\tau$.

(2) Inclusive photons at the $\psi(3096)$ and evidence for the production of direct photons.

(3) Inclusive proton and Λ^o production.

(4) Observation of the production of charmed baryons.

(5) The $\psi(3770)$ resonance.

(6) Decays of the charmed D mesons.

---

* Work supported primarily by the Department of Energy under contract numbers DE-AC03-76SF00515 and W-7405-ENG-48.

ISSN:0094-243X/80/590159-26$1.50

THE MARK II DETECTOR

The SLAC-LBL Mark II solenoidal magnetic detector is shown in end view in Fig. 1. The design goals were to build a detector which

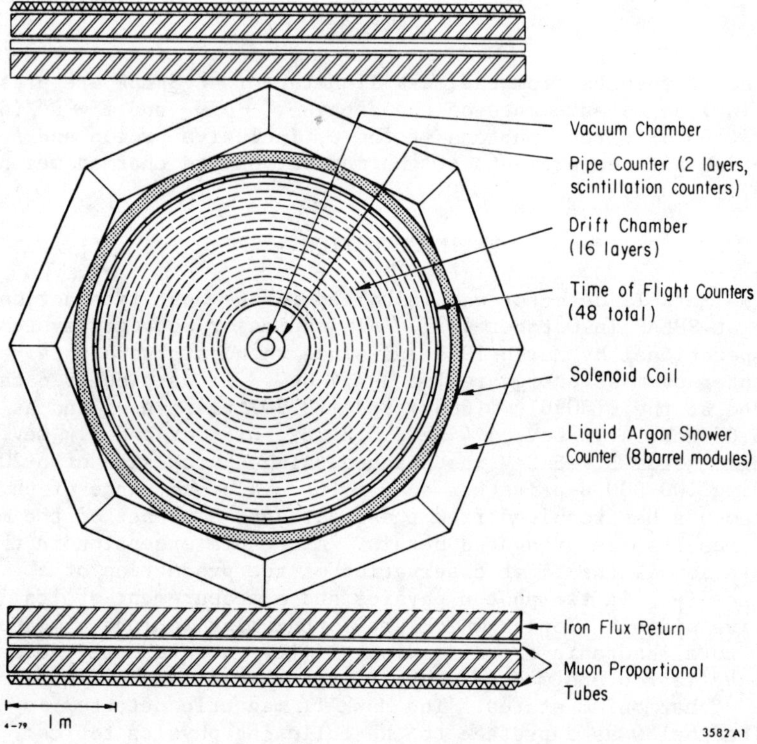

Fig. 1. End view of the SLAC-LBL Mark II detector.

tracked charged and neutral particles with good precision and efficiency over a large solid angle. In addition high priority was given to the identification of leptons with small hadron contamination and the clean separation of pions, kaons and protons over a wide range of momenta. The principal features of the detector are:

(a) Sixteen layers of drift chamber[3] which provide charged particle tracking over a solid angle of 75% of 4π. Azimuthal tracking information comes from 6 axial layers while the remaining 10 stereo layers, strung at $\pm 3^o$ to the beam axis, provide polar angle information. The rms momentum resolution for tracks constrained to the e^+e^- interaction point is $\sigma_p/p = [(0.005p)^2 + (0.0145)^2]^{\frac{1}{2}}$ where p is the track momentum in GeV/c. The two terms in the error parametrization correspond to the measurement error and the multiple scattering respectively. The tracking efficiency is greater than 95% for tracks whose momenta are above 100 MeV/c.

(b) A time-of-flight (TOF) system comprising 48 scintillation counters arranged azimuthally around the exterior of the drift chamber. This system has a resolution of 300 picoseconds for hadrons which translates into one standard deviation separation levels for momenta of 300 MeV/c for pions and electrons, 1.35 GeV/c for pions and kaons and 2.0 GeV/c for kaons and protons. The following method is used to obtain TOF particle identification for all the physics topics discussed below. The measured momentum and flight path are used to predict a flight time for each mass hypothesis (π, K and proton). A Gaussian weight is obtained for each hypothesis using the difference between the measured and predicted times. These weights, which are proportional to the probability that the measured time is compatable with the mass hypothesis, are summed and normalized to unity. Particles are labeled protons or kaons if their respective weight exceeds 0.5, otherwise the particle is called a pion. A similar scheme is used below 300 MeV/c to label tracks electrons or pions.

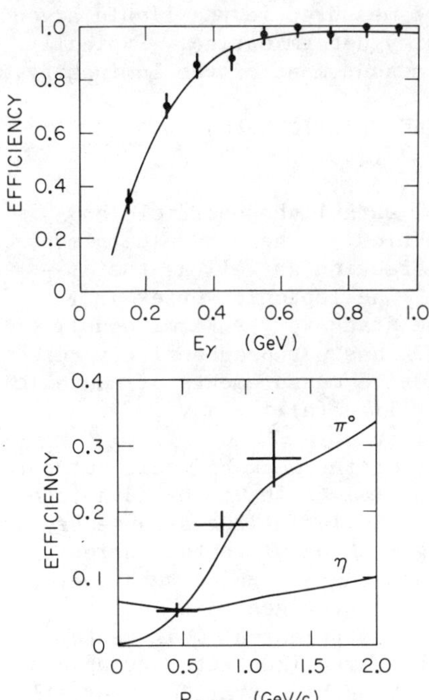

Fig. 2. (a) Photon efficiency for the liquid argon system. (b) π^o and η^o detection efficiency including geometrical effects and $B(\eta \rightarrow \gamma\gamma)$. The solid curves are Monte Carlo predictions.

(c) Eight lead-liquid argon shower counters[4] instrumented over 64% of 4π. These counters comprise readout strips parallel, perpendicular and at 45° to the beam direction and are 14 radiation lengths thick. The energy resolution has been measured to be $\sigma_E/E = 11.5\%/\sqrt{E}$ (E in GeV) for high energy electrons and photons. For energies below 500 MeV this resolution degrades slightly due to the 1.36 rℓ coil which precedes the counters. Angular resolution for high energy photons and electrons has been measured to be 3.5 mrads which in turn degrades to ~8 mrads below 500 MeV. The photon detection efficiency has been measured using the decay $\psi \rightarrow \pi^+\pi^-\pi^+\pi^-\pi^o$ and is shown in Fig. 2(a). The solid curve, which is the prediction of the Monte Carlo simulation program,[5] agrees well with the data. Figure 2(b) shows the resulting detection efficiency for π^o and η^o where all geometric effects are included as well as the branching ratio for $\eta^o \rightarrow \gamma\gamma$.

(d) Below 300 MeV/c electrons are identified using TOF as described earlier. Above 300 MeV/c cuts in TOF (below 500 MeV/c), total energy, longitudinal and transverse shower development are used to separate electrons and hadrons. The optimal cuts were established from pure samples of electrons, obtained from photon conversions and radiative Bhabhas, and pions obtained from the decay $\psi \rightarrow \pi^+\pi^-\pi^+\pi^-\pi^o$. The probability that a hadron is misidentified as an electron has been measured to be 7% below 500 MeV/c, 4% at 600 MeV/c and 2% at 800 MeV/c.

(e) Two layers of proportional tubes interleaved with steel are used for the identification of muons. The threshold momentum for the two layers is 700 MeV/c and 1.0 GeV/c respectively and the system, which covers 55% of 4π, has an efficiency for tagging muons above threshold in excess of 98%. The probability that a hadron is called a muon is 4% at 800 MeV/c, 11% at 900 MeV/c and 2% above 1 GeV/c.

(f) A small angle (25 mrad) tagging system is used to measure Bhabha scattering and hence provides a measure of the luminosity. Large angle Bhabha scattering events are measured in the liquid argon system as a cross-check on this luminosity determination. Typically we assign a 7% systematic error to the measurement of the luminosity.

MEASUREMENT OF THE BRANCHING FRACTIONS FOR $\tau^- \rightarrow \rho^-\nu_\tau$ AND $\tau^- \rightarrow K^{*-}(890)\nu_\tau$

The heavy lepton, τ^-, is now a well established particle and much is known about its intrinsic properties.[6] There have been many measurements of the leptonic branching fraction as well as the inclusive hadronic branching fractions.[6] The leptonic vertex in τ decays (see Fig. 3) seems to involve the standard V-A Fermi coupling[7] (although V+A has not been entirely ruled out). Inclusive measurements of hadronic final states like (a) $\tau^- \rightarrow \pi^-\nu_\tau$, (b) $\tau^- \rightarrow \rho^-\nu_\tau$ and (c) $\tau^- \rightarrow K^{*-}(890)\nu_\tau$[8] probe the coupling of the weak hadronic current and reactions (a)-(c) have the advantage of separately isolating the strangeness non-changing weak axial vector current, the strangeness non-changing weak vector current and the strangeness changing (i.e., Cabibbo suppressed) weak vector current. The Mark II detector group has measured the branching fractions for all three of these modes and we present below a detailed description of $\tau^- \rightarrow \rho^-\nu_\tau$[9] and $\tau^- \rightarrow K^{*-}\nu_\tau$ and briefly mention our preliminary result on $\tau^- \rightarrow \pi^-\nu_\tau$.

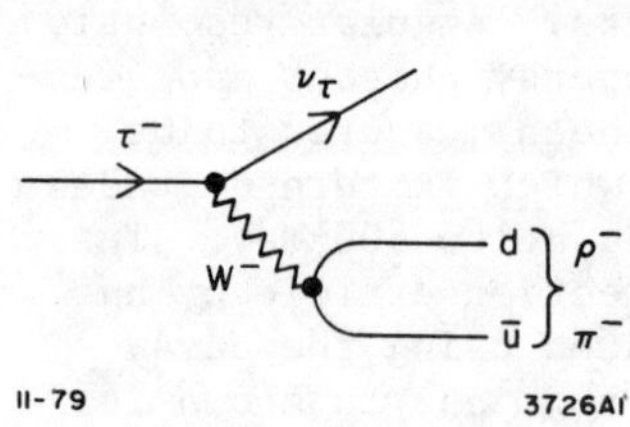

Fig. 3. The hadronic decay of the τ^-.

The data used in the $\tau \rightarrow \rho\nu$ study come from a scan in the center-of-mass energy range $4.5 \leq E_{cm} \leq 6.0$ GeV. This corresponds to an integrated luminosity of 3950 nb^{-1} or 11,500 produced $\tau^+\tau^-$ pairs. In searching for the decay $\tau^- \rightarrow \rho^-\nu_\tau$ we attempt to isolate events arising from the following decay sequence:

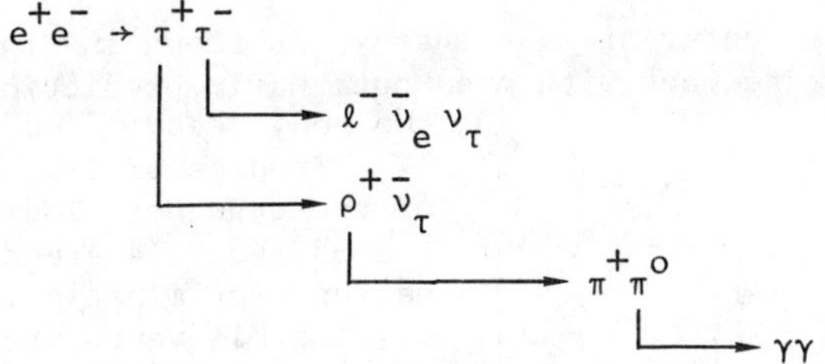

which results in two charged particles and two photons in the detector. Here ℓ^- represents either an electron or a muon and this lepton tag helps provide a clean signature for $\tau^+\tau^-$ pair production. The selected events have two oppositely charged particles, one a pion and the other a lepton, and two photons in the liquid argon shower counters with $E_\gamma \geq 100$ MeV. The two photon invariant mass ($M_{\gamma\gamma}$) spectrum is shown in Fig. 4 and the photon energies for the π^o candidates (those satisfying $80 \leq M_{\gamma\gamma} \leq 200$ MeV/c^2) were adjusted using the π^o mass as a single constraint. These π^o's were then combined with the charged pions to obtain the $\pi^\pm\pi^o$ invariant mass spectrum shown in Fig. 5. The data were fit to the sum of a smooth background and a Breit-Wigner resonance and the ρ mass and width obtained from this fit agree well with the $\rho^\pm$ resonance parameters. There are 85 $\rho^\pm\ell^\mp$ events in the signal region, 64 $\rho^\pm e^\mp$ and 21 $\rho^\pm\mu^\mp$. There are less $\rho\mu$ events because of the 700 MeV/c momentum threshold of the muon system. The spectrum for the variable $x_\rho = (E_\rho - E_{min})/(E_{max} - E_{min})$, where E_ρ is the measured energy of the ρ and E_{min} and E_{max} are the minimum and maximum energies for a ρ produced in the decay of a τ

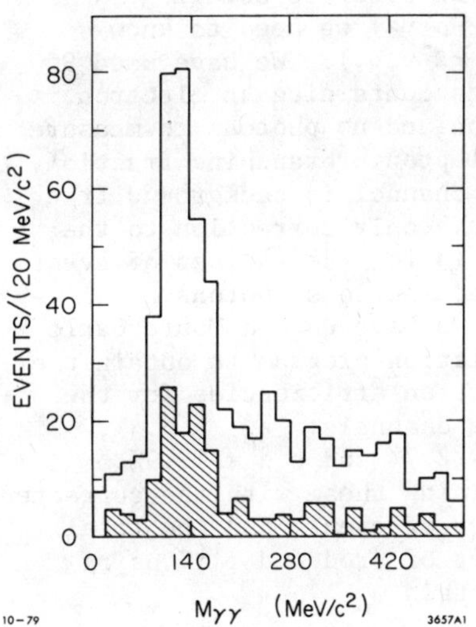

Fig. 4. Two photon invariant mass spectrum for $x^\pm\pi^\mp\gamma\gamma$ events where x represents either a pion or a lepton (ℓ). The shaded distribution is for $\ell^\pm\pi^\mp\gamma\gamma$ events.

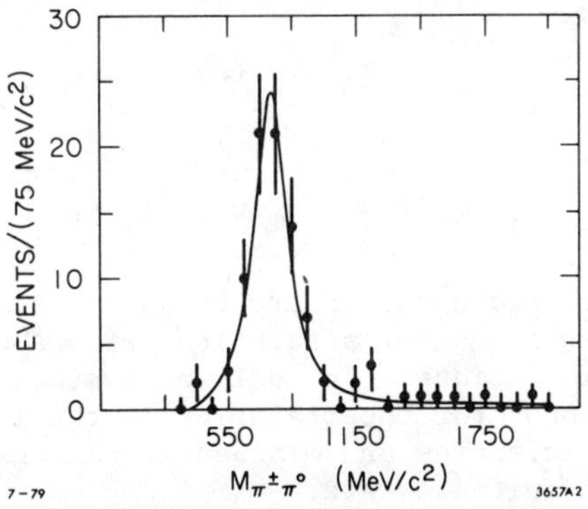

Fig. 5. $\pi^\pm\pi^o$ invariant mass distribution for $\ell^\pm\pi^\mp\pi^o$ events.

at the appropriate center-of-mass energy, is shown in Fig. 6. The data are in good agreement with the Monte Carlo prediction for the two body decay $\tau^- \to \rho^- \nu_\tau$.

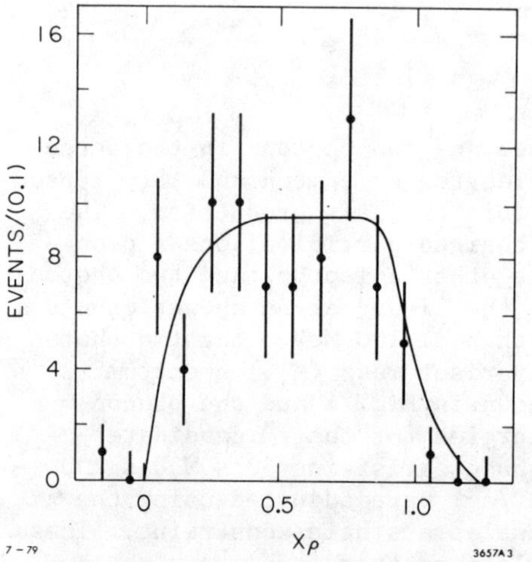

Fig. 6. The distribution $x_\rho = (E_\rho - E_{min})/(E_{max} - E_{min})$ for the 85 $\rho^\pm$ candidates. Here $x_\rho = 0$ (1) corresponds to the minimum (maximum) energy allowed for the ρ in the decay $\tau \to \rho\nu$. The solid line represents the prediction of the Monte Carlo simulation program.

Studies of like-sign events of the type $\rho^\pm \ell^\pm$ indicate that of the 85 events 4 are due to multihadron contamination. The remaining 81 events are assumed to be genuine τ decays and we correct for an estimated contribution from $\tau \to A_1\nu$ and $\tau \to (4\pi)\nu$ of 8.3 ρe events and 2.6 $\rho\mu$ events. In addition we have to correct for a 12.7% loss of events due to spurious photons.

In order to obtain $B(\tau^- \to \rho^- \nu_\tau)$ we need to know $B(\tau^- \to \ell^- \nu_\tau \bar{\nu}_\ell)$. We have used 95 events containing an electron, a muon and no photons to measure the leptonic branching fraction. This channel is background free and the only correction to the data is for the 6% loss of events due to spurious photons.

We have used a Monte Carlo simulation program to obtain the detection efficiencies for the three channels: $\varepsilon_{\rho e} = 6.4\%$, $\varepsilon_{\rho\mu} = 2.7\%$ and $\varepsilon_{e\mu} = 12.9\%$. Combining these with the corrected number of events and the total number of produced $\tau^+\tau^-$ pairs we obtain

$$B(\tau^- \to \rho^- \nu_\tau) B(\tau^+ \to e^+ \bar{\nu}_\tau \nu_e) = 0.042 \pm 0.009$$

$$B(\tau^- \to \rho^- \nu_\tau) B(\tau^+ \to \mu^+ \bar{\nu}_\tau \nu_\mu) = 0.033 \pm 0.010$$

and

$$\sqrt{B(\tau^- \to e^- \nu_\tau \bar{\nu}_e) B(\tau^+ \to \mu^+ \bar{\nu}_\tau \nu_\mu)} = 0.185 \pm 0.015$$

The errors quoted are based on the statistics of the uncorrected events and the statistical errors of the corrections and Monte Carlo calculations. In addition systematic errors have been included to account for uncertainties in the luminosity, the lepton tagging efficiencies and misidentifications and radiative corrections for the initial state.

Assuming μ-e universality in τ decays we further obtain

$$B(\tau^- \to \rho^- \nu_\tau) = (20.5 \pm 4.1)\%$$

$$B(\tau^- \to e^- \bar{\nu}_e \nu_\tau) = (18.5 \pm 1.5)\%$$

and

$$\frac{B(\tau^- \to \rho^- \nu_\tau)}{B(\tau^- \to e^- \nu_\tau \bar{\nu}_e)} = 1.11 \pm 0.23 \quad .$$

This ratio can be calculated using the conserved vector current hypothesis and $e^+e^- \to \rho^o$ data. In particular the prediction of Gilman and Miller[10] for this ratio is 1.2 in excellent agreement with the data. The result is also in agreement with the measurement of $B(\tau^- \to \rho^- \nu_\tau) = (24 \pm 9)\%$ of the DASP group.[11] The measurement presented here represents less than 20% of our data and we will soon have a sample of ~500 ρℓ events to upgrade this analysis. We also have obtained a preliminary result of $B(\tau^- \to \pi^- \nu_\tau) = (10.7 \pm 2.1)\%$.

In order to study the decay $\tau^- \to K^{*-}(890)\nu_\tau$ we have used all the Mark II data with $E_{cm} > 4.2$ GeV, which corresponds to an integrated luminosity of 15,300 nb^{-1} or 42,700 produced $\tau^+\tau^-$ pairs. For this analysis we search for one of the following two decay sequences:

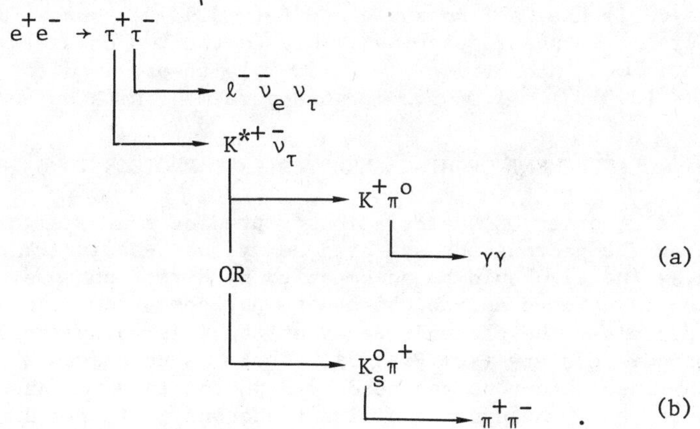

Again the lepton tag is used to isolate the τ channel and we require that the K^* decay either to (a) $K^\pm\pi^o$ or (b) $K^o_S\pi^\pm$. In case (a) the event selection is identical to the ρℓ events except that we require a charged kaon instead of a pion. In case (b) we require that there be four charged tracks with the appropriate charges and that two oppositely charged tracks, both pions, form a secondary vertex (distant from that formed by the $\pi^+\ell^-$ by at least 1 cm) and have a mass consistant with a K^o_S. Figure 7 shows the invariant mass spectrum for the $K^\pm\pi^o$ and $K^o_S\pi^\pm$ events selected as above. A clear peak is seen at the $K^*(890)$. There are 18 signal events with an estimated background of 3.4 events. We have used the detection efficiencies $\varepsilon_{K^+\pi^o\ell^-} = 1.3\%$ and $\varepsilon_{K^o_S\pi^+\ell^-} = 2.4\%$, the number of corrected events

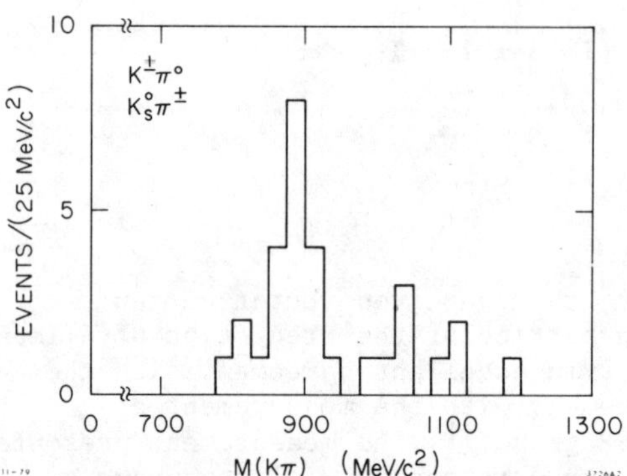

Fig. 7. $K_S^0\pi^{\pm}$ and $K^{\pm}\pi^0$ invariant mass distribution in candidate events for $\tau^- \to K^{*-}(890)\nu_\tau$.

and the number of produced $\tau^+\tau^-$ pairs to obtain

$$B(\tau^- \to K^{*-}(890)\nu_\tau) = 1.26 \pm 0.37 \pm 0.32$$

where the first error is statistical and the second error accounts for systematic effects. We interpret this result as evidence for a Cabibbo suppressed decay mode of the τ. Such a mode is expected and according to Tsai[12]

$$\frac{B(\tau \to K^*(890)\nu_\tau)}{B(\tau \to \rho\nu_\tau)} = \tan^2\theta_c \cdot f(M_\tau, M_\rho, M_{K^*}).$$

where θ_c is the Cabibbo angle and f (= 0.93) accounts for the difference in phase space available to the two channels. For $\tan^2\theta_c = 0.05$ and $B(\tau \to \rho\nu) = 20.5 \pm 4.1$, theory predicts $B(\tau \to K^*(890)\nu_\tau) = 1.0 \pm 0.2$ in good agreement with our measurement.

MEASUREMENT OF HIGH ENERGY DIRECT PHOTONS IN ψ DECAYS

First order QCD calculations[13] predict that a significant fraction of the hadronic decays of a heavy quark-antiquark resonance, such as the ψ, should be accompanied by direct photons. A direct photon is defined as one which does not come from π^0 or η^0 decay. QCD perceives the hadronic decay of the ψ as going through a 3 gluon intermediate state (see Fig. 8). There is no reason why one of the gluon lines cannot be replaced by a photon in which case hadronic decays of the ψ would contain direct photons. First order QCD predicts

$$B_\gamma = \Gamma(\psi \to \gamma gg)/\Gamma(\psi \to ggg) = (\alpha/\alpha_s)C(e_Q/e)^2$$

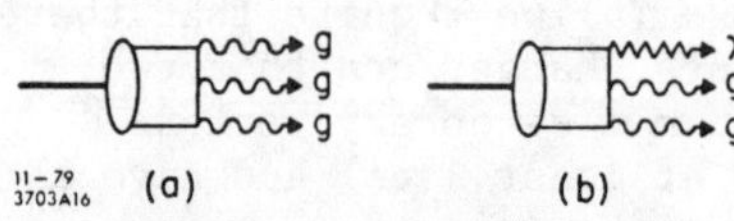

Fig. 8. (a) QCD model for the hadronic decay of the ψ and (b) the contribution to the direct photon production.

where $C = 36/5$ is a color SU(3) factor, e_Q is the charge of the charmed quark, and α_s is the strong coupling constant. For $\alpha_s = 0.18$,[14] one calculates $B_\gamma = 13\%$. If we equate the decay of ψ into hadrons with the 3 gluon decay and correct for second-order electromagnetic effects, the first order QCD prediction for the branching ratio for direct photons from the ψ is 8%. The photon momentum spectrum is predicted to be roughly linear, peaking at the beam energy.

The data used for this analysis comprises 280,000 hadronic events at the ψ. The events are required to have at least two charged particles and background due to Bhabha events, in which one of the electrons radiates, is removed. Photons are detected in the liquid argon shower counters. Neutral pions are reconstructed by combining pairs of photons each of which is required to have $E_\gamma \geq 150$ MeV. Photon pairs with an invariant mass between 75 and 200 MeV are considered to be π^o candidates and the candidate events are then corrected (as a function of π^o momentum) for background. The shape of the background was obtained by combining real photons with "pseudo-photons" from the same event. Pseudo-photons were created by assigning charged π's a π^o identity and letting them decay into two photons. The background was normalized to the two photon mass spectrum above 300 MeV/c^2.

The photons and π^o's obtained in this way are used to obtain inclusive photon and π^o spectra. The detected spectra are corrected using the known photon and π^o detection efficiencies and the trigger efficiency. The trigger efficiency is obtained from a sample of ψ' decays. Events of the type $\psi' \to \psi\pi^+\pi^-$ are selected which have two oppositely charged particles whose missing mass is consistant with the ψ mass. This sample of ψ events is identified solely by the π^+ and π^- and hence there is no trigger bias associated with the ψ. Trigger efficiencies as a function of photon and π^o momenta are obtained from these events as the fraction of ψ events which would have triggered the Mark II in the absence of the π^+ and π^-. The photon trigger efficiency varies from 70% at $x = 0.1$ to ~40% near $x = 1.0$ where for a photon of momentum p, $x = 2p/M_\psi$. The π^o trigger efficiency varies from 70% at $x = 0.3$ to ~50% at $x = 1.0$.

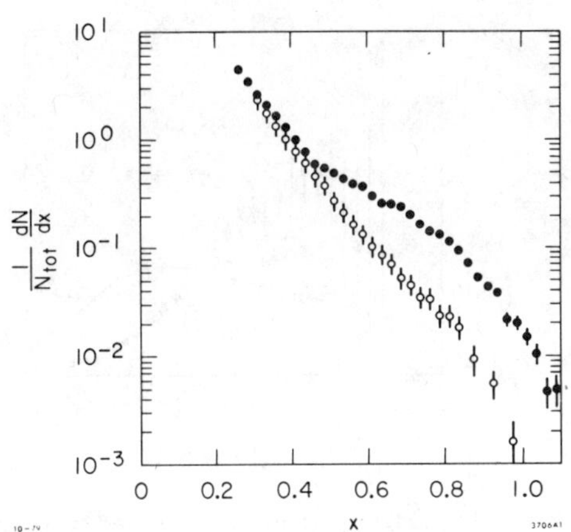

Fig. 9. Solid points show the inclusive photon momentum distribution. Open points show the expected photon momentum distribution based on measurements of π^o and η^o production. The error bars on the measured photons are statistical only; the error bars on the expected distribution include a ±20% systematic error.

The solid points in Fig. 9 show the inclusive photon spectrum after correction for the trigger efficiency and the photon detection efficiency. The distribution has been normalized to the total number of produced ψ's and the error bars are statistical only. Typical systematic errors are ±20%, approximately half of which are x dependent. Fig. 10 shows the π^o inclusive spectrum. Systematic errors of ±30% are not shown.

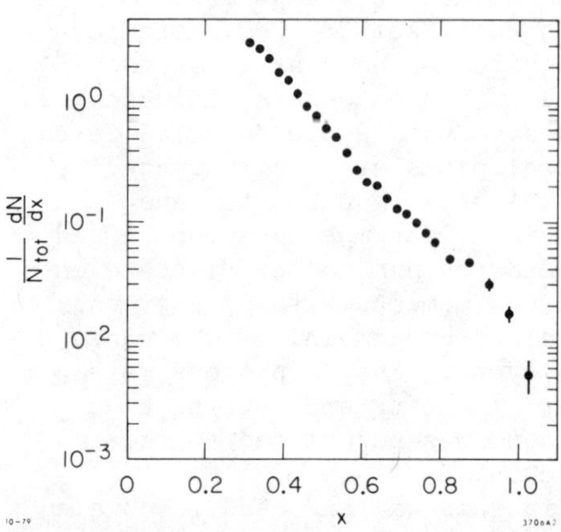

Fig. 10. Inclusive π^o momentum distribution. Systematic error are ±30% and are not shown on the figure.

We have used the π^o inclusive distribution to make a prediction for the expected photon x distribution. We have also accounted for photons arising from η^o decay. To determine the η^o contribution, fits have been made to the background subtracted two photon mass distributions. We have measured the ratio $R(p) = B(\psi \to \eta^o + x) \cdot B(\eta^o \to \gamma\gamma)/B(\psi \to \pi^o + x)$ as a function of momentum and find that R(p) is less than 0.1 in all momentum bins except p > 1.2 GeV where $R(p) = 0.16 \pm 0.06$. Using these estimates of η^o production and the measured π^o x distribution, we produce the expected photon inclusive distribution shown in Fig. 9 as open circles. The error bars include a ±22% systematic error. In Fig. 9 we see a clear excess of inclusive photons for $x \geq 0.5$.

Figure 11 shows the difference between the two distributions in Fig. 9 and is hence the excess of measured photons over and above those arising from π^o and η^o decays. Both statistical and systematic errors are shown although there is an additional ±17% systematic error in the subtracted distribution due to uncertainties in the efficiency corrections and knowledge of the direct photon angular distribution. Integrating the direct photon momentum distribution above $x = 0.6$ we obtain an inclusive rate for direct photon production of $(4.1 \pm 0.8)\%$. For $x > 0.6$ first order QCD predicts 5% in good agreement with the data. The shape of the direct photon x distribution predicted by first order QCD and convoluted with our detector resolution is shown in Fig. 11. It does not agree well with the observed shape, but it is important to point out that higher order QCD effects could well be comparable to those in first order. The predicted x distribution will also be softened by radiative

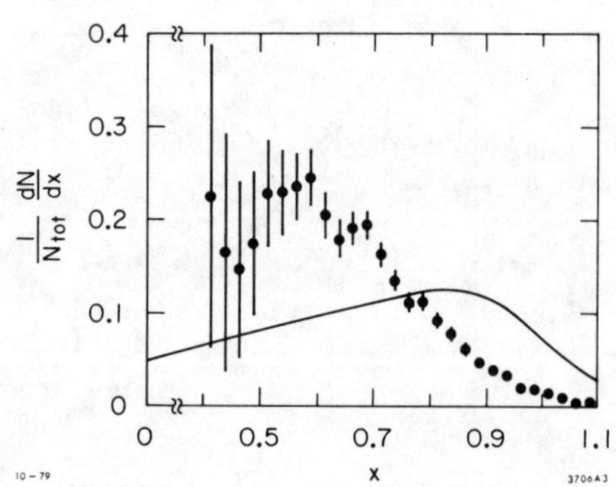

Fig. 11. Direct photon momentum distribution. The solid curve is the lowest order QCD prediction convoluted with the energy resolution.

effects and the mass of the final state hadrons which were not included in the QCD calculations.

The angular distribution for the photons with $x > 0.6$ has been fit to the form $1 + \alpha_\gamma \cos^2\theta$ where θ is the polar angle of the photon with respect to the beam direction. Approximately 25% of these photons are from π^o decay and allowing for an isotropic distribution for these photons we obtain $\alpha_\gamma = 0.18 \pm 0.18$ for the direct photons. First order QCD predicts $\alpha_\gamma \approx 0.3$ for $x > 0.6$.

As a check on possible systematic errors and trigger bias problems, the analysis presented above has been repeated using two other independent methods. The first method uses the ψ events obtained from $\psi' \to \psi\pi^+\pi^-$ as discussed earlier. The second involves using ψ data and requiring one photon in the liquid argon system and another arising from a photon conversion in the 0.06 radiation length of material which precedes the drift chamber. In the first case the π^+ and π^- provide the trigger and in the second the e^+ and e^- from the conversion provide the trigger. So both of these methods are free of the trigger bias, and in both cases the resulting photon and π^o distributions are consistant, within errors, with the primary method discussed above.

INCLUSIVE PROTON AND Λ^o PRODUCTION

A study has been made of the inclusive cross section for the production of protons and Λ^o's in the center-of-mass energy range 3.7-7.4 GeV.[15] A measurement of this nature is important in its own right. However the main motivation for this study was to look for threshold effects which might signify the onset of charmed baryon production.

Anti-protons ($\bar{p}$) were used for this inclusive measurement to avoid the contamination present in proton events from beam-gas scattering. All hadronic events with two or more charged prongs were used for the $\bar{p}$ analysis and TOF was used to identify $\bar{p}$'s with momenta ≤ 2.0 GeV/c. The $\bar{p}$ detection efficiency is estimated from a Monte Carlo model in which $\bar{p}$'s are generated according to the invariant cross section $E/4\pi p^2 d\sigma/dp \sim e^{-bE}$, where p is the momentum of the $\bar{p}$, E its energy and b is an adjustable slope parameter. The rest of the phase space is filled with a second nucleon and $\langle n \rangle - 2$ pions, where $\langle n \rangle$ is the mean particle multiplicity. The parameters b and $\langle n \rangle$ are adjusted at each center-of-mass energy so that the observed $\bar{p}$ momentum spectra are adequately reproduced. As a check on this $\bar{p}$ production model we have used it to obtain invariant cross sections as a function of momentum at various energies in the 4 GeV region and the resulting spectra agree well with those obtained by the DASP group.[16] The detection efficiency for $\bar{p}$'s varies slowly with center-of-mass energy, the average being 60%.

The Λ^o and $\bar{\Lambda}^o$ hyperons are identified by their $p\pi^-$ and $\bar{p}\pi^+$ decay modes. All hadronic events with three or more charged particles were used. The rms mass resolution for the Λ's is 3 MeV and background subtractions of $\lesssim 15\%$ are made. For the Λ's (not $\bar{\Lambda}$'s) we require that the total observed charge be $< +1$. This requirement reduces the beam-gas background substantially and results in an efficiency of ~75% for Λ's relative to $\bar{\Lambda}$'s. The detection efficiency is obtained from the same Monte Carlo model as the $\bar{p}$'s, and its varies from 10% at 3.7 GeV to 13% at 7.4 GeV.

The results are presented in Fig. 12 in the form of $R(p+\bar{p}) = 2\sigma_p/\sigma_{\mu\mu}$ and $R(\Lambda+\bar{\Lambda}) = [\sigma_\Lambda + \sigma_{\bar{\Lambda}}]/\sigma_{\mu\mu}$ (where $\sigma_{\mu\mu}$ is the μ-pair cross section) as a function of center-of-mass energy. These data have been corrected for acceptance and the Λ branching fraction into pπ. All errors are statistical; the estimated systematic errors are ±17% for $R(p+\bar{p})$ and ±27% for $R(\Lambda+\bar{\Lambda})$ and are dominated by the uncertainties in the production process. It is important to point out that $R(p+\bar{p})$ is not corrected for the contribution of protons from decays of Λ^o or Σ^o hyperons. The measurement of $R(p+\bar{p})$ agrees well in shape but lies ~25% above previous measurements.[16,17] The measurement of $R(\Lambda+\bar{\Lambda})$ is in poor agreement with the previous measurement;[17] the rate measured in the Mark II, which is about 2½ times higher than the previous measurement, is based on a cleaner Λ signal and more detailed efficiency studies. The data for both the $\bar{p}$'s and Λ's show an appreciable increase in baryon yield between 4.6 GeV and 5.2 GeV. The observed step sizes are $\Delta R(p+\bar{p}) = 0.31 \pm 0.06$ and $\Delta R(\Lambda+\bar{\Lambda}) = 0.10 \pm 0.03$ and the Λ/p ratio, corrected for the proton contribution from Λ's, is (41 ± 15)%.

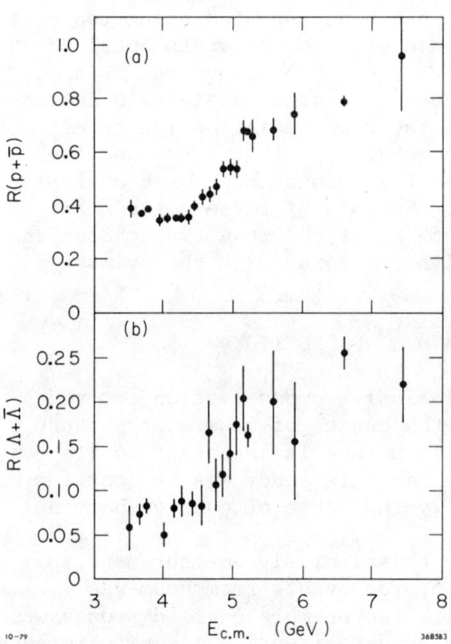

Fig. 12. (a) $R(p+\bar{p})$ as a function of E_{cm} and (b) $R(\Lambda+\bar{\Lambda})$ as a function of E_{cm}. Errors are statistical only.

OBSERVATION OF CHARMED BARYON DECAYS

If the rise in the baryon yield at 4.6 GeV is due to the onset of the production of charmed baryons, exclusive states with masses ~2.3 GeV should be seen decaying into final states containing one unit of baryon number and at least one unit of strangeness. Such weakly decaying states have been sought in invariant mass spectra and a prominent signal has been seen in the channel $pK^-\pi^+ + \bar{p}K^+\pi^-$.[15] All the Mark II data in the center-of-mass energy range 4.5-6.0 GeV are used for the analysis. These data correspond to an integrated luminosity of 9150 nb^{-1}, 5150 nb^{-1} of which is at 5.2 GeV. Charged kaons and protons are identified by the TOF system.

Figure 13 shows the invariant mass for various combinations of pKπ. A recoil mass cut of > 2.2 GeV has been applied to these data. Figure 13(a) shows the invariant mass of $pK^-\pi^+ + \bar{p}K^+\pi^-$ and a narrow signal of 39 ± 8 events is seen at a mass of 2.285 GeV/c^2 above a background of 20 events. The quantum numbers for these two states are those appropriate to the lowest lying charmed baryon, henceforth called Λ_c^+. Figure 13(b) shows the invariant mass of pKπ states whose quantum numbers are not appropriate for the Λ_c^+ and no structure is seen in these channels.

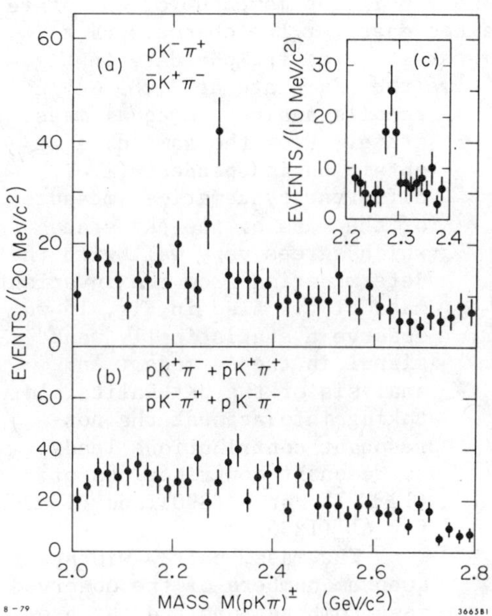

Fig. 13. (a) The combined $pK^-\pi^+$ and $\bar{p}K^+\pi^-$ invariant mass distribution for recoil masses > 2.2 GeV/c^2. Part of this plot is shown with finer bins in (c). (b) The $pK^+\pi^-$ and $pK^-\pi^-$ (plus charge conjugate states) mass distribution for recoil masses > 2.2 GeV/c^2.

A fit using a Gaussian plus a background shape obtained from Fig. 13(b) has been performed on the data in Fig. 13(a). The width obtained for the Gaussian is consistant with the detector resolution indicating that the pKπ state results from a weak decay. We obtain a mass of 2.285 ± 0.006 GeV/c^2 for the pKπ state, where the error is dominanted by systematic effects. We have also observed a signal in the channel $pK_s^o + \bar{p}K_s^o$ which is shown in Fig. 14. The signal corresponds to 10 ± 5 events and occurs at a mass which is in good agreement with that obtained from the pKπ signal.

A systematic error of 6 MeV/c^2 has been assigned to the mass of the pKπ state. The uncertainty in the magnetic field contributes 2 MeV/c^2 to the mass error. A 20% error in our correction for the proton and antiproton energy loss contributes 2 MeV/c^2. The

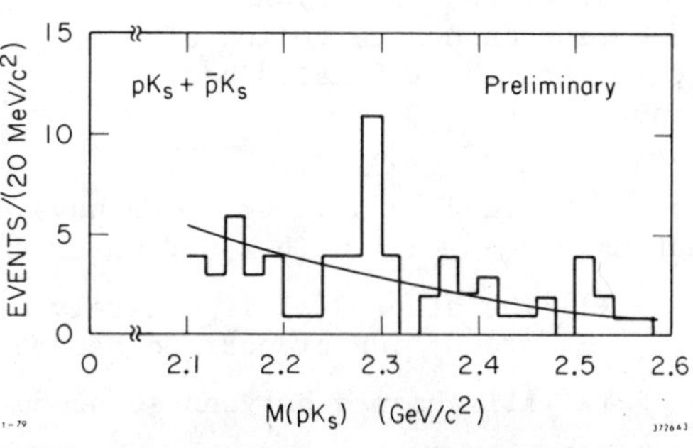

Fig. 14. Invariant mass for pK_s and $\bar{p}K_s$ for events with a recoil mass > 2.2 GeV/c^2.

absolute mass scale is checked using 20,000 reconstructed K_s^o decays whose mass agrees with the world average to within 0.3 MeV/c^2. Furthermore, in data taken many months apart, the D^o mass varies by at most 3 MeV/c^2.

If the observed state is produced in equal mass pairs then the energy of the state will be the beam energy (E_b), and its mass can be calculated as $M = \sqrt{E_b^2 - p^2}$, where p is the momentum of the state. Figure 15 shows this beam constrained mass for the channels $pK^-\pi^+$, pK_s^o and $\Lambda\pi^+$ and their charge conjugates. From the $pK\pi$ data in Fig. 15 we find that (26 ± 11)% of the time the $pK\pi$ state at 2.285 GeV/c^2 recoils against an equal mass state. From the same data we obtain an independent (i.e., different systematics) measure of the mass of the $pK\pi$ state which agrees very well with the determination from the invariant mass plot. Also in Fig. 15 we observe a statistically weak signal in the $\Lambda\pi$ mode. An analysis of the $pK\pi$ Dalitz plot taking into account the non-resonant contributions leads to resonant contributions of (12 ± 7)% for $K^*(890)$ and (17 ± 7)% for $\Delta^{++}(1236)$.

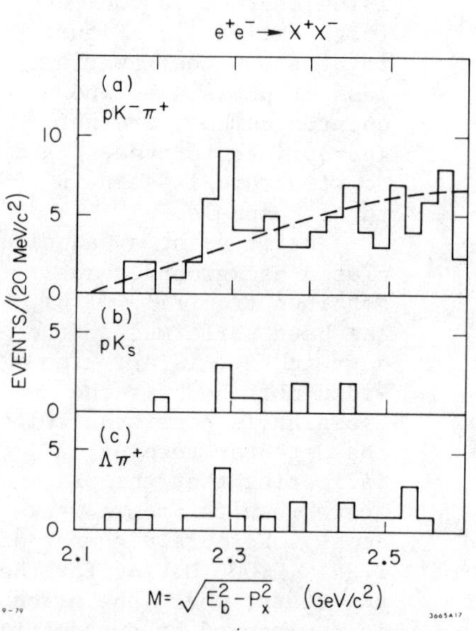

Fig. 15. Beam constrained mass is plotted to search for equal mass production $e^+e^- \to x^+x^-$ where $x^\pm$ decays to (a) $pK^-\pi^+$ or $\bar{p}K^+\pi^-$, (b) pK_s or $\bar{p}K_s$ and (c) $\Lambda\pi^+$ or $\bar{\Lambda}\pi^-$. The dashed curve in (a) represents the estimated background.

The mass, narrow width, quantum numbers of the observed mass combinations and the presence of associated production lead to the most obvious identification of this state as the Λ_c^+. The Λ_c^+ has been observed by several other experiments,[18] although the majority of these experiments have reported a mass near 2.26 GeV/c^2.

A Monte Carlo simulation program has been used to estimate the detection efficiency for the $pK\pi$ mode as (13.0 ± 2.5)% which along with the integrated luminosity provides a measure of $\sigma(\Lambda_c + \bar{\Lambda}_c) \cdot B(\Lambda_c \to pK\pi) =$ 0.037 ± 0.012 nb at 5.2 GeV. We have used the measured step size to estimate $\sigma(\Lambda_c + \bar{\Lambda}_c)$ at 5.2 GeV under the following assumptions:

(1) The step $\Delta R(p)$ is entirely due to the onset of pair production of charmed baryons.[19]

(2) All charmed baryons so produced cascade down to the Λ_c^+.

(3) The Λ_c^+ decays to a proton (as opposed to a neutron) (60 ± 10)% of the time.[20]

Hence

$$\sigma(\Lambda_c + \bar{\Lambda}_c) = \frac{\Delta R(p+\bar{p})}{0.6}\,\sigma_{\mu\mu}$$

$$= 1.7 \pm 0.4 \text{ nb}$$

and furthermore

$$B(\Lambda_c^+ \rightarrow pK^-\pi^+) = (2.2 \pm 1.0)\% \quad .$$

In addition from the 10 $K_S^o p$ events we obtain

$$\frac{B(\Lambda_c^+ \rightarrow p\bar{K}^o)}{B(\Lambda_c^+ \rightarrow pK^-\pi^+)} = 0.54 \pm 0.25$$

and the 6 $\Lambda\pi$ events in the constrained mass plot (Fig. 15) correspond to

$$\frac{B(\Lambda_c^+ \rightarrow \Lambda\pi^+)}{B(\Lambda_c^+ \rightarrow pK^-\pi^+)} = 0.35 \pm 0.20 \quad .$$

THE $\psi(3770)$ RESONANCE

The $\psi(3770)$[21,22] resonance lies just above the threshold for the production of D^+D^- and $D^o\bar{D}^o$, but below the threshold for DD^* production. It decays entirely to $D\bar{D}$ and provides a sample of kinematically well determined and relatively background-free D mesons. The Mark II group has performed a scan in the energy range 3.67-3.87 GeV in order to measure the $\psi(3770)$ resonance parameters. In addition 2850 nb^{-1} of integrated luminosity were obtained at the fixed energy of 3.771 GeV. The fixed energy running is useful for understanding both inclusive and exclusive properties of the D mesons and absolute branching fractions can be obtained because the scan data provides a measure of the $D\bar{D}$ production cross section.

The scan data is shown in Fig. 16(a) in the form of R which is the ratio of the hadronic cross section to the μ-pair cross section. The contribution from $\tau^+\tau^-$ pairs has been removed and the continuum (but not the ψ, ψ' and ψ'') has been radiatively corrected. The prominant features of Fig. 16(a) are the $\psi(3684)$ and its radiative tail plus a broader resonance centered at ~3.77 GeV, the $\psi(3770)$.

The data in Fig. 16(a) have been fit to a function which accounts for the resonance shape, the continuum and the radiative tails of the $\psi(3096)$, $\psi(3684)$ and the $\psi(3770)$ itself. The form of the resonance shape is a non-relativistic p-wave Breit-Wigner with an energy dependent total width[23] which takes into account the proximity of the resonance to the $D\bar{D}$ threshold. In particular

$$R(E_{cm}) = \frac{1}{\sigma_{\mu\mu}}\frac{3\pi}{M^2}\frac{\Gamma_{ee}\,\Gamma_{tot}(E_{cm})}{(E_{cm}-M)^2+\Gamma_{tot}^2(E_{cm})/4}$$

and

$$\Gamma_{tot}(E_{cm}) \propto \frac{p_+^3}{1+(rp_+)^2} + \frac{p_o^3}{1+(rp_o)^2}$$

where p_+ (p_o) is the momentum of the pair produced D^+ (D^o) and the mass of the resonance (M) its total width (Γ_{tot}) and its partial width to electrons (Γ_{ee}) are parameters in the fit. The total width is normalized at the peak of the resonance and the interaction length r is taken to be 2.5 fermis for this analysis. The resonance parameters obtained in the fit are found to be insensitive to the size of r.

The fit has a χ^2 of 12 for 17 degrees of freedom and is shown, after removal of the continuum and radiative effects, superimposed on the data in Fig. 16(b). The parameters obtained from the fit are:

$$M = 3763.7 \pm 5.1 \text{ MeV/c}^2$$
$$\Gamma_{tot} = 23.5 \pm 5.0 \text{ MeV}$$
$$\Gamma_{ee} = 276 \pm 50 \text{ eV} \quad .$$

The errors include systematic effects: the error in the mass is dominated by the 0.13% energy calibration of SPEAR. We also obtain the peak cross section $\sigma(3764) = 9.3 \pm 1.4$ nb.

Table I compares the Mark II results with those of DELCO[22] and the Lead Glass Wall[21] (LGW) experiment. The Mark II measurement of Γ_{tot} agrees well with the previous measurements while Γ_{ee} is intermediate between the two previous measurements. The mass we obtain appears 6-8 MeV/c^2 lower than the previous measurements, which seems to be within errors. However, we can remove the ring calibration error by

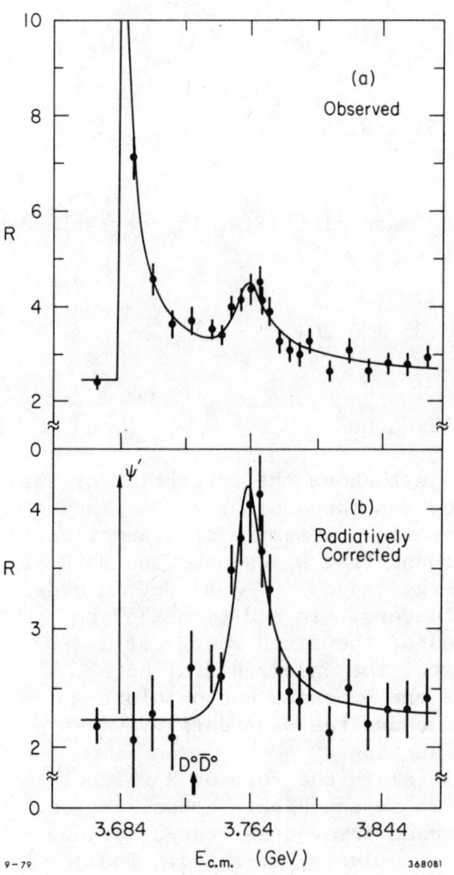

Fig. 16. The ratio of the hadron cross section to the μ-pair cross section in the region of the ψ(3770) resonance for (a) observed and (b) radiatively corrected for the ψ(3096), ψ(3684) and ψ(3771). The curve represents the fit to the data.

TABLE I

Measurements of the $\psi(3770)$ Resonance Parameters

Experiment	Mass MeV/c^2	Γ_{tot} MeV	Γ_{ee} eV	ΔM^* MeV/c^2
DELCO[22]	3770 ± 6	24 ± 5	180 ± 60	86 ± 2[24]
LGW[21]	3772 ± 6	28 ± 5	345 ± 85	88 ± 3
Mark II	3764 ± 5	24 ± 5	276 ± 50	80 ± 2

* ΔM is the mass difference between the $\psi(3684)$ and $\psi(3770)$.

measuring masses relative to the $\psi(3684)$ and this comparison is shown in the last column in Table I. The mass disagreement is now more significant.

The $\psi(3770)$ is typically associated with the $^3D_1$ state of charmonium and the Mark II values for the resonance parameters are in good agreement with the theoretical predictions.[25] In such charmonium models the relatively large leptonic width of the $\psi(3770)$ results from its mixing with the $^3S_1$ state, $\psi(3684)$. We obtain a mixing angle of $(20.3 \pm 2.8)^o$.

DECAYS OF CHARMED MESONS

The data presented below was obtained from the fixed energy running at 3.771 GeV, which corresponds to 49,000 hadronic events in the Mark II. From the scan data we measure $\sigma_{D\bar{D}}(3771) = 6.85 \pm 1.2$ nb and further obtain the inclusive $D^{\pm}$ and $D^o(\bar{D}^o)$ cross sections by assuming that the $\psi(3770)$ is a resonance of definite isospin (either 0 or 1) and that the decays of the $\psi(3770)$ are saturated by $D\bar{D}$. The assumption concerning isospin leads to the consequence of equal partial widths to the states $D^o\bar{D}^o$ and D^+D^- up to a phase space factor and the saturation of the decay channels by $D\bar{D}$ is reasonable in light of the narrow width of the nearby $\psi(3684)$. Hence we obtain

$$\sigma_{D^+}(3771) = 5.9 \pm 1.0 \text{ nb}$$

$$\sigma_{D^o}(3771) = 7.8 \pm 1.2 \text{ nb} \quad .$$

At the $\psi(3770)$ one is able to exploit the $D\bar{D}$ pair production by using the beam energy and the small D momentum to obtain invariant mass resolution of ~ 3 MeV/c^2. For any combination of particles which are candidates for D meson decay states we require that the energy of those particles be within 50 MeV of the beam energy and calculate the invariant mass using the beam energy (E_b):

$$M = \sqrt{E_b^2 - p^2}$$

where p is the momentum of the particle combination (~280 MeV/c). Figures 17 and 18 show the invariant mass for various $D^o(\bar{D}^o)$ and $D^{\pm}$

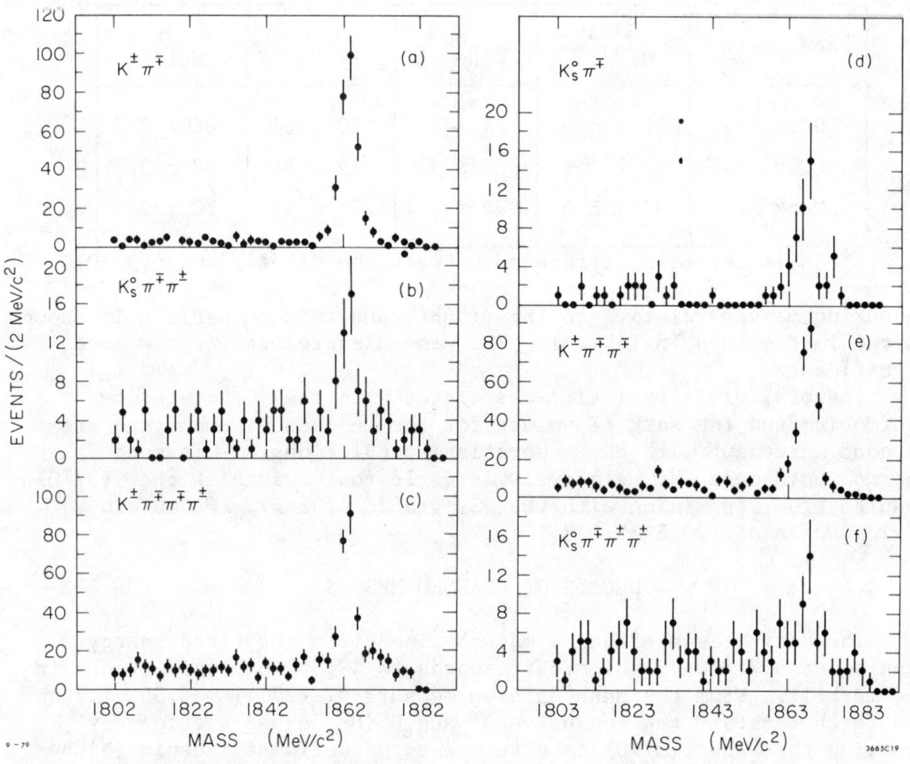

Fig. 17. Invariant mass spectra for various $D^o(\bar{D}^o)$ decay modes (a)-(c) and $D^{\pm}$ decay modes (d)-(f).

decay states. The low level of background in the $D^o \rightarrow K^-\pi^+$, $D^o \rightarrow K^-\pi^-\pi^+\pi^+$ and $D^+ \rightarrow K^-\pi^+\pi^+$ channels is evident and these channels will be used in a later section as clean tags of events containing D's to study inclusive D meson properties. Table II summarizes the results of the D branching fraction study along with a comparison with the results of the LGW experiment.[26] It is interesting to note that the rate for $D^o \rightarrow \bar{K}^o\pi^o$ is roughly the same as for $D^o \rightarrow K^-\pi^+$ which contradicts the theoretical prediction of color suppression.[27]

In addition to the Cabibbo favored decay channels discussed above, we have observed the Cabibbo suppressed decay modes $D^o \rightarrow K^+K^-$ and $D^o \rightarrow \pi^+\pi^-$. Figure 19 shows quark diagrams for the Cabibbo favored decay mode as well as the two Cabibbo suppressed modes, and those diagrams reflect the flavor preference for the $c \rightarrow s$ coupling over the $c \rightarrow d$ coupling. In Fig. 19, θ_A is the familiar Cabibbo angle (θ_c) whereas θ_B can be thought of as the charm analogue of θ_c. Both θ_A

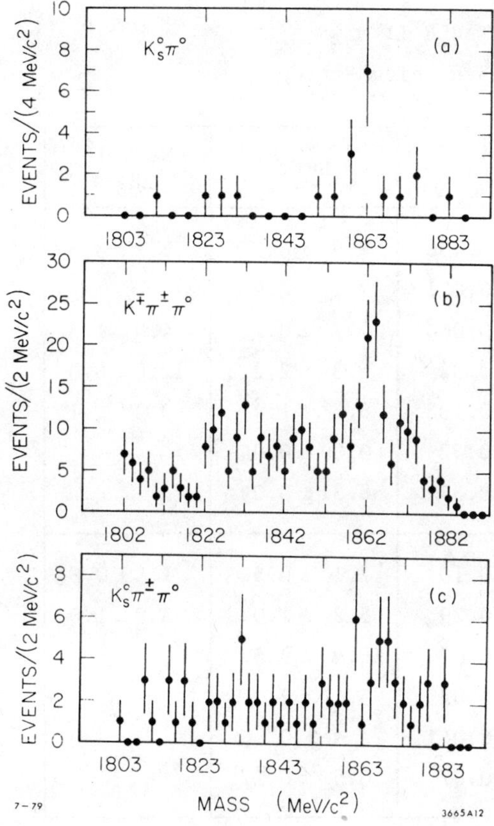

Fig. 18. Invariant mass spectra for the decays (a) $D^0 \to K_S^0\pi^0$, (b) $D^0 \to K^-\pi^+\pi^0$ and (c) $D^+ \to K_S^0\pi^+\pi^0$.

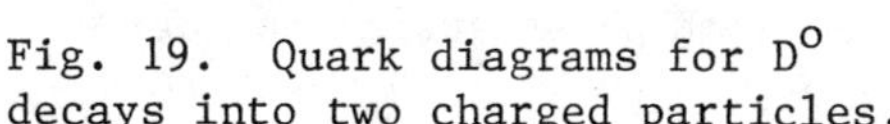
Fig. 19. Quark diagrams for D^0 decays into two charged particles.

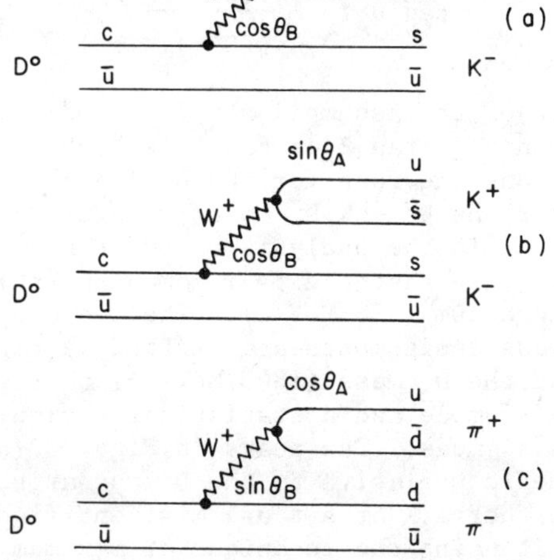

TABLE II

D-Meson Branching Ratios

Mode	# Events	ε	BR(%)	Mark I[29] BR(%)
$K^-\pi^+$	271 ± 17	0.436	2.8 ± 0.5	2.2 ± 0.6
$\bar{K}^o\pi^o$	9 ± 3	0.021	2.1 ± 0.9	
$\bar{K}^o\pi^+\pi^-$	39 ± 7	0.064	2.7 ± 0.7	4.0 ± 1.3
$K^-\pi^+\pi^o$	37 ± 9	0.028	6.3 ± 2.2	12.0 ± 6.0
$K^-\pi^+\pi^+\pi^-$	197 ± 16	0.133	6.7 ± 1.4	3.2 ± 1.1
$\pi^+\pi^-$	9 ± 4	0.52	0.09 ± 0.04	
K^+K^-	22 ± 5	0.37	0.31 ± 0.09	
$\bar{K}^o\pi^+$	37 ± 7	0.10	2.1 ± 0.5	1.5 ± 0.6
$K^-\pi^+\pi^+$	251 ± 17	0.29	5.2 ± 1.0	3.9 ± 1.0
$\bar{K}^o\pi^+\pi^o$	9 ± 4	0.004	16.4 ± 9.5	
$\bar{K}^o\pi^+\pi^+\pi^-$	22 ± 7	0.025	5.1 ± 2.0	
$K^-\pi^+\pi^+\pi^+\pi^-$	5 ± 3.5	0.041	< 2.0*	
$\bar{K}^oK^+$	6 ± 3	0.07	0.5 ± 0.27	

* 90% confidence limit.

and θ_B can be measured via the relations

$$\tan^2\theta_A = \frac{\Gamma(D^o \to K^+K^-)}{\Gamma(D^o \to K^-\pi^+)} \quad ; \quad \tan^2\theta_B = \frac{\Gamma(D^o \to \pi^+\pi^-)}{\Gamma(D^o \to K^-\pi^+)} \ .$$

Under the assumption $\theta_A = \theta_B$ and SU(3) invariance, one can predict $\tan^2\theta_A = \tan^2\theta_B = \tan^2\theta_C = 0.05$. Because of phase space considerations the rate for the $D^o \to \pi^+\pi^-$ mode is raised by 7% and the rate for the $D^o \to K^+K^-$ mode is lowered by 8%.

In the analysis of the three two body modes mentioned above we require that the pair momentum (p_D) be within 30 MeV/c of the expected momentum (288 MeV/c). The two body invariant masses for the three mass assignments are plotted in Fig. 20. One sees a prominent peak at the D_o mass (1863 MeV/c^2) in the $K^-\pi^+$ mode, a clear signal in the K^+K^- mode and a statistically significant excess of events in the $\pi^+\pi^-$ mode. The peaks in Figs. 20(a) and 20(b) which appear at ~±120 MeV/c^2 relative to the D mass arise from the decay $D^o \to K^-\pi^+$ in which either a K or a π are misidentified by TOF. The data in Fig. 20 is fit using the technique of maximum likelihood. Background distributions shown in Fig. 20 are obtained from sidebands in the p_D plot.

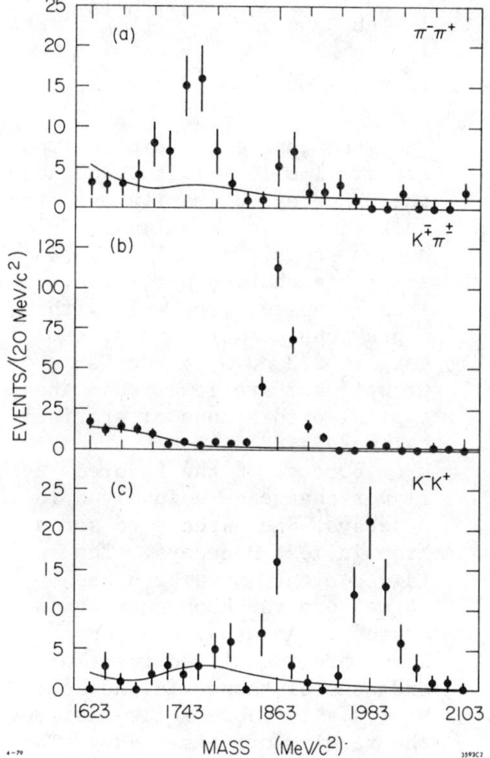

The results of the fit are 235 ± 16 $K^{\pm}\pi^{\mp}$ events, 22 ± 5 K^+K^- events and 9 ± 3.9 $\pi^+\pi^-$ events. Accounting for the relative efficiencies gives $\Gamma(D^o \to K^+K^-)/\Gamma(D^o \to K^-\pi^+) = 0.113 \pm 0.03$ and $\Gamma(D^o \to \pi^+\pi^-)/\Gamma(D^o \to K^-\pi^+) = 0.033 \pm 0.015$. The errors quoted include systematic effects the most prominent of which is the background estimate used for the fit. The results show that Cabibbo suppressed D decays exist and that they have roughly the expected magnitude.

Fig. 20. Invariant mass of two particle combinations with momenta within 30 MeV/c of the expected D^o momentum.

INCLUSIVE STUDIES OF D-MESON DECAYS

As we mentioned before, the relatively background-free channels $D^o \to K^-\pi^+$, $D^o \to K^-\pi^+\pi^-\pi^+$ and $D^+ \to K^-\pi^+\pi^+$ are used for inclusive studies. The observed D channel serves as a clean tag for $D\bar{D}$ production and we look to see what is produced in association with the tag. For these inclusive studies a sample of events comprising 283 $K^-\pi^+$ decays (with 17 ± 2 background), 211 $K^-\pi^+\pi^-\pi^+$ decays (with 31 ± 3 background) and 290 $K^-\pi^+\pi^+$ decays (with 33 ± 2 background) are used.

Figure 21(a) shows the observed charge multiplicity opposite the tagged events. These spectra have been corrected for the background contributions shown shaded in the plot. The distributions are not corrected for the $K_s \to \pi^+\pi^-$ decay which enters into the spectra in Fig. 21 as two charged particles. Using a Monte Carlo simulation program one obtains a matrix which relates the probability of observing j charged particles when i were produced. This matrix is used to "unfold" the produced charged multiplicity, the results of which are shown in Fig. 21(b). The average multiplicities are:

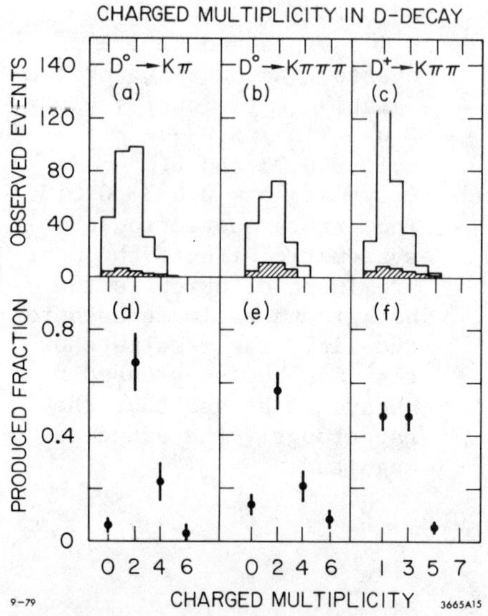

Fig. 21. The observed (a)-(c) and produced (d)-(f) charged multiplicity distributions obtained using the channels $D^o \to K^-\pi^+$, $D^o \to K^-\pi^+\pi^-\pi^+$ and $D^+ \to K^-\pi^+\pi^+$ as tags.

$$\langle n_{ch} \rangle_{D^o} = 2.46 \pm 0.14$$

$$\langle n_{ch} \rangle_{D^+} = 2.16 \pm 0.16 \quad .$$

The mean charged multiplicities are similar in magnitude because the D^o decays primarily into two charged particles while the D^+ decays roughly equally into one and three charged particles. These results agree well with a previous measurement of $\langle n_{ch} \rangle = 2.3 \pm 0.3$ by the LGW group[29] and are lower than the typical predictions of statistical models.[30]

Because of the favored flavor change $c \to s$ involved in D decays, one expects to see a kaon in most D decays. The tagged events have been used to measure the kaon content in D decay. As usual charged kaons are identified from TOF and neutral kaons via the decay $K_s \to \pi^+\pi^-$. Table III summarizes the results of this study. The backgrounds for charged kaons come from TOF misidentifications and for neutral kaons from random $\pi^+\pi^-$ pairs whose mass is consistent with a K_s. From Table III we can conclude that the Cabibbo suppressed decays ($\bar{D} \to K^+$) are seen at the expected rate. The rate for $D^+ \to K^-$

TABLE III

Kaon Content in D Decays

Decay Mode	# Tags	# Kaons	Background	BR(%)
$D^o \to K^-$		111	5.1 ± 1.2	56 ± 5.6
$\to K^+$	476 ± 23	20	5.8 ± 1.3	7.9 ± 2.6
$\to K^o$		15	5.1 ± 0.5	20.3 ± 8.5
$D^+ \to K^-$		21	2.2 ± 0.7	17.2 ± 4.1
$\to K^+$	257 ± 18	11	4.5 ± 1.6	5.6 ± 2.9
$\to K^o$		13	3.5 ± 0.4	44.0 ± 15.0

is suprisingly low. However, this measurement is in good agreement with (10 ± 7)% obtained by LGW.[29] Although the errors are large, particularly for the $D \to K^o$ rates, the inclusive fraction of both D^o and D^+ into kaons are lower than naive expectations.

We have studied semi-leptonic D decays using the electron content in tagged events. Here we must distinguish between "right sign" and "wrong sign" electrons - the "wrong sign" electron having the opposite sign of the kaon observed in the tagged D. The data are summarized in Table IV. The excess (observed events above background) of "right sign" events are corrected using the excess of "wrong sign" events in order to obtain the true branching fractions. We note that $B(D^+ \to e^+)$ is approximately three times as large as $B(D^o \to e^+)$. If we assume that the D^+ and D^o have the same semi-leptonic widths we obtain

$$\frac{\Gamma(D^o \to all)}{\Gamma(D^+ \to all)} = \frac{B(D^+ \to e^+)}{B(D^o \to e^+)} = \frac{\tau^+}{\tau^-}$$

Hence measurement of the ratio of the leptonic branching fractions for D^+ and D^o measures their relative lifetimes. Using the data in Table IV and a maximum likelihood method which incorporates the (Poisson) statistics of the observed events and the subtractions we obtain

$$\frac{\tau^+}{\tau^o} = 3.08 {+4.1 \atop -1.33} \quad .$$

DELCO has reported a similar trend in the relative lifetimes of the D^+ and D^o.[31] Correcting for phase space effects we also obtain

$$B(D \to e^+) = (9.8 \pm 3.0)\% \quad .$$

TABLE IV

Semi-Leptonic Decays of D^+ and D^o

Decay Mode	# Tags	# Electrons	Background	BR(%)
$D^+ \to e^+$	295 ± 18	38	15 ± 1	15.8 ± 5.3
$\to e$		4	3.9 ± 0.5	
$D^o \to e^+$	480 ± 23	36	19 ± 1	5.2 ± 3.3
$\to e$		19	12 ± 1	

CONCLUSIONS

The Mark II is able to make valuable contributions to the understanding of a broad range of physics topics. Data from the study of τ hadronic decays agree well with the theoretical predictions. The mode $\tau \to \rho\nu$ supports the CVC hypothesis and the decay $\tau \to K^*(890)\nu$ is evidence of the presence of a Cabibbo suppressed vector current in τ decays. The studies of D decay properties are in may cases statistics limited but there are some significant disagreements with the standard theoretical lore. In particular color suppression is not supported by the data, the fraction of both D^o and D^+ decaying to a kaon seems low, the D^o and D^+ have lifetimes differing by about a factor of 3 and there are still a large fraction of D decay modes unaccounted for. Cabibbo suppressed D decays have been observed at approximately the expected rate. A direct photon signal has been observed at the ψ, and the rate for these direct photons agrees well with the prediction of first order QCD calculations. However the shape of the direct photon momentum distribution is in poor agreement with this QCD prediction and better theoretical input is needed. Clear evidence for the charmed baryon Λ_c^+ has been seen for the first time in e^+e^- interactions. The Λ_c^+ is seen decaying into at least two decay channels and an absolute branching fraction is obtained for the decay $\Lambda_c^+ \to pK^-\pi^+$. Unfortunately it appears as though we have insufficient data to obtain a clear understanding of the charmed baryon spectroscopy and this task will have to be left to future experiments.

REFERENCES

1. Members of the SLAC-LBL collaboration: G. S. Abrams, M. S. Alam, C. A. Blocker, A. M. Boyarski, M. Breidenbach, C. H. Broll, D. L. Burke, W. C. Carithers, W. Chinowsky, M. W. Coles, S. Cooper, B. Couchman, W. E. Dieterle, J. B. Dillon, J. Dorenbosch, J. M. Dorfan, M. W. Eaton, G. J. Feldman, H. G. Fischer, M. E. B. Franklin, G. Gidal, G. Goldhaber, G. Hanson, K. G. Hayes, T. Himel, D. G. Hitlin, R. J. Hollebeek, W. R. Innes, J. A. Jaros, P. Jenni, A. D. Johnson, J. A. Kadyk, A. J. Lankford, R. R. Larsen, D. Lüke, V. Lüth, J. F. Martin, R. E. Millikan, M. E. Nelson, C. Y. Pang, J. F. Patrick, M. L. Perl, B. Richter, J. J. Russell, D. L. Scharre, R. H. Schindler, R. F. Schwitters, S. R. Shannon, J. L. Siegrist, J. Strait, H. Taureg, V. I. Telnov, M. Tonutti, G. H. Trilling, E. N. Vella, R. A. Vidal, I. Videau, J. M. Weiss, and H. Zaccone.

2. G. S. Abrams _et al_., Phys. Rev. Lett. _43_, 477 (1979).

3. W. Davies-White _et al_., Nucl. Instrum. Methods _160_, 227 (1979).

4. G. S. Abrams _et al_., to be published in IEEE Trans. on Nucl. Sci. _NS-27_, 1 (Feb. 1980); G. S. Abrams _et al_., IEEE Trans. on Nucl. Sci. _NS-25_, 1, 309 (1978).

5. EGS, Electromagnetic Shower Program, R. L. Ford and W. R. Nelson, SLAC-Report 210 (1978).

6. See for instance G. Feldman, Proceedings of the 19th International Conference on High Energy Physics, Tokyo, 1978, p. 777.

7. W. Bacino *et al.*, Phys. Rev. Lett. 42, 749 (1979).

8. The notation $\tau^- \to \rho^- \nu_\tau$, $\tau^- \to \ell^- \nu_\tau \bar{\nu}_e$ and $\tau^- \to \pi^- \nu_\tau$ imply also the charge conjugate reactions.

9. G. S. Abrams *et al.*, Phys. Rev. Lett. 43, 1555 (1979).

10. F. J. Gilman and D. H. Miller, Phys. Rev. D17, 1846 (1978).

11. R. Brandelik *et al.*, Z. Phys. C1, 233 (1979).

12. Y. S. Tsai, Phys. Rev. D4, 2821 (1971).

13. See for instance S. J. Brodsky *et al.*, Phys. Lett. 73B, 203 (1978).

14. This value of α_s is calculated from the ratio of the leptonic to the hadronic width of the ψ [see T. Appelquist and H. D. Politzer, Phys. Rev. Lett. 34, 43 (1975)]. The predicted direct photon contirubtion is included in the $\Gamma_{tot}(\psi)$ used in the calculation.

15. G. S. Abrams *et al.*, submitted to Phys. Rev. Lett. (1979), SLAC-PUB-2406 and LBL-9855 (1979).

16. R. Brandelik *et al.*, Nucl. Phys. B148, 189 (1979).

17. M. Piccolo *et al.*, Phys. Rev. Lett. 39, 1503 (1977).

18. E. G. Cazzoli *et al.*, Phys. Rev. Lett. 34, 882 (1976); A. M. Cnops *et al.*, Phys. Rev. Lett. 42, 197 (1979); C. Baltay *et al.*, Phys. Rev. Lett. 42, 1721 (1979); C. Angelini *et al.*, Phys. Lett. 80B, 428 (1979). Contributions to the EPS Conference on High Energy Physics, Geneva 1979; D. Drijard *et al.*, CERN-EP preprint (1979); K. L. Giboni *et al.*, CERN-EP preprint (1979); P. Schlein *et al.*, UCLA preprint (1979).

19. An upper limit of 0.5 nb (90% C.L.) on the production of charmed baryons in association with a charmed meson, $\bar{D}^o$ or D^-, had been obtained from a handscan of all events with a measured decay $D^o \to K\pi$.

20. This value is somewhat model dependent. The quoted value was estimated from our measurements of $\Delta R(p+\bar{p})$ and $\Delta R(\Lambda+\bar{\Lambda})$ and a simple isospin statistical model.

21. P. A. Rapidis *et al.*, Phys. Rev. Lett. 39, 526 (1977).

22. W. Bacino *et al.*, Phys. Rev. Lett. 40, 671 (1978).

23. Cool and Marshak, Advances in Particle Physics, Vol. 2, p. 193 (1968).

24. Private communication, Walt Bacino, UCLA.

25. E. Eichten *et al.*, Phys. Rev. D17, 3090 (1978); T. Appelquist *et al.*, Ann. Rev. Nucl. Sci. 29, 387 (1978).

26. I. Peruzzi et al., Phys. Rev. Lett. 39, 1301 (1977).

27. D. Fakirov, B. Stech, Nucl. Phys. B133, 315 (1978); N. Cabibbo and L. Maiani, Phys. Lett. 73B, 418 (1978).

28. G. S. Abrams et al., Phys. Rev. Lett. 43, 481 (1979).

29. V. Vuillemin et al., Phys. Rev. Lett. 44, 1149 (1978).

30. C. Quigg and J. L. Rosner, Phys. Rev. D17, 239 (1978).

31. J. Kirkby, 1979 International Symposium on Lepton and Photon Interactions at High Energy, Batavia, August 23-29, 1979; SLAC-PUB-2419.

RECENT RESULTS FROM THE CRYSTAL BALL DETECTOR AT SPEAR

Presented by Charles W. Peck
Representing the Crystal Ball Collaboration*

R. Partridge, C. Peck, and F. Porter**
California Institute of Technology, Pasadena, California 91125

W. Kollmann, M. Richardson, K. Strauch, and K. Wacker
Harvard University, Cambridge, Massachusetts 02138

D. Aschman, T. Burnett,*** M. Cavalli-Sforza, D. Coyne, and H. Sadrozinski
Princeton University, Princeton, New Jersey 08540

R. Hofstadter, I. Kirkbride, H. Kolanoski, A. Liberman, J. O'Reilly, and J. Tompkins
Stanford University (HEPL), Stanford, California 94305

E. Bloom, F. Bulos, R. Chestnut, J. Gaiser, G. Godfrey, C. Kiesling, and M. Oreglia
Stanford Linear Accelerator Center, Stanford, California 94305

ABSTRACT

Evidence for an η_c candidate with a mass of 2980 ± 15 MeV from the ψ' inclusive γ spectrum is presented. The corresponding region in the J/ψ inclusive γ spectrum is then studied and shown to be consistent with a broad Breit-Wigner enhancement superimposed upon a smooth background. Results from a detailed study of the two exclusive decay modes $\psi' \to \gamma\gamma\ \psi \to \gamma\gamma\ e^+e^-$ and $\psi' \to \gamma\gamma\ \psi \to \gamma\gamma\mu^+\mu^-$ are also given. These consist of measurements of the cascade branching ratios $BR(\psi' \to \gamma\chi \to \gamma\gamma\psi)$ for $\chi(3.55)$ and $\chi(3.51)$ and upper limits on this branching ratio for $\chi(3.41)$, $\chi(3.45)$, and $\chi(3.59)$. Finally, we give branching ratios for the two decays $\psi' \to \eta\psi$ and $\psi' \to \pi^\circ\psi$.

---

* Work supported in part by the U.S. Department of Energy under Contract numbers DE-AC-03-79ER0068(Caltech), DE-AC-03-76SF00515 (SLAC), and EY-76-C-02-3064(Harvard), and by the National Science Foundation, grant numbers PHY78-00967(HEPL), PHY78-07343(Princeton), and PHY75-22980(Caltech).

** Chaim Weismann Fellow.

*** Alfred P. Sloan Fellow, on leave from the University of Washington, Seattle, Washington.

ISSN:0094-243X/80/590185-18$1.50

INTRODUCTION

The properties of charmonium below the threshold for the productio of charmed mesons have been a subject of intense experimental and theoretical work in the five years since the J/ψ and ψ' were discovered. Of particular interest has been the question of the mass and width of the η_c and η_c', the $^1S_0$ partners of the two $^3S_1$ charmonium states, the J/ψ (3095) and the ψ' (3684). Theoretical estimates from both potential models(1) and dispersion relations(2) suggest that $\Delta m = m(^3S_1) - m(^1S_0)$ should be in the range -20 to +100 MeV. The only candidate (3,4) for the η_c has been the X(2.83) with Δm = 265 MeV. However, this large mas difference, the product(3) of branching ratios $BR(\psi \to \gamma\chi(2.83)) \cdot BR(\chi(2.83) \to 2\gamma)$, and the limits(8,11) on the branching ratio for $\psi \to \gamma\chi(2.83)$ made it difficult to understand within the conventional charmonium picture. Similarly, the two possible candidates for the η_c', the $\chi(3.45)$(5 and the $\chi(3.59)$(6) have Δm's of 234 MeV and 94 MeV respectively. Thus the second of these is the favored candidate for the η_c'. Confirmatory evidence of these states and measurement of their spins and parities are clearly important experimental problems.

In previously reported preliminary results(7,8,9) from the Crystal Ball detector at SPEAR, we have found no evidence for the $\chi(2.83)$ and we have shown an enhancement(9) at 640 MeV in the inclusive γ energy distribution from the ψ'. We interpret this enhancement as a candidate for η_c with a mass of 2980 ± 15 MeV. In this report, we show that this enhancement is consistent with our instrumental resolution and that it appears uniformly in all parts of our data sample. However, we have not been able to isolate a signal at 2980 MeV in any of the small set of exclusive channels we have investigated so far. We then turn to a study of the region in the inclusive γ distribution from the J/ψ corresponding to a missing mass of 2980 MeV and show that our preliminary data is consistent with a broad enhancement ($\Gamma = 20^{+15}_{-10}$ MeV) at the expected energy of about 110 MeV.

Next, we present our results from a detailed study of $\psi' \to \gamma\chi(3.55 \to \gamma\gamma\psi$, $\psi' \to \gamma\chi(3.51) \to \gamma\gamma\psi$, $\psi' \to \eta\psi$ and $\psi' \to \pi^\circ\psi$ and give branching ratios for them. However, we find only very weak or no evidence for γ cascade decays involving the $\chi(3.59)$, $\chi(3.45)$, or the well established $\chi(3.41)$. For these three cases, we give upper limits on the branching ratio product $BR(\psi' \to \gamma\chi) \cdot BR(\chi \to \gamma\psi)$.

THE APPARATUS

The Crystal Ball detector is a non-magnetic device optimized for the detection of γ's in e^+e^- annihilations. The three major subsystems of the apparatus which were used to obtain the results presented here are (in order outward from the interaction region): three layers of charged particle tracking chambers, the NaI(Tℓ) spherical shell of the ball itself with tunnel cutouts for the beams to enter and exit, and endcaps of magnetostrictive spark chambers and NaI covering most of the solid angle of these tunnels. The central charged particle trackin system consists of two gaps of magnetostrictive readout spark chambers covering 94% of the sphere about the interaction point, two planes of proportional wire chambers with 80% coverage, and finally, two gaps of magnetostrictive spark chambers with 71% coverage. The proportional wire chambers provided us with fast charged multiplicities for use in our trigger and the magnetostrictive chambers were our primary source of charged particle directions in offline analysis. The Crystal Ball itself is two hemispherical, hermetically sealed shells of NaI(Tℓ), each consisting of 336 optically isolated triangular tapering prisms. Each prism (or module) is viewed by a single 2" photomultiplier. The inner diameter of the NaI shell is 20" and the outer diameter, 52". Thus, the shell thickness in physical units is 16 electromagnetic radiation lengths and about 1 hadronic absorption length. The main ball covers 94% of the sphere. Figure 1 shows the geodesic dome geometry of the ball. The endcaps cover the holes in the ball required for the beams and complete the coverage to 98%.

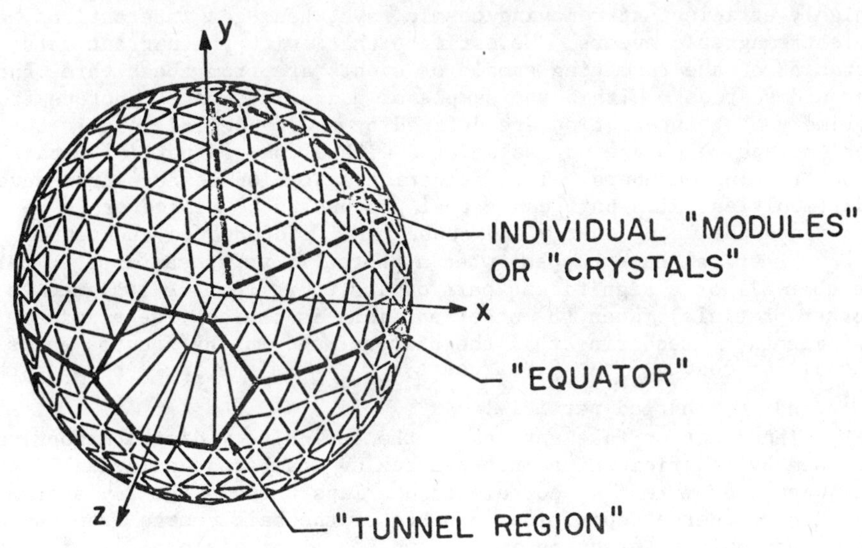

Figure 1 - The icosahedral geometry of the Crystal Ball. The full spherical shell is segmented into 720 triangular prisms. To provide beam entrance and exit tunnels, 48 of these are removed.

Although the ball and its endcaps cover all but 2% of the sphere about the interaction point with a highly sensitive photon and charged particle detector, it is uniform in its energy and angular resolutions only in a region which avoids the edge effects of the tunnels and their associated endcaps. Accordingly, in all the data presented here, the requirement is imposed that $|\cos\theta| < 0.9$, where θ is the angle between the beam and a particle whose energy and direction are important in the event class being considered (for the inclusive γ spectra, the more restrictive requirement $|\cos\theta|<0.85$ was used). It is important to notice, however, that the 98% coverage is crucial in event selection. In the part of the ball with uniform response, $|\cos\theta|<0.9$, the photon angular resolution has a standard deviation of 2° at 200 MeV and the FWHM energy resolution for electrons and photons is about $6.6\%/\sqrt[4]{E}$, with E in GeV; this includes uncompensated time dependent drifts, intercalibration errors between the modules, and the intrinsic resolution of NaI. The angular resolution for charged particles is 5mr. For further discussion of the apparatus and its triggering, see ref. 7.

THE INCLUSIVE γ SPECTRA

Figures 2, 4, and 5 show the inclusive γ energy distributions from the $\psi'(3684)$, $\psi''(3.77)$, and $J/\psi(3095)$ respectively. This section discusses the data selection used to obtain these spectra. We first define hadronic events by a series of cuts which can be shown to be highly efficient at removing cosmic rays, beam-gas interactions, and electromagnetic events. We estimate that, with the current cuts, less than 5% of the resulting sample of events are from these three background sources. Within the sample of hadronic events, photons from the primary e^+e^- interaction are defined by energy depositions in the ball or endcaps which are not associated with a charged particle track in the tracking chambers. This selection criterion suffers from several difficulties which both cause real photons to be missed or to be assigned the wrong energy and cause fake "photons" to be created.

The first case happens when a photon is very near another particle; either all or a significant part of its energy can be assigned to the other particle. When the other particle is charged, we eliminate these accidents by requiring that the photons used in the spectra satisfy the condition $\cos\theta_{\gamma-ch} < 0.85$, where $\theta_{\gamma-ch}$ is the angle between the photon and any charged particle.

Incorrect or fake entries in the inclusive γ distributions can be caused by inefficiencies in the tracking chambers, accidental overlaps between 2 or more γ's, correlated overlaps between the 2 γ's from a π° decay, energy deposition in parts of the ball remote from the site of a hadronic interaction in it, and K°, n, or $\bar{n}$'s produced in the e^+e^- annihilation. Inefficiency in the tracking system is known to be at the few percent level. It gives rise to a distortion in the inclusive γ spectra by creating a spurious enhancement centered at about 210 MeV, which is the minimum energy loss of a non-interacting charged particle penetrating the full shell.

Accidental overlaps between pairs of photons are included in the spectra as single photons and so create some distortion. The correlated overlaps of the two γ's from a π° decay can occur for $E_{\pi^\circ} \gtrsim 800$ MeV, and, for this reason, we believe that significant spectrum distortion may occur above this energy. Thus, pending improvements in our π° detection algorithms, we cut off the inclusive γ spectra at 1 GeV.

By a variety of mechanisms (n, γ, $\pi^\pm$, etc.), energy can be transported sufficiently far from a charged hadron which strongly interacts in the ball so that our energy assignment algorithms cannot associate it with the charged particle. This energy then becomes a fake photon. We believe that this distorts the spectrum most seriously for $E_\gamma < \sim 50$ MeV, and so we cut off inclusive spectra at this energy. Finally, the inclusive cross sections for n, $\bar{n}$, and K°_L are small and hadronic interactions of these particles are not expected to produce any significant spectrum distortion.

In order to enhance our sensitivity for observing monochromatic photons due to a radiative transition to a new state, we have paired as many photons as possible in each event into π°'s and then removed them from the inclusive spectra. For this purpose, a looser definition of photons than that discussed above was used. In particular, the $|\cos\theta|<0.85$ was not imposed on photons used in pairings. It is these resulting π°-subtracted spectra that are shown in Figs. 2-6. The spectra of the removed photons show no "holes" which would lead to spurious enhancements in the inclusive spectra. We think that a large part of the final smooth spectrum is due to γ's from π°'s which our pairing algorithm was unable to detect for any of a variety of reasons (overlaps, edge effects near tunnels, incorrect pairings, etc.).

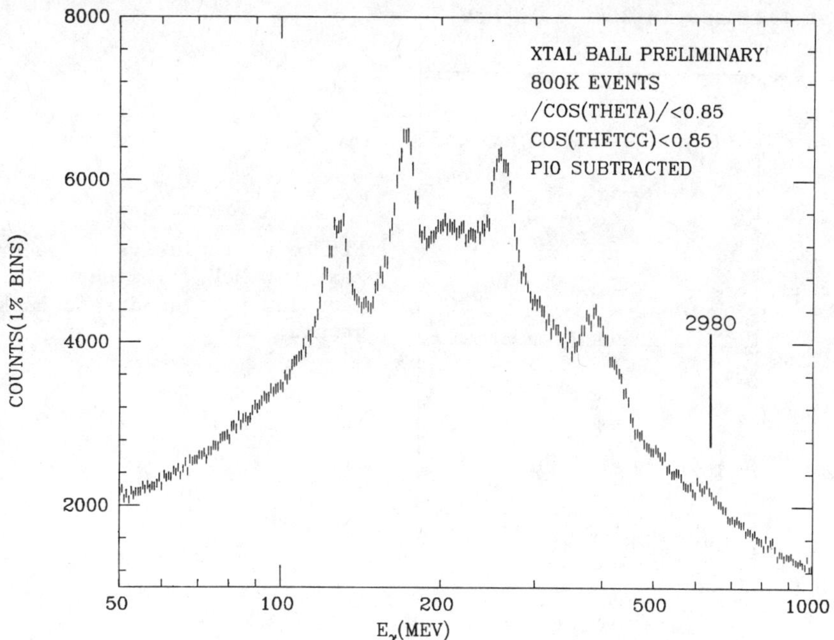

Figure 2 - Inclusive γ spectrum from $\psi'(3684)$ - 1% bins.

In the final $\psi'(3684)$ spectrum, Fig. 2, the three strong peaks at 129 MeV, 173 MeV, and 265 MeV are due to the monochromatic photons from $\psi' \to \gamma\chi$ for the three well-known states $\chi(3.55)$, $\chi(3.51)$, and $\chi(3.41)$ respectively. The broader peak centered at 390 MeV with a clear shoulder extending up to about 450 MeV is due to the Doppler broadened transitions $\chi(3.51) \to \gamma\psi$ and $\chi(3.55) \to \gamma\psi$ respectively. Finally, an enchancement at 640 MeV with a width comparable to our resolution is readily apparent and we interpret it as evidence for a new state with mass 2980 ± 15 MeV. In the $J/\psi(3095)$ distribution, Fig. 5, there is no evidence[8] for any enhancement below 1000 MeV with a width equal to our rcoolution.

$\psi' \to \gamma U(2.98)$

The portion of the ψ' inclusive γ spectrum near 640 MeV is shown in Fig. 3. This data has been fit to a quadratic background and a gaussian signal in a variety of ways. What is shown in Fig. 3 (and Fig. 6) is the result of a global fit to the ψ' and J/ψ data together and it will be discussed in more detail later. The gaussian parameters for the fit in Fig. 3 correspond to a missing mass of 2984 MeV with a statistical error of 2 MeV, a FWHM of 8.9%, and an area of 1600 ± 250 counts. Our expected resolution at this energy has a FWHM of 7.3% and a gaussian with this width also gives an acceptible fit. Thus, we conclude that the enhancement is quantitatively consistent with our expected resolution. At present we assign a systematic uncertainty of ±2% to our absolute energy calibration and conclude that the best fit mass is 2980 ± 15 MeV.

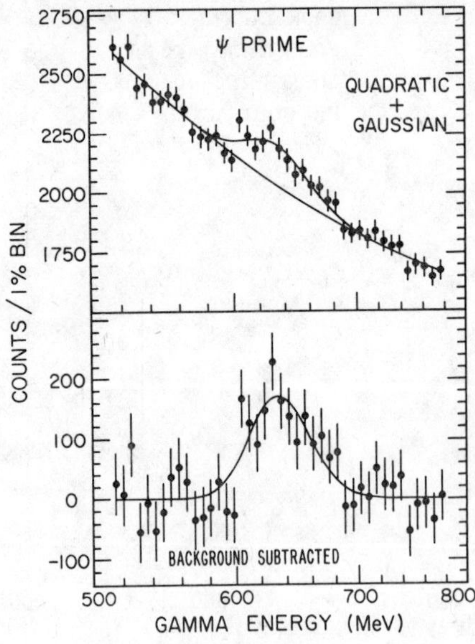

Figure 3

Fit to ψ' inclusive γ spectrum near 640 MeV. The curve is a gaussian plus quadratic background.

In order to show that this effect is uniformly persistent in all parts of our data sample, we have segmented the data in various ways. The results of independent fits to each of these portions of the data are shown in Table I. In each of these fits, the mean and area of the gaussian were left free but the width was held fixed at 8.9% FWHM. In all cases the observed number of counts in a portion of the data is statistically consistent with what would be expected from the size of the sample. Similarly, the same mean, to within the statistical uncertainty, is found in the three partitionings of the ball, but there is a systematic difference of 19 MeV between the means of the early and late data samples. We attribute this to uncompensated drifts in our overall system gain in these preliminary data; it is this systematic difference which leads us to assign a systematic uncertainty of ± 15 MeV to the mass of the U(2.98) at this time.

TABLE I

Results of independent fits in the 640 MeV region to different portions of the full ψ' inclusive γ spectrum. In each case "observed counts" is the area of the best fit gaussian for the portion of the sample being considered and "expected counts" is the gaussian area of the full sample multiplied by the relative size of that portion.

Portion of Data Sample	Observed Counts / Expected Counts	Mean of Gaussian (errors are statistical only) (MeV)
Full Data Sample	1.00	$633.8 \pm 1.7^{*}$
Top Half of Ball	1.15 ± 0.20	634 ± 3
Bottom Half of Ball	0.85 ± 0.18	638 ± 4
Left Half	0.84 ± 0.20	633 ± 3
Right Half	1.22 ± 0.14	638 ± 3
Forward Half	1.26 ± 0.18	633 ± 3
Backward Half	0.76 ± 0.20	640 ± 4
Early 1/3 of Data	1.22 ± 0.22	625 ± 3
Late 2/3 of Data	0.95 ± 0.10	644 ± 3

* This results from a global fit with a common missing mass in both the J/ψ and ψ' inclusive γ spectra.

In Table II, we show the results of cutting the ball into four θ ranges, where θ is the angle between a photon and the incident beam. This is compared with the angular distribution expected for a psuedo-scalar particle and is consistent with it. However, the data are consistent with many other spin-parity assignments and so this certainly does not constitute a determination of the spin-parity of the U(2.98). However, it does demonstrate that the effect is seen over the entire geometry of the detector.

TABLE II

Results of independent fits in the 640 MeV region of the ψ' inclusive γ spectrum for photons in different angular ranges. The θ is that between the photon and the incident beam. The mean and width of the fitting gaussian were held fixed and only its area varied in the fitting to give "observed counts". "Expected counts" is calculated from the gaussian area of the full sample and an assumed angular distribution of $(1 + \cos^2\theta)$.

Angular Range	Observed Counts/Expected Counts
$0 < \lvert\cos\theta\rvert < 0.21$	1.01 ± 0.38
$0.21 < \lvert\cos\theta\rvert < 0.43$	1.02 ± 0.37
$0.43 < \lvert\cos\theta\rvert < 0.64$	1.06 ± 0.33
$0.64 < \lvert\cos\theta\rvert < 0.85$	0.96 ± 0.27

To demonstrate that our apparatus does in fact have the expected sensitivity and resolution in the 600 MeV energy region, one would like to have a monochromatic line near this energy for direct calibration purposes; radiative corrections provide us with such a calibration line. At any energy $E_{cm} > m_\psi$ in e^+e^- annihilations, the radiative tail of the J/ψ will provide a source of monochromatic photons with energy $(E_{cm}^2 - m_\psi^2)/(2\,E_{cm})$. These are due to radiation from the incident electron or positron followed by formation of a J/ψ. The angular distribution and strength of the resulting monochromatic line is calculable from QED. Similarly, for $E_{cm} > m_{\psi'}$, a monochromatic line associated with the ψ' radiative tail will be produced. In general these lines are weak and can only be seen in a large exposure with relatively little π° background in the inclusive γ distribution. Thus at $E_{cm} = m_{\psi'}$, there are so many hadronic events that the small expected line at 542 MeV from J/ψ radiative corrections is not visible over the π° background. However, we have a 1770 event/nb exposure at the $\psi''(3.77)$ and the inclusive γ distribution from it is shown in Fig. 4. Using our estimated detection efficiencies for the cuts defining this distribution, we expect about 100 counts centered at

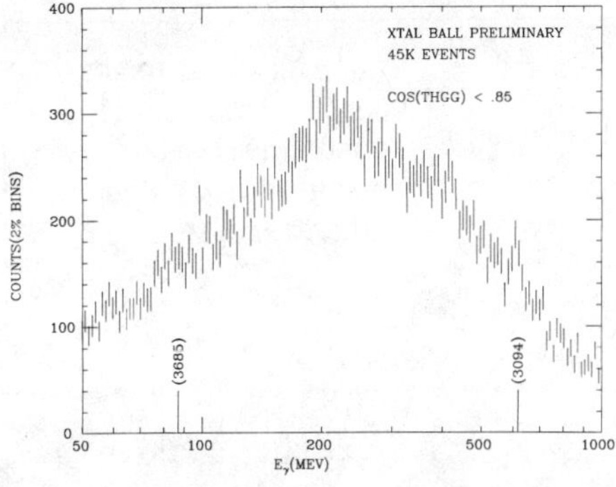

Figure 4

Inclusive γ spectrum at $\psi(3.77)$ - 2% bins.

617 MeV from J/ψ(3095) radiative corrections and about 70 counts centered at 86 MeV from the ψ'(3684). The enhancement near 620 MeV is consistent in energy, width, size, and angular distribution with the QED expectations and our experimental energy resolution. This result gives us assurance that our assumptions about the response of the ball to photons near 600 MeV in the ψ' data analysis were valid. The expected signal at 86 MeV is near the limit of our sensitivity and has not been intensively studied as yet. It should be noted that the 3770 MeV data, being in the continuum, are not nearly so rich in hadronic events as those at the ψ'. There are enough differences in the data analysis at these two energies so that the detection efficiency for photons near 600 MeV is not the same in the two cases. Thus, without further analysis, we cannot use our quantitative results from the ψ''(3.77) in analyzing the ψ' data.

From the fitted area of the gaussian in the ψ' inclusive γ distribution, the number of ψ''s in our sample, and an estimated detection efficiency of 55%, we obtain a preliminary value of the branching ratio for $\psi' \to \gamma U(2.98)$ of about 0.4%. This can be compared with recent theoretical estimates[10] of the branching ratio for the hindered M1 transition, $\psi' \to \gamma\eta_c(2980)$, of about 0.4%.

Preliminary work searching for exclusive hadronic decay modes of U(2.98) has begun. So far, we have studied only decay modes involving more than about 1 GeV of neutral energy to avoid biases due to our trigger. The mass spectrum for X in the reaction $\psi' \to \gamma X$, where X is $p\bar{p}\ \pi^\circ\pi^\circ$, $K^+K^-\pi^\circ\pi^\circ$, $\pi^+\pi^-\pi^\circ\pi^\circ$, $p\bar{p}\ \pi^\circ$, $K^+K^-\pi^\circ$, $K^+K^-\eta$, or $\pi^+\pi^-\eta$, does not show any convincing structure at 2980 MeV; nine events from these seven channels are observed in $\pm 1\sigma$ of 2980 MeV. From our estimated detection efficiencies for these channels, the size of the signal in the ψ' inclusive γ spectrum, and the assumption of no background, we can only conclude that the upper limit on the branching ratio into each of these channels is a few percent.

$J/\psi \to \gamma U(2.98)$

We consider next the inclusive γ distribution from the J/ψ(3095) shown in Fig. 5. As mentioned already, there is no evidence[8] for any resolution-limited enhancement above about 0.7% for $75 < E_\gamma < 1000$ MeV in this distribution. This does not rule out the possibility of broad

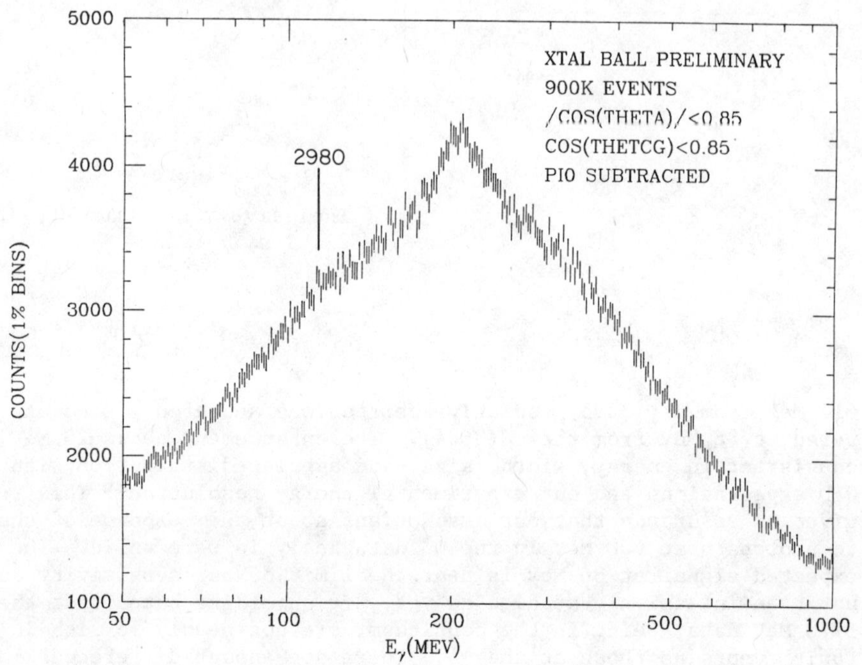

Figure 5 - Inclusive γ spectrum at $J/\psi(3095)$ - 1% bins.

resonances, however. Since we now have evidence for a state at 2980 MeV, we have investigated the region around 110 MeV in the ψ inclusive γ spectrum. Figures 6 and 3 show the result of a 9 parameter joint best fit to the ψ' data near 640 MeV and the ψ data near 110 MeV. The nine parameters are: 3 for a background quadratic for the ψ', a similar 3 for the ψ, the area of a gaussian signal in the ψ', the area of a Breit-Wigner signal in the J/ψ, and the mass of the assumed resonance. For the fit shown in Fig. 6, the natural line width, Γ, of the Breit-Wigner shape was fixed at 30 MeV. This joint fit has a χ^2 of 52 for 66 degrees of freedom. From this we conclude that the ψ data is consistent with a broad resonance, the U(2.98).

In order to extract the best information possible regarding Γ, the natural width of the U(2.98), we have folded a Breit-Wigner line shape with a gaussian of width reflecting our instrumental resolution; using this convolution as the shape of the signal, we have calculated the χ^2 of the resulting global fit as a function of Γ. The result is shown in Fig. 7. From this we conclude that the best value of $\Gamma(2.98)$ is 20^{+15}_{-10} MeV and the size of the U(2.98) signal in the J/ψ inclusive γ spectrum is 3500^{+4500}_{-1500} counts.

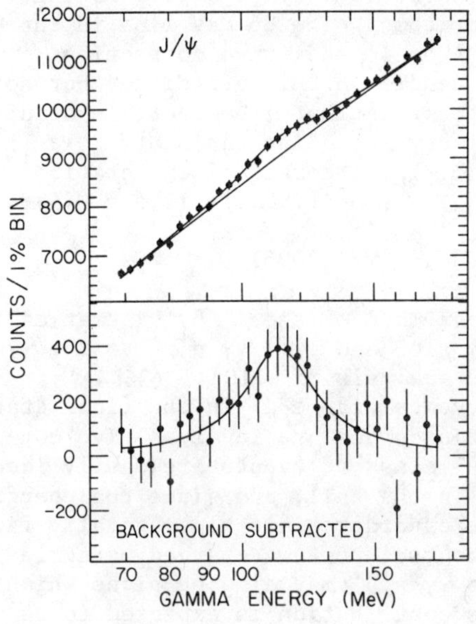

Figure 6

Fit to $\psi(3095)$ inclusive γ spectrum near 110 MeV. The curve is a Breit-Wigner shape with Γ = 30 MeV plus a quadratic background.

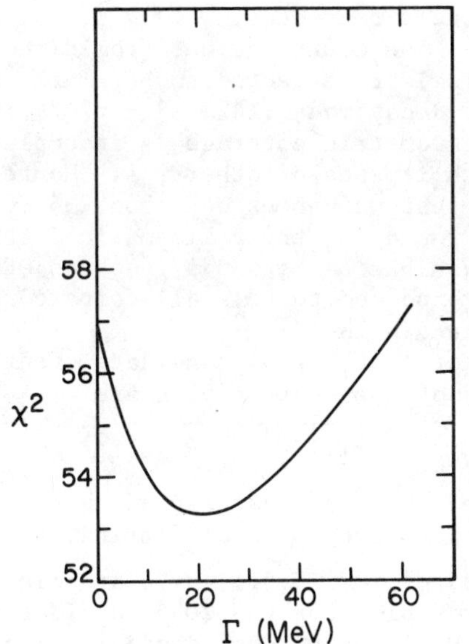

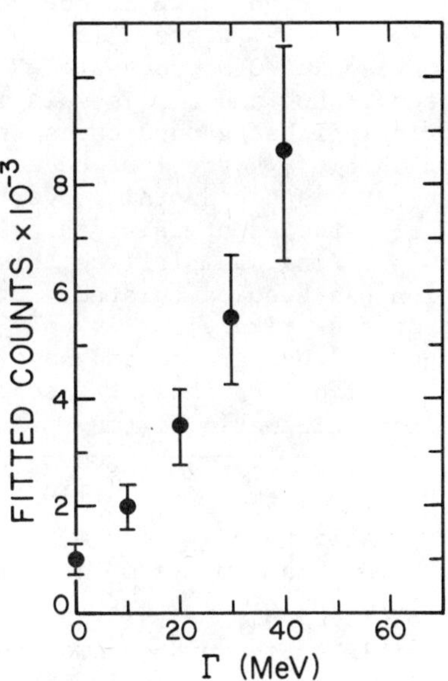

Figure 7 - (a) Goodness of the fit to the $J/\psi(3095)$ and $\psi'(3684)$ inclusive γ spectra near 110 MeV and 640 MeV respectively as a function of the width of the assumed Breit-Wigner resonance. (b) The fitted number of counts in $J/\psi \to \gamma U(2.98)$ as a function of the width of the assumed Breit-Wigner shape.

At this time, we are very uncertain about our detection efficiency in the 100 MeV region because it is very sensitive to the details of the π° subtraction algorithms. The 86 MeV line in the $\psi''(3.77)$ inclusive γ distribution from ψ' radiative correction photons will eventually give us a handle on this efficiency but not at the present time. Our current best estimate, however, is about 30% and from this we get a preliminary branching ratio of $J/\psi(3095) \rightarrow \gamma U(2.98)$ of about 1.5%. The theoretical expectation[(9)] for the branching ratio of $J/\psi(3095) \rightarrow \gamma\ \eta_c(2.98)$ is $(2.5 \pm 0.4)\%$.

THE DECAYS $\psi'(3684) \rightarrow \gamma\gamma\psi(3095)$

We have studied the above decay mode of the ψ' by kinematically fitting candidates for the $\psi' \rightarrow \gamma\gamma\ e^+e^-$ and $\psi' \rightarrow \gamma\gamma\ \mu^+\mu^-$ to the $\gamma\gamma\psi$ hypothesis. We have completed the analysis of 8.10^5 $\psi'(3684)$'s, our full data sample. Candidate events with e^+e^- in the final state are fully measured and so the fitting procedure involves five constraints. The muons in the $\gamma\gamma\ \mu^+\mu^-$ class of events are easily identified by a minimum ionizing signal in the ball, and since the energies of the muons are unmeasured, these candidates are kinematically fit with only three constraints. Of course, the two charged particles in the class of events we identify as $\gamma\gamma\ \mu^+\mu^-$ may also be pions which do not interact in the ball, but this contribution is expected to be very small.

The signals which one can expect in this decay are $\psi' \rightarrow \gamma\chi \rightarrow \gamma\gamma\psi$, $\psi' \rightarrow X\psi \rightarrow \gamma\gamma\psi$ where X is η or π°, the direct decay $\psi' \rightarrow \gamma\gamma\psi$, and the nonresonant QED process $e^+e^- \rightarrow \gamma\gamma\psi$. The contributions from direct decay, QED, and π° (isospin violating) are expected to be small. The principal background comes from the decay mode (16%) $\psi' \rightarrow \pi^\circ\pi^\circ\psi$ in which two γ's are lost because the geometric coverage is incomplete and/or γ's accidentally overlap with leptons or other γ's. Monte Carlo studies of this process show that we can expect about 35 events in our final sample from this background and their kinematic distribution has been calculated. Other possible backgrounds, such as QED processes like $e^+e^- \rightarrow \gamma\gamma\ e^+e^-$, are expected to be small compared to the $\pi^\circ\pi^\circ$ background and have not been included.

Figure 8a shows the scatter plot of the 2783 candidate events before kinematic fitting. The axes of the scatter plot are $m_{high} = \sqrt{m^2_{\psi'} - 2m_{\psi'} E_{\gamma min}}$ and $m_{low} = \sqrt{m^2_{\psi'} - 2m_{\psi'} E_{\gamma max}}$. The kinematic boundary for $\psi' \rightarrow \gamma\gamma\psi$ and the lines corresponding to $m_{\gamma\gamma} = m_{\pi^\circ}$ and $m_{\gamma\gamma} = m_\eta$ are also shown. Even without fitting, the two cascades $\psi' \rightarrow \gamma\chi(3.55) \rightarrow \gamma\gamma\psi$ and $\psi' \rightarrow \gamma\chi(3.51) \rightarrow \gamma\gamma\psi$ are clearly visible. Figure 8b shows the scatter plot for the 2056 candidates which survive kinematic fitting by having a χ^2 confidence level of greater than 0.5%. Above this value, the confidence level distribution is satisfyingly flat for both the e^+e^- and $\mu^+\mu^-$ events separately. A clear η band appears after fitting. Furthermore there is also an accumulation of points along the π° line. In fact, for $m_{high} > 3.58$ GeV, there are no events except those on the π° line; for $m_{high} < 3.48$ GeV, there are a few events between the π° and η bands.

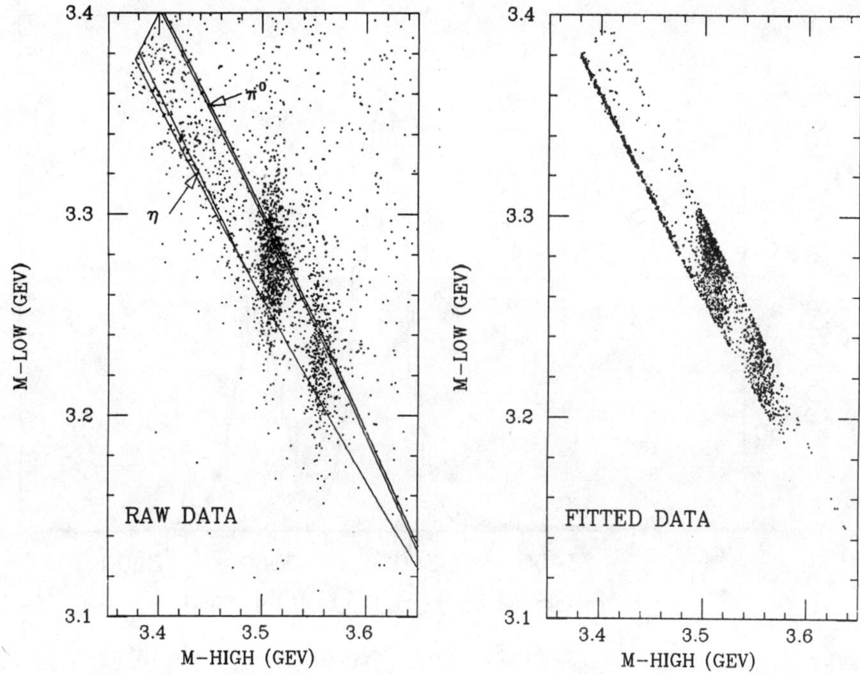

Figure 8 - Scatterplots of events of the two classes $\psi' \to \gamma\gamma e^+e^-$ and $\psi' \to \gamma\gamma\mu^+\mu^-$ before and after kinematic fitting to the hypothesis $\psi' \to \gamma\gamma\psi$. The kinematic boundary and lines of definite $m_{\gamma\gamma}$ are shown in Fig. 8(a).

This distribution of nonresonant events is consistent with the Monte Carlo calculations for the $2\pi°$ background and at present we attribute them all to this background. No attempt has been made to extract a signal for $\psi' \to \gamma\gamma\psi$ direct decays.

Figure 9 shows the projection on the m_{high} axis of the scatter plot of Fig. 8b, but with both the η ($m_{\gamma\gamma}$ > 525 MeV) and $\pi°$ ($110 < m_{\gamma\gamma} < 160$ MeV) removed. The η cut removed 411 events and the $\pi°$ cut, another 130. The curve through the data consists of three parts: two resonant contributions from the $\chi(3.55)$ and $\chi(3.51)$ whose amplitudes and means have been fitted and whose widths are the instrumental resolution, and the contribution from $\psi' \to \pi°\pi°\psi$ as a priori calculated. None of the parameters of the $\pi°\pi°$ were fitted; both its size and shape were obtained from the Monte Carlo calculation. There is clearly no evidence for the $\chi(3.59)$ or $\chi(3.45)$ and that for the $\chi(3.41)$ is weak. Until we have studied our backgrounds further, we choose to give only an upper limit for $\psi' \to \gamma\chi(3.41) \to \gamma\gamma\psi$. Tables III and IV give the results for these five cascade states.

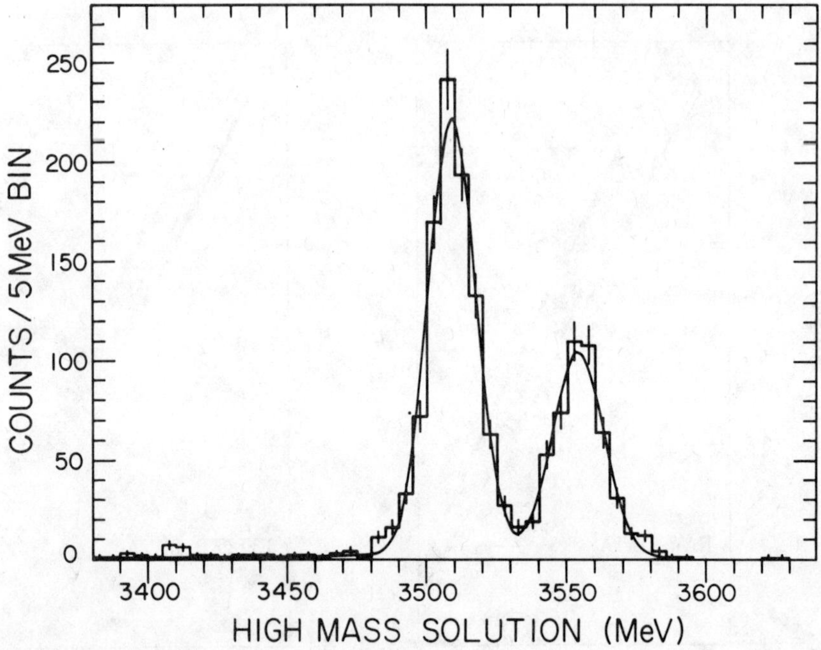

Figure 9 - Projection of fitted $\psi' \rightarrow \gamma\gamma\psi$ events on the high $(\gamma\psi)$ mass axis with the η and π° events removed.

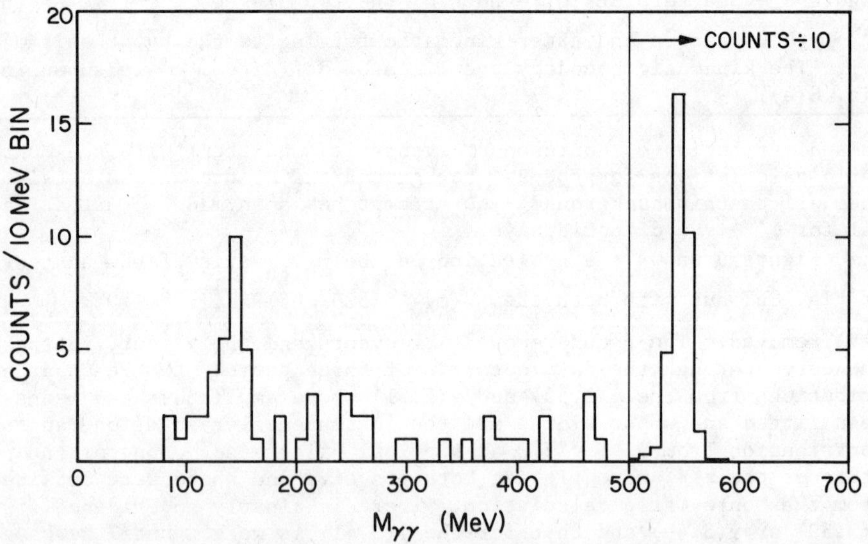

Figure 10 - Projection of fitted $\psi' \rightarrow \gamma\gamma\psi$ events on the $m_{\gamma\gamma}$ axis with the $\chi(3.55)$ and $\chi(3.51)$ cut out.

Finally, in Fig. 10 we show the projection of the scatter plot of Fig. 8b onto the $m_{\gamma\gamma}$ axis with the χ lines removed (the ranges 3480 < m_{high} < 3580 MeV and 3406 < m_{high} < 3416 MeV have been cut out). The η and the π° stand out clearly and from this distribution we obtain the branching ratios given in Table 5. No corrections to these branching ratios have been made for the direct QED processes $e^+e^- \to \eta\psi$ or $e^+e^- \to \pi^\circ\psi$, but from preliminary studies of these processes at the nearby c.m. energy of 3770 MeV, we believe that these corrections are well within our systematic errors. Finally, we have investigated the angular distribution of the η's in $\psi' \to \eta\psi$ and find that it is consistent with $(1 + \cos^2\theta)$ as expected for a pseudoscalar.

TABLE III

Results for cascade decays $\psi' \to \gamma\chi \to \gamma\gamma\psi$ for the two states $\chi(3.55)$ and $\chi(3.51)$. For both the fitted values of the masses and the branching ratios, the first error given is purely statistical and the second, systematic. For the masses, the systematic error is due to a 2% uncertainty in the absolute calibration and for the branching ratios, the dominant contribution to the systematic error is from uncertainty in the number of $\psi'(3684)$'s.

State	Observed Events	Fitted Mass (MeV)			$BR(\psi' \to \gamma\chi)\cdot BR(\chi \to \gamma\psi)$ (%)		
		Value	Stat.	Syst.	Value	Stat.	Syst.
$\chi(3.55)$	528	3553.9	± 0.4	± 2.7	1.13	± 0.05	± 0.19
$\chi(3.51)$	996	3509.2	± 0.3	± 3.6	1.95	± 0.06	± 0.33

TABLE IV

Results for cascade decays $\psi' \to \gamma\chi \to \gamma\gamma\psi$ for the three states $\chi(3.41)$, $\chi(3.45)$, and $\chi(3.59)$. In all three cases, the 90% confidence level upper limit on the number of observed events is calculated from a Poisson likelihood function with a calculated background due only to the reaction $\psi' \to \pi^\circ\pi^\circ\psi$.

State	Events (90% C.L. upper limit)	$BR(\psi' \to \gamma\chi)\cdot BR(\chi \to \gamma\psi)$ (90% C.L. upper limit)
$\chi(3.41)$	17.8	0.050%
$\chi(3.45)$	3.5	0.008%
$\chi(3.59)$	2.3	0.005%

TABLE V

Results for the two decays $\psi' \to \eta\psi$ and $\psi' \to \pi^\circ\psi$. The first error given is purely statistical and the second, systematic. The principal source of systematic error is due to uncertainty in the number of $\psi'(3684)$'s.

State	Observed Events	BR (%) Value	Stat.	Syst.
η	416	2.06	±0.10	±0.31
π°	22	0.08	±0.02	±0.02

DISCUSSIONS AND CONCLUSIONS

From previously reported preliminary results[7,8] we have seen no sign of the X(2.83) in the Crystal Ball, but we have found evidence[9] for an η_c candidate at 2980 ± 15. We have presented here further evidence in the $\psi'(3684)$ inclusive γ spectrum for this state by showing that it is a persistent feature in our data. Furthermore, we have shown that our inclusive γ distribution at the J/ψ is consistent with an enhancement at 2980 MeV with $\Gamma = 20^{+15}_{-10}$ MeV, but since we do not yet have an a priori understanding of the general shape of this inclusive γ distribution we cannot say that our ψ data demand the U(2.98). We have as yet found no evidence for this state in any exclusive channel. Finally, our experimentally determined branching ratios for $\psi' \to \gamma U(2.98)$ and $\psi \to \gamma U(2.98)$ are but poorly known at this time, but they seem fairly consistent with recent theoretical expectations[10].

From the exclusive decay mode $\psi' \to \gamma\gamma\psi$, we have obtained values for the cascade branching ratios $\psi' \to \gamma\chi \to \gamma\gamma\psi$ for the $\chi(3.55)$ and the $\chi(3.51)$ which are consistent with previously measured values[5,6,11,12,13]. Our upper limit on the branching ratio for the $\chi(3.45)$ is much lower than the smallest previously reported[13] limit of 0.12% by MARK II. For the $\chi(3.59)$, our limit is again much lower than the previously measured[6] value of (0.18 ± 0.06)%. And finally, our limit for the $\chi(3.41)$ is consistent with the smallest previously reported[13] value of (0.08 ± 0.08)% by MARK II. If we use the tentative J^P assignments of 2^+ for the $\chi(3.55)$ and 0^+ for the $\chi(3.41)$, we can, following Jackson[14], use our preliminary data[8] on $\psi' \to \gamma\chi_J$, J = 0,2, and the cascade branching ratios reported here to set a limit on the ratio $\Gamma(\chi_2 \to$ hadrons$)/\Gamma(\chi_0 \to$ hadrons$)$. We obtain an upper limit (90% confidence level) of 0.11. This upper limit is about 2.5 times smaller than the QCD expectation[14,15] of 4/15 for this ratio.

Our measurement of the branching ratio for $\psi' \to \eta\psi$ is about a factor of two smaller than the world average of $(3.7 \pm 0.2)\%$ from previous experiments[6,12,16]. Finally, our value for the branching ratio of the isospin forbidden decay $\psi' \to \pi^{\circ}\psi$ is consistent with the smallest previously reported[6] upper limit of 0.1% and is within the range estimated theoretically by various authors[17].

ACKNOWLEDGMENTS

We wish to acknowledge the important contributions of B. Beron and E. B. Hughes of HEPL to the construction of the Crystal Ball, and of the SPEAR and SLAC operations staff for making the accelerators perform as reliable and powerful tools for physics research.

REFERENCES

1. T. Appelquist and H. D. Politzer, Phys. Rev. Lett. 34, 43 (1975); T. Appelquist, A. De Rújula, H. D. Politzer, S. L. Glashow, Phys. Rev. Lett. 34, 365 (1975); H. J. Schnitzer, Phys. Rev. D13, 74 (1976); R. Barbieri, R. Kögerler, Z. Kunszt, R. Gatto, Nucl. Phys. B105, 125 (1976); H. J. Lipkin, H. R. Rubinstein, N. Isgur, Phys. Lett. 78B, 295 (1978); P. Minkowski, Phys. Lett. 85B, 231 (1979); E. Eichten, F. L. Feinberg, Phys. Rev. Lett. 43, 1205 (1979).
2. V. A. Novikov, L. B. Okun, M. A. Shifman, A. I. Vainshtein, M. B. Voloshin, V. I. Zakharov, Phys. Lett. 67B, 409 (1977); M. Shifman, A. Vainshtein, V. Zakharov, Nucl. Phys. B147, 450 (1979).
3. W. Braunschweig et al., Phys. Lett. 67B, 243 (1977); W. Bartel et al., Proc. of 1977 Int'l. Symposium on Lepton and Photon Interactions at High Energies, p. 117.
4. W. D. Apel et al., Phys. Lett. 72B, 500 (1978).
5. J. S. Whitaker et al., Phys. Rev. Lett. 37, 1596 (1976); W. Tanenbaum, et al., Phys. Rev. D17, 1731 (1978).
6. W. Bartel et al., Phys. Lett. 79B, 492 (1978).
7. E. D. Bloom, XIVth Recontre de Moriond, Les Arcs, France; March 11-23, 1979; Proceedings edited by Tran Thanh Van.
8. C. Kiesling, Proc. of the 1979 EPS Int'l. Conf. on High Energy Physics, Geneva, Switzerland, June 27-July 4, 1979 (to be published).
9. E. Bloom, Proc. of the 1979 International Symposium on Lepton and Photon Interactions at High Energies, Aug. 23-29, 1979, Batavia, Ill. (to be published).
10. E. Eichten, K. Gottfried, T. Kinoshita, K. D. Lane, T. M. Yan, CLNS-425 (June 1979) (to be published in Phys. Rev. D).
11. C. Biddick et al., Phys. Rev. Lett. 38, 1324 (1977).
12. R. Brandelik et al., Z. Phys., C1, 233 (1979).
13. J. M. Weiss et al., Proc. of the 1979 EPS Int'l. Conf. on High Energy Physics, Geneva, Switzerland, June 27-July 4, 1979 (to be published).
14. J. D. Jackson, CERN Report TH.2730-CERN; J. D. Jackson, XIIth Recontre de Moriond, Vol. 3, 75 (1977); M. S. Chanowitz and F. J. Gilman, Phys. Lett. 63B, 178 (1976).

15. R. Barbieri, R. Gatto and E. Remiddi, Phys. Lett. 61B, 465 (1976).
16. W. Tanenbaum et al., Phys. Rev. Lett. 36, 402 (1976).
17. G. Segre and J. Weyers, Phys. Lett. 62B, 91 (1976); N. Deshpande and E. Ma, Phys. Lett. 69B, 343 (1977); H. Genz, Lett. Nuovo Cimento 21, 270 (1978); R. Bhandari and L. Wolfenstein, Phys. Rev. D17, 1852 (1978).

REPORT ON THE STATUS OF THE CORNELL ELECTRON STORAGE RING

R. H. Siemann
Cornell University, Ithaca, N. Y. 14853

ABSTRACT

The status of the CESR project at the time of the Montreal meeting is discussed.

The last few months have been a time of tremendous progress at CESR (Cornell Electron Storage Ring); this progress has been shared by the storage ring itself and the two associated high energy physics experiments. This report is a summary of the status as of the Montreal APS meeting.

The principal CESR design parameters are given in Table I[1]. The design luminosity scales as E^2, and therefore in the B quark threshold and T regions it is about 4×10^{31} $cm^{-2}sec^{-1}$. We have concentrated on bringing the ring into operation in these interesting energy regions.

There are two interaction regions (Fig. 1); both are available for high energy physics experiments. In the south interaction

Table I CESR Design Parameters

Luminosity at 8 GeV	10^{32} cm^{-2} sec^{-1}
Number of Interaction Regions	2
Quadrupole Spacing at IR's	7 m
Number of Electrons per Bunch at 8 GeV	1.5×10^{12}
Circumference	768.4 m
Maximum Bending Radius	140.63 m
Minimum Bending Radius	31.65 m
Bending Radius in Normal Cells	87.89 m
RF Power at 8 GeV	2 MW
RF Frequency	499.7615 MHz

Table II CESR Performance as of October 25, 1979

Luminosity	2×10^{30} cm^{-2} sec^{-1}
Operating Energy Range	4.7 - 5.5 GeV
e^- current	13 mA
e^+ current	17 mA (61 bunches)
	5 mA (1 bunch)
Beam Lifetimes	3 hours

ISSN:0094-243X/80/590203-7$1.50

region there is a large, general purpose, magnetic detector, CLEO, and the experiment in the north region has been designed to study the photon spectroscopy from Υ' and Υ" decays.

The CESR vacuum chamber was closed on March 31, 1979, and operations began the following day[2]. Rapid progress was made through the various stages of e^- storage (April 13), e^+ storage (May 28), e^+ coalescing (June 18), and the first luminosity (August 18). The e^+ coalescing step is unique to CESR operation; in this process the positrons which are initially distributed among 61 r.f. buckets are combined into a single r.f. bucket. Our performance to date is summarized in Table II. The difference between the 61 and 1 bunch e^+ currents is due to losses in the coalescing process. These losses are worse than the factor of three suggested by the table, as the peak 61 bunch and 1 bunch currents were obtained using different injection procedures. Our coalescing efficiency is about a factor of five below the design

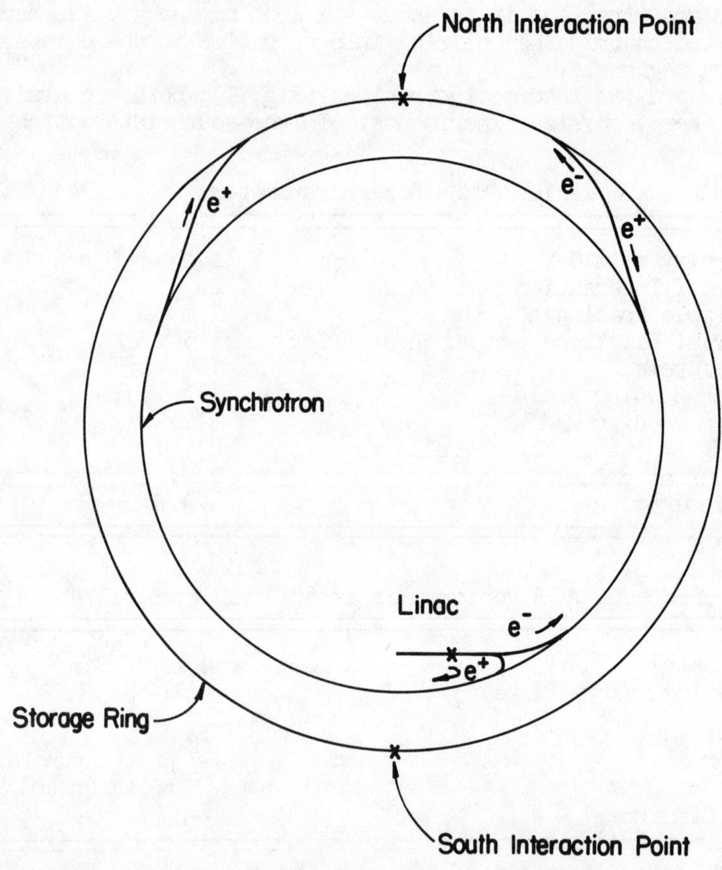

Fig. 1. Schematic drawing of the CESR accelerator complex.

value of 50%. Luminosities close to the record 2×10^{30} $cm^{-2}sec^{-1}$ have been obtained on several fills of the ring.

The present luminosity limit is due to a number of problems; these include positron injection rate, difficulities with energy ramping, and an incomplete understanding of the beam-beam interaction. Work is proceeding in all of these areas, and we hope for significant improvements in the luminosity by the end of the year.

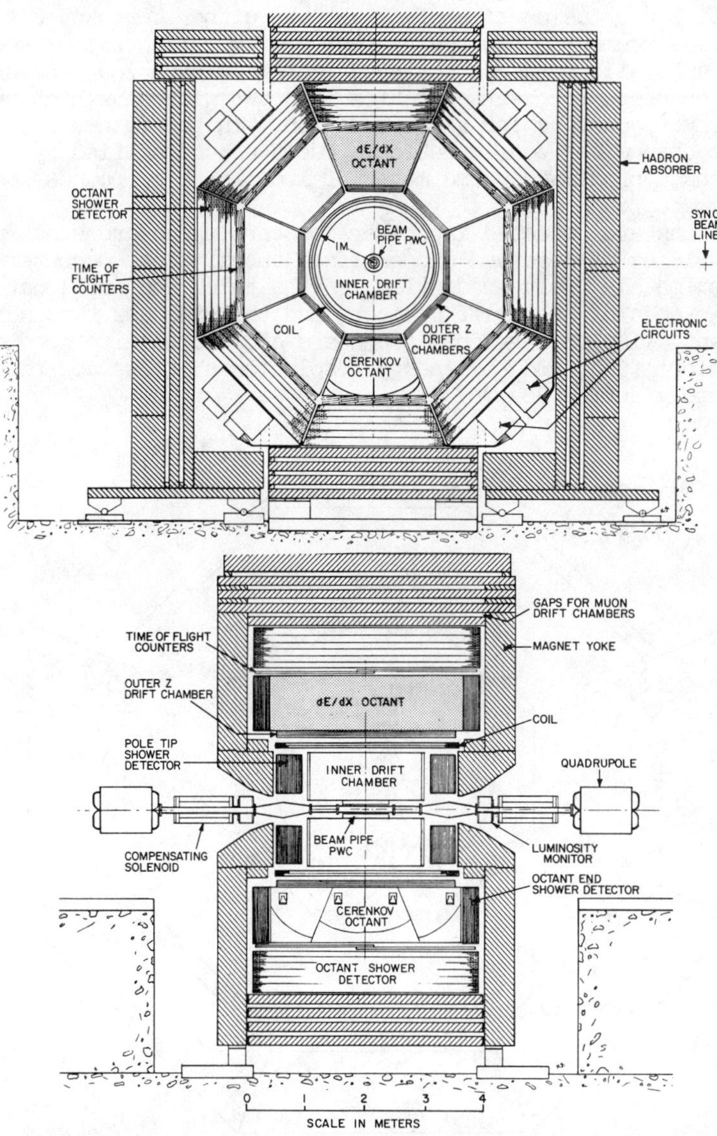

Fig. 2. End and side views of the CLEO detector.

The CLEO detector[3,4] is a general purpose magnetic detector (see Fig. 2). Charged particle momenta are measured with a seventeen cylinder drift chamber in a one meter radius, 3.7 kG solenoidal magnetic field. The z coordinates of a track (z being the coordinate axis parallel to the beam direction) are measured by the beam pipe proportional chamber, the central drift chamber, and a set of drift chambers outside the magnet coil. Charged particles are identified by using time-of-flight and either Cerenkov counters or dE/dx measurements. In the present configuration two of the eight octants will be dE/dx octants, four will have low pressure Cerenkov counters for π-e separation, and two will have high pressure Cerenkov counters for π-K separation. In the future we expect to change the configuration. Muons are identified by their penetrating approximately one meter of iron which surrounds the detector.

Photons are detected in shower counters which are constructed from sandwiches of lead and proportional tubes. These counters are approximately twelve radiation lengths thick and are located at the two ends of the drift chamber, and the side and two ends of each octant. The proportional tubes are crossed to give good coordinate information, and the energy resolution of these counters is approximately $20\%/\sqrt{E(\mathrm{Gev})}$.

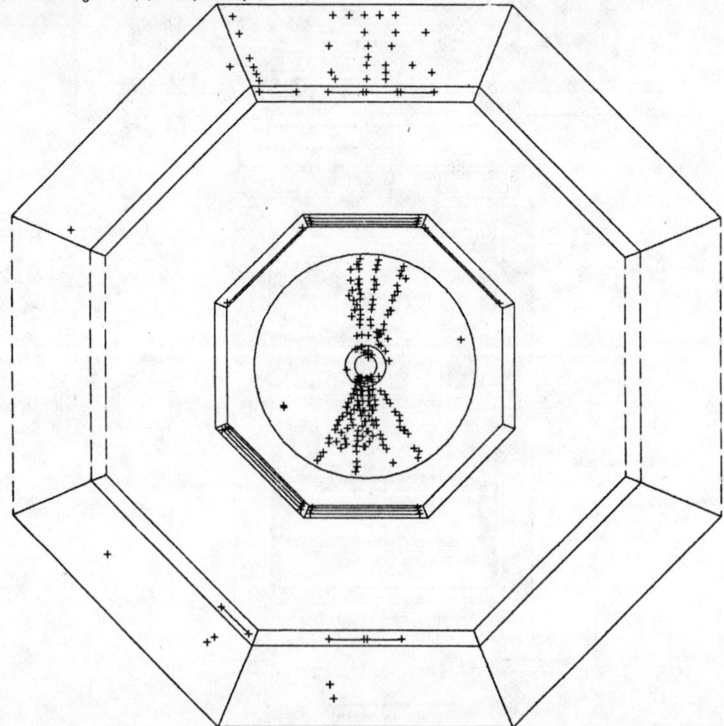

Fig. 3. Annihilation event seen with CLEO. The dashed lines indicate octants which were not installed when this event was observed.

At the time of the Montreal meeting CLEO had run for several days with single beams and almost a week with colliding beams. The single beam runs showed that both the drift chamber and the beam pipe proportional chamber were bothered by the flux of synchrotron radiation reaching them. Improved synchrotron radiation masking solved this problem. In addition, the drift chamber had a few wires which were prone to breakdown in the presence of radiation. These wires were identified and disconnected.

During the running time with colliding beams, various event triggers were tested, and the first sample of data was accumulated. From this sample, clear annihilation events were seen; Fig. 3 is an example.

Many problems exist before CLEO is fully operational. The outer detectors need to be completed, and a large body of software is, at present, untested on real data. Many of our needs have

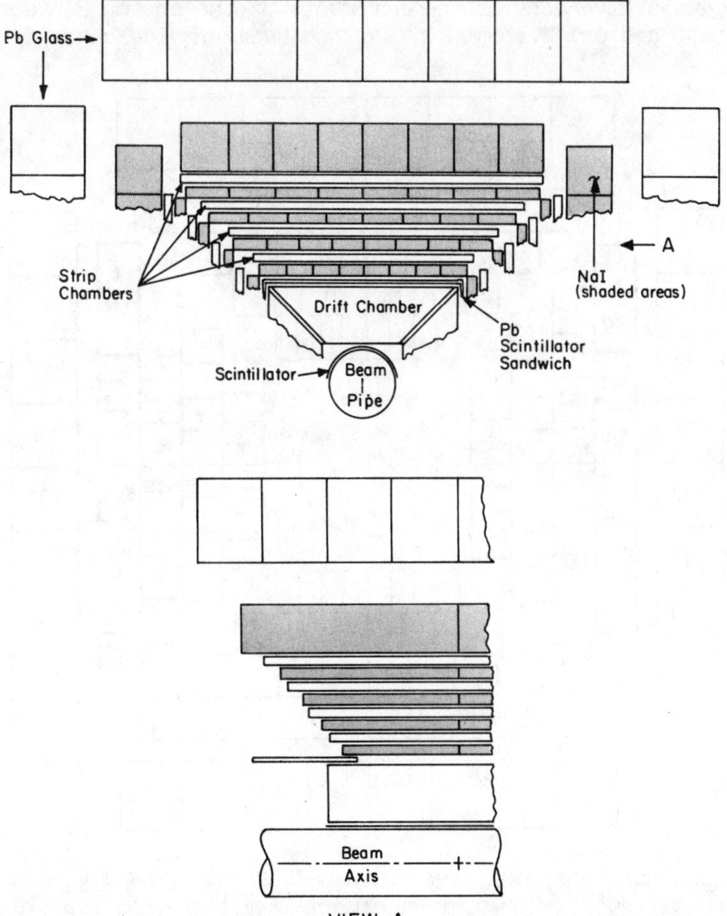

Fig. 4. Two views of the north interaction region detector.

become more clearly identified now that we have our first sample of data, and one can expect rapid progress towards the goal of a working experiment. However, even before this is done we will be starting our physics program; the first parts of this program can be pursued with only a part of the detector. Our goals will be to find the Υ and the Υ' to establish and then confirm our energy calibration, and then begin a search for the Υ''. Confirmation of the existence of this state and a measurement of its mass should be the first new results from CLEO. After these initial measurements we will be making extensive measurements in the Υ, Υ', Υ'' and $B\bar{B}$ continuum regions.

The experiment in the north interaction region emphasizes photon detection and energy measurement[5]. It consists of five layers of NaI (8.8 X_o for photons of normal incidence) followed by a lead glass array (7 X_o for photons of normal incidence). Chambers between layers give accurate measurements of the photon conversion coordinates, and drift chambers near the beam provide charged

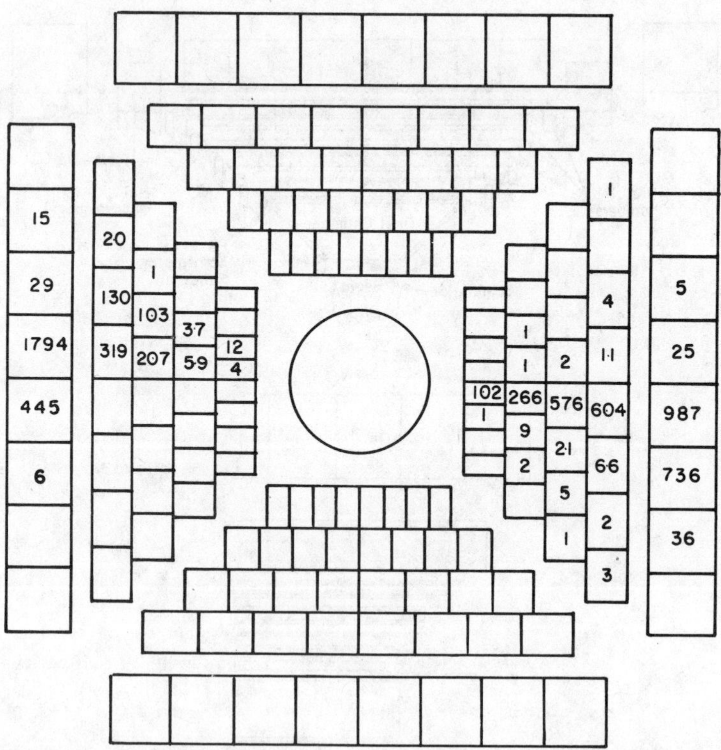

Fig. 5. Bhabha scattering event observed in the NaI portion of the north area detector. The numbers indicate energy deposited in each crystal in MeV.

particle information. Fig. 4 shows two views of one of the quadrants of this experiment.

During the running time when colliding beams were available, one half of the NaI was installed. Bhabha scattering events were clearly seen (Fig. 5); single beam runs confirmed these to be background free. Assembly of the detector is continuing and is expected to be complete by the end of the year.

We are looking forward to the next few months when we will be improving the accelerator performance, completing and understanding the detectors, and beginning a long and fruitful physics program.

This work has been supported in part by the National Science Foundation.

FOOTNOTES AND REFERENCES

1. More details about the CESR design are available in M. Tigner, IEEE trans. Nucl. Sci. NS-24, No. 3, 1849(1977).
2. The people who have contributed to the design and operation of CESR are: K. Berkelman, M. Billing, D. Cassel, J. DeWire, B. Gibbard, D. Hartill, J. Kirchgessner, D. Larsen, R. Littauer, B. McDaniel, R. Meller, N. Mistry, D. Morse, E. Nordberg, S. Peck, S. Peggs, L. Phillips, D. Rice, G. Rouse, J. Seeman, K. Shinsky, R. Siemann, R. Sundelin, R. Talman, M. Tigner, and E. Von Borstel.
3. The CLEO collaboration is: D. Andrews, K. Berkelman, R. Cabenda, D. Cassel, J. DeWire, R. Ehrlich, T. Ferguson, T. Gentile, B. Gibbard, M. Gilchriese, B. Gittleman, D. Hartill, D. Herrup, M. Herzlinger, D. Kreinick, D. Larsen, R. Littauer, B. McDaniel, N. Mistry, E. Nordberg, R. Perchonok, R. Plunkett, K. Shinsky, R. Siemann, A. Silverman, J. Smith-Kintner, S. Stone, R. Talman, G. Thonemann, M. Tigner, and D. Weber from Cornell University; C. Bebek, B. Finn, R. Haggerdy, J. Izen, R. Kline, W. Loomis, F. Pipkin, W. Tanenbaum, and R. Wilson from Harvard University; A. Sadoff from Ithaca College; D. Bridges from Lemoyne College; K. Chadwick, P. Ganci, H. Kagan, F. Lobkowicz, W. Metcalf, S. Olsen, R. Poling, C. Rosenfeld, G. Rucinski, E. Thorndike, and G. Warren from the University of Rochester; D. Bechis, J. Mueller, D. Potter, F. Sannes, P. Skubic, and R. Stone from Rutgers University; A. Brody, J. Cannellos, A. Chen, M. Goldberg, N. Horwitz, J. Kandaswamy, H. Kooy, P. Lariccia, G. Moneti, and R. Whitman from Syracuse University; and M. Alam, S. Csorna, R. Panvini and J. Poucher from Vanderbilt University.
4. More details about the CLEO detector are available in: B. Gittleman, Cornell Preprint CLNS-373 (1977) (unpublished), and E. Nordberg and A. Silverman, Cornell note CBX 79-6 (1979) (unpublished).
5. The north area collaboration is: T. Bohringer, F. Costantini, J. Dobbins, P. Franzini, K. Han, S. Herb, D. Kaplan, L. Lederman, G. Mageras, D. Peterson, E. Rice, and J. Yoh from Columbia University, and G. Finocchiaro, R. Giannini, J. Lee-Franzini, D. Schamberger, L. Spencer, M. Sivertz and P. Tuts from the State University of New York at Stony Brook.

SOURCES OF PROMPT LEPTONS IN HADRONIC COLLISIONS*

Arie Bodek**
Department of Physics and Astronomy
University of Rochester, Rochester, N. Y. 14627

ABSTRACT

Experimental evidence for various sources of prompt leptons produced in hadronic collisions is discussed. Emphasis is placed on sources of prompt single leptons and prompt multileptons. Such final states are expected from production and subsequent weak decays of short lived states. Hadronic charm production cross sections of 20 μb per nucleon at 400 GeV account for the observed single lepton signal as well as dileptons with missing energy (indicative of final state neutrinos). Limits of 40 μb/nucleon for bottom particle production in 400 GeV proton collisions are extracted from multimuon final state signatures. Various prompt dilepton sources (such as Drell-Yan and resonance production) are also summarized.

I. INTRODUCTION AND HISTORICAL PERSPECTIVE

The subject of prompt muon production in hadronic collisions has been investigated for a sufficiently long time to have been reviewed by two[1,2] Physics Reports articles. The first review article by Lederman[1] was published in 1976, and the second article by Craigie[2] was published in 1978. Therefore, I will not elaborate much on the historical development of the field or on the early data.

Investigations of prompt lepton production sources have resulted in several important discoveries and measurements. I will only mention a few highlights. The J/ψ was simultaneously discovered in e^+e^- collisions at SLAC[3] and in a hadronic dilepton experiment at BNL[4]. The Υ was discovered in a hadronic dilepton experiment at Fermilab[5], and high energy dimuon data has been used to extract nucleon[6] and pion[7,8], structure functions. Recently, new results on prompt single muon production[9,10], prompt neutrino production[11] and prompt dileptons with missing energy[12] have determined for the first time the level of charm production in hadronic collisions. In addition, limits on bottom particle production have been extracted[13] from data on prompt multimuon final states, and that technique looks promising as a tool for experiments to measure bottom particle production cross sections[14]. Multimuon final states can also be used to enhance bottom state signals in multiparticle spectrometer data[15] or in emulsions[16]

In this review I will discuss dilepton sources only briefly in Appendix II. Dileptons are reviewed in detail by A.J.S. Smith[17] at

*Work supported in part by the U.S.Department of Energy under Contract No. EY-76-C-02-3065.
**Alfred P. Sloan Foundation Fellow.
ISSN:0094-243X/80/590211-36$1.50

this conference. I will concentrate instead on prompt leptons that come predominantly from the production and subsequent weak decay of short lived states (i.e. charm and bottom states).

Historically[1,2] early data on prompt lepton production has been obtained by small acceptance single arm experiments. This prompt lepton data has been presented and lepton/pion yields. The advantages of that approach were:

1. Data from various energies (s), X_F, P_T and several nuclear targets could be plotted on the same linear scale (lepton/pion ≈ 10^{-4}).
2. Some experimental uncertainties in acceptance would cancel if both lepton and pion data were measured in the same apparatus.
3. In prompt muon experiment, the π yield can also be extracted from the π decay muons, and thus the apparatus acceptance would also cancel in the ratio.

The disadvantages of the lepton/pion measurements were that they told us little about the sources of prompt leptons. The experiments were valuable in establishing the <u>existence</u> of a prompt muon signal in various s, X_F, P_T regions, and provided the motivation for the design of second generation experiments to study the phenomenon in detail. These experiments were designed to investigate particular final states. For example, the study as to whether prompt leptons come from single or pair sources required a large acceptance detector to identify second muons as well as a high density detector to minimize pion and kaon decays[9,10]. Such a detector is also suitable for multimuon studies[12,13]. On the other hand, detailed studies of di-lepton final states require high resolution. Such resolutions were obtained in two arm spectrometers[4,5,6] and in open geometry large acceptance spectrometers[7,8].

Prompt muons are <u>defined</u> to be muons originating directly at the <u>primary vertex or</u> originating from the decays (e.g. weak or electromagnetic) of <u>short lived</u> states. Short lived is somewhat experiment dependent but is typically defined as particles with lifetime $(C\tau)\lesssim 1$ cm. Using this criteria we find that all strange particles are relatively long lived (see Appendix I).

<u>Direct electrons</u> is a term used to designate electrons that are not only prompt (i.e. from short lived states) but also do not originate from Dalitz decays such as $\pi_o \rightarrow \gamma e^+e^-$ and $\eta \rightarrow \gamma e^+e^-$. Dalitz decay contributions are usually not subtracted from prompt <u>muon</u> signals because the π^o does not decay to muons and the muonic Dalitz decays of the η and other particles are very small (see Table III).

The major sources of background in direct electron experiments are photon conversions, Dalitz decays and Ke3 decays. The major sources of backgrounds for prompt muon experiments are pion and kaon two body decays. The sources of non-prompt leptons for both types of experiments are discussed in Appendix I.

The contribution of pion and kaon decay muons is determined by measurements of the muon rate as a function of target and absorber density (ρ). The decay probability of pions and kaons is linearly proportional to the mean free path before interaction. This path is proportional to $1/\rho$. Therefore, the intercept of the prompt muon rate plotted versus $1/\rho$ at $1/\rho=0$ is the prompt muon signal. If the target and absorber are a single unit, then the π decay muon rate would extrapolate to zero because the density is changed for pions from primary and secondary collisions at the same time[9,10].

Early prompt muon experiments[18,19] have used a different method. They used a fixed target and a movable absorber. The fake prompt signal due to interactions in the absorber had to be subtracted using Monte Carlo calculations[18].

The expected weak decay sources of prompt leptons are shown in Table I. These are mostly single lepton sources. Wherever experimental values for the lifetimes and branching ratios are not available, I have inserted a theoretical estimate designated with a (TH). Table II lists the particles that yield dileptons from electromagnetic sources, and have equal branching ratios to electron and muon pairs. Table III lists short lived sources which have different branching ratios to electrons and muons (mostly Dalitz decays). Here, wherever the branching ratio was not measured, I have included a rough estimate (RE) by using the measured photon decay mode as an input. Table IV lists sources of prompt multileptons that come from double weak decays and bottom particle cascade decays. Here a lot of the branching ratios are theoretical estimates. The non-prompt background sources are listed in Appendix I.

II. THE TOTAL PROMPT MUON RATE

Recently, the total prompt muon production rate for almost the entire forward hemisphere has been measured[10]. The experiment (Fermilab experiment E595; Caltech-Fermilab-Rochester-Stanford) was done in the Fermilab N5 line using 350 GeV diffracted protons. There are some preliminary results from a short test run performed in Feb. 1979. Since these are first results from this experiment, I will discuss the apparatus and analysis in more detail.

II.1 Experimental Apparatus (Caltech-Fermilab-Rochester-Stanford)

The experiment was performed at low intensities, i.e. about 1×10^4 350 GeV protons per pulse. The incident protons interacted in a target-calorimeter consisting of 49 (76×76 cm^2) steel plates (a total of 2.44 m of steel). The plates were independently mounted on rails so that the interplate spacing (and thus the calorimeter density) can be varied. The calorimeter was followed by a muon range detector consisting of eighty-eight 3 m $\times$ 3 m $\times$ 5.08 cm steel plates interspersed with forty-two 3 m $\times$ 3 m $\times$ 3.2 cm liquid scintillation counters with wave-shifter light collectors[30], and twenty-two 3.2 $\times$ 3.2 m^2 spark chambers with magnetostrictive readout[31]. The amount of light for each muon in the scintillation counters (15 photoelectrons) was sufficient to allow discrimination between one and two muons by use of counter pulse heights alone. The range detector was followed by a 3.5 m magnetized steel toroidal muon spectrometer.

The trigger required a proton interaction in the calorimeter in coincidence with a muon that penetrated at least 5.75 m of steel. Because of the large size of the range detector, this trigger effectively selected all muons of momentum $P_\mu > 8$ GeV, corresponding to almost the entire forward hemisphere in the center of mass (Fig. 2).

Most muons satisfying this trigger were due to the decay of pions and kaons. This non-prompt background was measured by uniformly expanding the first 38 plates (1.68 m of steel) of the calorimeter, thereby proportionally increasing the mean path length and decay probability of hadrons in this region. Data were taken at three different densities: fully compacted, expanded by a factor of 1.75, and expanded by a factor of 2.5. The mean density of the compacted calorimeter is about 3/4 that of steel since there are gaps between plates where scintillation counters are mounted.

II.2 Analysis

The muon event rate (i.e. events with at least one muon with $P_\mu > 8$ GeV) per interacting proton exhibits a linear dependence on inverse density, ρ (Fig. 3). The intercept at $1/\rho=0$ of $(3.95 \pm 0.40) \times 10^{-4}$ is the prompt rate. However, to get the total rate, events with two muons both satisfying the requirement $P_\mu > 8$ GeV must be counted twice. This rate, determined using the scintillation counters in the range detector after corrections for accidentals[32], is $(0.94 \pm 0.01) \times 10^{-4}$, giving a total prompt muon rate of $(4.89 \pm 0.40) \times 10^{-4}$ per interacting proton. This rate has been corrected for a density independent background ($8 \pm 4\%$) from decays occurring downstream of the expanded region. This contribution was determined by a measurement in which only the downstream portion of the calorimeter was expanded.

The above quoted rate is for a thick target and includes contributions from secondary and tertiary hadron and photon interactions. To extract the contribution of the first collision a shower development calculation was performed. The calculation was based on radial scaling parametrization of particle production data from hydrogen[33] modified for nuclear effects. The contribution from secondary and tertiary interactions was calculated[34] to be $(1.60 \pm 0.2) \times 10^{-4}$, and subtracted from the total prompt muon rate. The remaining rate of $(3.29 \pm 0.45) \times 10^{-4}$ is the prompt muon rate for primary collisions. We can express this rate as a $(\mu^+ + \mu^-)/(\pi^+ + \pi^-)$ ratio by dividing by the number of π's (with $P_\pi > 8$ GeV) produced by eacy primary proton collision. The number of π's of 3.55, extracted from the measured non-prompt muon rate and the shower development calculation[35], yields $\mu/\pi = (0.93 \pm 0.13) \times 10^{-4}$. The errors here are statistical. However, fairly conservative assumptions[36] yield an upper limit of 1.38×10^{-4} and a lower limit of 0.78×10^{-4}. The measurement is consistent with the trend of other μ/π data[37] at higher values of X_F as shown[38] in Fig. 4b. In a similar X_F and P_T region, ISR 30° data[39] (Fig. 4a) indicate that the average[40] e/π is between 3.2 and 4.8×10^{-4}. The data are in agreement with the trend established by another[41] ISR experiment at 90° although that experiment does not extend to low values of P_t (see Fig. 4a).

II.3 Discussion

There is about a factor of 4 difference between the 30° ISR e/π

results (Baum et al[39]) and the Fermilab μ/π data (Ritchie et al[10]). Although the two experiments are in a different energy range and looked at electrons and muons respectively, it is instructive to investigate possible reasons for the different results.

1. Ritchie et al. used a steel target, (p-Fe) while the ISR 30° data is for p-p. If all the difference were attributed to nuclear effects (p-Fe vs pp), then the A dependence of prompt muons at small X_F and small P_T would have to be $A^{0.43}$ (i.e. much less than the inelastic cross section which rises as $A^{0.70}$). Such an unusual A dependence is unlikely. For example, the A dependence of π production (for π's with $P_\pi > 8$ GeV) is $A^{0.79}$ yielding only a 30% difference between p-Fe (3.55 π's) and pp[33] (2.45 π's) interactions. Also, at higher X_F($X_F \gtrsim 0.1$) the A dependence for $\mu^+\mu^-$ pairs has been measured[42] to be $\sim A^{0.70}$ for low $M_{\mu\mu}$ and $A^{1.0}$ for high $M_{\mu\mu}$.
2. The center of mass energy at the ISR ($\sqrt{s}$=53 GeV) is larger than that of the Fermilab 350 GeV data ($\sqrt{s}$=23.7 GeV). This difference initially appeard unimportant since a low energy[43] experiment also indicated a large e/π ratio. However, the experimental situation at low energies is somewhat unclear since more recent low energy experiments[44] at SLAC and ANL indicate a <u>small</u> e/π ratio. We conclude that s-dependent effects cannot be ruled out.
3. Baum et al. measured e/π versus the μ/π measured by Ritchie et al.

It is possible that the source of low P_T electrons does not yield a corresponding rate of low P_T muons. This would be the case if the low P_T electrons were from low mass pairs (e.g. due to the muon-electron mass difference) rather than from, for example, charm decays. The ISR experimenters, based on their $e\mu$ measurements, have concluded[45] that charm production[46] is not the source of the large low P_T electron rate. Since that experiment vetoes on very low mass e-pairs, they conclude that the source of direct electrons at low P_T are e^+e^- pairs with m_{ee}>0.1 GeV. At the low P_T region there is a large contribution from Dalitz decays which must be calculated and subtracted. The Dalitz decay contribution (see Table III) may be larger than expected[48]. For example, the production cross section for η mesons at low P_T has not been measured, and high P_T measurements[49] (P_T>1 GeV) indicate a large η cross section ($\eta/\pi \simeq 0.5$). Therefore, a source such as copious η production at low P_T at ISR energies could account for the difference between the two experiments.

II.4 Single Muons and Charm Production

There are also some preliminary results regarding the contribution of prompt <u>single</u> muons to the total prompt muon rate. A preliminary separation of single muon and dimuon events (using the scintillation counters only) after corrections for misidentified dimuon events, yields a prompt single muon rate of $(1.05 \pm 0.5) \times 10^{-4}$, this rate indicates that at 350 GeV (30±15)% of the prompt muons come from single muon sources. The prompt single muon rate can be used to obtain a charm production cross section. The acceptance for charm production is large (39±1%) and rather insensitive to model assumptions because muons from almost the entire forward hemisphere are detected. Using a semileptonic branching ratio of 8% and a linear A dependence one obtains a charm cross section of 22 ± 9 μb/nucleon.

The A dependence of prompt single muons originating from charm production is expected to be similar to the A dependence of ψ production (i.e. linear). On the other hand, the dimuon rate is dominated by low mass sources with an $A^{2/3}$ dependence. Therefore, the relative contribution of single vs. 2μ sources to the total prompt muon rate is expected to depend on the target material. High A targets (e.g. Fe) are expected to have a larger prompt single muon contribution than low A targets (e.g. H_2,Be). Similarly, the energy dependence of charm production is expected to be steeper than that for ρ meson and other low mass dimuon pairs. Therefore, prompt single muons are expected to contribute more to the promp muon rate at higher incident energies.

III. INVESTIGATION OF THE CHARM PRODUCTION CONTRIBUTION

In the previous section, it was mentioned that the contribution of prompt single muons to the total prompt muon rate is ≃30 ± 15%. Several years ago, it was pointed out by Bourguin and Gaillard[49] that the P_T dependence of muon from charm decays does not fall as steeply as the P_T dependence of muons from ρ and low mass dimuons. Similarly, because of the very steep P_T dependence of pions, it was expected that the background from π decay muons would be small for P_T values of around 1 GeV. I will now discuss the results of Fermilab experiment E379 (Caltech-Stanford)[9] which concentrated on the P_T region near 1 GeV.

III.1 Apparatus (Caltech-Stanford)

This experiment was the first to establish a prompt single muon signal[9] and reported on the observation of a prompt 1-μ signal in the moderately high P_T ($0.8<P_T^{\mu^+}<2.5$ GeV) and low X_F ($10<E^{\mu^+}<60$ GeV) region produced by 400 GeV p - N interactions. Approximately equal production cross-sections for 1-μ and 2-μ final states were found in this kinematic region. Evidence for the observation of missing energy (indicative of final state neutrinos) in association with hadronically produced $\mu^+\mu^-$ pairs, was presented and related to the observed single muon signal[12]. The calorimeter used in the missing energy investigation was the same one described in the previous section. The major experimental difficulties in the measurement of a prompt single muon signal come from:

1. Separating the π,K decay from the prompt signal.
2. Distinguishing 1μ from 2μ events.

The experimental approach taken was:

1. Vary the target-absorber density to determine the π,K decay background.

2. Maximize signal/background by triggering on a μ^+ in the high P_T region ($P_T \simeq 1$ GeV).
3. Greatest possible acceptance for identifying 2nd muons in di-muon events.
4. "Detection" of final state neutrinos by tagging missing energy in the final state.

The experimental apparatus is shown in Fig. 5.

The experiment was performed in the Fermilab N5 beam with 400 GeV protons at typical intensities of $3\text{-}5 \times 10^5$/sec. The primary elements of the detector (Fig. 5) were a fine-grained target-calorimeter of variable density (energy resolution of 3.5% at 400 GeV), a muon identifier (MI), and a toroidal muon spectrometer.

The data were taken with a high-P_T trigger, which required a coincidence of both a beam and a muon trigger component. The muon component required the muon to remain in the same quadrant throughout the toroid system by requiring the appropriate coincidence of counters C, S2, ACR, T4 (which were divided into quadrants) and S1, T2, T3, (divided into half-planes). This requirement preferentially selected muons with high P_T ($P_T^{\mu+} > .8$ GeV).

The beam component required an incident proton to pass through counters B0 and B1 (7.6×7.6 cm and 5.1×5.1 cm) and to interact within the first 10 plates of the calorimeter. To reject any background from upstream interactions, triggers were vetoed by the presence of any additional particles in the beam or halo counters within 95 nanoseconds of the trigger. Further beam information was provided by the pulse height of the trigger counters and by the incident proton's trajectory and momentum, as measured by a spectrometer immediately upstream of the calorimeter. Interactions satisfying the beam trigger alone were scaled, and one out of each 2^{16} was recorded to provide a control sample of interactions without any muon requirement.

III.2 Analysis

The experiment collected data at three different densities (keeping the mean interaction point fixed in space): fully compacted, expanded by a factor of 1.5, and expanded by a factor of 2. The mean calorimeter density in the compacted configuration was 3/4 that of steel due to the gaps (1.3 cm) between plates. After all software cuts, the rates in each density configuration were normalized to the beam trigger rates and plotted as shown in Fig. 6. As expected, the 2-μ rate is flat, and the 1-μ rate shows a linear increase with the effective pion interaction length. The 1-μ slope measures the rate from non-prompt decays, and the intercept of $(10.5 \pm .5)10^{-6}$ at infinite density is the raw prompt 1-μ signal.

To obtain the true prompt single muon rate, the raw prompt 1-μ rate had to be corrected for several background sources:

a) $\mu^+\mu^-$ events with a low energy μ^- which ranged out in the calorimeter or muon identfier. A Monte-Carlo calculation using the measured $\mu^+\mu^-$ distributions gave a correction of 10±2% (systematic errors included) of the raw prompt 1-μ signal. This component was subtracted from the 1-μ signal and added to the 2-μ signal.

b) Muons from decays of pions and kaons in the unexpanded part of the calorimeter (after plate 25). A Monte-Carlo simulation of the hadron shower, which reproduced the mean shower profile measured in the experiment, gave a correction of 8±3% of the measured decay rate. This corresponds to 16±6% of the prompt 1-μ signal.

c) A subtraction of 20±10% of the prompt 1-μ signal due to second order variation in the acceptance with density. These arise because, although the mean interaction point stays fixed, multiple scattering effects and production by secondaries move downstream when the calorimeter is expanded. Since the toroid hole subtends a larger angle for particle originating downstream, this yields a reduction of 4±2% in the acceptance of the expanded relative to the compacted configuration. This correction was obtained from the $\mu^+\mu^-$ events (which should be constant with density).

III.3 Single Muon Results

After all corrections, the measured prompt 1-μ rate was $(5.8 \pm 1.5)\times10^{-6}$ per incident proton and the 2-μ rate was $(5.9\pm.2)\times10^{-6}$; the errors are largely systematic. These rates indicate that in the P_T region near 1 GeV prompt single muons account for 50±13% of all prompt muons.

III.4 Two Prompt Muons with Missing Energy

The total observed energy spectrum for $\mu^+\mu^-$ events ($E_{tot}=E_{\mu^+}+E_{\mu^-}+E_{calorimeter}$) is shown in Fig. 7. The dashed curve shown for comparison is the E_{tot} spectrum exhibited by beam interactions without final state muon. There is a pronounced enhancement of missing energy events for $m_{\mu^+\mu^-}<2.4$ GeV. We observe 227 $\mu^+\mu^-$ events with missing energy in excess of 45 GeV. An estimate of the double π,K decay background is provided by the 5 <u>observed like sign</u> dimuon events with large missing energy. Monte Carlo calculation of $K\bar{K}$ production and double decay also yields a background of 5 events. Also, since the toroid spectrometer is instrumented with acrylic calorimetry counters, we can rule our catastrophic muon energy loss in the steel as significant source of background. We conclude that all backgrounds are unlikely to contribute more than 10% of the observed $\mu^+\mu^-$ with missing energy signal.

III.5 Interpretation

To estimate a charm production cross-section from these data, we have assumed that all the signal comes from the semileptonic decays $D\rightarrow K\mu\nu$ (60%) and $D\rightarrow K^*\mu\nu$ (40%) with a total semileptonic branching ratio of 8%. The inclusive D cross-section was assumed to increase linearly with the atomic number A of the nucleus and was parameterized as

$$E\frac{d^3\sigma}{dp^3} = C\,(1-x_F)^\beta\, e^{-\alpha p_t} \quad \text{(for inclusive D production)} \qquad (1)$$

The single muon data were consistent with values in the range α=2.0-3.5 GeV^{-1} and β>3. Varying α and β over these allowed ranges yields charm cross-sections in the range 15-75 μb/nucleon. For β=3 and α=2.5, the acceptance for the produced μ^+'s was 2.8% and the cross-section for D production was $\sigma_{D\bar{D}}$=35±9 μb/nucleon. This model, in which the two charmed states are uncorrelated gives an acceptance of 0.22% for the $\mu^+\mu^-$ events with 45 GeV of missing energy and yields a charm cross-section of 17±3 μb. However, the $\mu^+\mu^-$ mass and momentum distributions do not fit this model. The 25% error in the charm cross section extracted from the single muon results and the 20% error in the cross section extracted from the 2μ results include experimental statistical and systematic errors (mostly systematic) but do not include model dependent undertainties in acceptance or branching ratios.

In order to include the expected correlation between the D and $\bar{D}$ state we assume a $D\bar{D}$ model production model

$$E\frac{d^3\sigma}{dp^3} = \frac{C}{M^3}\,(1-x_F)^\beta\, e^{-\alpha p_t}\, e^{-M/\sqrt{s}} \quad \text{(for } D\bar{D} \text{ production)} \qquad (2)$$

and calculated the fraction of $D\bar{D}$ double muonic decays which satisfy our trigger requirement, give 2 μ's that pass the muon cuts, and yield a measured missing energy in excess of 45 GeV. Here the kinematic variables in the above cross-section equation refer to the composite $D\bar{D}$ system (and $\sqrt{s}$=27.4). The acceptance was rather insensitive (to ±30%) to variations in α between 1.5 and 3.0 GeV^{-1} and γ between 0.0 and 17.5. For α=1.3, β=6 and γ=20 we obtain an acceptance of 0.27% yielding a charm cross section of 14±3 μb. Using this same model we obtain a charm cross-section of 21±5 μb from the single muon data. Changing β from 2.96 to 6.0 changes the $\mu^+\mu^-$ acceptance from 0.39% to 0.24%. (See Table 5.)

III.6 Discussion

In general the charm cross sections extracted from the 2μ data are lower than those extracted from the single muon data (see Table 5). Although for the best fit model parameter the extracted cross sections are consistent (14±3 and 21±5) closer agreement between the two sets of data is obtained with an average branching ratio of 5% which gives $\sigma_{charm}\simeq$35 μb/nucleon. In view of the recently reported difference between the D^o and D^+ semileptonic branching ratios, (see Table I). This may be indicative of large D^o/D^+ production ratio. It is theoretically expected that $\sigma_{D^o}\simeq 3\sigma_{D^+}$ (see Rosner, ref. 23). It is expected[23] that the fragmentation of charmed quarks would lead to final state charm particles in the following ratios: $1D^+$, $3D^o$, (0.7 to 2) F^+, and (0.9 to 2)Λ_c. Since the D^o,F and Λ_c have a smaller ratio than the charged D, average hadronic branching ratios of order 5% are not unexpected.

The observation of a signal in both the single μ and 2μ plus missing energy signatures indicates that short lived hadronically produced states with branching ratio to muons of order 5% are produced with a total cross section of about 30 μb. Such a branching ratio indicates that the observed signal comes from charm rather than a new particle with a large muonic branching ratio such as $\tau^+\tau^-$ production ($\tau \rightarrow \mu\nu\nu$ BR is 17.5%). Theoretically it is expected that the cross section for $\tau^+\tau^-$ pairs is equal to the $\mu^+\mu^-$ Drell-Yan cross section for $\mu_{\mu\mu}$>4 GeV, i.e. a $\tau^+\tau^-$ production cross section of order 0.1 nb/nucleon. This cross section is much less than the observed 20 μb. Similarly, the observation of 2μ plus missing energy signal indicates that the muons do not originate from a copious production of a new particle which has a branching ratio to muons which is as small as those of the strange hyperons. The strange hyperon contribution is removed via the density extrapolation since the hyperons are relatively long lived. (See Appendix I.)

III.7 Comparison with Beam Dump Experiments

A detailed discussion of beam dump experiments can be found in the review talk by H. Wachsmuth[11] at the Batavia Lepton Photon (1979) Conference. Early beam dump results have indicated rather large cross sections (40 to 400 μb).[11] Recent runs yield smaller cross sections of order 15 μb which are in agreement with the prompt single muon results. The recent experiments have careful checks on upstream beam scraping and have data at different densities. The new data provided additional statistics to the small bubble chamber event sample. The 15 μb cross section assumes a linear A dependence and includes a correction for charm production by secondary interactions. The acceptance is model dependent. A production model for D mesons of the form

$$\frac{E d^3\sigma}{d^3 p} \sim (1 - |x|)^n e^{-bp_t}$$

was used with n=3 and b=2. The cross sections are rather insensitive to b for b>2 but can vary[50] by a factor of 5 if n is changed from n=1 to n=5.

IV. SEARCH FOR BOTTOM STATES

Recently, there have been theoretical calculations, based on gluon vector-dominance models[51], which indicated large bottom particle production cross sections. Cross sections of the order of 200 nb for π^-N collisions at 150 GeV and 850 nb for p-p collisions at 400 GeV were predicted. Furthermore, experimental evidence has been reported[15] for the production of a heavy state (with a mass of 5.3 GeV) in π^-N collisions at 150 and 175 GeV. The size of the observed experimental cross section was large (≃200 nb). The decay mode[25] ($\psi K\pi$) of the state make it tempting to associate this state with the predicted B meson, i.e. a bound state of a b and $\bar{d}$ quark.

Bottom particle production is expected to lead to multimuon

final states (as shown in Table 4). Such final states were investigated using the same detector described in the previous section (Caltech-Stanford).

IV.1 Bottom Production Analysis

Three particular signatures of $B\bar{B}$ production and decay were investigated

1) $pN \to \psi\mu X$, i.e., a final state containing 3 muons, 2 of which reconstruct to a Ψ mass (2.6 GeV < M < 3.6 GeV) arising from $\bar{B}(B) \to \mu^+\nu X$, or $\bar{B}(B) \to \bar{D}(D)X$ with $D \to \mu\nu X$.

2) $pN \to \mu\mu\mu X$, a 3μ final state without any mass restriction arising from the following decay chain (or it charge conjugate):

$B \to D\mu^-\nu$ with $D \to \mu^+ X$

$\bar{B} \to \bar{D}\mu^+\nu$ with $\bar{D} \to$ hadrons

or, alternatively,

$B \to D\mu^-\nu$ with $D \to \mu^+\nu X$

$\bar{B} \to \bar{D}X$ with $\bar{D} \to \mu^-\nu X$

3) $pN \to \mu^+\mu^+X$, arising from

$B \to DX$ with $D \to \mu\nu^+ X$

$\bar{B} \to \bar{D}\mu^+\nu$ with $\bar{D} \to$ hadrons

The experimental features which make the experimental setup particularly suited for bottom particle investigations are:

1. The trigger required one high p_t muon ($p_t \gtrsim 0.8$ GeV) which enhanced muons from ψ and B particle decays.
2. The high density of the apparatus ($= \frac{4}{3}\rho_{Fe}$) minimized muons from π and K decay.
3. Background multimuon events arising from an extra muon unassociated with the interaction were eliminated by running the experiment at moderately low intensities (5×10^5 protons/sec) and vetoing triggers if there were any additional beam particles in beam or halo counters within ±95 nsec of the trigger. All muons were required to be in time with the interaction and to extrapolate to a vertex inside the calorimeter.
4. The probability for other background muons (primarily from π and K decay in the hadron shower) was directly measured by triggering the apparatus on random proton interactions without any muon requirement. In addition, the contribution of π- and K-decays to multimuon events was checked by comparing multimuon rates at different calorimeter densities.
5. Signal/background could be enhanced by the requirement of missing energy associated with bottom particle decays or the requirement of a high p_t muon associated with such decays.

The efficiencies for the detection of specific final states were calculated using the following assumptions about $B\bar{B}$ production and decay modes:

a) B's are produced via the reaction

$$p + Fe \rightarrow B + \bar{B} + \ldots$$

with the invariant cross section given by

$$E\frac{d^3\sigma}{dp^3} = \frac{1}{M^3}(1-x)^{\alpha}\, e^{-\beta p_t}\, e^{-\gamma M/\sqrt{s}}$$

where x_F, p_t and M refer to the compound $B\bar{B}$ system (with M $\gtrsim$ 10.6 GeV), and $\alpha = 3.0$, $\beta = 2.2$, and $\gamma = 15$.

b) The cross-section follows linear A dependence.

c) The leptonic decay mode of the B proceeds via

$$B \rightarrow D\mu\nu$$

d) The nonleptonic decay mode of the B proceeds via

$$B \rightarrow D\pi\pi$$

e) All the ψ final states resulting from the B decay are represented by

$$B \rightarrow \psi K\pi$$

The detection efficiencies are only mildly dependent on these production and decay assumptions. However, the determination of $B\bar{B}$ production cross-sections from multimuon signatures is obviously dependent on the branching ratio assumptions. The assumptions used were $B \rightarrow \psi X = 3\%$, $B \rightarrow D\mu\nu = 10\%$, $B \rightarrow DX = 100\%$, and $D \rightarrow \mu\nu K$ (or K*) = 8% (with a K/K* ratio of 1.5). It was assumed, for simplicity, that the charged and neutral states have equal semileptonic branching ratios.

The trigger efficiencies, as well as the final acceptance after all analysis cuts on the muons, were calculated for each final-state category using a Monte Carlo simulation of the apparatus. These efficiencies, with the assumed branching ratio into each final state, are summarized in Table 6. Included, as subcategories, are the effects of additional cuts in p_t and missing energy which serve to enhance signal/background. The sensitivity (i.e., partial cross-section per detected event), given in the fourth column, is the product $\sigma_{pN}/(N_p\varepsilon)$, where σ_{pN}=13 nb/nucleon is the p-N cross-section on iron nuclei, N_p=1.01×10^{10} is the total number of interacting protons (after dead time correction and cuts), and ε is the acceptance for detecting the final state muons. The inclusive $B\bar{B}$ cross-section per event in the last column is obtained from the partial cross-section/event using the branching ratio in column 5.

Table 7 summarizes the number of detected events in each category, the estimated background, and the resulting signals. Only data from the compacted density are used here, since backgrounds are smaller. Backgrounds arise from a single uncorrelated muon accompanying a ψ, $\mu^+\mu^-$, or μ^+ final state, and were calculated from the production rate of single muons which satisfied all analysis requirements. A μ^+ rate of $(2.5 \pm .6) \times 10^{-4}$/interaction and a μ^- rate of $(5.5 \pm 1.1) \times 10^{-4}$/interaction were measured in this experiment using the sample of random proton interactions taken throughout the run.

The difference between μ^+ and μ^- rates was due to acceptance; low energy μ^+'s were focussed into the toroid hole and then failed the analysis cuts. The backgrounds in coincidence with a ψ or $\mu^+\mu^-$ were estimated to be ≈15% smaller because of the reduction in hadron shower energy due to the production of the dimuon pair. In the case of $\psi\mu$ final states, for example, the background of 3.0 events is the product of 4399 recorded ψ's times an estimated rate of 6.8×10^{-4}/interaction for an accompanying uncorrelated μ^+ or μ^-.

As a check of these background estimates, multimuon events taken at three different densities are compared in Fig. 8. The curves give the shape expected if the extra muons were entirely due to π and K decay backgrounds. It is evident that most, if not all, of the events are due to this background.

IV.2 Conclusion

None of the categories recorded in Table 8 shows any evidence of a positive signal (with the possible exception of the single event which survives the p_t cut in category 1a). We conclude from the limits in column 5 that $\sigma_{B\bar{B}} \leq 40$ nb. This is much smaller than the 850 nb predicted by gluon vector dominance model[51] and is also smaller than the 200 nb indicated[15] in 150 GeV π^-N interaction. The limit is in agreement with QCD calculations[55] indicating a $B\bar{B}$ cross section of 10 nb in 400 GeV pp collisions.

IV.3 Discussion

The disagreement with the predictions of the gluon vector dominance model[51] is probably not surprising in view of the fact that the same model[51] predicts a charm cross section of 170 μb in 400 GeV proton collisions. This charm cross section is also larger than the 20 μb indicated by the single muon results reported here. It may be possible to change the gluon distribution such as to reduce both the charm and bottom particle production cross sections, and bring the theory closer to agreement with the data. However, in any model dominated by gluon interactions high energy protons (e.g. 400 GeV) would be more effective in producing bottom states than low energy pions (e.g. 150 GeV). On the other hand protons would be less effective than pions if quark-antiquark annihilation processes were important. Such is the case in T production processes where pions are more effective than protons. Figure 9 shows T production cross section from J. Badier et al.[52]. Using the measured[52] value for $\sigma_T \times (BR \to \mu^+\mu^-)$ of (1.5 ± 0.5)pb/nucleon in 200 GeV π^-N collisions and extrapolating to 150 GeV we obtain $\sigma_T\ (BR \to \mu^+\mu^-) \approx (1 \pm 0.5)$. The corresponding cross section for 400 GeV protons[53] is $0.5 \pm .24$ pb/nucleon. Therefore, if bottom particle cross sections were proportional to the T cross section, then 150 GeV π^- would be about twice as effective as 400 GeV protons in producing botton states (note the large errors, the ratio is 2 ± 1.4).

One can make the observation that the relation between ρ and π production is

$$\frac{\sigma_\pi}{\sigma_\rho} \simeq 10$$

and for strange particles

$$\frac{\sigma_K}{\sigma_\phi} \simeq 15$$

and for charm (400 GeV protons)

$$\frac{\sigma_{charm}}{\sigma_\psi} \simeq 100$$

where we have used a charm cross section of 20 μb and a ψ cross section[54] (measured in the same experiment) of $\sigma_\psi \cdot BR(\psi \to \mu\mu) = 17$ nb with $BR(\psi \to \mu\mu) = 0.07$. If $\sigma_{B\bar{B}}/\sigma_\Upsilon \approx 100$ we obtain (using[27] $BR(\Upsilon \to \mu\mu) = 2.3 \pm 1.4\%$ from Table II). $\sigma_{B\bar{B}} \approx 2$nb/nucleon which is consistent with the 40 nb limit.

In the case of ψμ final states a more direct comparison with the data of Barate et al. can be made. One would expect $\sigma(B\bar{B} \to \psi K\pi) \simeq 3\sigma(B\bar{B} \to \psi\mu)$ since $BR(B \to \psi K\pi)/BR(B \to \psi X) \simeq 0.5$ and $[BR(B \to \mu X) + BR(B \to DX) BR(D \to \mu X)] \simeq 0.18$. Barate et al. quote $\sigma_B \cdot BR(B \to \psi K\pi) = 2$ nb. This implies $\sigma_{B\bar{B}} BR(B\bar{B} \to \psi\mu) \cdot BR(\psi \to \mu\mu) = 90$ pb (using $BR(\psi \to \mu\mu) = 0.07$), for 150 GeV π^-, which is considerably larger than the measurements (see Table VII) of (-8 ± 12) pb and (22 ± 22) pb for 400 GeV protons.

In the analysis it was assumed for simplicity that the charged and neutral states have equal semileptonic branching ratio of 10%. We have assumed 10% rather than the free quark model value of 14% to account for some enhancement of the hadronic modes. The ψμ signature is sensitive only to the <u>average</u> semileptonic BR of the B^o and B^+. It is even less sensitive because some of the extra muons come from the final state D decay. However, if the muonic branching ratio of the B^+ is greater than that of the B^o in analogy with the recently observed[20,21,22] difference between D^+ and D^o branching ratios, then the 3μ final states would be suppressed and the $\mu^+\mu^+$ final states would be slightly enhanced due to the sign correlation between B and D in the cascade decays. A difference between the B^o and B^+ branching ratio is not unexpected in view of recent theoretical investigations[24] of the reasons for the D^o, D^+ lifetime difference.

V. CONCLUSIONS

The experimental evidence for prompt single muons, prompt neutrinos and prompt dimuons with missing energy indicates that about 30% of prompt muons come from single sources in 400 GeV p-Fe collisions. The data indicates that the source of such single leptons is a particle with a semileptonic branching ratio of ∿5% and with a hadronic production cross section of ∿20 nb. Charm production is the most likely explanation since all other <u>known</u> lepton sources are ruled out.

A charm cross section of 20 μb is smaller than the 170 μb predicted by gluon vector dominance models[51]. It is somewhat larger than the few μb originally predicted by QCD calculations[63]. It can accomodated within QCD models if the effective mass of the charm quark is taken to be[56] about 1.3 GeV rather than 1.8 GeV. It may also be accomodated if non-perturbative diagrams are included.[55]

The multimuon final states can be studied to look for bottom particle production signatures. Limits of 40 nb per nucleon are obtained from the data. These are smaller than the 850 nb predicted by gluon vector dominance models[51] but are consistent with the 10 nb obtained from QCD calculations.[55] The 40 nb limit for 400 GeV pp interaction is smaller than the 200 nb indicated by recent reports[15] of a peak in the $\psi K\pi$ mass spectrum for 150 GeV π^-N collisions.

The backgrounds in the multimuon signature are small (if p_t and missing energy cuts are applied). This indicates that with additional data a signal of the order of 10 nb (predicted by QCD) may be observable.

The measurement of hadronic production cross sections for heavy quark final states (i.e. charm, bottom) is important theoretically. It is very rare that hadronic cross sections can be calculated theoretically. However, QCD calculations of the production cross section of charm and bottom states are more feasible because of the heavy mass of the final state quarks. These calculations involve both q-$\bar{q}$ and gluon processes. Therefore, the measurement of charm and bottom cross sections provide information on quark and gluon distributions. In this sense, these measurements compliment Drell-Yan and deep inelastic experiments which yield information mostly on quark distributions within the nucleon.

APPENDIX I SOURCES OF NON-PROMPT LEPTONS

The backgrounds to prompt lepton experiments originate from long lived states. Aside from the conversion of real photons to e^+e^- and $\mu^+\mu^-$ pairs, all other non-prompt sources such as π,k decays, yield single leptons in the final state (aside from uncorrelated double π decays).

AI.1 Non-linearities (muons)

Table VIII shows the sources of non-prompt muons. These are dominated by the π and k decay contribution. These backgrounds are eliminated using the density extrapolation method. For a particle of lifetime $c\tau$ the decay length in the laboratory system is $\Lambda d = c\tau\gamma$ where $\gamma = E/M$. The interaction length in steel Λ_i is about 20 cm.

The muon decay rate from any particle as a function of density is

$$R(\rho) = \sigma \times (BR) \times \frac{\Lambda_i}{\Lambda_d} \frac{\rho_{Fe}}{\rho} \left[\frac{1}{1 + \left(\frac{\Lambda_i}{\Lambda_d}\right)\left(\frac{\rho_{Fe}}{\rho}\right)} \right]$$

where σ is the production cross section for π,K etc. and BR is the branching ratio to muons. The dependence on $1/\rho$ is exactly linear in the limit $\Lambda_i/\Lambda_d \to 0$.

Table 9 shows the ratio for all non prompt muon sources with a typical lab energy necessary to yield 8 GeV muon. As can be seen, the non-linear term is very small for π's and K's which dominate the non-prompt muon rate. For the hyperons, the non linear term is somewhat larger, but the hyperon branching ratios to muons is small and the production cross sections for high energy hyperons is also small[57]. The hyperons must be high energy in order that their 3 body decay yield an 8 GeV muon.

AI.2 Fake Intercept (muons)

In addition to small non-linearity arising from the finite lifetime of pions and kaons, there is an additional fake signal which arises from the fact that the density is non uniform. The target is composed from steel plates with air gaps. Pions attenuate in the steel plates before they reach the gaps. A simple calculation yields a fake signal which is a fraction $1/12(\ell/\Lambda_i)^2$ of the π,K decay. Here $\Lambda_i \simeq 20$ cm is the π,K interaction length and ℓ is the plate thickness. In a typical experiment[9,10] $\ell \simeq 3.8$ cm. So the fake signal is 0.3% of the π,K decay contribution (i.e. very small).

In addition, there could be a non prompt intercept due to decays occurring in non-expanded regions, i.e. upstream and downstream interactions. This has been discussed in the text.

AI.3 Non-direct Electron and Non-prompt ν_e Sources

The background for direct electron experiment are e^+e^- pairs from real photon conversions and Dalitz decays (see Table III). In addition, there are electron sources that come from long lived sources. These must be calculated and subtracted. They are important in beam dump experiments[11] [see H. Wachsmuth CERN/EP 79-125 (1979)]. These are listed in Table 10. The Ke3 and the Σ^- decays are most important.

APPENDIX II SOURCES OF PROMPT DILEPTONS

This will only be a brief summary since dileptons have already been reviewed by Smith at this Conference. The dilepton spectrum consists of a continuum plus resonances.

1. The resonance contributions tend to follow a scaling law. The scaling function $F_R(\tau)$ is similar to that of Drell-Yan production, but is not identical to the Drell-Yan function because gluon-gluon processes contribute also. Each resonance (listed in Table II e.g. $\rho,\omega,\phi,\psi,\psi',\Upsilon$ etc.) cross section could be described[55] as

$$\left(\frac{d\sigma}{dM_{\mu\mu}}\right)_R = A(M_{\mu\mu}-M_R)\ \frac{F_R(\tau)}{M_R^3}\ \Gamma_R(\mu\mu)$$

where M_R is the resonance mass, $\tau = M_R^2/S$, $\Gamma_R(\mu\mu)$ is the leptonic width, S is the square of the center of mass energy, $F_R(\tau)$ is a universal function for the resonances and $A(M_{\mu\mu} - M_R)$ is the resonance shape (i.e. a Breit-Wigner).

2. The high mass continuum ($M_{\mu\mu} > 4$ GeV) is described by the contribution of initial state quark-antiquark annihilation Drell-Yan processes, i.e.

$$\frac{d\sigma}{dM_{\mu\mu}} = \frac{F'_{DY}(\tau, M_{\mu\mu})}{M_{\mu\mu}^3}$$

where $F'(\tau, M_{\mu\mu})$ is related to the target and projectile structure functions. The scaling violations (i.e. $F'(\tau, M_{\mu\mu})$ not a function of τ only) at high energies are small for p-p collisions[6]. For π^-N collisions[7,8] there is a question of a normalization[8] factor of 1.5 to 2.0 between data and theory or a possible A dependence problem. The normalization problem may be related[62] to second order QCD (see review by Smith, this Conference).

3. In addition to the Drell-Yan continuum, there is a hadronic source of dimuons at low masses $1 < M_{\mu\mu} < 1$ GeV. One interpretation of this contribution is that it is due to final state quark-antiquark annihilation of $q\bar{q}$ pairs which have been hadronically produced. The Rochester-NSF-BNL collaboration has extracted[59] the ratio of this contribution to the Drell-Yan contribution by looking at the ratio of π^+N and π^-N dimuon cross sections. For $1 < M_{\mu\mu} < 4$ GeV they obtain

$$\frac{d\sigma}{dM_{\mu\mu}} = \left.\frac{d\sigma}{dM_{\mu\mu}}\right|_{DY} \times \frac{1}{\gamma(M_{\mu\mu})}$$

where $\gamma(M_{\mu\mu}) \to 1$ for $M_{\mu\mu} \to 4$ GeV. This ratio would be target mass dependent. This is because the Drell-Yan pairs have an $A^{1.0}$ target mass dependent and the hadronic $\mu\mu$ pairs have an $A^{2/3}$ dependence. This may be the reason why the A dependence of $\mu\mu$ pairs changes from $A^{2/3}$ at low $M_{\mu\mu}$ to $A^{1.0}$ at high $M_{\mu\mu}$. The dependence of $\gamma(M_{\mu\mu})$ on $M_{\mu\mu}$ for a Cu target[59] is shown in Fig. 10.

4. There is some uncertainty about the magnitude of the scaling violations in dimuon production. The p-p data indicates small scaling violations[5]. This is expected because the high energy data has rather limited Q^2 range and accurate lower energy pp data is not available. Scaling violations in p-p dimuon production are expected to be small because most of the contribution to the cross section is from the antiquarks in the low X region where the scaling violations are not large. On the other hand, π^-N collisions involve antiquarks at large X where the structure function exhibit large scaling violations. A comparison of the π^- structure functions extracted from low energy BNL data[9] with higher energy data shows agreement with the QCD predictions[9,61]. This agreement must be viewed with caution because the high energy data has the normalization uncertainty of 1.5

to 2.0. Also, the extraction of the structure functions from low energy data involve the previously discussed correction for non Drell-Yan hadronically produced pairs. This hadronic contribution is subtracted before the structure functions are extracted from the low energy data.

5. The mean p_T of $\mu^+\mu^-$ pairs is large at high energies, possibly due to gluon corrections. There are attempts to extract the primodial p_T distribution of the quarks from lower energy $\mu^+\mu^-$ data[60] where the gluon contribution is expected to be smaller.

6. At very low masses $M_{\mu\mu} < 1$ GeV, there are large Dalitz decay contributions in addition to the Drell-Yan and the $\mu^+\mu^-$ pairs of hadronic origin. Since this region is dominated by non Drell-Yan pairs, the A dependence is about $A^{2/3}$.

RECORD OF QUESTIONS

Name: M. S. Tannenbaum, Rockefeller University

I'd like to comment on the ISR single electron data. There are 2 experiments, CCRS at 90° (Busser et al), and Baum et al 30°. For the CCRS experiments, we vetoed on π^o Dalitz decay and also tried to measure low mass e^+e^- pairs with μ_{ee}>400 MeV/c^2. We concluded that the single electrons could not be explained by a low mass e^+e^- continuum unless the mass dependence was M^{-1} or flatter. A continuum falling like 1/M could occur from Dalitz decay or real single photon production, and a γ/π^o ratio of ∿10% for p_T>1.3 GeV/c could have explained the single lepton signal. However subsequent experiments showed that γ/π^o in this p_T range was $\lesssim$1%, thus ruling out this production mechanism. The fact that part of this single e signal might be explained by charm was first discussed by Boorquin and Gaillard and by Hinchliff and Llewelyn Smith.

Bodek: I agree, the CCRS data and the Fermilab data are both consistent with a significant charm contribution

Name: Neville W. Reay, Ohio State University

You ran with 400 GeV protons on target, Goliath ran with 165 GeV pions on target. Near threshold, the rise of pion and proton production cross-sections are considerably different. Can you comment on how you expect your limits on $B\bar{B}$ production should compare with the Goliath signal on $B\bar{B}$ production?

Bodek: As I discussed in the talk, gluon-vector-dominance model predict a larger cross section for 400 GeV pp than for 150 GeV π^-p (850 nb vs. 200 nb). Our data is not consistent with that. On the other hand if $q\bar{q}$ processes are important, pions would be more effective. For example, the ratio of Υ production cross section for 150 GeV π^-p to Υ production by 400 GeV protons is only 2±1.4. Therefore, our limit of 40 nb as compared to Goliath's 200 nb is about 2σ from that ratio. This indicates the need to do $B\bar{B}$ searches and multimuon experiments with pion beams.

Name: Jack Sandweiss, Yale University

1. How many pions of energy greater than 150 GeV are produced by the proton interactions in your experiment?

2. How sensitive are your limits on the $B\bar{B}$ production cross sections to the assumed branching ratios for the B system?

Bodek: 1. Each proton interaction produces secondary protons and pions which reinteract. We have not subtracted the contribution of these secondary interaction from our $B\bar{B}$ limits. However, there are many more secondary protons with E>150 GeV than secondary pions. For each incident proton there are 0.4 secondary and tertiary protons with E>150 GeV and only 0.018 secondary and tertiary pions ($\pi^+ + \pi^-$).

2. As discussed in the text, our sensitivity to the semi-leptonic branching ratio of the B particles affects the conclusion from the ψ_μ data in less than a proportional dependence. We assume that the $B \to X_{\mu\nu}$ branching ratio is 10% and the $B \to D \to X_{\mu\nu}$ is 8%. If the $B \to X_{\mu\nu}$ branching ratio were 5% then our limit would increase by less than a factor of 2. The limits from the 3μ and $\mu^+\mu^-$ final state are more sensitive to the difference between the B^oB^+, D^o and D^+ branching ratios due to the sign conditions in the cascade decays. If the B^o semileptonic branching ratio was smaller than the B^+ then the 3μ final states would be suppressed and the $\mu^+\mu^+$ final state would be slightly enhanced.

REFERENCES

1. L. M. Lederman, Physics Reports C, 1976 (149-181).
2. N. S. Craigie, Physics Reports 47, 1978 (1-148).
3. J. E. Augustin et al., Phys. Rev. Lett. 33, 1406 (1974).
4. J. J. Aubert et al., Phys. Rev. Lett. 33, 1404 (1974).
5. S. Herb et al., Phys. Rev. Lett. 39, 252 (1977).
6. J. A. Appel et al., Phys. Rev. Lett. 40, 435 (1978).
7. C. B. Newman et al., Phys. Rev. Lett. 25, 1523 (1970) (Chicago-Princeton collaboration).
8. J. Bodier et al., EPS Intern. Conf. on High Energy Physics (Geneva 1979); R. Barate et al., Phys. Rev. Lett. 43, 1541 (1979).
   D. McCal et al., Phys. Lett. 85B, 432 (1978) (Rochester-NSF-BNL collaboration).
9. K. W. Brown et al., Phys. Rev. Lett. 43, 410 (1979)(Caltech-Stanford collaboration).
10. J. Ritchie et al., Rochester Preprint UR-717 (Oct. 1979) "Prompt Muon Production at Small X_F and P_T in 350 GeV p-Fe Collisions" [Caltech-Fermilab-Rochester- Stanford collaboration].
11. P. C. Boseti et al., Phys. Lett. 74B, 143 (1973); P. Alibran et al., ibid 134, T. Hansl et al., ibid 139; for more recent results see H. Wachsmuth, Invited talk - Lepton Photon Conference, Batavia, 1979, also available as CERN-EP/79-115 Oct. 1979 and H. Wachsmuth CERN/EP 79-125, Oct. 1979. See also Ref. 50.

12. A. M. Diament-Berger et al. Phys.Rev.Lett. 43, 1774 (1979).
13. A. M. Diamant-Berger et al., "Search for Possible Signatures of Bottom-Quark States Produced in 400 GeV p-Fe Interactions."
14. Fermilab proposal E595, Caltech-Fermilab-Rochester-Stanford Collaboration, A. Bodek, spokesman.
15. R. Barate et al., "Possible Observation of a New Meson at 5.3 GeV." Paper #184, submitted to the International Photon and Lepton Conference, Batavia, IL. 1979.
16. "An Experiment to Observe Directly Beauty Particles Selected by Muonic Decay in Emulsion" CERN Proposal #223 PPESP/79/41 (Oct. 1979) P. Musset, spokesman.Bori, Brussels, CERN, UC Dublin, UC London, Open University, Rome, Turin collaboration.
17. A.J.S. Smith, "Lepton Pair Production in Hadronic Collisions", this conference.
18. K. J. Anderson et al., Phys. Rev. Lett. 37, 803 (1976).
19. J. P. Boymond et al., Phys. Rev. Lett. 33, 112 (1974).
20. N. W. Reay, this conference.
21. J. Kirkby, Invited talk, Lepton-Photon Conference, Batavia, IL. (1979) (DELCO results).
22. J. Dorfan, this conference (Mark II results).
23. J. L. Rosner, Proceedings of the Cosmic Ray and Particle Physics-1978 (Bartol Conference) p. 297, T. K. Gaisser editor.
24. K. Jagannathan and V. S. Mathur, Rochester preprint UR-728 (Oct. 1979).
25. H. Fritzsch, Physics Letters 80B, 343 (1979).
26. C. Bricman, "Review of Particle Properties" Phys. Lett. 75B, 1 (1978).
27. H. Meyer, Invited talk, Lepton-Photon Conference, Batavia, IL. 1979.
28. Yu. B. Bushnia et al., Phys. Lett. 79B, 147 (1978); R. I. Dzhelyadin, Phys. Lett. 84B, 143 (1979); R. I. Dzhelyadin et al., IFVE 79-121.
29. C. H. Lai and C. Quigg, Fermilab FN-296 (Sept. 1976); N. S. Craigie and D. Schildknecht, Nucl. Phys. B118, 311 (1977), C. Quigg and J. Jackson, UCRL 18487 (1968).
30. B. C. Barish et al., IEEE, Transactions on Nuclear Science, BS25, No. 1, p. 532 (Feb. 1978).
31. The range detector consists of half of the neutrino target used in Lab E neutrino experiments. (E356, E616 Caltech-Fermilab-Rochester-Rockefeller).
32. The accidental 2μ rate originates from uncorrelated decays of two pions. It is determined from the square of the single muon rate at each density and subtracted.
33. F. E. Taylor, et al., Phys. Rev. D14, 1217 (1976).
34. The parametrization used was $\mu/\pi = A(s)B(X_F)\times10^{-4}$, with $A(s)=(-1.91+0.88 \ln\sqrt{s})$ for $\sqrt{s}>15.44$, $A(s)=0.5$ for $\sqrt{s}<15.44$, and $B(X_F)=(1-2X_F)$ for $X_F > 0.4$ and $B(X_F)=(1-X_F)/3$ for $0.4 < X_F < 1.0$. This form was used with the hadron spectra from secondary and tertiary interactions and yielded a contribution of 1.22×10^{-4} prompt muons per interaction. An additional contribution of

0.38×10^{-4} from photon conversions was calculated using π^o spectra.

35. The shower calculation correctly predicted the observed non-prompt muon rate of 18×10^{-4} per interaction in the compacted configuration. A contribution of 7.2×10^{-4} per interaction comes from the decay of 3.55 pions and 0.39 kaons (with P_π>8 GeV) produced in the primary collision. An additional 10.5×10^{-4} originates from the 2.7 pions and 0.3 kaons produced in secondary and tertiary collisions.
36. The upper limit is obtained by dividing the prompt μ rate from all generations (primaries, secondaries and tertiaries) by the 3.55 primary π's only. The lower limit is obtained by assuming the $\sqrt{s}$ dependence of μ/π is flat (i.e., same for primaries and secondaries). The prompt μ rate from all generations is then divided by the 6.25 π's from all generations.
37. J. G. Branson et al., Phys. Rev. Lett. 38, 457 (1977), these authors have also expressed the results of H. Kasha et al., Phys. Rev. Lett. 36, 1007 (1976), on μ^+/π^+ and μ^-/π^- in terms of $\mu^++\mu^-/\pi^++\pi^-$ using π^+/π^- data; K.W.B. Merritt et al., (to be published). Preliminary data presented by M.H.Shaevitz in Proceedings of the 3rd International Conference on New Results in High Energy Physics, Vanderbilt (1978) p. 138.
38. A 10% correction has been applied yielding $\mu/\pi=(1.03\pm0.14)\times10^{-4}$ at $X_F=0.03$. The value of 0.93×10^{-4} is an average over the forward hemisphere and includes contribution from larger value of X_F (Fig. 2).
39. M. Barone et al., Nucl. Phys. B132, 29 (1978);L. Baum et al., Phys. Lett. 60B, 485 (1976).
40. An average value of 4.0×10^{-4} was obtained by taking a fit to the ISR data (Fig. 4a) $e/\pi=1.42\times10^{-4}$ for P_T 1.06 GeV, $e/\pi=(1.5/P_T)\times10^{-4}$ for $0.25\leqslant P_T\leqslant1.06$ and $e/\pi=6.0\times10^{-4}$ for $P_T\lesssim0.25$ GeV, and folding it against the P_T spectrum of pions with P_π>8 GeV. The range of $(3.2 \text{ to } 4.8)\times10^{-4}$ is obtained by using extreme assumptions about the behavior of e/π for P_T<0.2 (i.e. the three curves shown in Fig. 4a).
41. F. W. Busser et al., Nucl. Phys. B113, 189 (1976).
42. J. G. Branson et al., Phys. Rev. Lett. 38, 1334 (1977).
43. E. W. Beier et al., Phys. Rev. Lett. 37, 1117 (1976).
44. J. Ballam et al., Phys. Rev. Lett. 41, 1207 (1978); Y. Makdisi et al., Phys. Rev. Lett. 41, 367 (1978).
45. L. Baum et al., Phys. Lett. 77B, 337 (1978); 68B, 279 (1977).
46. Large charm baryon cross sections have been reported by ISR experiments (see A.Kernan, Lepton-Photon Conference, Batavia 1979 and ref. 47) but these measurements are at large X_F as compared to the small X_F of Ref. 39.
47. K. L. Giboni et al., Phys. Lett. 85B, 437 (1979); W. Lockman et al., ibid 443; D. Drijard et al., 85B, 452 (1979).
48. W. M. Geist, CERN/EP 79-78 (July 1979).
49. E. Amaldi et al., Nucl. Phys. B158, 1 (1979).
50. H. Wachsmuth, Proceedings of the Topical Conference on Neutrino Physics, Oxford, July 197 , p. 233, A. G. Michette, P. B. Renton, editors.

51. H. Fritzsch and K. H. Streng, Phys. Lett. 78B, 447 (1978).
52. J. Badier et al., Phys. Lett. 86B, 98 (1979).
53. D. Garelick et al., Phys. Rev. D18, 945 (1978); J. K. Yoh et al., Phys. Rev. Lett. 41, 684 (1978).
54. E. J. Siskind et al., CALT-68-665 (1979) to be published in Phys. Rev. D.
55. F. Halzen, Proceedings of the Cosmic Ray and Particle Physics-1978 (Bartol Conference) Delaware, p. 261, T. K. Gaiser,ed.).
56. C. Carlson and R. Suaya, SLAC PUB 2212 (71).
57. H. Kichimi et al., Phys. Lett. 72B, 411 (1978); M. Bourquin et al., Nucl. Phys. B153, 13 (1979); K. Doroba, Fermilab TM-818 (1979).
58. D. A. Dicus and J. R. Letaw, ORO 3992-368 Aug. 1979, Univ. of Texas, Austin preprint (Leptonic decays of the Ω^-).
59. C. Reece et al., Phys. Lett. 85B, 427 (1979); D.McCal et al., Phys. Lett. 43, 410 (1978), (Rochester-BNL-NSF collaboration).
60. W. J. Metcalf et al., Rochester preprint UR-726 (1979).
61. A. Donnachie and P.V. Landshoff, Nucl. Phys. B112, 233 (1976).
62. J. Ellis, invited talk, Lepton-Photon Conference, Batavia, Oct. 1979 (available as Ref. TH.2744-CERN).
63. J. Babcock, D. Sivers, S. Wolfram, Phys. Rev. D18, 162 (1978).

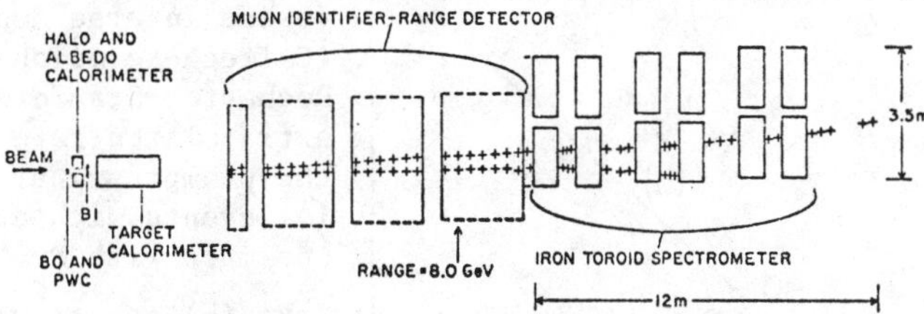

Fig. 1 Plan view of the Caltech-Fermilab-Rochester-Stanford apparatus. The beam is incident from the left. A dimuon event is shown to illustrate the spark chamber locations. In addition to the 22 3m×3m spark chambers, the 360 ton range detector contains 42 3m×3m scintillation counters. (Ref. 10.)

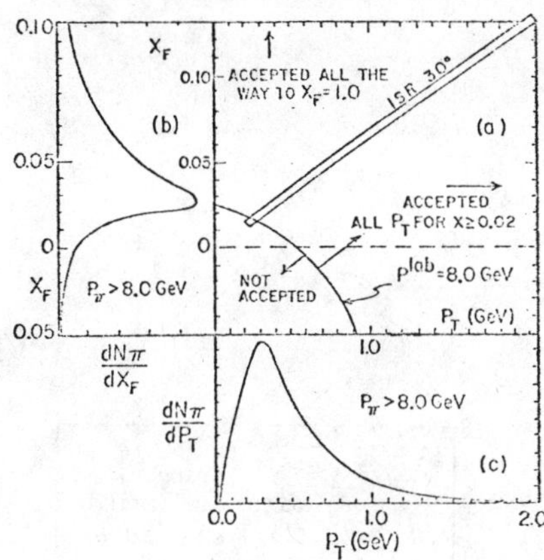

Fig. 2 The acceptance of the Caltech-Fermilab-Rochester-Stanford experiment covers almost the entire forward hemisphere in the center of mass, except for a small region around the point $X_F=0, P_T=0$. The region covered by the ISR 30° direct electron experiment is shown for comparison. A calculated spectrum of pions produced in 350 GeV P-Fe interactions with $P_\pi > 8$ GeV is shown versus X_F and P_T. (Ref. 10.)

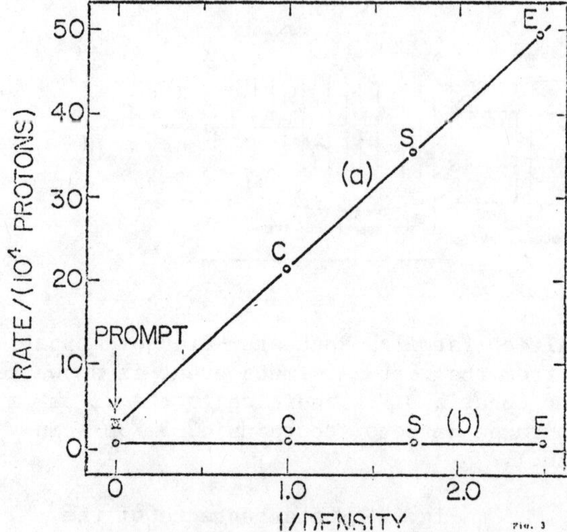

Fig. 3 The event rates versus inverse density. (Caltech-Fermilab-Rochester-Stanford) The extrapolated rate is the prompt signal. (a) events with at least one muon with $P_\mu > 8$ GeV; (b) events with two muons each with $P_\mu > 8$ GeV. (Ref. 10.)

Fig. 4 The μ/π ratio measured in the Caltech-Fermilab-Rochester-Stanford experiment (p-Fe at 350 GeV) compared to (a) ISR p-p direct electron data at small X_F vs. P_T (Refs.39 and 41), and (b) prompt muon data at small P_T vs. X_F (Ref. 37). The data of Kasha et al. (p-Cu) is at 400 GeV and of Branson et al. (p-Fe) is at 200 GeV. The experiment of K.W.B. Merritt et al. (p-Fe, 400 GeV) was performed using an earlier version of our apparatus.

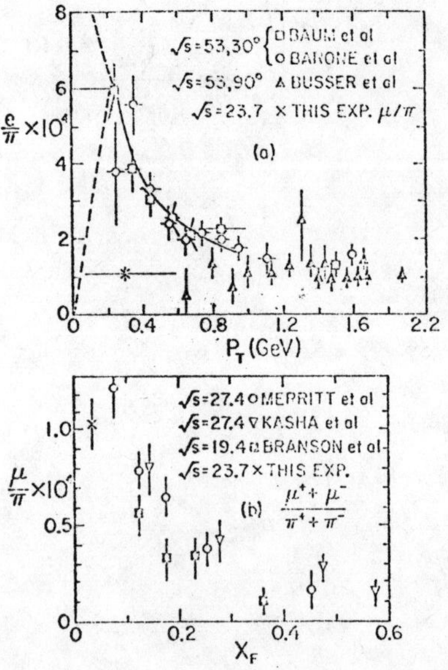

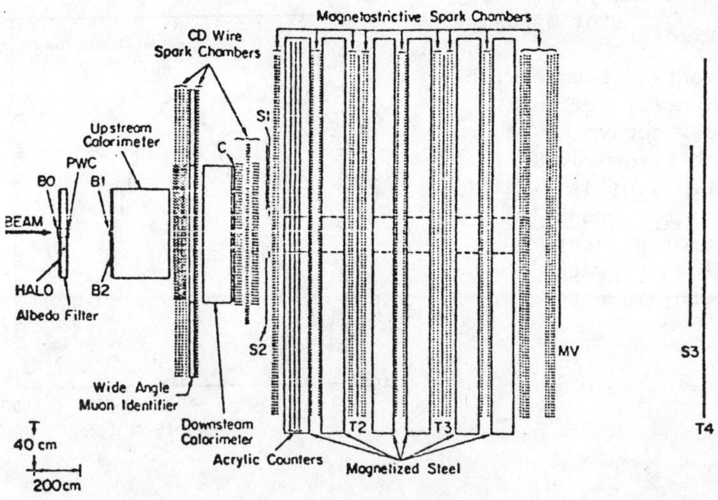

Fig. 5 The Caltech-Stanford apparatus for detection of prompt muons at high P_T. The toroidal geometry of the magnets is used to trigger on high P_T muons by requiring muons to remain in the same toroid quadrant (Ref. 9).

Fig. 6 Prompt single μ rate and 2μ rate versus inverse density for muon with $P_T \gtrsim 0.8$ GeV (Caltech-Stanford, Ref. 9).

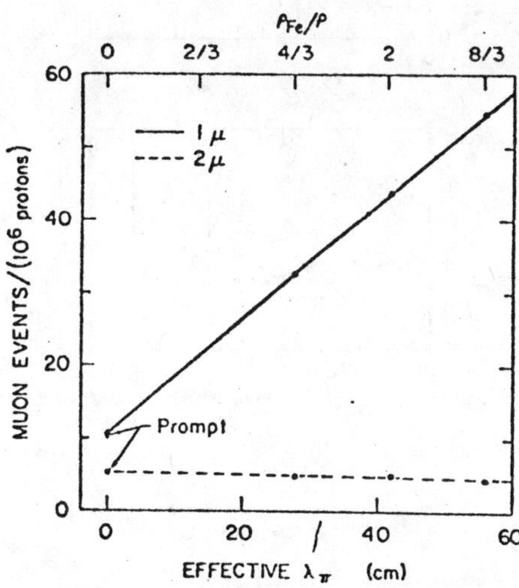

Fig. 7 The total observed energy $E_{\mu^-} + E_{\mu^-} + E_{calorimeter}$ for opposite sign dimuon events. Events with large missing energy originate from double weak decays of charm states (Caltech-Stanford, Ref. 12). The dashed line is a fit to data taken simultaneously for unbiased proton interactions in the calorimeter (i.e. without any muon requirements).

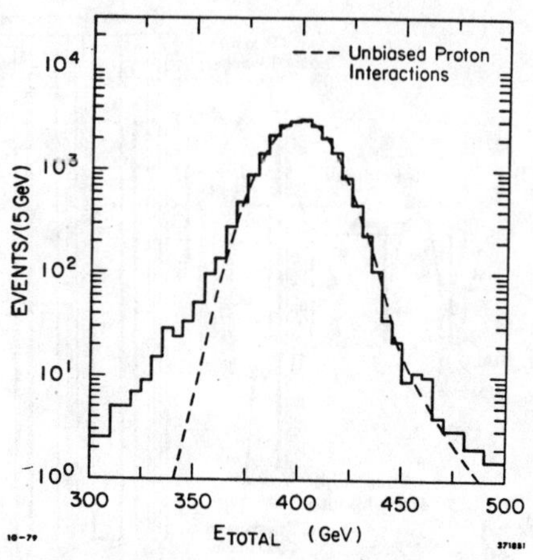

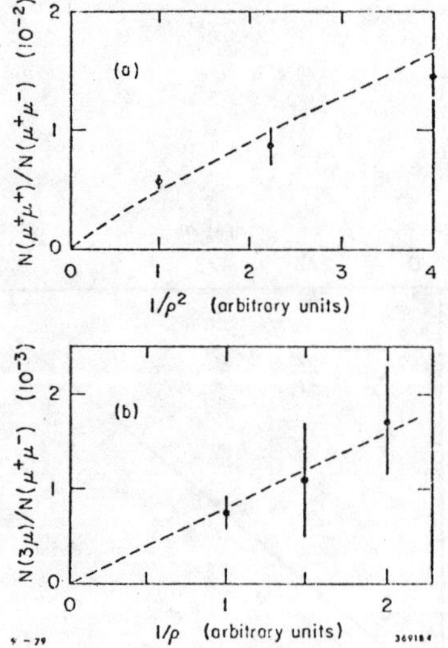

Fig. 8 Multimuon rates versus inverse density ($1/\rho$)(Caltech-Stanford, Ref. 13).
a) $\mu^+\mu^+/\mu^-\mu^-$ events vs. $1/\rho^2$. The curve is the expected distribution if all same sign events were due to a single muon (from charm or π decay) with an additional accidental π decay muon.
b) $3\mu/\mu^+\mu^-$ vs. $1/\rho$. The curve is the expected distribution if all the events were due to a $\mu^+\mu^-$ pairs from electromagnetic sources accompanied by an additional accidental π decay muon.

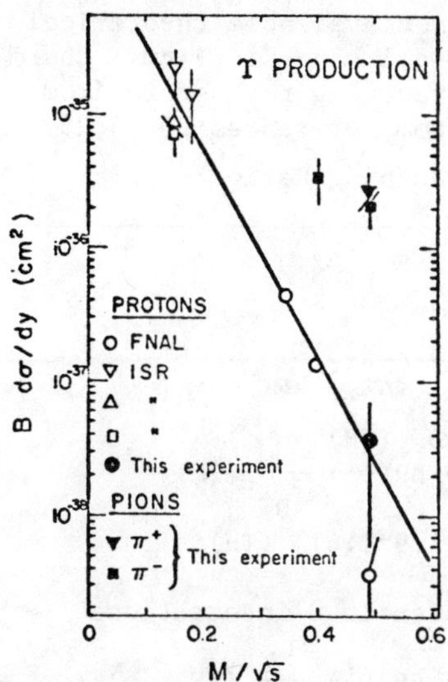

Fig. 9 Υ production cross section for protons and pions (Ref. 52).

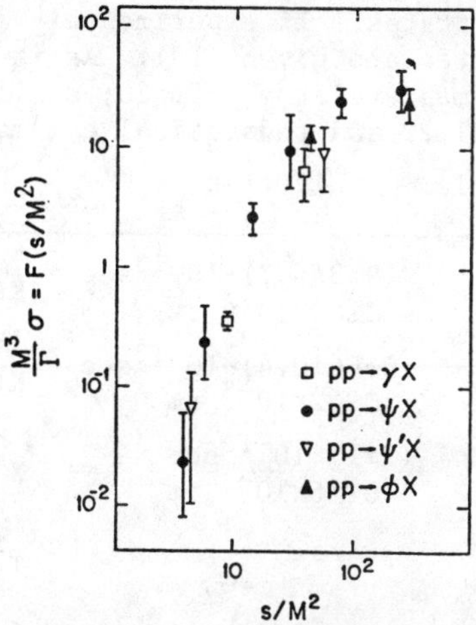

Fig. 10 The figure demonstrates that the quantity m^3/Γ-σ is a universal function of s/m^2 for different flavor bound states. m, Γ and σ are respectively the mass, direct hadronic width and production cross section of the flavor bound state (Ref. 55).

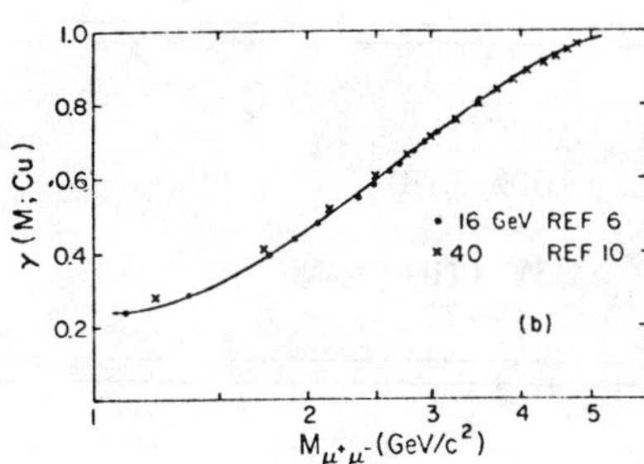

Fig. 11 The fraction of Drell-Yan $\mu^-\mu^-$ pairs for a Cu target as a function of the $\mu^+\mu^+$ mass. (Rochester-BNL-NSF, Ref. 59)

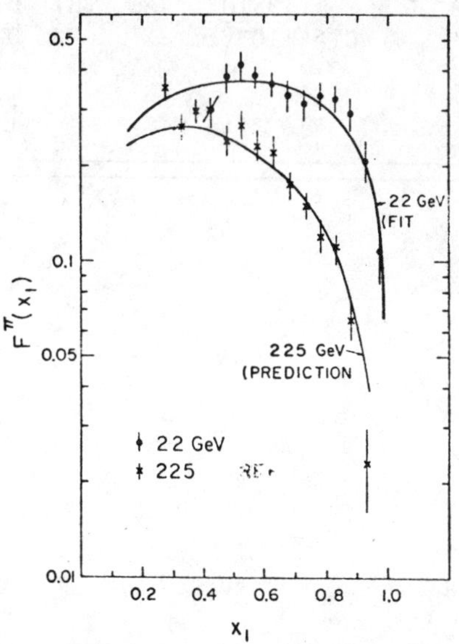

Fig. 12 Pion structure function extracted from low and high energy data (Rochester-BNL-NSF, Ref. 59 and Chicago-Princeton, Ref. 7).

Table 1: Sources of prompt leptons from weak decays. These yield mostly single leptons with equal branching ratios to $X_{\mu\nu}$ and $X_{e\nu}$ final states. If experimental values are not available theoretical estimates are given (TH). As is indicated all have $C\tau \ll 1$ cm. Charm lifetimes are from Ref. 20; semileptonic branching ratios are from Ref. 21,22 and theoretical estimates are from References 23,24,25.

Particle	Lifetime	Decay Mode	Branching Ratio
Charm:			
$D^{\pm}$	$(9.3\pm5.2)\times10^{-13}$sec $C\tau\simeq3\times10^{-2}$ cm	$K(K^*)\mu\nu$	$\begin{cases}15.8\pm5.3\% \\ 23 \pm 6 \ \%\end{cases}$
D^{o}	$(0.66\pm0.4)\times10^{-13}$sec $C\tau\simeq2\times10^{-3}$ cm	$K(K^*\mu\nu)$	$(5.2\pm3.3)\%$ $<4\%$
$F^{\pm}$	$(2\pm1)\times10^{-13}$sec $C\tau\simeq6\times10^{-3}$ cm	$X_{\mu\nu}$	$\begin{cases}=D^{o} \text{ (TH)} \\ = BR(D^{+}) \dfrac{\tau_F}{\tau_{D^+}} \text{ (TH)}\end{cases}$
		$\tau^{\pm}\nu_{\tau}$	$(1.9\pm1.6)\%$ (TH)
Λ_C	$(0.55\pm0.33)\times10^{-13}$sec $C\tau\simeq1.6\times10^{-3}$cm $(2 \text{ to } 4)\times10^{-13}$sec (TH)	$\Lambda_C X_{\mu\nu}$	seen in ν interactions $= BR(D^{+}) \dfrac{\tau_{\Lambda}}{\tau_{D^+}}$ (TH)
$\tau^{\pm}$	2.9×10^{-13}sec (TH) $C\tau\simeq9\times10^{-3}$cm		$(17.5\pm1.7)\%$
Bottom			
$B^{\pm},B^{o}$	$\tau_B>1.3\times10^{-15}$sec (TH) $C\tau>4\times10^{-5}$cm	$D(D^*)\mu\nu$ DX └ $\mu\nu$ ψX └ $\mu^+\mu^-$	$\sim10\%$ (TH) 100% (TH) └→8% 3% (TH) └→7%

Table 2: Electromagnetic sources of prompt dileptons which are expected to have equal branching ratio to e^+e^- and $\mu^+\mu^-$ pairs (note $\Gamma = 1$ ev $\Rightarrow \tau \simeq 10^{-15}$ sec). Most data is from ref. 26. Data on the Υ family is from ref. 27.

Particle	Γ, Full width (MeV)	BR to (e^+e^- or $\mu^+\mu^-$)
γ_v virtual photon	(for $m_{\mu\mu} \gtrsim 4m_\mu$)	($e^+e^- = \mu^+\mu^-$ away from $2m_\mu$ threshold)
η(549)	(0.85 ± 0.12) KeV	$(2.2 \pm 0.8) \times 10^{-5}$
ρ(770)	(155 ± 3) MeV	$(4.3 \pm 0.5) \times 10^{-5}$
ω(783)	(10.1 ± 0.3) MeV	$(7.6 \pm 1.7) \times 10^{-5}$
φ(1020)	(4.1 ± 0.2) MeV	$(3.1 \pm 0.1) \times 10^{-4}$
ψ(3100)	(67 ± 12) KeV	(7 ± 1)%
ψ'(3685)	(228 ± 56) KeV	(0.9 ± 0.1)%
ψ"(3772)	(28 ± 5) MeV	$(1.3 \pm 0.2) \times 10^{-5}$
ψ"(4415)	(33 ± 10) MeV	$(1.3 \pm 0.3) \times 10^{-5}$
Υ(9460)	$\Gamma_t \simeq 57$ KeV, $\Gamma_e \simeq (1.35 \pm 0.14)$ KeV	(2.3 ± 1.4)%
Υ'(10010)	$\Gamma_t \simeq$? , $\Gamma_e = (0.32 \pm 0.13)$ KeV	seen
Υ"(10410)	$\Gamma_t \simeq$? , $\Gamma_e = 2/3\ \Gamma_e(\Upsilon')$	seen

Table 3: Short lived electromagnetic sources of lepton pairs that have different branching ratios for electrons and muons. Muons from such sources are typically not subtracted in prompt muon experiments but must be subtracted in direct electron experiments because π^o and η Dalitz decays dominate the measured electron spectra. Branching ratios are from ref. 26 and ref. 28. Wherever branching ratios measurements are not available I have inserted theoretical estimates from ref. 29 (TH) or rough estimates (RE) using $BR(Xee)/BR(X\gamma) \approx 10^{-2}$ and $BR(X\mu\mu)/BR(X\gamma) \approx 10^{-3}$.

Source		Decay Mode	BR	
γ_v(virtual photon)		$2m_e < m_{ee} < 4m_{\mu\mu}$	μ,e difference due to threshold factors	
π_o(135)		$\gamma\gamma$	(98.85±0.05)%	
	e^+e^-	$\gamma\ e^+e^-$	(1.15±0.05)%	
		$e^+e^-e^+e^-$	3.32×10^{-5}	
η(549)		$\gamma\gamma$	(38 ± 1)%	
		$\pi^+\pi^-\gamma$	(4.89 ± 0.13)%	
		$\pi^o\gamma^-\gamma$	(3.1 ± 1.1)%	
	e^+e^-	$\gamma\ e^+e^-$	(0.50 ± 0.12)%	
		$\pi^+\pi^-e^+e^-$	(0.1 ± 0.1)%	
		$\pi^o\gamma\ e^+e^-$	$\sim 3\times10^{-4}$	(RE)
	$\mu\ \mu$	$\gamma\ \mu^+\mu^-$	$(1.5 \pm 0.75)\times10^{-4}$	
		$\pi^+\pi^-\mu^+\mu^-$	$\sim 5\times10^{-5}$	(RE)
		$\pi^o\gamma\ \mu^+\mu^-$	$\sim 3\times10^{-5}$	(RE)
ρ(770)		$\pi^o\gamma$	$(2.4 \pm 0.7)\times10^{-4}$	
		$\pi^o\ e^+e^-$	$\sim 2.4\times10^{-6}$	(RE)
		$\pi^o\ \mu^+\mu^-$	$\sim 2.4\times10^{-7}$	(RE)
ω(738)		$\pi^o\ \gamma$	(8.8 ± 0.5)%	
		$\pi^o\ e^+e^-$	8×10^{-4}	(TH)
		$\pi^o\ \mu^+\mu^-$	$(9 \pm 5)\times10^{-5}$	

Table 3 (continued)

Source		Decay Mode	BR	
$\phi(1020)$		$\eta\gamma$	$(1.6 \pm 0.6)\%$	
		$\pi^o\gamma$	$(0.14 \pm 0.05)\%$	
	e^+e^-	ηe^+e^-	1.7×10^{-4}	(TH)
		$\pi^o e^+e^-$	1.25×10^{-5}	(TH)
	$\mu^+\mu^-$	$\eta \mu^+\mu^-$	6.7×10^{-6}	(TH)
		$\pi^o \mu^+\mu^-$	1.25×10^{-6}	(TH)
$\eta'(958)$		$\rho^o\gamma$	$(29.8 \pm 1.7)\%$	
		$\omega\gamma$	$(2.1 \times 0.4)\%$	
		$\gamma\gamma$	$2.0 \times 0.3\ \%$	
	e^+e^-	ρe^+e^-	2×10^{-3}	(TH)
		ωe^+e^-	$\sim 2 \times 10^{-4}$	(RE)
		γe^+e^-	3.6×10^{-4}	(TH)
	$\mu^+\mu^-$	$\rho \mu^+\mu^-$	3×10^{-4}	(RE)
		$\omega \mu \mu$	2×10^{-5}	(RE)
		$\gamma \mu \mu$	$(8 \pm 4) \times 10^{-5}$	

Table 4 : Sources of prompt multilepton final states. These basically come from simultaneous pair decays or from cascade decays of botton states. Most estimates for branching ratios come from squares of branching ratios listed in Table I. Branching ratios for electron and muon final states are expected to be the same.

	Final State	Source	Branching Ratio
Opposite sign dimuons	$\mu^+\mu^-\nu\bar{\nu}$	a. double charm decay e.g. $D\bar{D}$	$(\sim 8\%)^2$
		b. double $B\bar{B}$ decay	$(\sim 10\%)^2$
	$\mu^+\mu^-\nu\bar{\nu}\nu_\tau\bar{\nu}_\tau$	c. double $\tau^+\tau^-$ decay	$(17.5\%)^2$
Same sign dimuons	$\mu^+\mu^+\nu\,\nu$ or $\mu^-\mu^-\nu\nu$	a. double decay bottom cascade e.g. $\{B\to XD\to X\mu\nu;\ \bar{B}\to X\mu\nu\}$	$2\times(10\%)\times(8\%)$
Trimuons	$(\psi\to\mu^+\mu^-)\mu\nu$	a. double bottom decay $\{B\to\psi\to\mu\nu;\ \bar{B}\to\mu X$ or $\bar{B}\to XD\to\mu\}$	$2\times3\%\times7\%\times(10\%+8\%)$
		b. $\psi DD\to\psi$ + extra μ	$2\times7\%\times8\%$
	$\mu^+\mu^-\mu^\pm\nu\nu\nu$	a. bottom cascade $B\to\mu D\nu$; $\bar{B}\to\mu\bar{D}X$, $\bar{D}\to\mu\nu X$	$2\times10\%\times10\%\times8\%$
4-muons	$\mu^+\mu^-\mu^+\mu^-\nu\nu\nu\nu$	a. bottom cascade $B\to\mu DX$, $D\to\mu\nu X$; $\bar{B}\to\mu\bar{D}X$, $\bar{D}\to\mu\nu X$	$(10\%)^2\times(8\%)^2$
	$(\psi\to\mu^+\mu^-)(\psi\to\mu^+\mu^-)$	$B\to X\psi\to\mu\mu$; $\bar{B}\to\bar{X}\psi\to\mu\mu$	$(3\%)^2\times(7\%)^2$

Table 5: Acceptance calculations (ε) for different charm production models and the extracted charm cross section using the 1μ and 2μ plus missing energy data. The correlated $D\bar{D}$ production models (1a-1f) fit the data with model 1a yielding the best fit. The uncorrelated production models (2a,2b) give poor fits to the data.

					1μ data		2μ data	
Model	α	β	γ	K/K^*	ε(%)	σ(μb)	ε(%)	σ(μb)
1a	1.3	6	20	1.5	4.6	21.0	.23	14.1
1b	1.3	2	20	1.5	4.6	20.9	.40	7.9
1c	2.7	6	20	1.5	3.1	31.1	.17	19.4
1d	1.3	6	30	1.5	4.1	23.9	.20	16.0
1e	2.2	3	15	1.5	4.1	23.8	.33	9.7
1f	2.2	3	15	.67	3.5	28.0	.26	12.2
2a	2.5	3	--	1.5	2.8	35.2	.19	16.9
2b	2.5	5	--	1.5	2.4	40.8	.12	27.4

Table 6: Calculated accpetance and sensitivity for $B\bar{B}$ decay final states.

	Final State	Trigger Acceptance	Acceptance After Cuts	$BR.\sigma_{B\bar{B}}$ Per Event	BR	$\sigma_{B\bar{B}}$ Per EVENT
1a.	Ψμ	65%	16%	8 pb	$7.6x10^{-4}$	10.7 nb
b.	$\Psi\mu(P_t^\mu > 1.4)$	65%	5.6%	23 pb	$7.6x10^{-4}$	30.3 nb
2a.	3μ	43%	6.0%	22 pb	$2.9x10^{-3}$	7.6 nb
b.	3μ, $E_\nu > 30$ GeV	43%	2.7%	48 pb	$2.9x10^{-3}$	16.6 nb
3a.	$\mu^+\mu^+$, $E_\nu > 30$ GeV	37%	8.5%	15 pb	$8.0x10^{-3}$	1.9 nb
b.	$\mu^+\mu^+$, $E_\nu > 30$ GeV	37%	4.0%	33 pb	$8.0x10^{-3}$	4.1 nb

Table 7: Measured events and cross section limits for each $B\bar{B}$ decay final state.

	Final State	Number of Events	Background	Events	BR $\sigma_{B\bar{B}}$ (pb)	$\sigma_{B\bar{B}}$ (nb)	$\sigma_{B\bar{B}}$ 90% C.L.(n.b)
1a.	$\psi\mu$	2	3.0 ± 0.5	$-1.0^{+2.7}_{-1.4}$	-8^{+22}_{-12}	-11^{+29}_{-15}	< 26
b.	$\psi\mu$ with $p_t^{\mu} > 1.4$ GeV/c	1	.06 ± .04	$.9^{+2.3}_{-.9}$	22^{+53}_{-20}	28^{+70}_{-26}	<116
2a.	3μ	19	17 ± 3	$2.0^{+6.2}_{-5.3}$	44^{+136}_{-117}	15^{+47}_{-40}	< 76
b.	3μ with $E_{\nu} > 30$ GeV	1	.7 ± .4	$.3^{+2.4}_{-1.0}$	14^{+115}_{-48}	5^{+40}_{-17}	< 53
3a.	$\mu^+\mu^+$	92	70^{+17}_{-17}	22 ± 20	330 ± 300	41 ± 38	< 90
b.	$\mu^+\mu^+$ with $E_{\nu} > 30$ GeV	8	6 ± 2	$2.0^{+4.5}_{-3.4}$	66^{+149}_{-112}	8^{+19}_{-14}	< 31

Table 8 Non-prompt sources of muons (from ref. 26 and ref. 58)

Particle	(Mass MeV)	$c\tau$(cm)	Decay mode	BR
π	(139.6)	780.4	$\mu\,\nu$	100%
			$\mu\,\nu\,\gamma$	$(1.24 \pm 0.25) \times 10^{-4}$
K^+	(493.7)	370.9	$\mu\,\nu$	(63.5 ± 0.16)%
			$\mu\,\nu\,\pi^o$	(3.2 ± 0.09)%
			$\mu\,\nu\,\gamma$	$(5.8 \pm 3.5) \times 10^{-4}$
K^o_L	(497.7)	1554	$\pi\,\mu\,\nu$	(26.8 ± 0.5)%
Λ	(1115)	7.89	$p\,\mu\,\nu$	$(1.57 \pm 0.35) \times 10^{-4}$
Σ^-	(1197)	4.45	$n\,\mu^-\nu$	$(0.45 \pm 0.04) \times 10^{-3}$
Ξ^o	(1314)	8.69	$\Sigma^+\mu\,\nu$	$<3.3 \times 10^{-3}$ (Exp)
				$<3.3 \times 10^{-4}$ (RE)
Ξ^-	(1321)	4.96	$\Lambda\,\mu\,\nu$	$(3.5 \pm 3.5) \times 10^{-4}$
Ω^-	(1672)	2.47	$\Xi^o\mu^-\nu$	(0.5×10^{-2}) (TH)
			$\Xi^{o*}\mu^-\nu$	3.6×10^{-6} (TH)

Table 9 Non linearity terms in density extrapolations from finite lifetime of particles (assuming a 20 cm interaction length Λ_i) for $\Lambda d \equiv c\tau\gamma$.

Particle	Energy	γ	Λd(cm)	$\Lambda_i/\Lambda d$
π	12 GeV	86	6.7×10^4	3×10^{-4}
K	12 GeV	24	0.9×10^4	2×10^{-3}
K_L^o	20 GeV	40	6.2×10^4	3×10^{-4}
Λ	69 GeV	62	490	4×10^{-2}
Σ^-	61 GeV	51	227	9×10^{-2}
Ξ^o	94 GeV	71	617	3×10^{-2}
Ξ^-	76 GeV	58	288	7×10^{-2}
Ω^-	71 GeV	42	104	0.20

Table 10 Non-prompt sources of single electron and electron neutrinos (from Ref. 26 and Ref. 58)

Particle	Cτ (cm)	Decay mode	BR	
μ	6.59×10^4	$e\,\bar{\nu}\nu$	98.6%	
		$e\,\bar{\nu}\nu\gamma$	1.4%	
π	780.4	$e\nu$	$(1.267 \pm 0.023) \times 10^{-4}$	
K	370.9	$e\nu\pi_o$	$(4.82 \pm 0.05)\%$	
K_{oL}	1554	$\pi e\nu$	38.8%	
		$\pi e\nu\gamma$	1.3%	
N	2.75×10^{13}	$pe^-\nu$	100%	
Λ	7.89	$pe^-\nu$	$(8.07 \pm 0.28) \times 10^{-4}$	
Σ^+	2.40	$\Lambda e^+\nu$	$(2.02 \pm 0.47) \times 10^{-5}$	
Σ^-	4.45	$ne^-\nu$	$(1.08 \pm 0.04) \times 10^{-3}$	
		$\Lambda e^-\nu$	$(0.60 \pm 0.06) \times 10^{-4}$	
Ξ^o	8.69	$\Sigma^+e^-\nu$	$<1.1 \times 10^{-3}$	
Ξ^-	4.96	$\Lambda e^-\nu$	$(0.69 \pm 0.18) \times 10^{-3}$	
		$\Sigma^o e^-\nu$	(0.8×10^{-4})	(TH)
Ω^-	2.47	$\Xi^o e^-\nu$	0.7×10^{-2}	(TH)
		$\Xi^{o*} e^-\nu$	(0.8×10^{-5})	(TH)
		(Xev)	$\sim 10^{-2}$	(EXP)

OBSERVATION OF HADRONIC CHARM PRODUCTION IN A HIGH RESOLUTION STREAMER CHAMBER EXPERIMENT †

J. Sandweiss, T. Cardello, P. Cooper, R. Kellogg⍭, D. Ljung≠, T. Ludlam#, R. Majka*, P. McBride, P. Nemethy*, L. Rosselet**, A. J. Slaughter, H. D. Taft, L. Teig, L. Tzeng
Yale University, New Haven, Ct. 06520

S. Ecklund and M. Johnson
Fermi National Accelerator Laboratory, Batavia, Il. 60510

INTRODUCTION

The recent discovery of charm and the theoretical expectation that charmed particles should have lifetimes of about 10^{-13} sec led to considerable effort to directly observe the decays of these particles.

As has been demonstrated at this conference(1), bubble chamber, and especially, emulsion techniques have been used to accomplish this goal. However, neither of these techniques have the possibility of being triggered on electronically selected events. In particular, the study of hadronic charm production benefits greatly from triggerability because, in contrast to charm production by neutrino interactions, the hadronic charm production rate is a fraction of the order of 10^{-3} of the total interaction rate.

We have developed a high resolution streamer chamber and have used it to study hadronic charm production by incident 350 GeV protons interacting with the nuclei of the chamber gas (Ne 90%, He 10%). The chamber has been described elsewhere(2) and only a brief summary of its properties will be given in the following.

EXPERIMENTAL ARRANGEMENT

A cutaway view of the streamer chamber is shown in figure 1, and the position of the chamber and muon filter used in the experiment is shown in figure 2. The experiment was set up in the M-1 beam line at FNAL. The beam was tuned to 350 GeV positive particles and consisted primarily of protons.

The beam was defined by the small counter B1. Upstream interactions were vetoed by requiring that the hole counters VH1 and VH2 not count for a good beam particle. The counter B2, which covered the full aperture of the entrance window assembly (about 3 cm x 3 cm), was used to reject interactions in B1 by requiring an output pulse height consistent with the traversal of a single

† Research supported in part by the Department of Energy
⍭ Currently at University of Maryland
≠ Currently at Fermi National Accelerator Laboratory
Currently at Brookhaven National Laboratory
* Currently at Lawrence Berkeley Laboratory
** Currently at CERN
*** Currently at Stanford Linear Accelerator Center

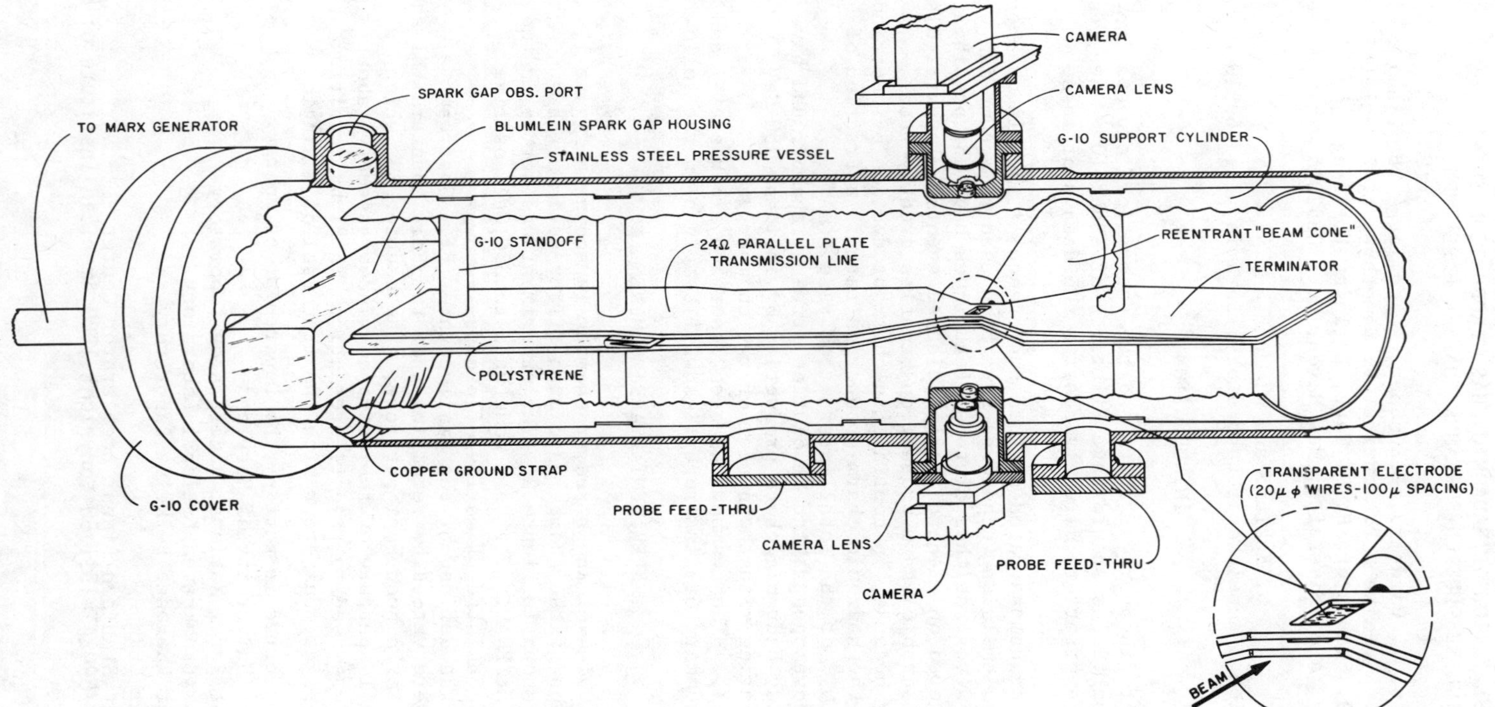

Fig. 1 A cutaway view of the High Resolution Streamer Chamber

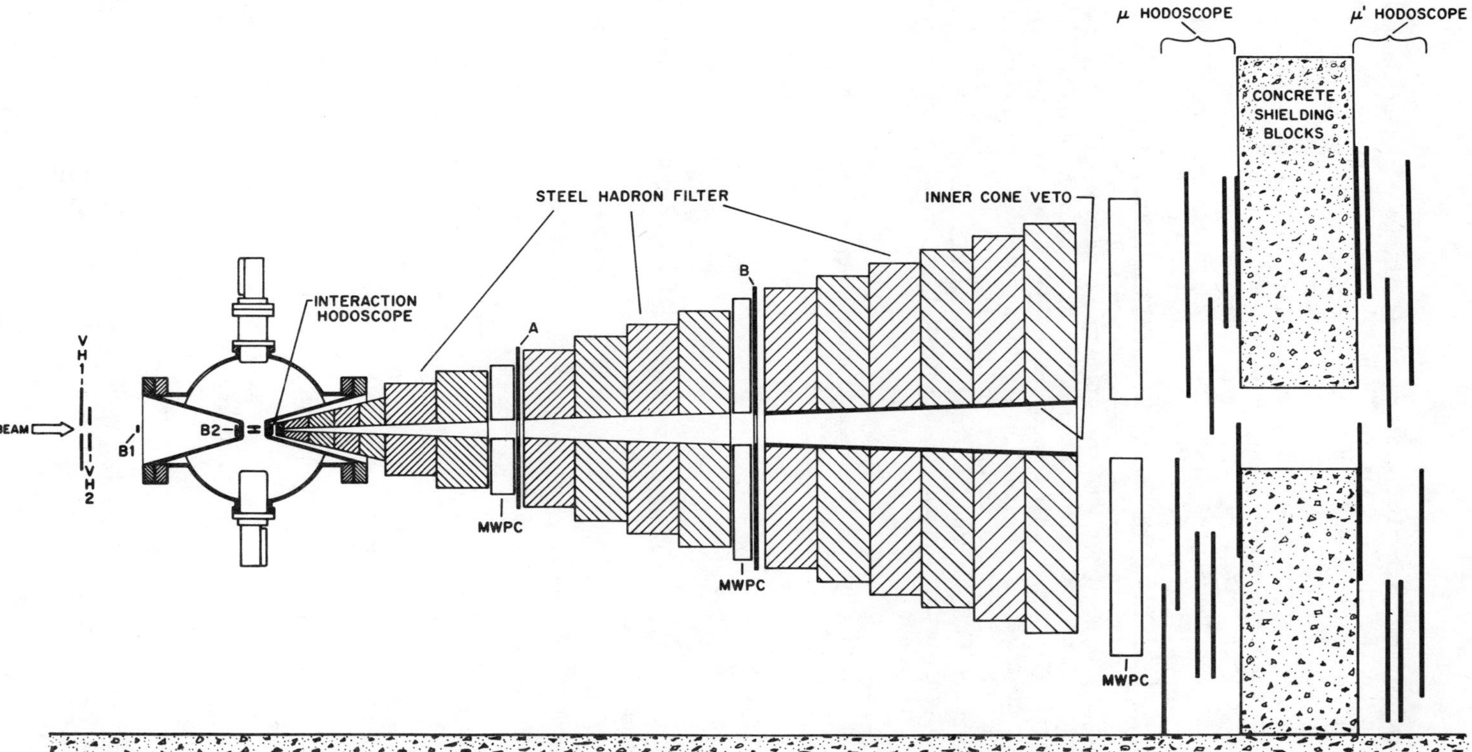

Fig. 2 Layout of the experiment

minimum ionizing particle. Interactions of the incident beam particles in the chamber gas (or windows) were detected by requiring 2 or more counts in a small 8 counter hodoscope located just behind the exit beam window.

A "muon" trigger required, in addition to a good beam particle and an interaction signal, that one or more of the counters behind the muon filter count. It was also required that suitable counters in the μ and B hodoscopes also fire so as to be consistent with one or more penetrating muons. Finally, in order to improve our rejection of hadronic punch through triggers, we also required that there not be a count in the inner cone veto counter.

The "muon" trigger was the primary trigger used in the data collection. For background studies we also took some data with an "interaction" trigger which required a good beam, an interaction and no inner cone veto. For the interaction trigger, the muon counters were ignored. For each trigger mode, all of the scintillation counters were latched and recorded on magnetic tape.

Table I summarizes the properties of the chamber as it was used in the experiment.

Table I Streamer Chamber Parameters

Gas	90% Ne, 10% He @ 24 atmosphere
Electric field	333 kV/cm
Pulse width	.5 ns
Image Intensifier	ITT Model F4112, optical gain 2500
Gap height	.45 cm
Length of visible region in beam direction	4.0 cm
Width of visible region	3.0 cm
Streamer diameter	≈ 50 μm
Track width in space	150-200 μm
Precision of measurement of track coordinates	40 μm (in space)

The track width of 150-200 μm was due to the diffusion of the primary ionization electrons prior to the application of the high voltage pulse. In our next run, improvements in the pulsing system allowing an increased chamber pressure and a smaller delay before applying the high voltage will reduce the width by about a factor of two.

RESULTS

Table II summarizes the E-490 (I) data sample, collected during the running period June to August 1978, on which the present analysis is based. The large ratio of non fiducial to fiducial interactions is due to the fact that the interaction detector was sensitive to interactions in the beam windows and in the non visible gas regions as well as to interactions in the approximately 3.0 cm of fiducial volume. It is interesting to note that this data sample corresponds to only about 40 hours of good beam time. Much of the time during the run, which was our first run with the high resolution chamber, was devoted to set up and debugging of the various experimental systems.

Table II E-490 (I) data sample

Incident beam	350 GeV protons
Incident flux per pulse	$(.5 \text{ to } .8) \times 10^6$
Average number of triggers per pulse	~ 1
Number of Fiducial Interactions with full muon trigger	1062
Number of fiducial interactions with interaction trigger	255
Ratio of non fiducial to fiducial triggers	10/1

In order to estimate charm cross sections it will be important to evaluate the performance of the trigger system. For this purpose we scaled and recorded the ungated rates for the following trigger categories:

$$\text{Full muon trigger} : B\cdot I\cdot \overline{CV}\cdot Mu$$

$$\text{Interaction trigger} : B\cdot I\cdot \overline{CV}$$

$$\text{Interaction rate} : B\cdot I$$

where B represents the pattern of VH1,VH2, B1 and B2 consistent with a good incident beam particle, I represents 2 or more hits in the 8 counter interaction hodoscope, and Mu represents the pattern of muon hodoscope hits consistent with one or more muons traversing the muon filter.

From these rates we can calculate two important ratios, averaged over the entire run.

$$\frac{B\cdot I\cdot \overline{CV}\cdot Mu}{B\cdot I} = .454 \times 10^{-3}$$

$$\frac{B\cdot I\cdot \overline{CV}}{B\cdot I} = .273$$

The first of these implies that the muon requirement rejects all but 1 in 2200 hadronic interactions and the second can be used to estimate the loss of charmed triggers due to the cone veto requirement. In particular, if the accompanying hadrons in charm production are similar to hadrons produced in ordinary hadronic interactions, we would expect 27.3% of charm production events to survive the cone veto requirement. As we shall see these numbers imply that our muon trigger events are between 15 and 50 times richer in charm production than a similar sample of "raw" interaction trigger events.

"Typical" pictures are shown in figures 3 and 4 both of which contain neutral V^o's. Figure 3 shows a heavily ionizing nuclear fragment track which is typical of about 60% of all interactions in the gas. Figure 3 also shows a background (post trigger) beam track whose narrower width clearly demonstrates the effect of diffusion of the primary electrons.

Each of the fiducial interactions was measured on our Yale image plane measuring system which has a resolution (in space) of 12 μm. All events in which the measured tracks failed to make a good fit to the hypothesis that the event contained only one vertex were examined by physicists and a sample of events in which one or more tracks clearly did not originate at the primary (production vertex) was obtained. This sample could have been obtained by visual scanning alone since each event in it has a clearly visible "miss distance". However, by flagging the events via measurement we believe we have maximized our scanning efficiency for detecting the charm decays. All of the following refers to this sample of events.

Because of the relatively large track width we do not expect to see all the features of a charm production event. We are sensitive only to one or two of the charm decay tracks which are produced at the largest angle relative to the incident beam direction. We have found it convenient to analyse our data in terms of an ℓ, θ_D plot. The distance ℓ is the distance in space (neglecting dip angles) which the charmed particle traveled before decaying and θ_D is the projected angle of the decay track. When the line of flight of the charmed particle cannot be observed we assume it to be along the beam direction. These definitions are illustrated in figure 5. The boundary separating the "charm" region and the "strange particle" region was chosen so that for charmed particle lifetimes of 10^{-12} or less there would be a negligible number of charmed particle decays in the strange particle region. It is worth noting that this boundary is determined solely by kinematics and does not depend on assumptions about the dynamics of hadronic charm production.

Figure 6 shows the data obtained from the muon trigger samples. We also require that the projected production angle of the decaying track be within 13° of the incident beam direction for an event to be considered a charm candidate. In the short decay cases we, naturally, do not apply this production angle requirement. Figure 6 shows that we have 10 such charm candidates. We have investigated the possible sources of background in this sample of charm candidates. In principle, delta rays, secondary interactions, and strange particle decays can all contribute background events in the charm region.

Fig. 3 A "typical" picture. The 350 GeV proton interacts in the chamber gas to produce a forward jet, three wide angle tracks and a neutral Vee. The full length of the picture is 4.0 cm in space.

Fig. 4 Another "typical" picture. The heavily ionizing track is a nuclear fragment. The picture also contains a neutral Vee and an unrelated post trigger beam track. The full length of the picture is 4.0 cm in space.

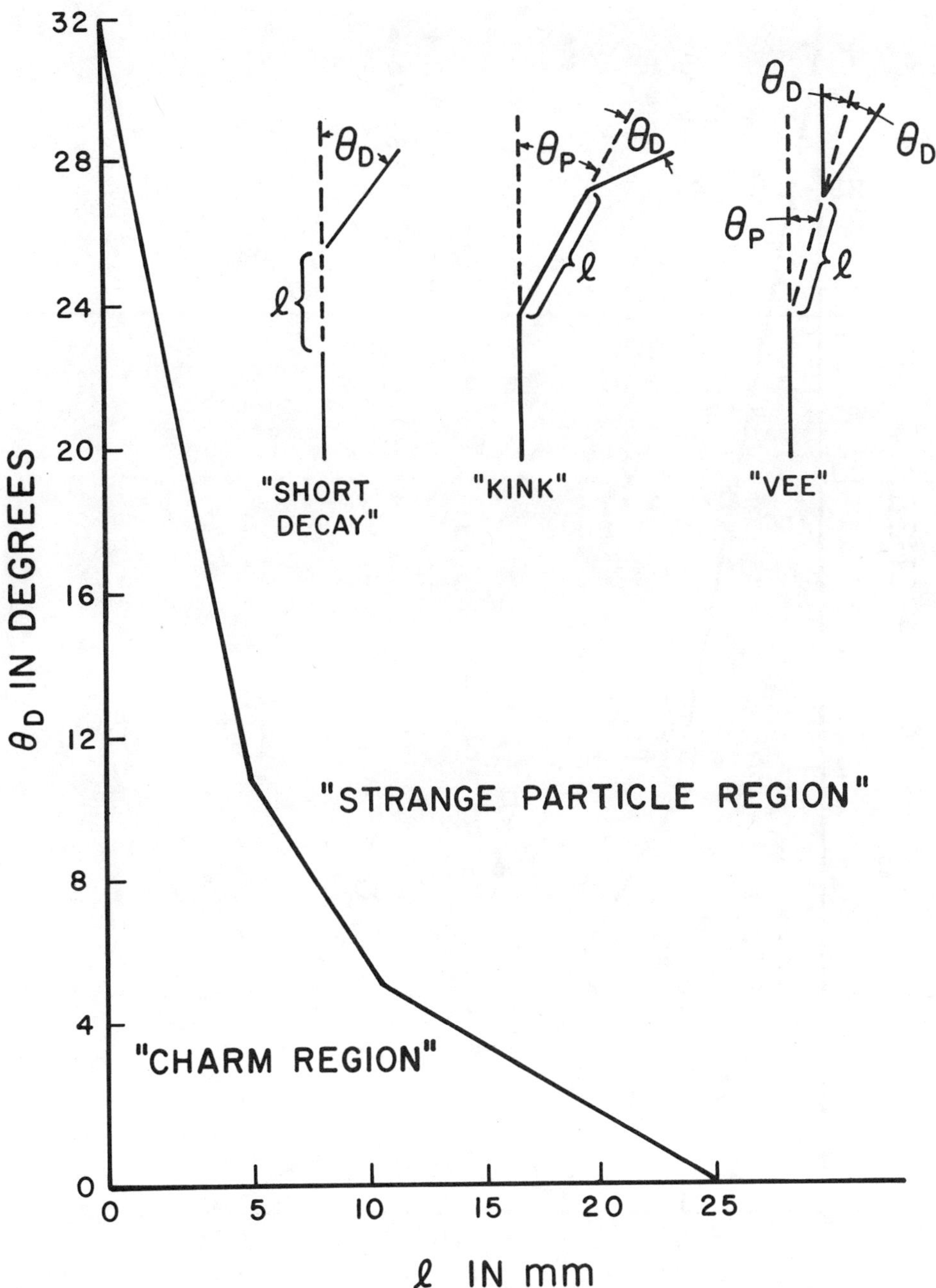

Fig. 5 The ℓ, θ plot and the definitions of event categories

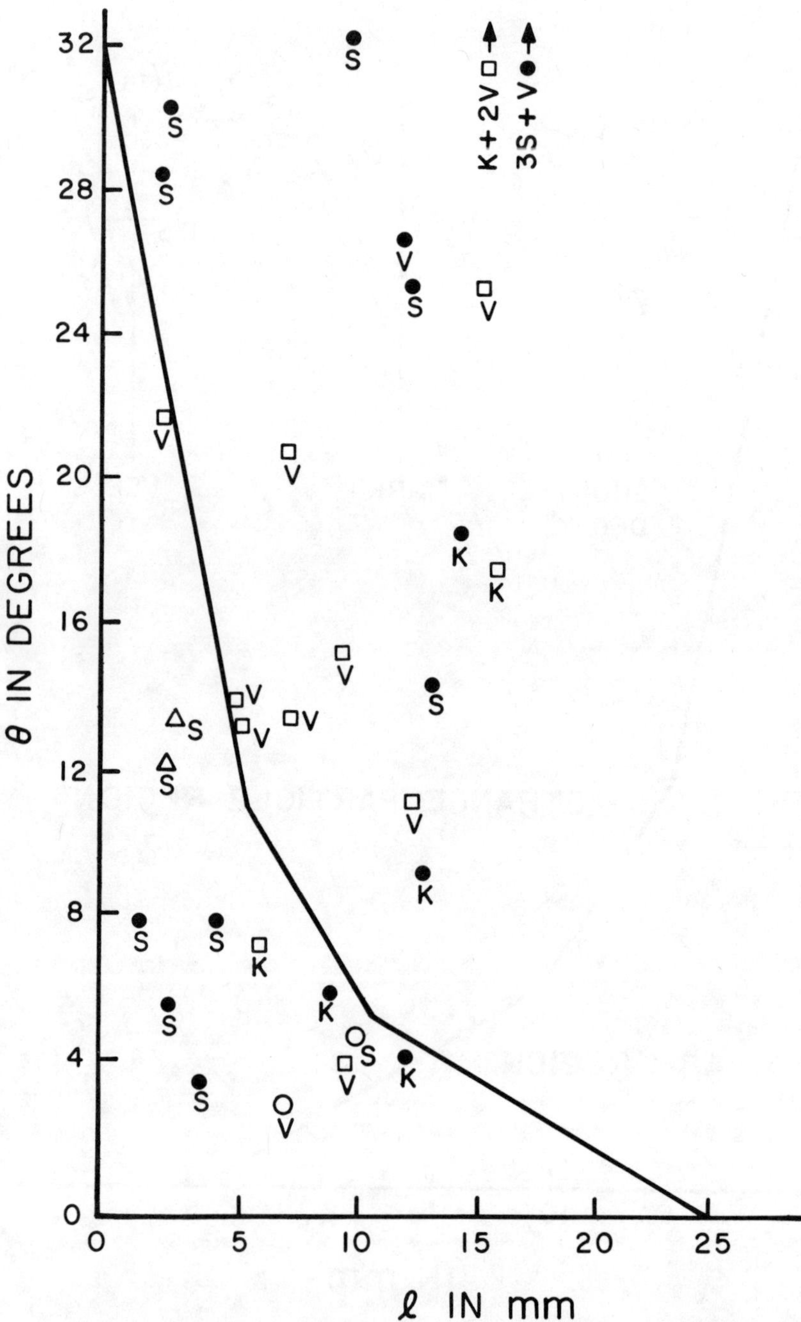

Fig. 6 Data from the muon trigger sample. The solid dots are short decay events, the squares are events with projected production angle ≥13°, the two open circles are events for which the decay track is the only possible muon and the triangles are events for which the decay track is a possible but not unique muon.

Quantitatively, delta rays give a negligible contribution and secondary interactions give less than 0.2 of a background event. Strange particles, however, are a potentially serious source of background. From the events observed in the strange particle region, we can estimate the strange particle contribution to the charm region in a straightforward way. This method will not adequately predict the background due to strange particles with momenta above about 2 GeV/c because at typical decay angles for these particles their potential path for decay is largely inside the charm region. We have estimated the contribution of these "fast" strange particles using data from a bubble chamber study of 205 GeV/c protons on hydrogen.[3] The bubble chamber data is also in good agreement with the number of events we have found in the strange particle region of the ℓ, θ_D plot. In carrying out the estimates of the fast strange particle and interaction background we have made certain simplifying assumptions which have the effect of overestimating the background, so that the resulting numbers are overestimates. A summary of the charm signal statistics is given in Table III.

Finally, we note that, based on the scintillator latches, for two of the charm candidate events, the decay track is the only track which could have been the trigger muon. In two other cases the decay track was a possible but not unique muon.

Table III charm signal summary

Trigger category	Strange particle events	Charm candidates
Muon trigger	26	10
Interaction trigger	8	1

Backgrounds to charm candidates

	Muon trigger	Interaction trigger
"Slow" strange particles	$1.07 \pm .38$	$.26 \pm .09$
Fast strange particles	$\leq .85$	$\leq .20$
Secondary interactions	$\leq .18$	$\leq .04$
Total	$\leq 2.1 \pm .4$	$\leq .5 \pm .1$

We conclude that we have observed a definite signal in the muon trigger events which cannot be explained by background. Clearly, we have not proven, in this experiment, that the signal is due to charm production but the muon association and the lifetime range involved make it a reasonable and self consistent conclusion.

In the following we shall assume that the signal is due to charm production and shall discuss the cross section and lifetime values which result.

In order to interpret our data we have carried out a Monte Carlo calculation in which we have generated charmed particles and their decays according to the model described below. The Monte Carlo program simulated the trigger system and for events which would have triggered our apparatus a tracing was generated of the charmed particle tracks, and their decay product tracks, just as they would have appeared in a chamber photograph. Each of these events were overlaid on a succession of typical pictures, from the charm candidate and strange particle region data, and examined by a physicist to determine whether or not the event would have been selected as a charm candidate. In this way, averaging over a suitable number of generated events for each assumed production and decay model, and over a suitable number of "typical" pictures (about 20), the net detection efficiency was determined for each of the assumed models.

Our charm production model assumed that D, $\bar{D}$ pairs were produced in an uncorrelated fashion with the following Feynman $X(X_F)$ and transverse momentum ($P_\perp$) dependence.

$$\frac{d^2\sigma}{d(P_\perp^2)dX_F} = Ae^{-9.94X_F^2 - 2P_\perp}$$

Our results are essentially unchanged if we use an X_F distribution of the form $(1-X_F)^{2.9}$ as suggested by studies of prompt single muons.[(4)] Similarly, including a correlation term of the form[(4)]

$$\frac{d\sigma}{dM} \sim M^3 \, {}^{-.55M}$$

where M = effective mass of the D, $\bar{D}$ pair in GeV, also has little effect on our results. The physical reason for this is the fact that with our current resolution we are essentially detecting only one of the charmed particles and are thus largely insensitive to the characteristics of the inclusive charm production.

We have assumed that the states D^+D^-, $D^+\bar{D}^o$, $D^o\bar{D}^o$and D^oD^- are produced with equal probability. For the purely hadronic decays of the D mesons we have used phase space with the multiplicities adjusted so that average number of charged decay particles from both charged and neutral D mesons is 2.3.

Finally, we have taken two different models for the semi-leptonic branching ratios and the ratio of D^o to $D^\pm$ lifetimes.

The first assumes

$$BR(D^o) = BR(D^\pm) = 10\%$$
$$\text{and} \quad \tau(D^o) = \tau(D^+)$$

The second assumes

$$BR(D^\pm) = 23\%$$
$$BR(D^o) = 0\%$$
$$\text{and} \quad \tau(0^o) = \frac{1}{5.8} \tau(D^\pm)$$

The second case was suggested by recent data from SPEAR.[5] The use of 0% rather than 23/5.8 = 4% was for computational simplicity and is not significant at our current level of statistics. The relationship between the D^o and $D^{\pm}$ lifetimes and their semileptonic branching is a consequence of the basic theory of charm which we assume in this analysis.

Figure 7 shows the results of our analysis for the two assumed cases. We have assumed that charm events are lost because of the cone veto requirement at the same rate as ordinary hadronic events. With this assumption, the Monte Carlo derived efficiency for detecting charm, and the measured rejection factor for ordinary hadronic interactions, we can determine an enhancement factor F for our muon trigger pictures. That is, each charm event observed should be given a weight of 1/F. The charm production cross section is then determined via

$$\sigma_c = 30 \text{ mb} \times \frac{N_{charm}}{F\, N_{total}}$$

where N_{charm} is the number of charm events observed and N_{total} is the total number of interactions observed.

The Monte Carlo calculations for each case were done for two assumed lifetimes for the $D^{\pm}$, 5×10^{-13} sec and 10^{-12} sec, and these are shown as the points with error flags in figure 7. In this general region the relationship between production cross section and assumed lifetime which would give our observed signal is approximately linear as has been indicated in the figure.

Although our event sample is too small to make real lifetime measurements we can set limits on the average lifetime of the charmed particles we observe. From the distribution of events in the charm particle region we can say that the average lifetime must be greater than 10^{-13} seconds. If the lifetime were that short, all of the events would have clustered at the lower "rim" of our scanning efficiency on the ℓ, θ plot. On the other hand, if the lifetime were greater than 2×10^{-12} seconds, a sufficient number of events would have occurred in the "strange particle" region so that our method of estimating the strange particle background would have led us to believe that the charm region events were all background.

SUMMARY

1. We have observed short lived particles, produced in association with muons by hadronic interactions. The most reasonable interpretation is that of charmed particle production.
2. The average lifetime of the charmed particles produced in our experiment must lie between 10^{-13} seconds and 2×10^{-12} seconds.
3. If, as suggested by references (1) and (5), the lifetime of the charged D is $\sim 10^{-12}$ seconds, we estimate the production cross section to lie between 20 and 50 μb per nucleon for 350 GeV incident protons. Figure 7 gives a more complete description of the cross section versus lifetime region implied by our experiment.

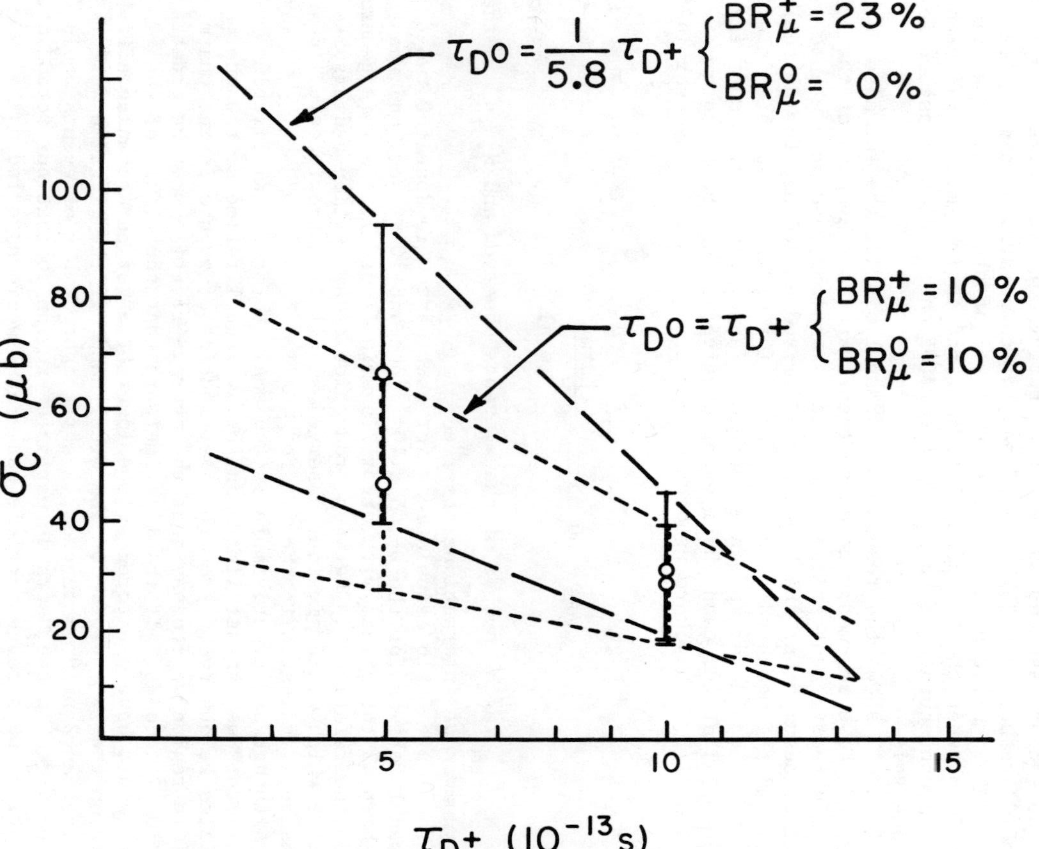

Fig. 7 The relationship between charm production cross section and lifetime of the $D^{\pm}$ implied by our experiment, for two assumptions listed. See text for additional explanation.

ACKNOWLEDGEMENTS

The development and utilization of the High Resolution Streamer Chamber would not have been possible without the skill and dedication of many people. We are especially appreciative of the support and assistance of the research division and meson laboratory division of FNAL in the entire process of building the chamber and setting up and running the experiment.

REFERENCES

1. N. W. Reay "Measuring Charm Lifetimes", proceedings of the Symposium of the Division of Particles and Fields, Montreal 1979.
2. R. Majka, et. al. "Design and Performance of a High Resolution Streamer Chamber", submitted to Nuclear Intruments and Methods, and J. Sandweiss. "The High Resolution Streamer Chamber", Physics Today, October 1978.
3. D. Ljung, Private Communication.
4. K.W. Brown, et. al. Physical Review Letters 43, 410 (1979)
5. J. Kirkby, "Recent e^+-e^- Results in the Energy Range 3 to 9 GeV", presented at the 1979 International Symposium on Lepton and Photon Interactions, Fermi National Accelerator Laboratory August 1979.

HADRON PHYSICS AT HIGH p_T

Michael J. Tannenbaum
The Rockefeller University
New York, New York 10021

ABSTRACT

Selected measurements of hadron physics at high p_T are reviewed. The inclusive single particle cross section for $p_T > 7.5$ GeV/c can be interpreted in terms of the scattering of constituents inside the proton via a force law similar in form to that governing Rutherford Scattering. The overall event structure contains two high p_T jets with a net transverse momentum imbalance K_T which increases with increasing p_T. The average fragmentation momentum transverse to the jet axis is $\langle j_T \rangle \simeq 600$ MeV/c which shows no evidence of changing with the jet p_T. Other topics discussed include direct single photon production.

INTRODUCTION

The last two years have seen many new developments in hadron physics at high p_T. I shall not attempt to describe them all; rather I shall concentrate on a few topics. For full coverage of all the issues, I can recommend the proceedings of two conferences last year at which major new results were presented that ushered in a new era in high p_T physics[1,2]. There are also many good reviews of the "classical" period of high p_T physics (1972-1977)[3,4] .

INCLUSIVE CROSS SECTION

One of the principal measurements in high p_T physics is that of the inclusive single particle cross section. Typically, I shall be concentrating on the reaction

$$p + p \to \pi + \text{anything}$$

for which the invariant cross section will be given

$$E \frac{d^3\sigma}{dp^3} = \frac{d^3\sigma}{p_T dp_T dy d\phi}$$

where p_T, ϕ and y are the transverse momentum, azimuthal angle and rapidity of the detected particle. The rapidity is related to the longitudinal momentum, or polar angle

$$y = \ln\left[(E + p_L)/\sqrt{p_T^2 + m^2}\right] \simeq -\ln \tan \theta/2$$

ISSN:0094-243X/80/590263-47$1.50

For simplicity, all quantities are taken in the center-of-mass system, of total energy $\sqrt{s}$. Two other useful variables are the scaled transverse and longitudinal momenta

$$x_T = 2p_T/\sqrt{s}$$
$$x_F = 2p_L/\sqrt{s}\ .$$

In general, most of the experiments have been performed near 90^0 in the C.M. system.

The state of the field at the end of the "classical era" is summarized in a nutshell by the data of the Chicago-Princeton Group[5] (Figure 1) in which inclusive π^+ and π^- production in p-p collisions is plotted as a function of x_T for three values of incident proton laboratory energy. This data, as well as most other data[6] in the p_T range 2 to 7 GeV/c for collisions with $\sqrt{s}$ between 20 and 60 GeV could be parameterized by the scaling form

$$E\,\frac{d^3\sigma}{dp^3} = \frac{1}{p_T{}^n}\ F(x_T)$$

with $n \simeq 8.6$ and $F(x_T) = A(1-x_T)^m$ with $m \simeq 10.6$. This scaling form, which appeared to be universal, provided an adequate description of the data of this era. The scaling law was originally suggested by models[7-11] in which particle production at large p_T is the result of hard scattering of hadronic constituents at small distances. The parameter n is related to the types of constituents and the force law governing their scattering. For instance, electromagnetic scattering of point constituents[7] or point-like vector gluon exchange[7,12] would give n=4; while quark-meson scattering via the exchange of a quark[8] would give n=8 as apparently observed.

The constituent scattering picture is also supported by another feature of the Chicago-Princeton data[5] (Figure 2). The ratio of inclusive π^+ to π^- production as a function of x_T is shown for (a) proton-proton collisions and (b) proton-neutron collisions. The particles produced at 90^0 at low x_T are roughly charge symmetric,

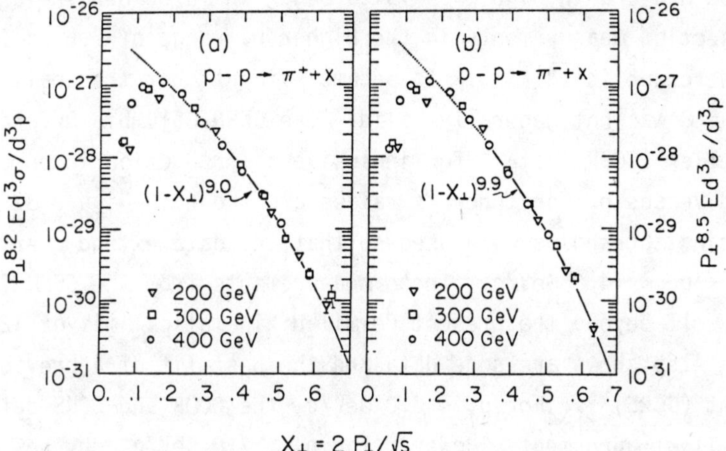
(a)
p - p → π⁺+x
(b)
p - p → π⁻+x
(1-X⊥)^9.0
(1-X⊥)^9.9
200 GeV
300 GeV
400 GeV
P⊥^8.2 Ed³σ/d³p
P⊥^8.5 Ed³σ/d³p
X⊥ = 2 P⊥/√s

FIG. 1

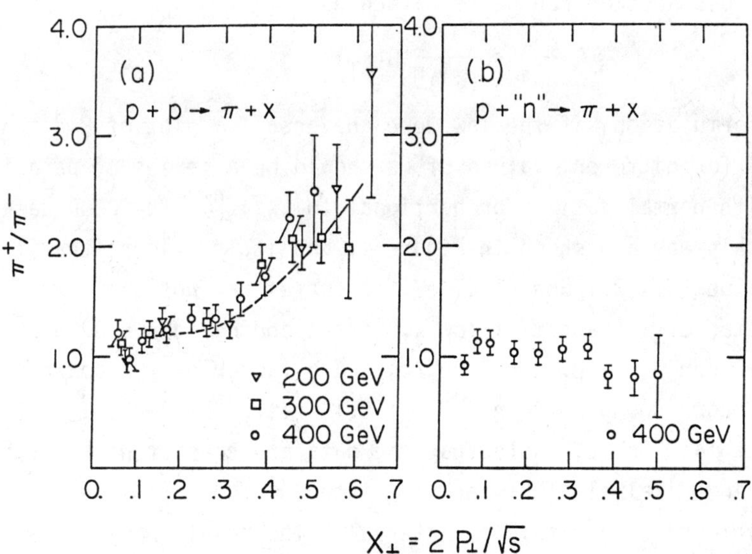
(a)
p + p → π + x
(b)
p + "n" → π + x
π⁺/π⁻
200 GeV
300 GeV
400 GeV
400 GeV
X⊥ = 2 P⊥/√s

FIG. 2

while at high x_T they exhibit the charge of the incident particles.

The new era in high p_T physics began in 1978 when inclusive cross section measurements in the higher p_T range of 7.5 to 14.0 GeV/c were published[13,14]. It became evident that the scaling law with $n \cong 8$ was not generally valid. The CERN-Columbia-Oxford-Rockefeller (CCOR) data[14] for inclusive π^0 production in pp collisions plotted versus p_T for three $\sqrt{s}$ values are shown in Figure 3. Several triggering thresholds were used so that the data extend over a range of $3.5 \leq p_T \leq 14.0$ GeV/c. These measurements from the CERN ISR extend well beyond the present Fermilab kinematic limit of 12.75 GeV/c. Also shown are some data and the best fit of a previous experiment (CCRS)[15]. For $p_T < 6.0$ GeV/c, the CCOR and CCRS data are in excellent agreement. However for $p_T > 7.0$ GeV/c, the new data are systematically above the extrapolation of the CCRS fit, $p_T^{-8.6}\,(1-x_T)^{10.6}$, shown by the dashed line.

CCOR was able to take the analysis one step further, and to study the range of validity of the scaling law. The scaling form for the cross section can be rewritten as

$$E\,\frac{d^3\sigma}{dp^3} = \frac{1}{(\sqrt{s})^n}\,G(x_T)$$

In this formulation, if the law were universal, a plot of $E\,d^3\sigma/dp^3$ versus x_T for different values of $\sqrt{s}$ should be a series of parallel curves with normalizations proportional to $(\sqrt{s})^{-n}$. The CCOR data plotted this way are shown in Figure 4; and it is evident that for the $\sqrt{s}$ values of 53.1 and 62.4 GeV the curves are not parallel. They are a factor of ~ 4 apart at low x_T corresponding to $n = 8.1 \pm 0.4$ fitted between $3.5 \leq p_T \leq 6.5$ GeV/c; and a factor of ~ 2 apart at higher x_T corresponding to $n = 5.1 \pm 0.14$ fitted between $7.5 \leq p_T \leq 14.0$ GeV/c. Note that the data are consistent with the curves being parallel at higher x_T, since n shows no sign of changing within errors over this p_T interval. The results are summarized in Figure 5 where n is plotted as determined at individual x_T values for the two lowest and two highest $\sqrt{s}$ values. It is evident that in terms of x_T, n is not universal as a function of $\sqrt{s}$. Thus,

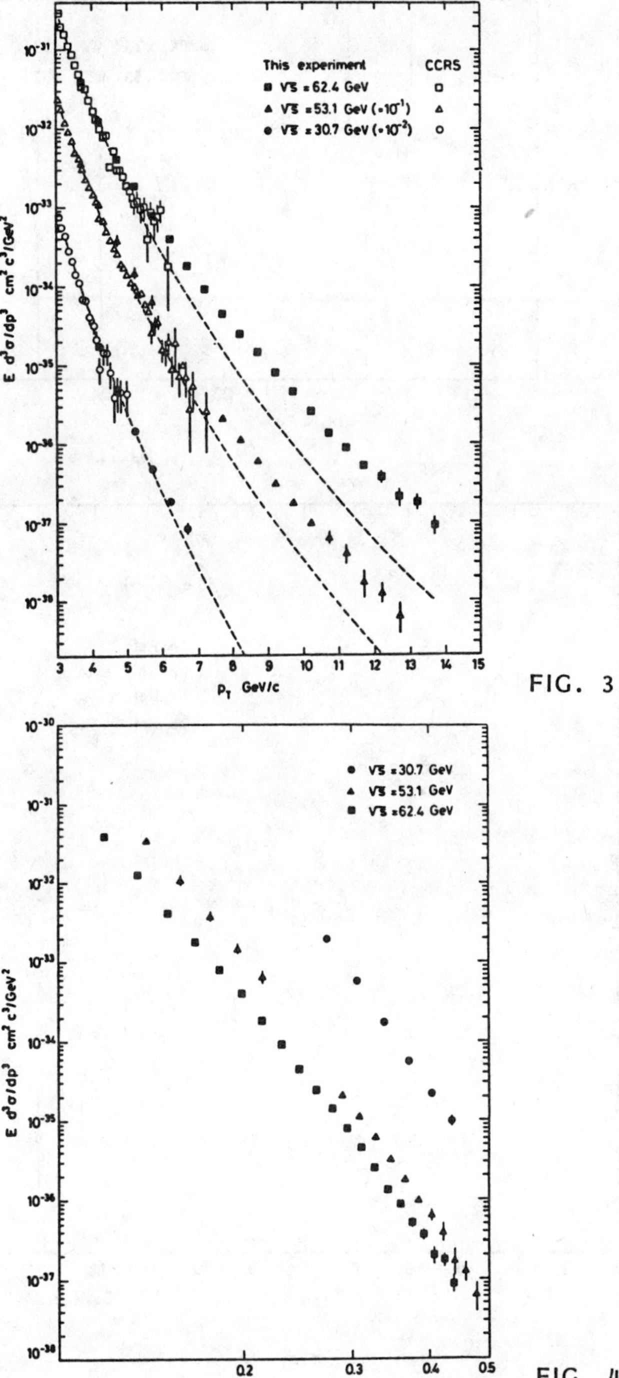
This experiment
CCRS
√s = 62.4 GeV
√s = 53.1 GeV (×10⁻¹)
√s = 30.7 GeV (×10⁻²)
E d³σ/dp³ cm² c³/GeV²
p_T GeV/c
√s = 30.7 GeV
√s = 53.1 GeV
√s = 62.4 GeV
x_T

FIG. 3

FIG. 4

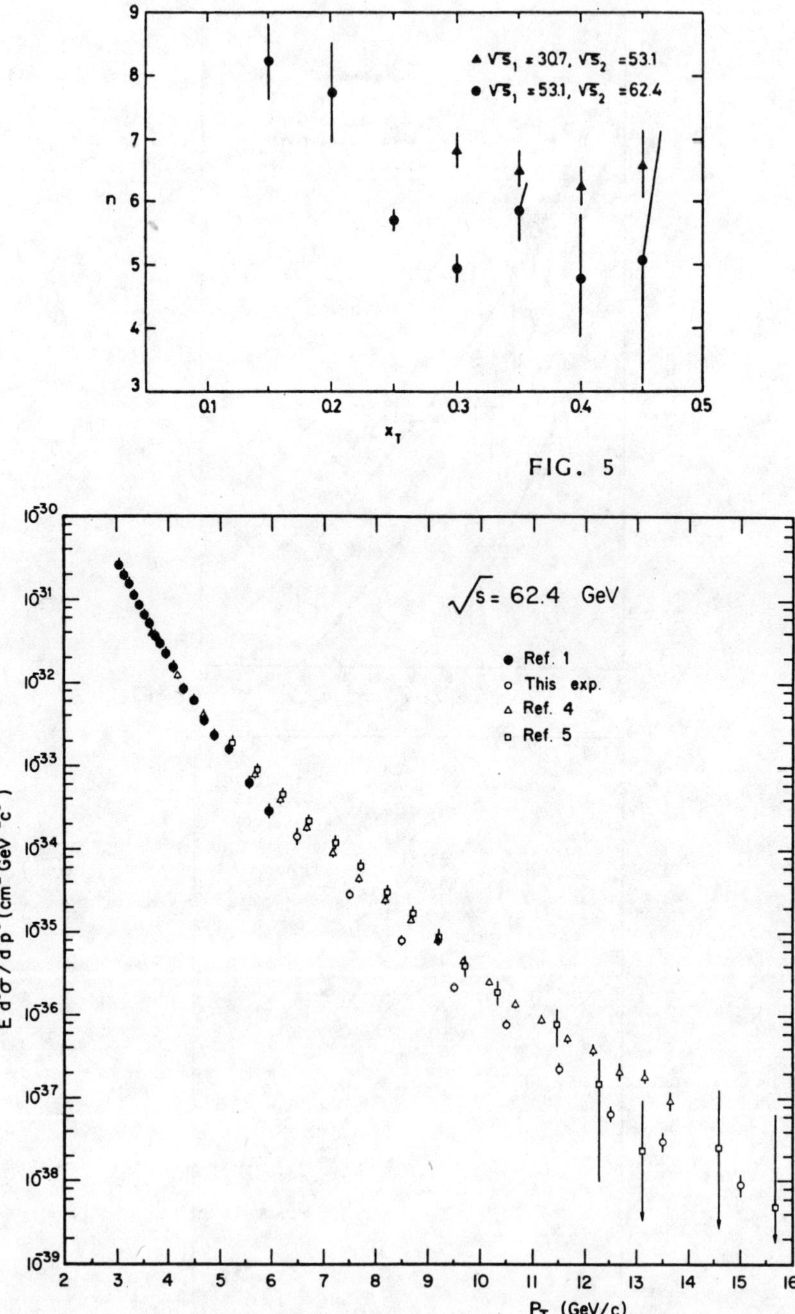
▲ √s₁ = 30.7, √s₂ = 53.1
● √s₁ = 53.1, √s₂ = 62.4
n
x_T
√s = 62.4 GeV
● Ref. 1
○ This exp.
△ Ref. 4
□ Ref. 5
$E\,d^3\sigma/dp^3$ ($cm^2\,GeV^{-2}c^3$)
P_T (GeV/c)

FIG. 5

FIG. 6

I am led to the conclusion that p_T is the important variable, and that the change of n by 3.0 ± 0.4 between the lower and higher p_T regions is indicative of a transition to a new realm of physics. Including all systematic errors[16] the new scaling form valid for $53.1 \leq \sqrt{s} \leq 62.4$ GeV and $7.5 \leq p_T \leq 14.0$ GeV is

$$E\,\frac{d^3\sigma}{dp^3} \simeq p_T^{-5.1 \pm 0.4}\,(1-x_T)^{12.1 \pm 0.6}$$

It is hopefully not just fortuitous that exactly this behavior for n and the inclusive cross section has been predicted by constituent scattering models based on Quantum Chromo Dynamics[17-21]. Essentially, most things in QCD are analogous to QED so the expectation is that n will approach 4 asymptotically. In the non-asymptotic p_T range where the measurements are available, n is expected[12] to be more like 5.5. It seems to me that QCD has to be taken seriously. In this context, it is worth pointing out that inclusive high p_T production in hadron collisions directly measures the QCD force law between constituents in analogy to the way that Rutherford scattering directly relates to Coulomb's law. Deeply inelastic muon and neutrino scattering[22], e^+e^- annihilation[23], and other aspects of high p_T hadron physics test QCD in a different way by measuring higher order effects like "Bremsstrahlung" and "Radiative Corrections."

Regardless of the theoretical interpretation, the experimental conclusions are solid. Another ISR experiment published this year has confirmed the CCOR analysis. The invariant cross sections of the Athens-Brookhaven-CERN-Syracuse collaboration[24] for inclusive π^0 production in p-p collisions at $\sqrt{s}$ = 62.4 GeV are shown in Figure 6 (circles) together with the data of the two other ISR experiments CSZ[13] (squares) and CCOR[14] (triangles). To the untutored observer, the situation looks like a mess since the data of the different groups appear to differ both in shape and absolute value. However, all groups are uncertain in their absolute p_T scale so that it is always dangerous to compare absolute cross sections. The quoted p_T scale uncertainty is $\pm 5\%$ for CCOR and $\pm 2\%$ for A^2BCS. Fortunately, the parameter n, extracted as in Figure 4, is insensitive to the absolute p_T calibration as can be readily seen from the log-log plot. The n values extracted by A^2BCS from their data at $\sqrt{s} = 52.7$ and 62.4 GeV are shown in Figure 7 as a function of x_T. To my eyes, this figure is virtually identical to Figure 5 and its conclusions are the same: for $p_T < 6.5$ GeV/c, n is in the range 8 to 9 while for $p_T > 7$ GeV/c, n decreases toward a value of 4 to 5.

While it is clear that the data at high p_T favor the QCD interpretation, one unresolved issue in pp collisions is whether the n=8 behavior at intermediate p_T values is a conspiracy[17], or whether it is evidence for the Constituent Interchange Model reaction:

$$\pi + \text{quark} \rightarrow \text{quark} + \pi$$

via the exchange of a quark[8]. An especially good testing ground for CIM[8] effects is in pion-induced high p_T production. The above reaction would imply a strong leading particle effect since the incident π acts like a constituent. For QCD models, my naive expectation would be that the reactions

$$\pi^+ + p \rightarrow \pi^\pm + x \quad \text{and} \quad \pi^- + p \rightarrow \pi^\pm + x$$

should be similar to

$$p + p \rightarrow \pi^\pm + x \quad \text{and} \quad n + p \rightarrow \pi^\pm + x$$

Data on this subject has been published by the CIT-UCLA-FNAL-UICC-Indiana Group at Fermilab[25]. The ratio of inclusive production of π^+ to π^- as a function of p_T is shown in Figure 8 for pp, π^+p, π^-p and k^-p reactions at $\sqrt{s} = 19.4$ GeV. Evidently, the π^+/π^- ratio observed in π^+p collisions is very similar to that observed in pp collisions, at intermediate values of p_T. The QCD-like model of Ref. 11 predicts the solid line labeled "FF" while a CIM prediction due to Chase and Stirling[25] is the dashed line, which disagrees with the data. This result has been confirmed by a Chicago-Princeton group at Fermilab[26] who plot the ratio of inclusive π^- to π^+

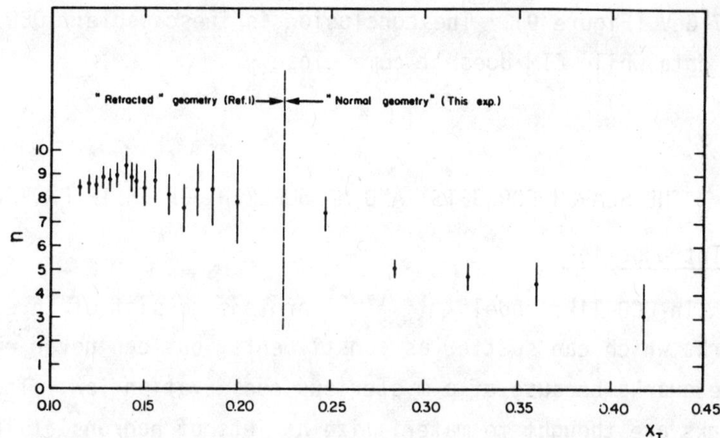

" Retracted " geometry (Ref. I)
" Normal geometry" (This exp.)
n
x_T
0 1 2 3 4 5 6 7 8 9 10
0.10 0.15 0.20 0.25 0.30 0.35 0.40 0.45

FIG. 7

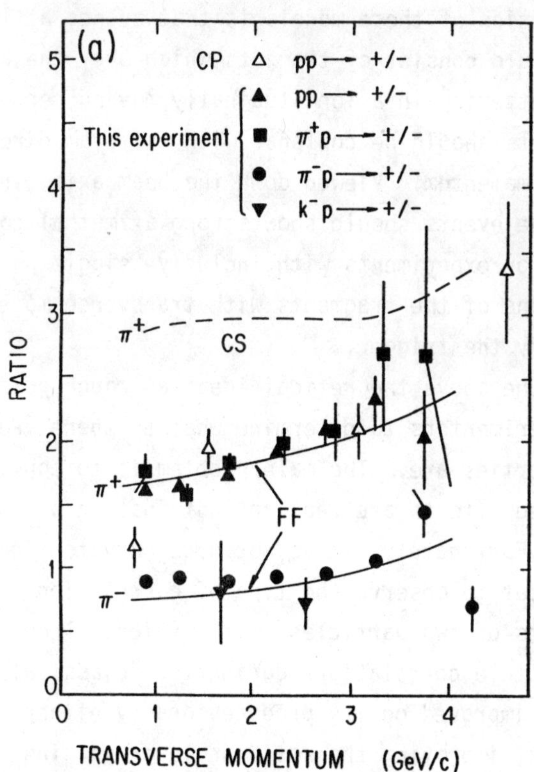

(a)
CP △ pp → +/-
This experiment
▲ pp → +/-
■ π⁺p → +/-
● π⁻p → +/-
▼ k⁻p → +/-
π⁺
CS
π⁺
FF
π⁻
RATIO
TRANSVERSE MOMENTUM (GeV/c)
0 1 2 3 4 5
0 1 2 3 4

FIG. 8

production as a function of p_T for $\pi^- p$ reactions at $\sqrt{s}$ = 19.4 and 23.7 GeV (Figure 9). The conclusion is inescapable: QCD describes the data while CIM doesn't come close.

THE SEARCH FOR JETS, AND MEASUREMENT OF THEIR PROPERTIES

a) Introduction

In QCD-like models[7-12,17-21] protons consist of three valence quarks which can scatter as constituents, but can never emerge as free quarks because of a mysterious conservation law. The scattered quarks are thought to materialize as jets of hadrons at large transverse momenta, while the "spectator" quarks continue in the beam directions, and aslo rematerialize as jets (Figure 10). A qualitative prediction of these models is that events at large transverse momenta should consist of two jets which obey the kinematics of elastic scattering in a longitudinally moving center-of-mass frame[27]. Thus the jets should be coplanar with the beam direction and balance transverse momentum. Viewed down the beam axis, in azimuthal projection, the events should show strong azimuthal correlations (Figure 11). For experiments with inclusive single particle high p_T triggers, one of the fragments with transverse momentum denoted p_{Tt} will satisfy the trigger.

With the above theoretical ideas as rough guidelines, it is up to the experimenters to determine whether there are jets and what their properties are. The main problem is to convince yourself that any observed effects are real and not just an artifact of a too-small detector. Fortunately, it is not necessary to find and reconstruct an entire jet to observe the type of correlation sketched above. Measurements of two particles will suffice. There has been much work on two particle correlations during the "classical" period . Each experiment improved on its predecessors by either adding momentum measurement, improving the solid angle or raising the trigger p_{Tt}. However, very substantial progress has been made during the past two years.

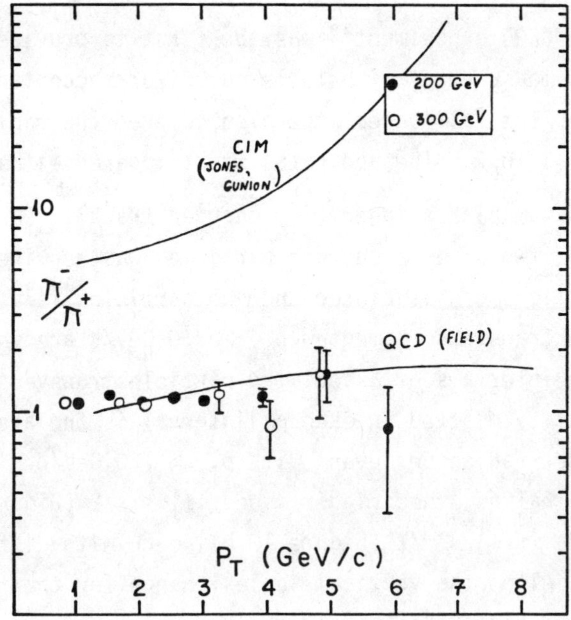

FIG. 9

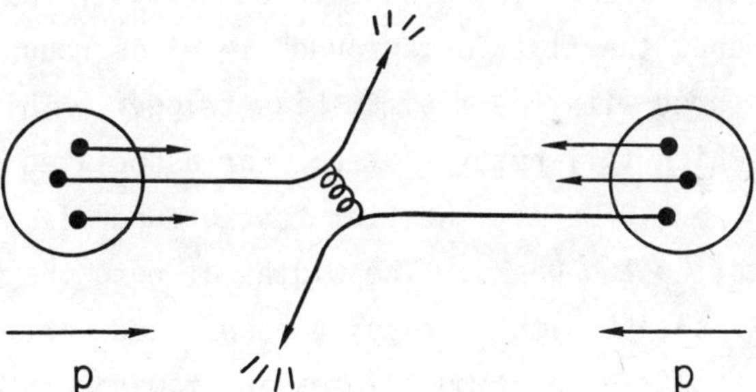

FIG. 10

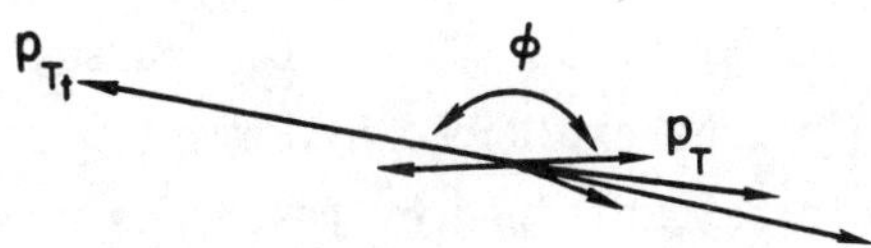

FIG. 11

b) Azimuthal Correlations

The CCOR experiment[28] was the first to provide charged particle momentum measurement with full and uniform acceptance over the entire azimuth. The full acceptance also covered the rapidity interval $-0.7 \leq y \leq +0.7$. The apparatus was triggered either by a high p_{Tt} π^0 or a minimum bias trigger (any charged track). In both cases a vertex of two or more charged particles was required. The azimuthal distributions of associated charged particles relative to a π^0 trigger with transverse momentum $p_{Tt} > 7.0$ GeV/c are shown in Figure 12 for five intervals of associated particle transverse momentum p_T. The quantity plotted in each p_T interval is the average number of tracks per radian per event with $p_{Tt} > 7.0$ GeV/c. The plot is split into two halves, the trigger side, $|\phi| \leq \pi/2$ (Figure 12a), and the away side $|\pi-\phi| \leq \pi/2$ (Figure 12 b), each with different vertical scales. Also, the vertical scales change for the five bands of associated transverse momentum.

In all cases, strong correlation peaks on top of a flat "background" level are observed on both the trigger and away sides. In each p_T band, the flat "background" level is roughly equal to the value measured with the minimum bias trigger. This is an interesting point to which I'll return later. For associated particle transverse momenta $p_T > 1.0$ GeV/c, the flat "background" is negligible in comparison to the two peaks. The widths of both the trigger and away side peaks shrink with increasing associated particle transverse momentum. This huge azimuthal peaking toward and away from a high p_{Tt} trigger seems to me to correspond to the naive two-jet picture. It seems farfetched to invoke momentum conservation to explain such strong peaking of particles with transverse momenta as low as 0.3 GeV/c relative to a 7 GeV/c trigger particle in a collision with a total available energy of 62.4 GeV. Of course, the peaking on the trigger side can not be explained by momentum conservation. In principle it could be explained by resonance production except that it is known that resonances contribute less than 10% of this effect[29].

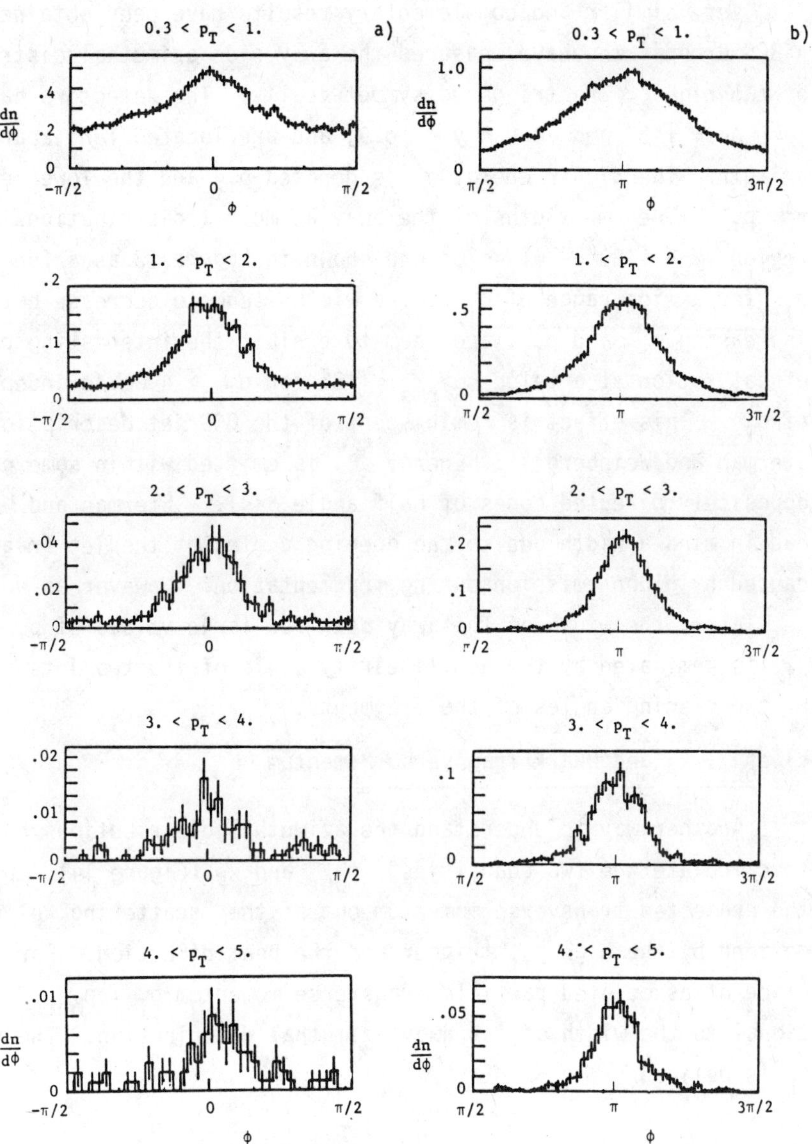

FIG. 12

Very similar and complementary results have been obtained by the A^2BCS Group[30] who have measured the away side azimuthal distribution of two high p_T π^0's triggered symmetrically. The detectors have aperture $\Delta\phi \simeq \pm 45^0$ and $-0.9 \leq y \leq +0.9$, and are located 180^0 apart in azimuth. The higher energy π^0 is denoted p_{T_1} and the lower energy π^0, p_{T_2}. The rms widths of the away azimuthal distributions for the region $\alpha \equiv |180^0 - \phi| < 70^0$ are shown in Figure 13 as a function of p_{T_2} for a wide range of p_{T_1}. The widths tend to decrease both with increasing p_{T_1} and p_{T_2}, and seem to exhibit the interesting property of saturation at a value $\langle\alpha\rangle_{rms} \simeq 13^0$ for $p_{T_2} > 4$ GeV/c independently of p_{T_1}. This effect is reminiscent of the QCD jet description of Sterman and Weinberg[31] : "Energy E is emitted within some pair of oppositely directed cones of half angle $\delta \ll 1$." Sterman and Weinberg had in mind a width due to the opening angles of the jet fragments caused by gluon emission during fragmentation. However as we shall see later, the width of the away peak for large values of p_{T_1} and p_{T2} is dominated by the acollinearity angle of the two jets and not by the opening angles of the fragments.

c) p_{out}, x_E and Quark Transverse Momentum

Another way to understand the azimuthal distributions has been to calculate the two quantities[32] p_{out} and x_E (Figure 14). p_{out} is the projected transverse momentum out of the "scattering" plane defined by the high p_{Tt} trigger and the beam direction. For any slice of associated particle transverse momentum p_T , p_{out} is proportional to the width of the away azimuthal distribution. The variable x_E is defined

$$x_E = \left| \frac{p_x}{p_{Tt}} \right|$$

so that x_E is the associated particle transverse momentum <u>projected</u> to the trigger axis and <u>scaled</u> by the trigger transverse momentum. If it should turn out that an away jet existed which exactly balanced p_{Tt} , then x_E would be the fragmentation variable, Z.

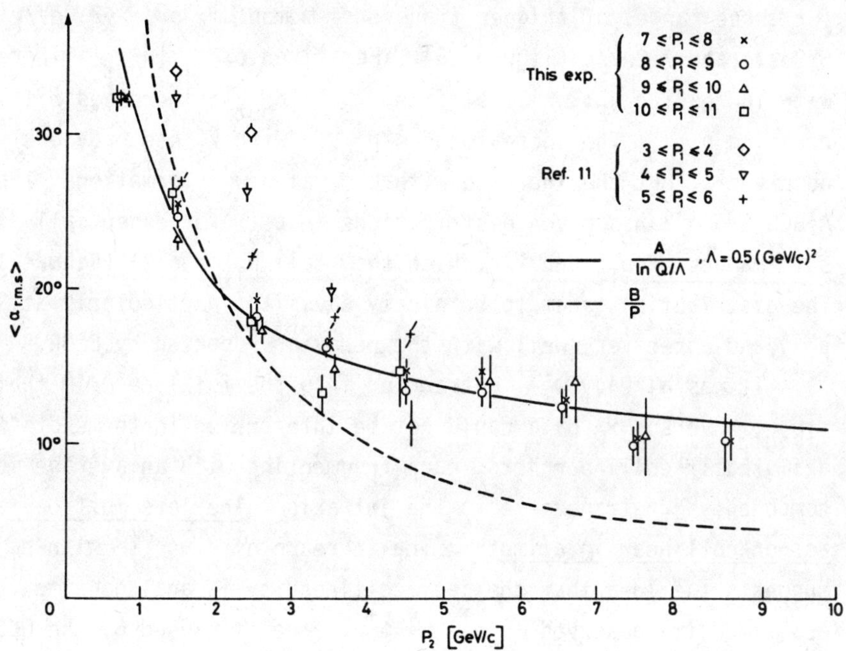

This exp.
7 ≤ P1 ≤ 8 ×
8 ≤ P1 ≤ 9 o
9 ≤ P1 ≤ 10 △
10 ≤ P1 ≤ 11 □
Ref. 11
3 ≤ P1 ≤ 4 ◇
4 ≤ P1 ≤ 5 ▽
5 ≤ P1 ≤ 6 +
A / ln Q/Λ , Λ = 0.5 (GeV/c)²
B / P
<α r.m.s>
30°
20°
10°
0
1
2
3
4
5
6
7
8
9
10
P2 [GeV/c]

FIG. 13

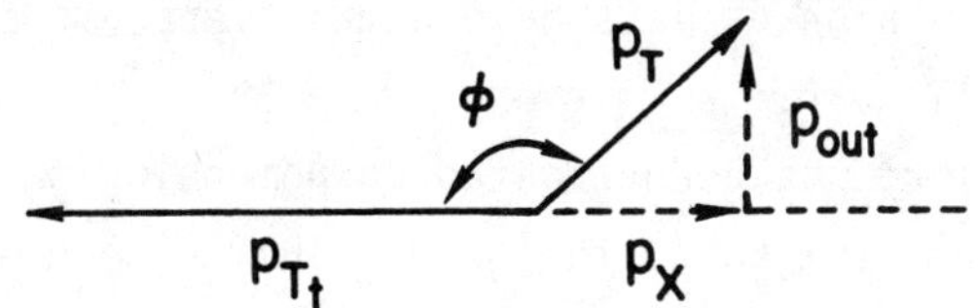

ϕ
pT
pout
pTt
px

FIG. 14

The CCOR data[28] are displayed as a plot of $<|p_{out}|>$ versus x_E for three ranges of trigger transverse momentum, $p_{Tt} > 3$ GeV/c, 5 GeV/c and 7 GeV/c (Figure 15). For fixed p_{Tt}, $<|p_{out}|>$ increases with increasing x_E; while at fixed x_E, $<|p_{out}|>$ increases with increasing p_{Tt}. The increase of $<|p_{out}|>$ with x_E had previously been observed[33], but the increase with p_{Tt} was new information. The A^2BCS Group[30] has given distributions in p_{out} for essentially the same ranges of p_{Tt} and x_E (which they call p_{T_1} and Z) (Figure 16). The distributions seem to be nicely Gaussian, particularly at low p_{T_1}, and agree very well with the behavior reported by CCOR.

It was already well entrenched into the folklore[33,34] that the $<|p_{out}|>$ versus x_E data could <u>not</u> be interpreted in terms of two azimuthally collinear jets, each fragmenting with an average momentum component $<j_T>$ transverse to the jet axis. <u>The jets must be taken as non-collinear in azimuth</u>. The increase of $<|p_{out}|>$ with p_{Tt} suggests the idea that the jet acollinearity is analogous to the acollinearity observed in $e^+e^- \to \mu^+\mu^-$, and is caused by the QCD analog of initial state radiation (Figure 17). The longitudinal momentum taken away by the initial state gluon radiation is not readily observable, since the constituents in a proton have a longitudinal momentum distribution to begin with. This effect is incorporated into the structure functions and is the explanation of scaling violations[22]. The transverse momentum taken away by the gluons goes near the beam direction and is thus missing from the central region. Therefore, the high p_T jets have a transverse momentum imbalance; they are acollinear.

This effect has been treated phenomenologically by assigning a transverse momentum k_T to the quarks inside each proton. The quark transverse momentum is thought to be composed of two parts[34], the "intrinsic" part due to the quark wave function inside the proton and the "effective" part due to QCD "radiative corrections." This same phenomenology allows the jet parameters to be determined from the two particle azimuthal correlations.

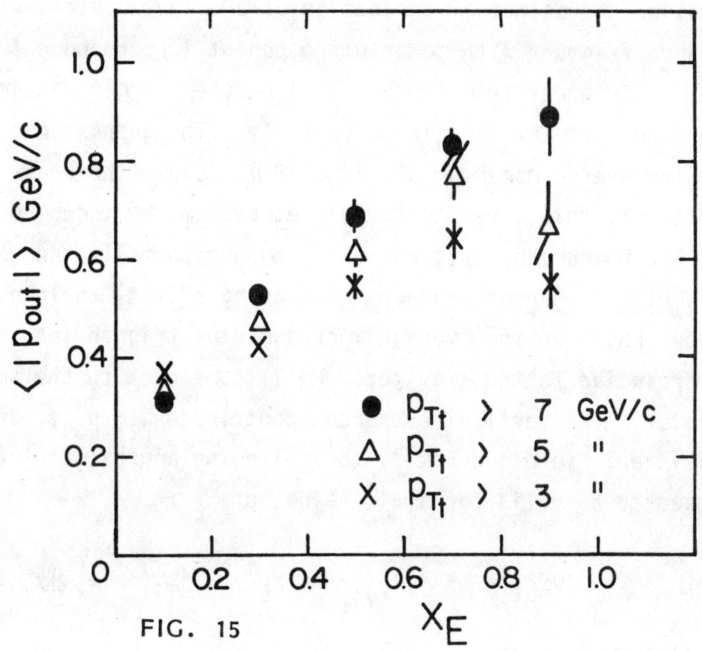
⟨| p_{out} |⟩ GeV/c
1.0
0.8
0.6
0.4
0.2
0
0.2
0.4
0.6
0.8
1.0
X_E
● p_{Tt} › 7 GeV/c
△ p_{Tt} › 5 "
× p_{Tt} › 3 "

FIG. 15

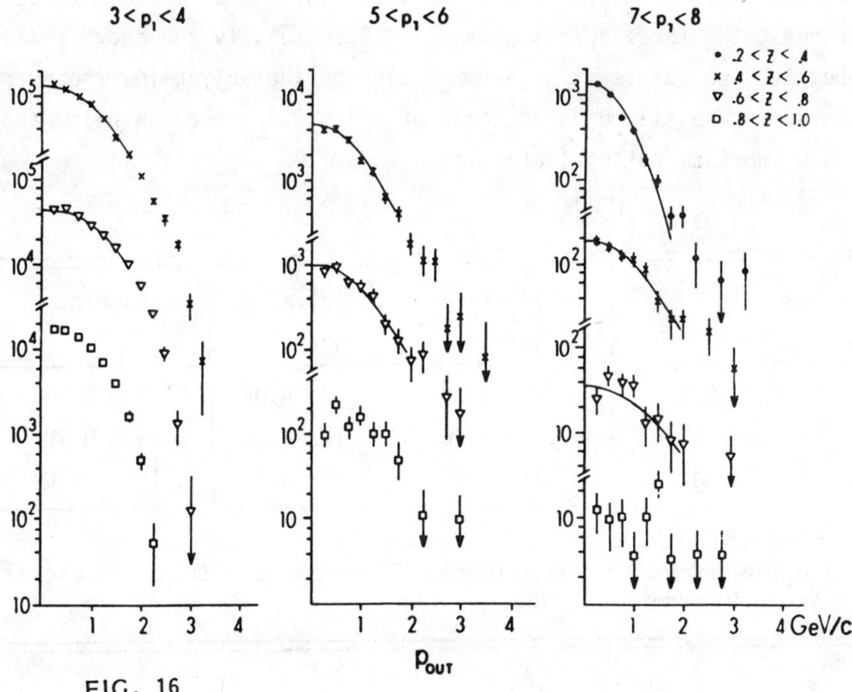
3 < p_1 < 4
5 < p_1 < 6
7 < p_1 < 8
• .2 < Z < .4
× .4 < Z < .6
▽ .6 < Z < .8
□ .8 < Z < 1.0
p_{OUT}
GeV/c

FIG. 16

Consider the azimuthal projection of a di-jet system, with each jet having a fragment with momentum component j_T perpendicular to the jet axis. A fragment from one jet will be the trigger particle, and from the other jet the associated particle. The quarks in each proton have transverse momentum k_T (Fig. 18a). Since there are two colliding protons, this gives to the di-jet system two independent components, k_T, which can be taken as acting horizontally and vertically. The horizontal component produces a smearing effect, analogous to resolution, which on the average increases the trigger jet transverse momentum relative to the away jet. We'll come back to the smearing effect later. The vertical component contributes to p_{out}, and according to Feynman, Field and Fox[34], the following approximate relation is supposed to be valid for small values of x_E :

$$<|p_{out}|>^2 = <|j_{T\phi}|>^2 + x_E^2 \left(<|j_{T\phi}|>^2 + <|k_T|>^2\right)$$

The CCOR data[28,35], which span the full x_E range, are plotted in this way, $<|p_{out}|>^2$ versus x_E^2, in Figure 18b. For all three p_{Tt} ranges, the three points with $x_E^2 \leq 0.3$ satisfy the above relation but the two points at higher x_E^2 lie considerably below the extrapolation. The fitted parameters for $x_E^2 < 0.3$ are given below and show two interesting features:

	$<\|j_{T\phi}\|>$ GeV/c	$<\|k_T\|>$ GeV/c	$<\|k_T\|>/p_{Tt}$
p_{Tt} > 3 GeV/c	0.352 ± 0.003	0.77 ± 0.05	0.223 ± 0.014
5 GeV/c	0.336 ± 0.003	1.12 ± 0.04	0.195 ± 0.008
7 GeV/c	0.321 ± 0.003	1.34 ± 0.03	0.166 ± 0.004

Note the errors are statistical. There is in addition an overall systematic error of ± 10%.

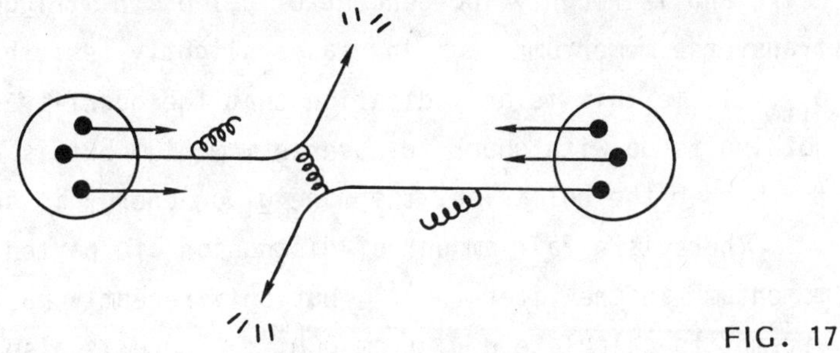

FIG. 17

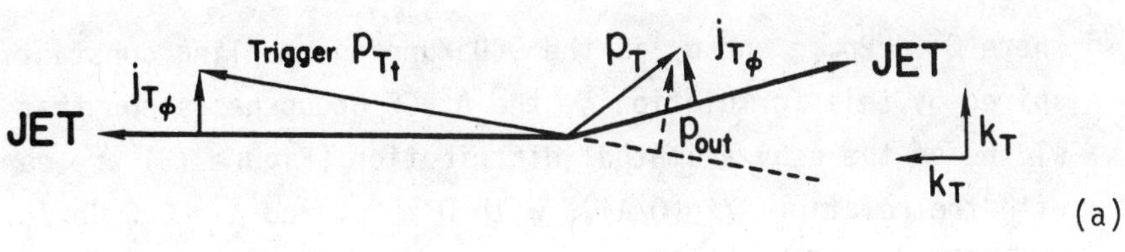

(a)

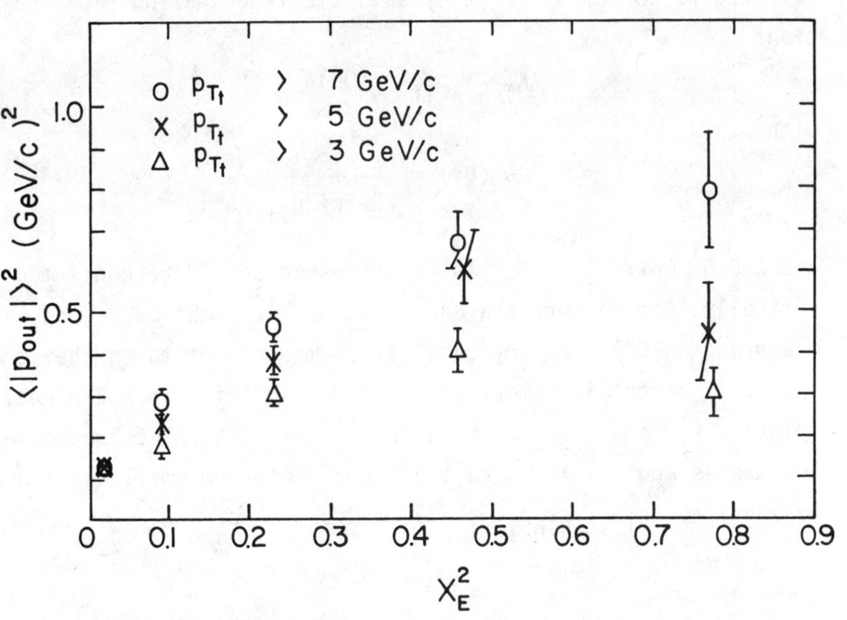

FIG. 18

(b)

The mean absolute value of the ϕ projection of j_T is equal to ~ 0.35 GeV/c and is roughly independent of p_{Tt} of the trigger. The "quark transverse momentum" $\langle k_T \rangle$ increases slightly less than linearly with p_{Tt}. To me this is an indication that the quantity $\langle k_T \rangle$ really has nothing to do with quark transverse momentum but is a pure QCD effect: the harder the collision, the more gluon energy is shaken off.

There is a fair amount of discussion of "parton transverse momentum" in the literature[36]; but only recently has there been an attempt to calculate p_{out} from QCD[37]. There is also the suggestion[38] that

$$\langle p_{out}{}^2 \rangle \sim \alpha_s(Q^2)\, Q^2$$

where $Q^2 = 2p_{Tt}{}^2$ and α_s is the QCD running coupling constant. Inspired by this formulation[39], the A^2BCS Group has shown that the widths of the away azimuthal distribution (Figure 13) are consistent with the relation $A/\ln(Q/\Lambda)$, with $Q = 2p_T$ and $\Lambda = 0.5$ GeV/c. Also, the CCOR data for the net transverse momentum of the di-jet system (assumed to be $\langle K_T \rangle = \sqrt{2}\, \langle k_T \rangle$) satisfies the amazing empirical equation:

$$\langle K_T \rangle = p_{Tt}\, \alpha_s(Q^2)$$

with

$$\alpha_s(Q^2) = \frac{12\pi}{25\, \ln(2p_{Tt}{}^2/\Lambda^2)}$$

and $\Lambda = 0.52$ GeV/c[40]. If this relationship should be confirmed theoretically, and be more than an accident, it might be a good way of measuring $\alpha_s(Q^2)$. In any case, it seems evident to me that the acollinearity effect is almost purely QCD in origin since the real intrinsic $\langle k_T \rangle$ has been measured in $\mu^+\mu^-$ production at Brookhaven[41] at low masses and low $\sqrt{s}$ where the QCD effects are small and found to be equal to $\langle k_T \rangle_{INTRINSIC} = 0.32 \pm 0.04$ GeV/c.

d) The k_T Component in the Scattering Plane: How to see it -- How to avoid it --

The in-plane component k_T produces a smearing effect, analogous to resolution, which is supposed[17] to be the principal contribution to the change of n from ~ 5 to ~ 8 at intermediate p_T. Smearing always flattens a steeply falling spectrum; and the steeper the spectrum, the larger is the relative effect. In order to overcome the k_T smearing, symmetric triggers 180^0 apart in azimuth have been used (Figure 19). For two jets selected to be roughly collinear, the in-plane component of transverse momentum imbalance should be due entirely to k_T.

$$\vec{k}_T = \vec{p}_{T_1} + \vec{p}_{T_2} \qquad \text{OR}$$

$$k_T = |\vec{p}_{T_1}| - |\vec{p}_{T_2}|$$

The average transverse momentum

$$p_{TAVG} = \frac{|\vec{p}_{T1}| + |\vec{p}_{T2}|}{2}$$

should be relatively independent of k_T. For back-to-back single particles, the situation is more complicated since each particle has only a fraction of the jet momentum; but, similarly, the effect of k_T cancels when using the average transverse momentum of the two particles.

The Stony Brook-Columbia-Fermilab Group[42] has given the six-fold differential two particle inclusive cross section for back-to-back $\pi^+\pi^-$ pairs as a function of x_{TAVG} (Figure 20). The cross section has been computed within the interval $|\vec{p}_{T_1} + \vec{p}_{T_2}| < 1.1$ GeV. The solid lines are a QCD calculation[43] which depends on $\langle k_T \rangle$ only to the extent that the data are given at $\phi_1 - \phi_2 = 180^0$, the peak of the azimuthal distribution (e.g., see Figure 12b). The height of the peak depends on the assumed width since the area, which is independent of $\langle k_T \rangle$, has been calculated. To avoid this complication, CCOR[44], for opposite side π^0 pairs at $\sqrt{s} = 62.4$ GeV, has integrated the six-fold two particle cross section over the relative azimuthal angle (Figure 21).

The agreement with the QCD prediction[43] is impressive if not slightly fortuitous because of the ± 5% experimental uncertainty in the absolute transverse momentum scale.

The SCF Group[42] has also given the distribution in $p_T' = |\vec{p}_{T_1}| - |\vec{p}_{T_2}|$ for back-to-back $\pi^+\pi^-$ for various values of $m' = 2\ p_{TAVG}$ (Figure 22). The distributions are relatively flat, but do show a drop-off for large values of p_T' with a FWHM ≃ 1.5 p_{TAVG}. The distribution in exactly the same quantity, $p_{Tx} = |\vec{p}_{T_1}| - |\vec{p}_{T_2}|$, for back-to-back jets has been obtained by the Wisconsin-Pennsylvania-Lehigh-Fermilab Group[45] in an experiment using two back-to-back calorimeters. A jet is defined as the vector sum of all momenta in a solid angle of ∿ 1½ steradians. A fiducial cut selects jets pointed at the central $\pm 10^0$ in c.m.s. polar and azimuthal angles in each calorimeter. The distribution in p_{Tx} for a range of p_{TAVG} between 2 and 3 GeV/c is shown in Figure 23 which dramatically illustrates that jets and single particles behave differently. The jets show a strong peaking about $p_{Tx} = 0$ with a Gaussian behavior. This implies that the two jets balance transverse momentum apart from the smearing, k_T, which can be obtained directly from the width of the Gaussian after two corrections (Figure 24). The corrections are from two sources: Instrumental resolution and uncollected jet fragments. The latter correction is model dependent since it depends on the assumed parameters of the jets, particularly the fragmentation function and the fragmentation transverse momentum $\langle j_T \rangle$. I note as a warning that the $\langle j_T \rangle$ used is 30% lower than the value measured for hadron collisions[46].

The corrected results for $\langle k_T \rangle_{rms}$ are shown in Figure 25 as a function of x_{TAVG}. At a given $\sqrt{s}$, the WPLF values of $\langle k_T \rangle_{rms}$ increase with increasing p_{TAVG}; but this effect is not very compelling for $\sqrt{s}$ below 27.4 GeV. The WPLF data also show that at fixed x_{TAVG}, $\langle k_T \rangle_{rms}$ increases with increasing $\sqrt{s}$, which is new information. In the absence of quantitative theoretical predictions, the WPLF Group give their own empirical relation:

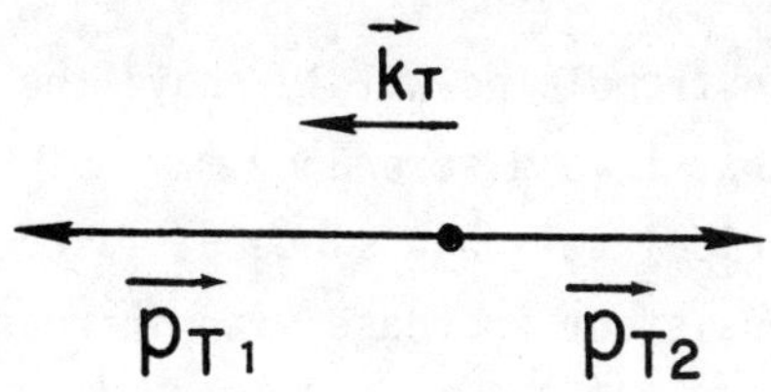

FIG. 19

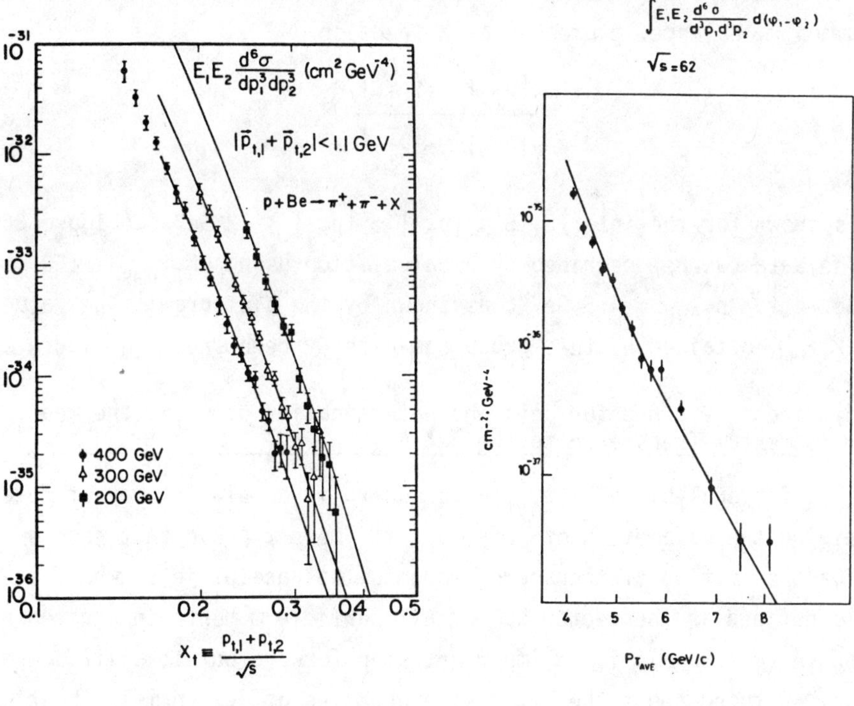

FIG. 20

FIG. 21

$$\langle k_T{}^2\rangle = (0.0144 \pm 0.0011)\, p_{TAVG} \sqrt{s} + 0.075 \pm 0.060$$

which unfortunately extrapolates wildly above the ISR measurements (Figure 25). Note that this discrepancy may not be due entirely to the form of the empirical relation since at fixed p_T, the WPLF values of $\langle k_T\rangle_{rms}$ show a consistent increase with increasing $\sqrt{s}$, while the CCOR results at $\sqrt{s}$ = 62.4 GeV, at a given p_T, are below the values given by WPLF at $\sqrt{s}$ = 27.4 GeV.

An ISR experiment has also measured the transverse momentum balance of jets. The CERN-Saclay Group[47] used a back-to-back π^0 pair trigger, and then reconstructed jets using all π^0 and charged particles within the solid angles of their two detectors of aperture $\Delta\phi \simeq \pm 15^0$ $\Delta\theta \simeq \pm 40^0$ each. The azimuthal aperture was inadequate, but a correction for fragments lost in this projection could be determined empirically from the distributions observed in the polar projection. The result for the number of events versus their in-plane transverse momentum balance, expressed as a fraction

$$\alpha = \frac{|\vec{p}_{T1}| - |\vec{p}_{T2}|}{|\vec{p}_{T1}| + |\vec{p}_{T2}|}$$

is shown for the interval $8 < |\vec{p}_{T_1}| + |\vec{p}_{T_2}| < 12$ GeV/c (Figure 26). The data can be described by a calculation using $\langle k_T\rangle_{rms}$ = 1.0 ± 0.1 GeV/c for $p_{TAVG} \simeq 4.5$ GeV/c as shown by the histogram. The value of $\langle k_T\rangle_{rms}$ obtained by this group has already been shown in Figure 25.

e) Direct Evidence for Jets and Determination of $\langle j_T\rangle$, the Mean Momentum Transverse to the Jet Axis

The analysis of $\langle|p_{out}|\rangle$ versus x_E is an elegant way of determining the parameters of jets, but it doesn't prove that jets exist. The observation of transverse momentum balance of jets, where a jet is defined as the vector sum of all particle momenta in a given solid angle ($\sim$1.5 sr), is an important step forward but it still doesn't say anything about the internal properties of jets namely their

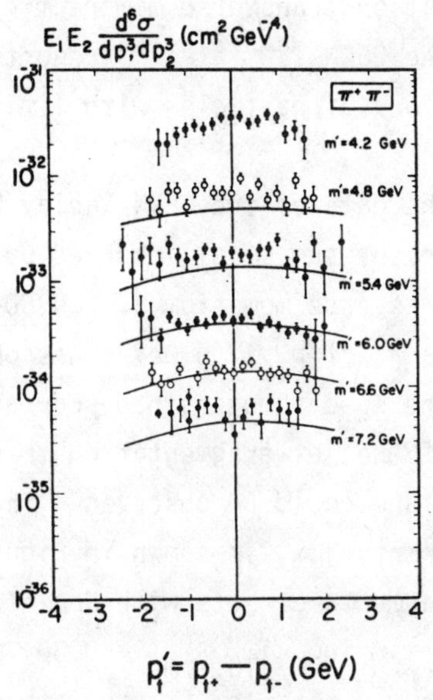

FIG. 22

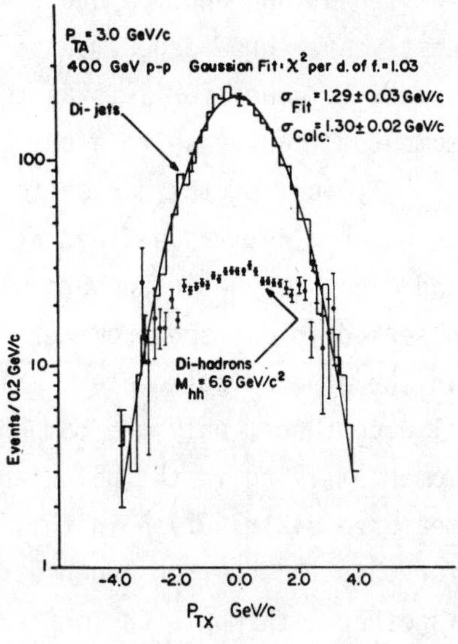

FIG. 23

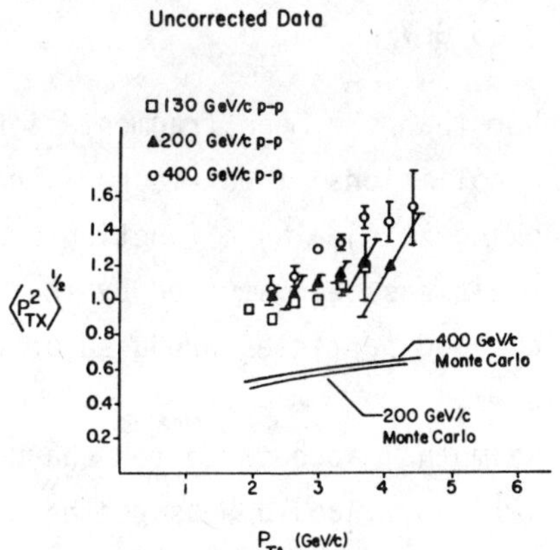

FIG. 24

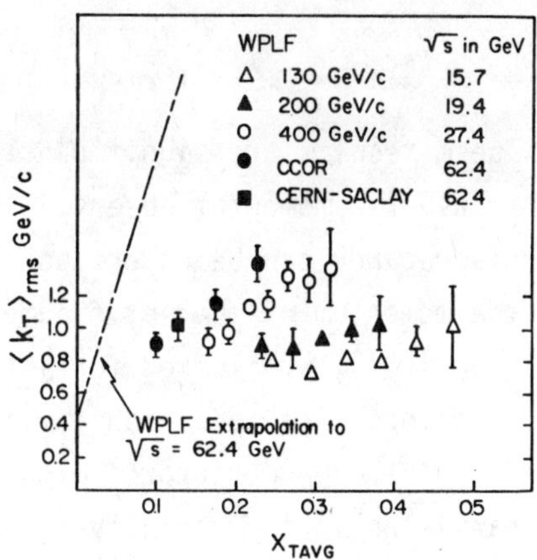

FIG. 25

fragmentation function and fragmentation transverse momentum $\langle j_T \rangle$. Last year, three experiments from the CERN ISR [28,47,48] reported direct evidence for jets as collections of particles with limited momentum transverse to a common axis.

Typical of these results are the data of the CERN-Saclay Group[47] in which a jet was defined as the vector sum of all charged particles and electromagnetic showers with transverse momentum $p_T > 0.80$ GeV/c observed in one spectrometer, when a π^0 of $p_{Tt} > 5$ GeV/c was observed in the other spectrometer. Since the spectrometers had poor azimuthal acceptance, only the component of the jet fragmentation transverse momentum lying in the polar angle plane could be observed. The measured distribution in this component, $j_{T\theta}$, is shown in Figure 27 for both the charged and neutral fragments of jets with $\Sigma \vec{p}_T > 3$ GeV/c, together with Monte Carlo predictions using $\langle j_{T\theta} \rangle_{rms} = 0.30$, 0.45 and 0.60 GeV/c. The CCOR Group[28] showed that the fragmentation was isotropic about the jet axis, so that the CERN-Saclay Group could infer the total fragmentation transverse momentum from the one measured projection. Their results are

$$\langle j_T \rangle = 0.55 \pm 0.05 \text{ GeV/c}$$

OR

$$\langle j_T \rangle_{rms} = 0.62 \pm 0.06 \text{ GeV/c}$$

These results are significantly higher than the mean fragmentation transverse momentum observed in e^+e^- collisions[23]; but much of the discrepancy can be understood by looking at $\langle j_T \rangle$ as a function of the momentum of the jet fragments. This has been studied by the CCOR Group[28,35], who looked for jets of charged particles produced produced opposite to a high p_{Tt} π^0 trigger.

The CCOR detector[28] with full azimuthal acceptance for charged particles in the rapidity interval $|y| < 0.7$, could measure the azimuthal distributions of charged particles produced in association with a high p_{Tt} π^0 trigger (Figures 12a and 12b). These distributions indicate that the jet-like peaks on the same side and opposite to a π^0 trigger are contained in azimuthal regions of width $\Delta\phi = \pm 60^0$

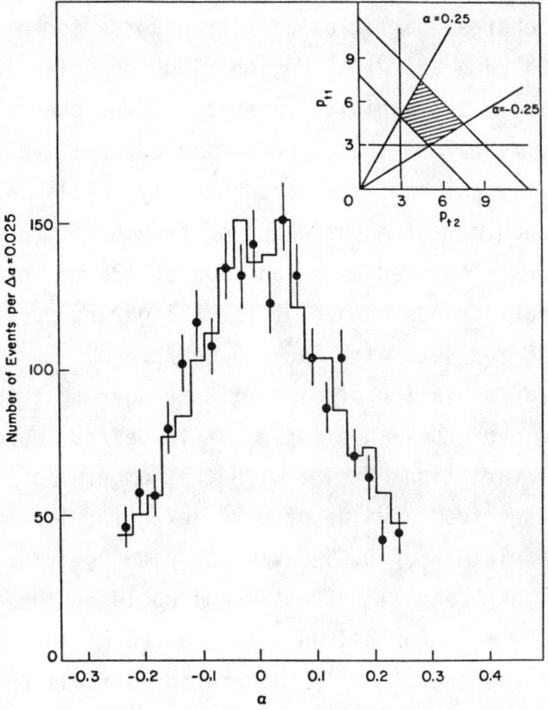

Number of Events per Δα = 0.025
α
α = 0.25
α = -0.25
p_{t1}
p_{t2}

FIG. 26

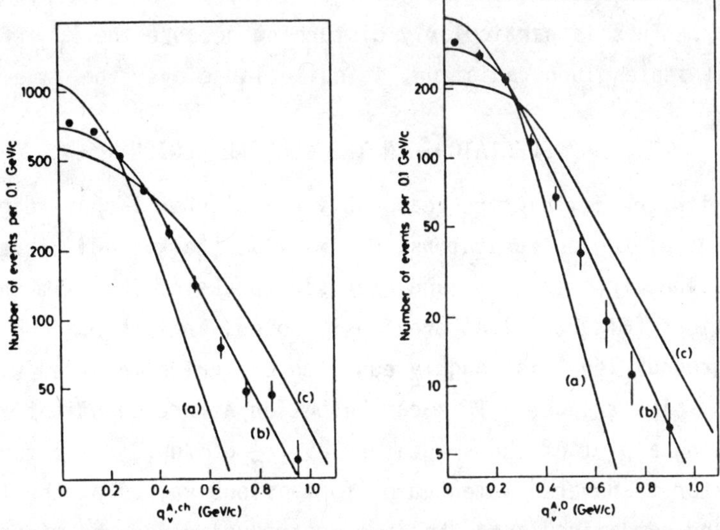

Number of events per 0.1 GeV/c
$q_v^{A,ch}$ (GeV/c)
$q_v^{A,0}$ (GeV/c)
(a)
(b)
(c)

FIG. 27

about the trigger axis. Thus, an away jet was defined as the vector sum of all charged particles with transverse momentum $p_T > 0.30$ GeV/c in the $\pm 60^0$ wide azimuthal region, 180^0 opposite in azimuth to the triggering π^0. The rapidity interval of the charged particles was fixed at $|y| < 0.7$, so that a centering cut was made on the rapidity of the jet, $|y_{JET}| < 0.3$, in order to try to eliminate edge effects. $\langle j_T \rangle$, the mean momentum component of fragments transverse to the jet sum vector was computed as a function of the momentum (p_T) of the fragments for various ranges of the triggering p_{Tt}. This is shown in Figure 28 for jets with $\Sigma \vec{p}_T > 3$ GeV/c.

No clear variation of $\langle j_T \rangle$ with triggering transverse momentum is observed over the range $3 \leq p_{Tt} \leq 11$ GeV/c. However, there is a substantial variation of $\langle j_T \rangle$ with the momentum of the jet fragments: $\langle j_T \rangle$ increases from a value of 0.34 GeV/c for low momentum (p_T) particles to a plateau of 0.65 GeV/c for $p_T > 2$ GeV/c. This behavior is similar to the "seagull" effect[49] and could be the explanation of the difference in $\langle j_T \rangle$ for hadron jets (quoted at the plateau value) and jets in e^+e^- annihilation[23] (integrated over all charged particle momenta). One disquieting feature of the hadron jet data is the apparent lack of final state gluon emission[23] which should be indicated by a broadening of the jets, or increase in $\langle j_T \rangle$, with increasing p_{Tt}. This is particularly disturbing because the k_T effect, or initial state gluon radiation, is quite large over the same p_{Tt} range.

SPECTATORS IN THE CENTRAL REGION

With the jet picture reasonably established I shall digress to the central region spectators, or the flat "background" level which is distinct from the same and away side peaks of the azimuthal distributions (Figure 12).As previously noted, in each p_T band the flat "background" level is roughly equal to the value measured with the minimum bias trigger. This can be put on a more quantitative basis by making a plot of the spectator level $\frac{1}{\sigma}\, d^2\sigma/dp_T d\phi$ as a function of spectator transverse momentum p_T for various values of the triggering p_{Tt}, and comparing it to the same distribution for the minimum bias

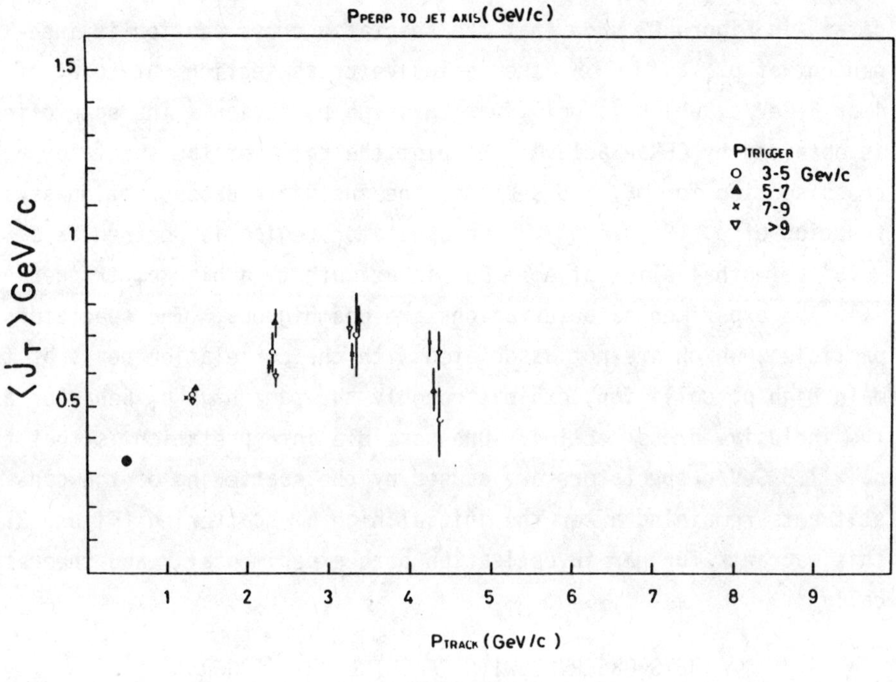
PPERP TO JET AXIS (GeV/c)
⟨ J_T ⟩ GeV/c
PTRIGGER
3-5 Gev/c
5-7
7-9
>9
PTRACK (GeV/c)

FIG. 28

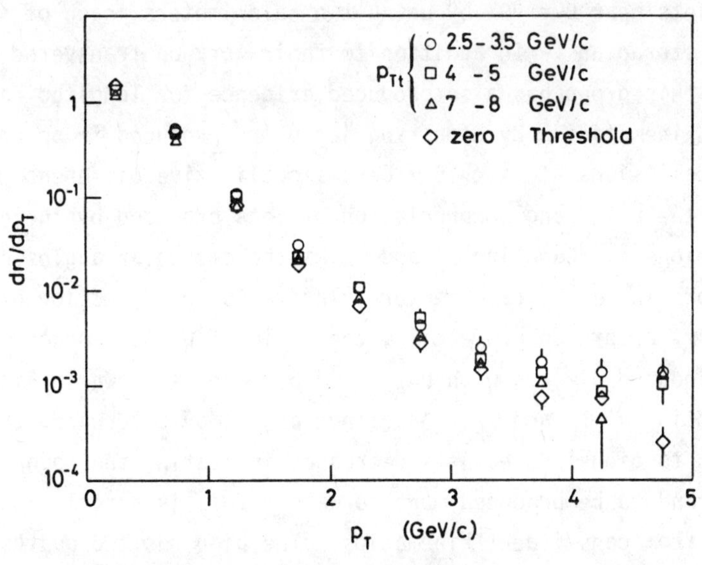
p_{Tt}
2.5-3.5 GeV/c
4 -5 GeV/c
7 -8 GeV/c
zero Threshold
dn/dp_T
p_T (GeV/c)

FIG. 29

zero threshold trigger, i.e., the inclusive cross section. The CCOR data[28] in Figure 29 show that the spectator cross section is independent of p_{Tt} and follows the inclusive cross section out to p_T of 4 or 5 GeV/c, which is well into the high p_T region. The same effect is obtained by CERN-Saclay[47] who plot the ratio of the spectator cross section for $p_{Tt} > 5$ GeV/c to the inclusive cross section as a function of p_T (Figure 30). The spectator region is defined as a $\pm 15^0$ azimuthal slice at $\Delta\phi = 90^0$ in azimuth to a high p_T trigger.

The experimental observations are unambiguous. The spectator particles, which are not associated with the correlation peaks of the main high p_T collision, exhibit roughly the same high p_T behavior as the inclusive cross section. One possible interpretation is that the $p_T > 1.5$ GeV/c spectators are caused by the scattering of the constituents remaining after the initial high p_T scattering (Figure 31). This warrants further investigation both experimentally and theoretically[50].

JETS OBSERVED WITH CALORIMETER TRIGGERS, and EVIDENCE FOR CONSTITUENT KINEMATICS

As previously mentioned, the WPLF Group studied azimuthally back-to-back jets near $\theta = 90^0$ by using two calorimeters, each of solid angle 1½ steradians. In addition to their work on transverse momentum balance, this group has also produced evidence for longitudinal constituent kinematics. By comparing jet pairs produced in pp collisions and π^+p collisions at $\sqrt{s} = 15.7$ GeV, a qualitative difference could be found in the polar angle correlation of jets produced by incident protons or pions[53]. Denoting θ_L and θ_R as the cms polar angles of the jet vectors in each calorimeter relative to the direction of the incident π^+ or proton projectile, the ratio of proton induced di-jets to pion induced di-jets with $p_{TAVG} \simeq 2.6$ GeV/c is shown in Figure 32 as a function of θ_L and θ_R. As either angle moves forward, the ratio of proton to pion induced jets decreases indicating that pion induced di-jets tend to be produced more forward. This is exactly what is expected from constituent kinematics. The pion has two quarks (actually a quark and an antiquark) but the proton has three quarks. Thus,

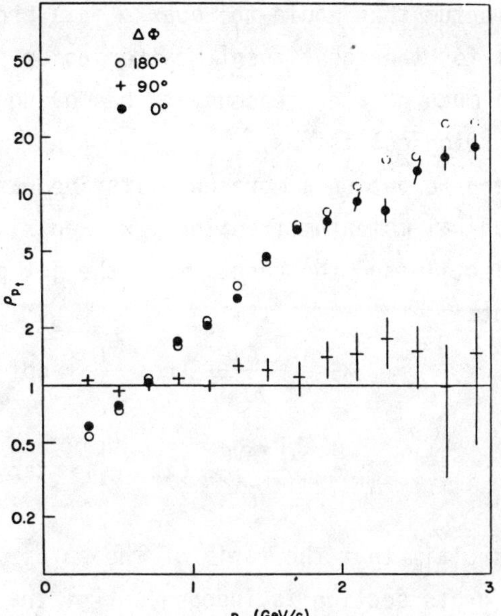

FIG. 30

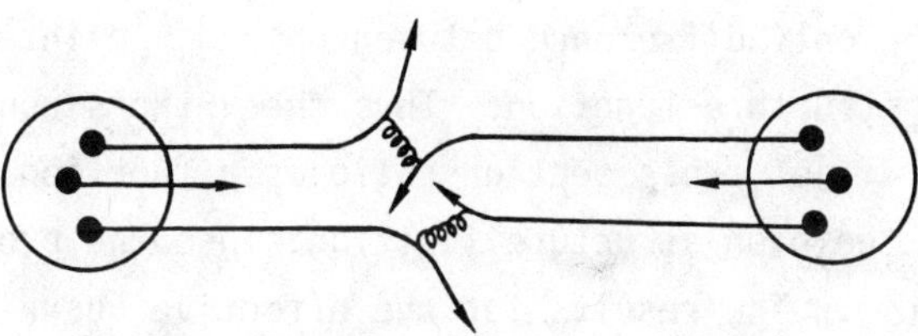

FIG. 31

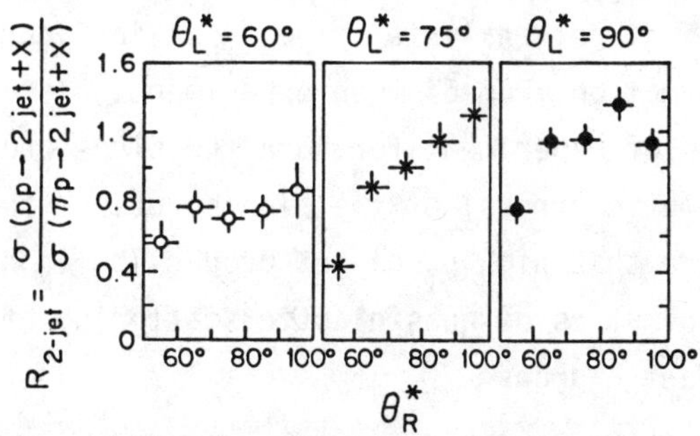

FIG. 32

each quark in a pion will carry a larger fraction of the total longitudinal momentum than would the quarks in a proton. Consequently the c.m. system for the constituent scattering in pion-proton collisions will have a much greater tendency to be moving forward than it would in proton-proton collisions.

This can be put on a more quantitative basis[54] by calculating the longitudinal momentum fractions, x_1 and x_2, of the constituents in the projectile and the target from the jet polar angles and transverse momenta:

$$x_1 = \frac{p_{TAVG}}{\sqrt{s}} \left(\cot \frac{\theta_L}{2} + \cot \frac{\theta_R}{2}\right)$$

$$x_2 = \frac{p_{TAVG}}{\sqrt{s}} \left(\tan \frac{\theta_L}{2} + \tan \frac{\theta_R}{2}\right)$$

The authors claim that the ratio of the pp di-jet cross section to the π^+p di-jet cross section is independent of the target x_2 and any other kinematic variables and depends almost entirely on x_1, the projectile momentum fraction. This observation agrees with QCD-type models in which the only difference between pp and π^+p interactions is the projectile structure function. Thus the authors conclude that the proton to pion di-jet cross section ratio as a function of x_1 can be used to extract the pion structure function since the proton structure function is known[22]. The results for the effective quark-plus-antiquark structure function of the pion

$$f^{EFF}_{(q+\bar{q}),\pi} \equiv \frac{f_{(q+\bar{q}),p}}{\sigma_p/\sigma_\pi}$$

from the WPLF di-jet analysis[54] are shown in Figure 33 together with similar information from di-muon experiments[55,56]. The agreement of the two types of experiment for both the slope and magnitude (within quoted systematic errors) of the pion structure function provides the first evidence that high p_T di-jet production in hadron collisions obeys the kinematics of quasielastic scattering of the constituents in the incident hadrons.

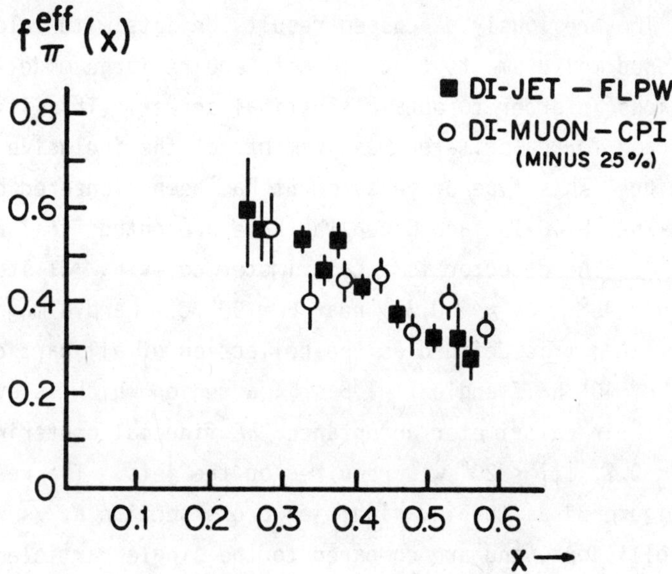

FIG. 33

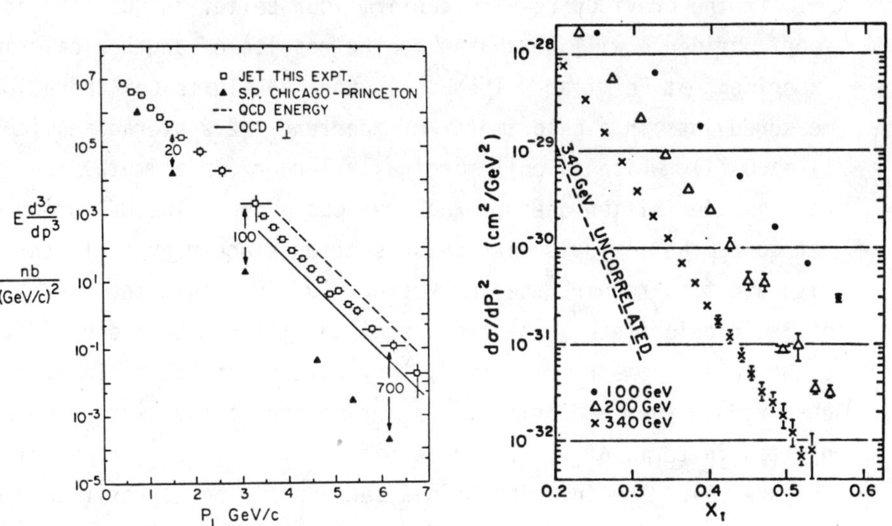

FIG. 34

FIG. 35

JET CROSS SECTION STUDIES WITH CALORIMETERS

The previously discussed results on jets with calorimeters have stressed measurements that did not require large model-dependent corrections in order to obtain the final answer. This is definitely not the case for calorimeter measurements of the inclusive jet cross section. This type of measurement has been pioneered by the Caltech-UCLA-FNAL-UICC-Indiana Group who have presented final results this year[57]. The detector is a calorimeter covering ~ 1 steradian ($\Delta\phi \simeq \pm 45^0$, $\Delta y \simeq \pm 0.35$) near $\theta = 90^0$ c.m. and a magnetic spectrometer. A jet is defined as the collection of all particles within a cone of 40^0 half angle (~ 1.5 sr), a region which is evidently larger than their calorimeter acceptance. A fiducial centering cut of $|y| < 0.2$, $|\phi| < 20^0$ was required on the jets. The results are shown in Figure 34 as the invariant jet cross section at $\sqrt{s} = 19.4$ GeV in pp collisions, and are compared to the single particle $(\pi^+ + \pi^-)/2$ cross sections[5] (triangles). The jet cross section is considerably larger than the single particle cross section at the highest p_T, and seems to be impressively in agreement with a QCD prediction.

If the previous results confirm your belief in QCD-like jets, your confidence will be shaken by the results of another calorimeter experiment at Fermilab. The Washington-FNAL-Tufts Collaboration[58] measured jets in a calorimeter of aperture ~ 2.2 steradians ($\Delta\phi = \pm 47^0$, $\Delta y = \pm 0.74$), which is only marginally larger (in azimuth) than the 40^0 cone jet definition used by the previous group. The WFT Group does not quote the inclusive jet cross section since they claim that even with a cut to require the jet vectors to lie within the central 0.2 sr of their calorimeter, the cross section will not be independent of the calorimeter geometry. Instead, they give the total yield of jets observed in their calorimeter as a function of the vector sum p_T plotted in terms of x_T for three different $\sqrt{s}$ values of pp collisions (Figure 35). The invariant cross section can be approximately obtained by multiplying the yield by a factor of 3. The most striking feature of the WFT data in Figure 35 is that the invariant cross sections are roughly 20 times larger at $\sqrt{s} = 19.4$ GeV than the CCFII results in Figure 34. Note that the WFT Group do claim to observe a jet-like correlation, since their jet yield is well above their calculation of what would be expected from a collection of uncorrelated particles having the measured single particle inclusive behavior.

It is difficult for the outsider to attempt to reconcile these results since the true jet cross section must be derived from the experimental data using an extensive model-dependent Monte Carlo calculation. Not surprisingly, the CCFII Group used the model of Feynman, Field and Fox[34]. The calculation used by the other group was not clearly documented. Two major problems are involved. There is the obvious problem of jet acceptance and undetected fragments; this effect certainly depends on the assumed parameters of the jets. Another problem claimed by the experimenters[59] is the lack of separation of the central region from the forward region. Thus, as the calorimeters get larger and larger, they might tend to pick up fragments of the forward spectator jets in addition to the high p_T jet. This problem does not seem to be evident in the jets observed at the ISR in the central rapidity region, $|y| < 0.7$, perhaps because of the higher $\sqrt{s}$. Nevertheless, it is claimed[59] that this lack of separation could be the explanation of the apparently contradictory jet cross section results. One could hope that the experimenters would make more of an effort to derive the jet parameters from their own data, or at least give some indication of the sensitivity of their conclusions to the parameters of the jets, particularly $\langle j_T \rangle$[60].

To end this section on a more positive note, I show the CCFII results[57] for the ratio of jet production at a given p_T by different particles incident on the same target (hydrogen). The jet acceptance cancels out when taking the ratio (Figure 36). The p_T spectra are the same for incident particles containing the same number of valence quarks. The pion induced p_T spectrum is harder than the proton induced spectrum since the pion has fewer, and therefore higher momentum, valence quarks.

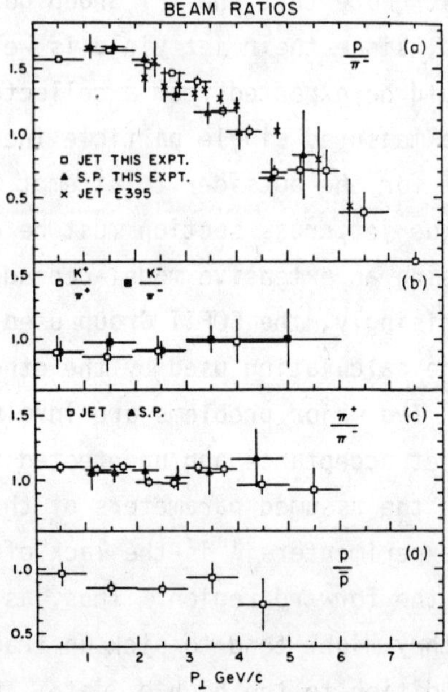

BEAM RATIOS
(a)
p/π⁻
JET THIS EXPT.
S.P. THIS EXPT.
JET E395
(b)
K⁺/π⁺
K⁻/π⁻
(c)
JET
S.P.
π⁺/π⁻
(d)
p/p̄
P⊥ GeV/c

FIG. 36

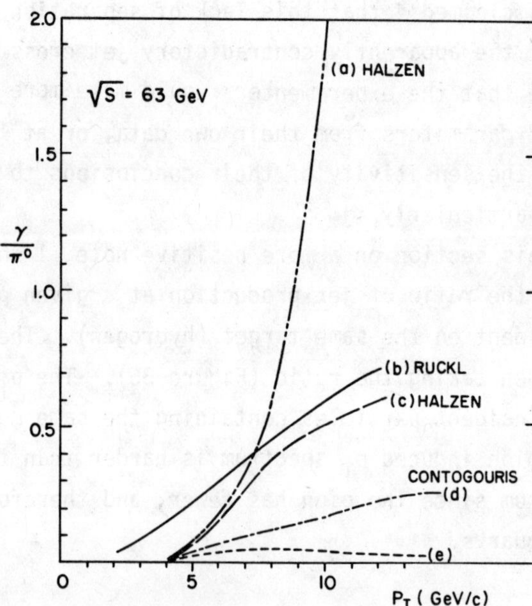

√S = 63 GeV
(a) HALZEN
(b) RUCKL
(c) HALZEN
CONTOGOURIS
(d)
(e)
γ/π⁰
P_T (GeV/c)

FIG. 37

DIRECT SINGLE PHOTON PRODUCTION

Direct photon production at high p_T has recently become a very active subject because of the prediction[61] of the QCD "compton effect", i.e., the constituent reaction quark + gluon → quark + γ. The beauty of this reaction as a hadronic probe is that γ-rays are allowed to emerge freely from inside a hadron. Thus, no rigamarole about jets and fragmentation is required. The high p_T QCD γ-rays from hadron collisions will emerge cleanly unaccompanied by any other same-side particles. In principle, the only unknown quantity is the gluon distribution inside a hadron. Thus it would seem that the calculations of direct photon production[62] should be reliable. As shown in Figure 37, however, there is nearly a two order of magnitude variation in the calculations at high p_T.

In all fairness, it should be pointed out that there are at least three reasons for the wide variation of the different calculations. i) The QCD calculations predict only the inclusive direct γ cross section whereas the experiments tend to quote γ/π^0. The ratio γ/π given in the QCD calculations is the QCD γ calculation divided by the experimentally measured inclusive π^0 cross section, where available, or by the extrapolation of the π^0 cross section to higher p_T[62a]. ii) Scaling structure functions were used originally, but it was pointed out by the McGill group[62d,e] that scale-violation effects were important. iii) Finally, the shape of the gluon distribution at the reference momentum transfer is unknown and, therefore, several possibilities have been tried. In Figure 37, the difference between curves (d) and (e) is the choice of the gluon distribution, $(1 - x)^4$ in (d) and $(1 - x)^5$ in (e); the difference between curves (d),(e) and the rest is the use of non-scaling structure functions for curves (d) and (e); the difference between curve (a) and the rest is an incorrect extrapolation of the π^0 inclusive cross section for $p_T > 7$ GeV/c in curve (a).

The experimental side of this issue is exceedingly difficult. The overriding problem is a background of apparent single γ-rays typically at a level of 20% or more of the π^0 signal. This background has nothing to do with direct γ-rays but originates from the following sources: i) the decays $\pi^0 \to \gamma\gamma$ or $\eta^0 \to \gamma\gamma$ in which a) either one γ is outside the detector or below threshold or b) the two γ-rays are not geometrically resolved; ii) neutral hadrons like $\bar{n}$, n, $K^0{}_L$, etc.

Four groups have presented data on direct photon production this year[63]; and there was a very good session on this topic at the 1979 Lepton-Photon Symposium[64] to which the reader is referred for details. The only group making a strong claim to see a significant direct photon signal is A^2BC from the CERN ISR[65] whose data for γ/π^0 are shown in Figure 38 for three $\sqrt{s}$ values (a) 31 GeV, (b) 53 GeV, (c) 63 GeV. The quoted γ/π^0 ratio approaches 30% for $p_T \simeq 6$ GeV/c. However, it must be stressed that this is not the inclusive γ/π^0 ratio but the ratio for γ's and π's unaccompanied by any other particle within a solid angle of 0.26 steradians. The inclusive ratio is lower than that shown in Figure 38 by an amount still being evaluated. However, one clue to the answer is to be found in another paper by the same group[66] in which it is claimed that the average γ/π^0 cross section ratio for p_T between 7 and 11 GeV/c is at most $\gamma/\pi^0 = 0.08 \pm 0.18$.

The γ/π^0 data from two other similar experiments are shown in Figures 39 and 40. The Brookhaven-CERN-Rome group[67] looked for γ's and π's unaccompanied by any other γ-rays. Their result for γ/π^0 at $\sqrt{s} = 53$ GeV is shown in Figure 39. The curve (Ref. 62c) might tempt one to believe that a signal existed, but the average ratio of γ/π^0 for p_T between 3 and 5 GeV/c is $\gamma/\pi^0 = 0.020 \pm 0.008$. The Fermilab-Johns Hopkins group[67] quote a γ/π^0 ratio of $\gamma/\pi = 0.070 \pm 0.025$ averaged over the p_T interval 1.5 to 4.0 GeV/c where neither the γ's nor the π's are accompanied by any charged particles. Their data are shown in Figure 40 for two different $\sqrt{s}$ values compared to some theoretical curves (a) (Ref. 62b), (b) (Ref. 62c). The shaded region indicates the range of the systematic errors.

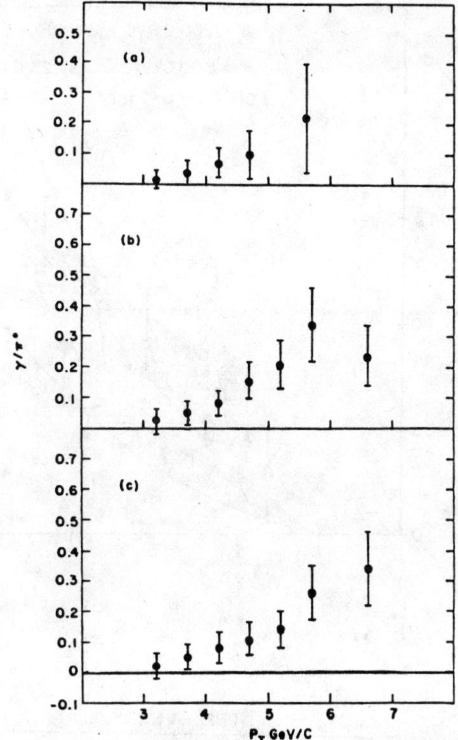
0.5
0.4
0.3
0.2
0.1
(a)
0.7
0.6
0.5
0.4
0.3
0.2
0.1
(b)
γ/π°
0.7
0.6
0.5
0.4
0.3
0.2
0.1
0
-0.1
(c)
3
4
5
6
7
P_T GeV/C

FIG. 38

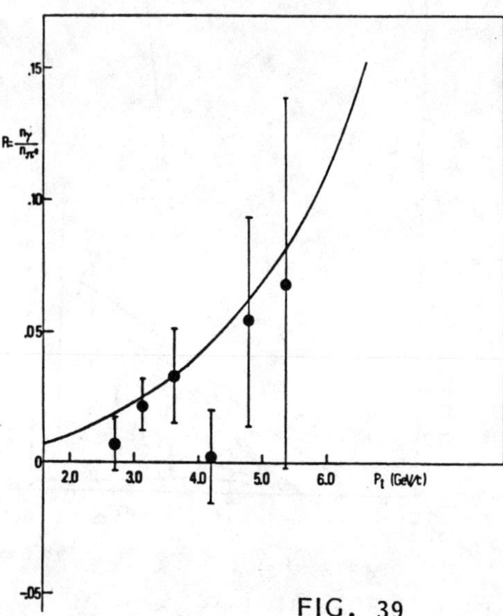
R= n_γ / n_π°
.15
.10
.05
0
-.05
2.0
3.0
4.0
5.0
6.0
P_t (GeV/c)

FIG. 39

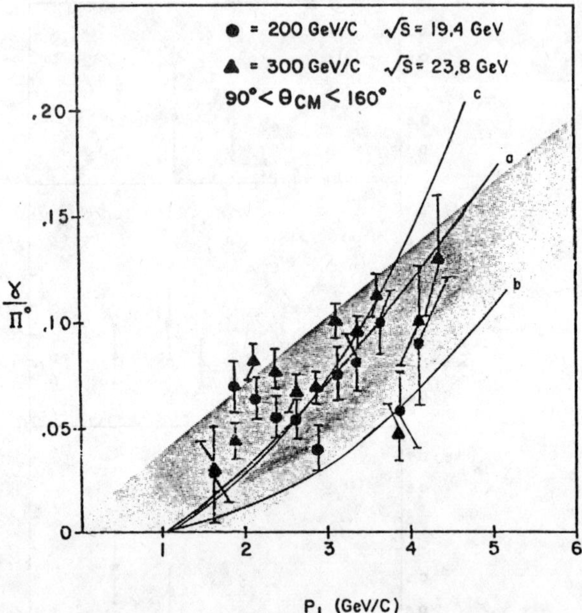
● = 200 GeV/C √S = 19.4 GeV
▲ = 300 GeV/C √S = 23.8 GeV
90° < θ_CM < 160°
c
a
b
.20
.15
.10
.05
0
γ/π°
1
2
3
4
5
6
P⊥ (GeV/C)

FIG. 40

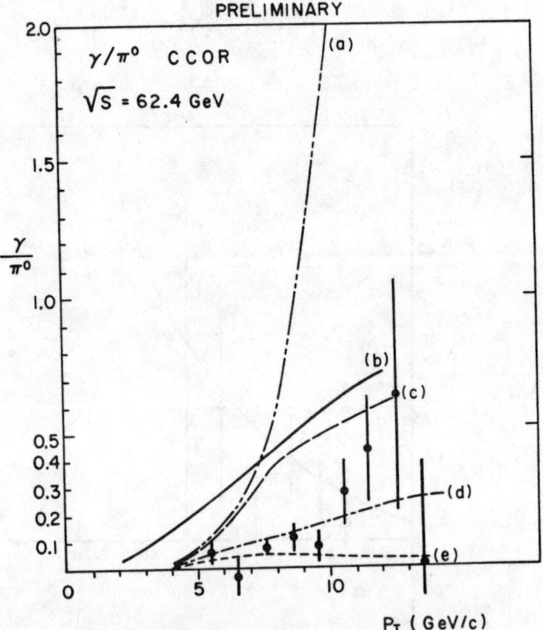
PRELIMINARY
γ/π° CCOR
√S = 62.4 GeV
(a)
(b)
(c)
(d)
(e)
2.0
1.5
1.0
0.5
0.4
0.3
0.2
0.1
γ/π°
0
5
10
P_T (GeV/c)

FIG. 41

The CCOR group is unable to geometrically resolve single γ rays and π^0 and therefore resorts to a completely different method of analysis than the other three experiments. The statistical division of a given sample into the fraction of single γ-rays and the fraction of $\pi^0 \to \gamma\gamma$ is accomplished by measuring the average conversion probability for the sample in a $1X_0$ thick converter. The preliminary CCOR result[64] for inclusive γ/π^0 is shown in Figure 41 compared to the predictions of Figure 37. The authors are hesitant to claim the existence of a single γ-ray signal from these data. Nevertheless, it is clear that the data exclude all the calculations except (d) and (e).

At the present time, the experimental situation in direct photon production is still a bit confusing. However, this is a new and difficult field. It is hoped that continued work will eventually clarify the situation and lead to quantitative measurements of the gluon structure function of the proton.

CONCLUSION AND APOLOGIES

Given the constraints of time and space, I have tried to make a selection of subjects that covered the underlying issues of high p_T physics and which reflected the latest thinking. Consequently, I have left out many important topics. I list some of them here, with references, as a guide to the interested reader:

i) x_E Scaling[32,33,48,28,42,30,47]

ii) Fragmentation Functions and Final State Particle Composition[5,47,57,48,29,66,69,70,71]

iii) Charge Correlations and Quantum Number Correlations[48,42,47,72,71,73,74]

iv) Correlation of High p_T Particles to Forward Spectators[51,52,72]

v) High p_T Production at Forward Angles[75]

REFERENCES

[1] Proc. Copenhagen Meeting on "Jets in High Energy Collisions," Physica Scripta 19 (1979).

[2] Proc. of the 19th International Conference on High Energy Physics, Tokyo 1978, edited by S. Homma, M. Kawaguchi and H. Miyazawa (Physical Society of Japan, 1979).

[3] M. Jacob and P.V. Landshoff, Phys. Reports 48, 285 (1978); S. Ellis and R. Stroynowski, Rev. Mod. Phys. 49, 753 (1976); D. Sivers, S. Brodsky and R. Blankenbecler, Phys. Reports 23C, 1 (1976). Also, the proceedings of the following annual or biennial conferences usually contain timely review articles: International Conference on High Energy Physics, American Physical Society Division of Particles and Fields, European Physical Society Conference on High Energy Physics, Rencontre de Moriond, International Colloquium on Multiparticle Reactions.

[4] M.J. Tannenbaum, "High p_T Processes - Experimental Aspects", in Proceedings of the Meeting in Marseilles on Hadron Physics at High Energies, June 1978, edited by C. Bourrely, J.W. Dash and J. Soffer (Centre de Physique Theorique C.N.R.S. - F-13288, Marseille, France).

[5] D. Antreasyan et al., Phys. Rev. Lett. 38, 112 (1977).

[6] Detailed citations of the experimental work of this era are given in Refs. 3 and 4.

[7] S.M. Berman, J.D. Bjorken, J. Kogut, Phys. Rev. D4, 3388 (1971); S.M. Berman and M. Jacob, Phys. Rev. Lett. 25, 1683 (1970); J.D. Bjorken, Phys. Rev. D8, 4098 (1973).

[8] R. Blankenbecler, S.J. Brodsky and J.F. Gunion, Phys. Rev. D6, 2652 (1972); Phys. Lett. 42B, 461 (1972); Phys. Rev. D12, 3469 (1975); ibid. D18, 900 (1978). See also, S.J. Brodsky and G.R. Farrar, Phys. Rev. Lett. 31, 1153 (1973); Phys. Rev. D11, 1309 (1975).

[9] J. Kogut, D.K. Sinclair and L. Susskind, Phys. Rev. D7, 3637 (1973); J. Kogut, G. Frye and L. Susskind, Phys. Lett. 40B, 469 (1972).

[10] S.D. Ellis and M.B. Kislinger, Phys. Rev. D9, 2027 (1974).

[11] R.D. Field and R.P. Feynman, Phys. Rev. D15, 2590 (1976); R.P. Feynman, *Photon-Hadron Interactions,* (Benjamin, Reading, Mass. 1972); R.P. Feynman, R.D. Field and G.C. Fox, Nucl. Phys. B128, 1 (1977); Phys. Rev. D18, 3320 (1978).

[12] R.F. Cahalan, K.A. Geer, J. Kogut and L. Susskind, Phys. Rev. D11, 1199 (1975).

[13] A.G. Clark et al., Phys. Lett. 74B, 267 (1978).

[14] A.L.S. Angelis et al., Phys. Lett. 79B, 505 (1978).

[15] F.W. Büsser et al., Nucl. Phys. B106, 1 (1976).

[16] The absolute scale uncertainty has no effect on n; the relative normalization uncertainty cancels when comparing n in the lower and higher p_T regions.

[17] R.D. Field, Phys. Rev. Lett. 40, 997 (1978).

[18] R. Cutler and D. Sivers, Phys. Rev. D16, 679 (1977); ibid. D17, 196 (1978).

[19] B.L. Combridge, J. Kripfganz and J. Ranft, Phys. Lett. 70B, 234 (1977).

[20] A.P. Contogouris, R. Gaskell and S. Papadopoulos, Phys. Rev. D17, 2314 (1978).

[21] J.F. Owens, E. Reya and M. Gluck, Phys. Rev. D18, 1501 (1978); J.F. Owens and J.D. Kimel, ibid. 3313 (1978).

[22] J. Steinberger, these proceedings.

[23] B. Wiik, these proceedings.

[24] C. Kourkoumelis et al., Phys. Lett. 84B, 271 (1979).

[25] C. Bromberg et al., Phys. Rev. Lett. 43, 561 (1979).

[26] N. Giokaris et al., submitted to the EPS International Conference on High Energy Physics, Geneva 1979, by B.G. Pope.

[27] e.g., see the discussion by J. Bjorken et al., in Proceedings of the 1971 International Symposium on Electron and Photon Interactions at High Energies, edited by N.B. Mistry (Laboratory of Nuclear Studies, Cornell University, Ithaca, N.Y., 1972), pp.296-7.

[28] A.L.S. Angelis et al., Physica Scripta 19, 116 (1979).

[29] H. Bøggild, Proc. of the XIVth Rencontre de Moriond, edited by J. Tran Thanh Van, Vol. I, p. 321; also, see M.J. Tannenbaum, ibid, p. 351 (1979).

[30] C. Kourkoumelis et al., Phys. Lett. 85B, 147 (1979); and C. Kourkoumelis et al., "Correlations of High Transverse Momentum π^0 Pairs Produced at the CERN ISR," CERN-EP/79-36 (to appear in Nucl. Phys. B).

[31] G. Sterman and S. Weinberg, Phys. Rev. Lett. 40, 1436 (1977).

[32] P. Darriulat et al., Nucl. Phys. B107, 429 (1976).

[33] M. Della Negra et al., Nucl. Phys. B127, 1 (1977).

[34] see FFF, Reference 11.

[35] A.L.S. Angelis, "Correlations of Charged Particles with a High p_T Trigger," (submitted to Phys. Lett. B).

[36] e.g., see the review by S. Matsuda in Reference 2.

[37] Z. Kunst and E. Pietarinen, "Production of Three Large p_T Jets in Hadron-Hadron Collisions," DESY 79/34, June 1979 (to be published).

38 A.P. Contogouris, R. Gaskell and S. Papadopoulos in Proc. of the IX International Symposium on Multi-Particle Dynamics, Tabor, Czechoslovakia, 3-7 July 1978.

39 Also, see A. De Rujula in Ref. 2.

40 The actual fit parameters are $\Lambda = 0.52^{+0.37}_{-0.22}$ GeV/c and the constant 25 which is supposed to equal $33-2n_f$ = 25.8 ± 4.8, or n_f = 3.6 ± 2.4.

41 W.J. Metcalf et al., "The Magnitude of Parton Instrinsic Transverse Momentum," Report COO-3065-249, The University of Rochester, Rochester, NY (submitted to Phys. Lett.)

42 H. Jöstlein et al., Phys. Rev. D20, 53 (1979).

43 R. Baier, J. Engels and B. Petersson, "Symmetric Pairs at Large Transverse Momenta as a Test of Hard Scattering Models," University of Bielefeld, Preprint BI-TP 79/10, April 1979 (submitted to Nucl. Phys. B).

44 A.L.S. Angelis et al., submitted to the E.P.S. Conference, Geneva, Switzerland, July 1979.

45 M.D. Corcoran et al., "A Study of Parton Transverse Momentum Using Jets From Hadron Interactions," University of Wisconsin Preprint (1979) - (submitted for publication).

46 Direct measurements of $\langle j_T \rangle$ in hadron collisions are discussed in the next section.

47 A.G. Clark et al., "Large Transverse Momentum Jets in High-Energy Proton-Proton Collisions," CERN Preprint EP/79-74, July 1979 (submitted to Nucl. Phys. B).

48 M.G. Albrow et al., Nucl. Phys. B145, 305 (1978).

49 C. del Papa et al., Phys. Rev. D15, 2425 (1977); G. Hanson, Proc. of the XIIIth Rencontre de Moriond, edited by J. Tran Thanh Van, Vol. II, p. 15 (1978); J. Bell et al., Phys. Rev.D19, 1 (1979); W.A. Loomis et al., ibid 2543 (1979).

50 Note that detailed studies of forward spectator effects have been presented by the CCHK Group (Ref. 51) and the BFS Group (Ref. 52).

51 E.E. Kluge et al., in Ref. 1; R. Sosnowski, in Ref. 2.

52 M.G. Albrow et al., Nucl. Phys. B135, 46 (1978).

53 M.D. Corcoran et al., Phys. Rev. Lett. 41, 9 (1978).

54 M. Dris et al., Phys. Rev. D19, 1361 (1979); M.D. Corcoran et al., "Evidence That High p_T Jet Pairs Give Direct Information on Parton-Parton Scattering," Preprint COO-088-105 (submitted for publication).

55 C.B. Newman et al., Phys. Rev. Lett. 42, 951 (1979).

56 A.J.S. Smith, these proceedings.

57 C. Bromberg et al., Phys. Rev. Lett. 43, 565 (1979).

58 V. Cook et al., "Large-Solid-Angle-Detector Rates, Energy Dependence, and Profiles of High p_T Events in pp Collisions," Fermilab-PUB-79/68-EXP, August 1979 (submitted to Phys. Lett.).

59 G. Fox, private communication.

60 Actually the WFT Group made an attempt in this direction by showing that the vertical profiles of their jets were twice as wide as they expected for spear-like jets (G. Hanson, Ref. 49).

61 H. Fritzsch and P. Minkowski, Phys. Lett. 69B, 316 (1977).

62 a) F. Halzen and D.M. Scott, Phys. Rev. Lett. 40, 1117 (1978).
b) R. Ruckl, S.J. Brodsky and J.F. Gunion, Phys. Rev. D18, 2469 (1978).
c) F. Halzen and D.M. Scott, Phys. Rev. D18, 3378 (1978).
d) McGill Group (Ref. 62e). I would like to thank Prof. Andreas Contogouris for making this information available to me prior to publication.
e) A.P. Contogouris, S. Papadopoulos and M. Hongoh, Phys. Rev. D19, 2607 (1979).

63 Earlier results are reviewed by M.J. Tannenbaum (Ref. 29).

64 Proc. of the 1979 International Symposium on Lepton and Photon Interactions at High Energies, Fermilab, Batavia, Ill., August 1979 (to be published--T.B.W. Kirk, editor).

65 M. Diakonou et al., "Direct Production of High p_T Single Photons in pp Collisions at the CERN ISR," BNL 26608, August 1979 (submitted to Phys. Lett.)

66 C. Kourkoumelis et al., Phys. Lett. 84B, 277 (1979).

67 E. Amaldi et al., Phys. Lett. 77B, 240 (1978); Nucl. Phys. B150, 326 (1979); Phys. Lett. 84B, 360 (1979).

68 R.M. Baltrusaitis et al., "A Search for Direct Photon Production in 200 and 300 GeV/c Proton-Beryllium Interactions," Fermilab-PUB-79/38-EXP, August 1979 (submitted to Phys. Lett.)

69 A.G. Clark et al., "Experimental Study of the Fragmentation of Large Transverse Momentum Jets in High Energy p-p Collisions," CERN Preprint, July 1978, (submitted to the 1978 Tokyo Conference).

70 A. Chilingarov et al., "Production of High Transverse Momentum Low Mass Electron-Positron Pairs in High Energy p-p Collisions," CERN Preprint, August 1978, (submitted to the 1978 Tokyo Conference).

71 M.G. Albrow et al., "Studies of Proton-Proton Collisions at the CERN ISR With an Identified Charged Hadron of High Transverse Momentum at 90^0.III Jet-like Structures," CERN-EP/79-56, June 1979, (submitted to Nucl. Phys. B)

72 D. Drijard et al., "Quantum Number Effects in Events with a Charged Particle of Large Transverse Momentum," (Part 1: Leading Particles in Jets), CERN/EP/PHYS 78-14 Rev., July 1978, (submitted to Nucl. Phys. B).

73 D.A. Finley et al., Phys. Rev. Lett. 42, 1028 (1979).

74 A.G. Clark et al., "Large Transverse Momentum Jets in High-Energy p-p Collisions," CERN Preprint, August 1978, (submitted to the 1978 Tokyo Conference).

75 R.M. Baltrusaitis et al., "Inclusive π^0 Production Over Large x_T and x_F Ranges in 200, 300, 400 GeV/c Proton-Beryllium Interactions," Fermilab-PUB-79/37-EXP, June 1979, (submitted to Phys. Rev. Lett.)

DISCUSSION

Chih Kwan Chen, Purdue University: Would you please mention the charge multiplicity in the away side jet in correlation with the species of the trigger particles?

Tannenbaum: CERN-Saclay has measured this with π^0 (Ref. 47). Other places to look would be BFS (Refs. 52, 48, 71) or CCHK (Refs. 72, 51, 33).

L. Clavelli, University of Bonn: Has anyone measured, or would it be worthwhile to measure, the constituent c.m. energy by measuring the total invariant mass of the outgoing jets?

Tannenbaum: Yes, definitely. This is a key issue. The WPLF data (Ref. 54) is a first step in this direction.

THE QCD PHENOMENOLOGY OF DEEP-INELASTIC SCATTERING*

L.F. Abbott
Brandeis University, Waltham, MA 02254

ABSTRACT

The role of higher-twist effects and corrections of non-leading order in α_s in the QCD phenomenology of deep-inelastic scattering is discussed. The possibility of higher-twist effects accounting for the anomalously large ratio of longitudinal to transverse electroproduction cross sections is examined.

QCD makes definite predictions[1] about the Q^2-dependence of structure functions (or their moments) measured in deep-inelastic scattering. In principle, these predictions can be compared with the experimental data to test the validity of the theory.[2,3] However, the leading-order results of QCD are subject to corrections coming from non-leading twist operators[4] and from higher-order terms in the QCD perturbation expansion. In this talk, I will discuss the role which these two types of corrections play in tests of QCD at present accelerator energies. I will also discuss the possible role of higher-twist effects in accounting for the large ratio of longitudinal to transverse cross sections measured at SLAC. Most of the experimental data I will be using come from SLAC-MIT electroproduction experiments[5] and my results are taken from a recent analysis of this data done in collaboration with Bill Atwood and Mike Barnett.[2]

To begin, let us recall exactly what the predictions of QCD are for deep-inelastic structure functions. The predictions are most easily stated in terms of the moments of the structure functions defined by

$$M_N(Q^2) = \int_0^1 dx\, x^{N-1}\, F(x,Q^2) \tag{1}$$

For a flavor non-singlet structure function like xF_3 or $F_2^p-F_2^n$, QCD tells us that these moments can be expressed in the following form:[6]

$$M_N(Q^2) = \frac{A_N}{(\ln Q^2/\Lambda^2)^{d_N}} \left\{ 1 + \sum_{i=1}^{\infty} B_{iN}(\ln Q^2/\Lambda^2) \left[\frac{N\mu_{iN}^2}{Q^2}\right]^i + \sum_{j=1}^{\infty} \left[\frac{1}{\ln Q^2/\Lambda^2}\right]^j \sum_{k=0}^{j} C_{jk}^{(N)} [\ln\ln Q^2/\Lambda^2]^k \right\} \tag{2}$$

The expression $A_N/(\ln Q^2/\Lambda^2)^{d_N}$ is the leading-twist, leading-order prediction of QCD. The first sum inside the curly bracket represents corrections coming from higher-twist effects.[4] It is a definite prediction of QCD that the dependence $[1/Q^2]^i$, which is determined by

*Talk presented at 1979 APS Conference, Division of Particles & Fields.

ISSN:0094-243X/80/590311-13$1.50

simple power counting, is only modified by logarithms of Q^2 expressed here by the function $B_{iN}(\ln Q^2/\Lambda^2)$. (In a fixed-point theory, for example, this Q^2-dependence would be modified by powers of $1/Q^2$ and we would have no way of knowing what the leading term in the operator-product expansion was.) The second sum (actually a double sum) inside the curly bracket of Eq. (2) represents the higher-order corrections in the perturbation expansion. These terms are down by powers of $1/\ln Q^2/\Lambda^2$ relative to the leading-order terms. At present, we know the values of d_N, $C_{10}^{(N)}$, $C_{11}^{(N)}$, $C_{21}^{(N)}$, and $C_{22}^{(N)}$ in Eq. (2) from perturbative QCD calculations.[1,7] In principle, all the coefficients C_{jk} are calculable although already the calculation of C_{10} involved a monumental effort by the authors of Ref. 7. Also, again in principle, all the functions $B_{iN}(\ln Q^2/\Lambda^2)$ are calculable although no such calculation has yet been attempted. However, the constants A_N and μ_{iN} are not presently calculable, even in principle. This is because these coefficients are related to matrix elements of local operators taken between nucleon states and thus they depend in detail on the structure of the nucleon which cannot be determined in perturbative QCD. Finally, the parameter Λ is of course not a calculated number but must be determined experimentally.

Although the QCD predictions for deep-inelastic scattering are most easily stated in terms of the moments, in actual phenomenological analyses it is often more instructive and otherwise advantageous to work directly with the structure functions themselves. QCD also makes predictions for the Q^2- evolution of the structure functions.[8] These predictions are obtained essentially by taking the Mellin transform of Eq. (2). Throughout this talk I will discuss the theory in terms of moments but when I actually show the data, I will frequently show the QCD predictions for the structure functions themselves instead. I hope this will not cause confusion - it should be remembered that most of the statements made about moments apply equally well to discussions of the structure functions directly and vice-versa.

For structure functions which are not purely flavor non-singlet, the momentum distribution of the gluons inside a nucleon enters into the QCD analysis. For $x \gtrsim .4$, the role of the gluons in the QCD predictions is very small. For $x \lesssim .1$, the evolution of the gluon distribution, as predicted by QCD, is highly erratic and unstable probably indicating the breakdown of the QCD perturbation series in the low-x region. Thus, the gluon distribution can only be reliably determined by examining data in the region $.1 \lesssim x \lesssim .4$. On the basis of SLAC-MIT electroproduction data,[5] we have not been able to determine a gluon distribution and have instead assumed a distribution for gluons of the form $(1-x)^5$. Changing the factor of 5 from 4-7 has little effect on the results I will report below.[2]

A logical procedure for testing QCD[2,3] is to take the leading prediction from Eq. (2),

$$M_N(Q^2) \approx \frac{A_N}{(\ln Q^2/\Lambda^2)^{d_N}} \tag{3}$$

and compare this with the data varying the A_N and Λ to obtain the best possible fit. (Equivalently, one can use the leading-order predictions for the Q^2-evolution of $F(x,Q^2)$ and compare this with the data directly.) In doing this, one is assuming that the terms inside the curly bracket of Eq. (2) are well approximated by the value 1. Is this a valid approximation?

First, let us consider the effects of higher-twist terms. To do this, we will ignore the higher-order perturbative corrections and in fact, for simplicity, let us only consider the first term in the higher-twist sum, the term which goes like $1/Q^2$. As a final simplification, I will ignore the logarithmic dependence on Q^2 coming from $B_{iN}(\ln Q^2/\Lambda^2)$ which is small compared to the $1/Q^2$ dependence of this term. Thus, we will consider the simplified result

$$M_N(Q^2) \simeq \frac{A_N}{(\ln Q^2/\Lambda^2)^{d_N}} \left\{ 1 + \frac{N\mu^2}{Q^2} \right\} \tag{4}$$

(The factor of N multiplying the term μ^2/Q^2 is suggested by quark counting arguments.)[9,10] Our question about the role of higher-twist effects then comes down to this - what is the value of μ^2 in Eq. (4)? Of course, the simplifications made to reduce Eq. (2) to Eq. (4) are only for the purposes of discussion and the remarks I will make apply to all of the higher-twist terms in the complete expression of Eq. (2).

There are two relevant mass scales that could affect the value of μ. These are the mass scale of perturbative QCD, Λ, and the nucleon mass, m_p. Therefore, we must ask whether μ is of order Λ or of order m_p. If we define the moments as in Eq. (1), then undoubtedly μ is of order m_p. This is because the kinematic target-mass dependences have not been taken into account in Eq. (1). However, this can be done by using the Nachtmann definition[11] for the moments in place of Eq. (1) (or equivalently, for the structure functions, by going to the ξ-scaling scheme of Georgi and Politzer.)[11] As can be seen in Fig. 1 (for N=4 moment) the change from the ordinary definition of Eq. (1) to the Nachtmann form involves a considerable change in the value of the moment when $Q^2 \lesssim 4$ GeV^2. If we take the Nachtmann definition, then the predicted forms of Eqs. (2) - (4) apply equally well for Nachtmann moments and we must now ask what value μ takes when kinematic dependences on m_p have been removed. This is a difficult question because, as is shown in Fig. 2, there is a considerable amount of resonance production and elastic scattering, coming from the region where the final-state hadronic mass W is less than 2 GeV, contributing to the moments for $Q^2 \lesssim 10$ GeV^2. These processes involve typical hadronic mass of order 1 GeV and so one might assume that μ was of order 1 GeV or m_p. However, one can argue on the basis of Bloom-Gilman duality,[10,12] that on the average the Nachtmann moments (or the ξ-variable) account for elastics and resonance production. In Fig. 3, the x-scaling predictions of QCD for F_2, shown by the solid lines, and the ξ-scaling predictions, indicated by dashed lines, are compared with the data[5] in the resonance region. The data clearly show resonance peaks and we have indicated the elastic scattering contribution by adding an extra data bin from x=1 to x=1.04. The area under the data point in this bin is equal to the area under the elastic peak at x=1 in the original

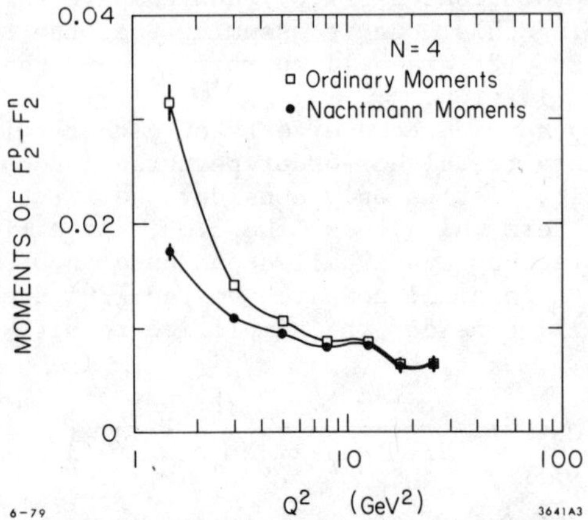

Fig. 1. A comparison of ordinary moments and Nachtmann moments from the SLAC data of Ref. 5. Curves are drawn connecting the data points to help guide the eye. Target-mass effects appear to be large for $Q^2 \lesssim 4$ GeV^2.

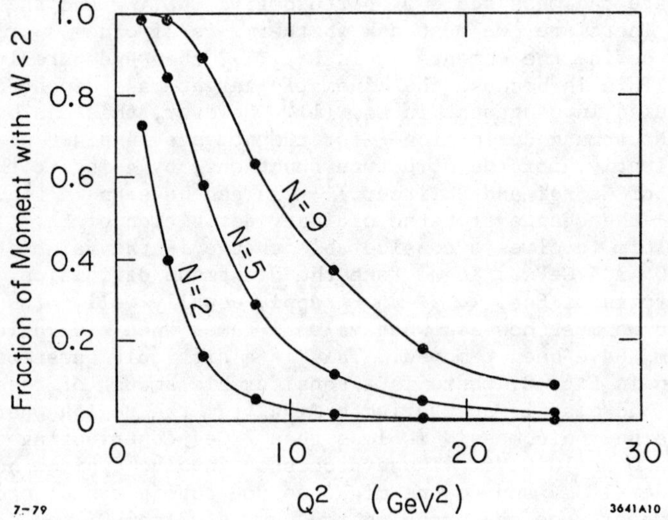

Fig. 2. The fraction of the Nachtmann moments (for N = 2,5,9) which comes from the resonance region (W < 2 GeV). The contributions at relatively large Q^2 are still quite significant. The data are from Ref. 5 with error bars not shown.

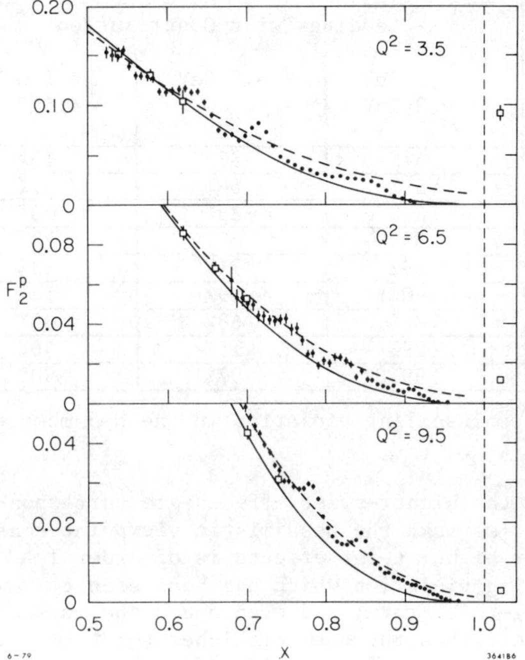

Fig. 3. $F_2(x,Q^2)$ for the proton. The solid (dashed) curve is the x (ξ) scaling prediction of QCD. Elastics are shown in extra bins from x=1 to 1.04 where the areas under the data points in these bins are equal to the areas under the elastic spikes at x=1 in the original data. All data are from SLAC-MIT (Ref. 5). The square points have W>2 GeV and are a compilation of SLAC data. The dots indicate some of the data in the resonance region.

data. Note that the ξ-scaling curve seems to go through the middle of the resonance peaks and that in overshooting the data near x=1 seems to account in some way for the elastic contribution. On the basis of this agreement one might speculate that all hadronic mass scales of order m_p have been eliminated by using the Nachtmann or Georgi-Politzer formalisms[11] and therefore μ is of order Λ.

In order to indicate the role played by one's assumptions about the magnitude of higher-twist effects in tests of QCD, let us consider three possible cases. The first is the rather optimistic view that μ=Λ=.3 GeV. Secondly, let us consider the more moderate case, μ=Λ=.5 GeV. Finally, for the pessimists, we will consider μ=1 GeV and Λ=.5 GeV. In Table 1, I have shown what percentage of the total scaling violation in the N=5 moment for various Q^2 ranges is coming from the leading-twist term of Eq. (3) if the actual Q^2-dependence is as in Eq. (4) with μ and Λ given the indicated values. I have also shown the accelerators for which the various Q^2 ranges are accessible. Note that if μ is of order 1 GeV then from the table we see that at all of the present accelerator Q^2 ranges less than 50% of the scaling violation seen is actually coming from the leading-twist QCD term. If μ is of order 1 GeV then the dominant source of Q^2-dependence in the N=5 moment is, in fact, higher-twist effects. We also see from the table that even if μ≈Λ anywhere from 10%-35% of the scaling violation seen above Q^2=5 GeV^2 at present accelerators originates from higher-twist effects. Thus, in detailed comparisons of the leading-twist predictions with experimental data, higher-twist terms represent a theoretical uncertainty of at least 10%-35% which should not be ignored. If

Accelerator	Q^2 Range (GeV^2)	Leading-Twist Contribution μ =.3 GeV Λ =.3 GeV	μ =.5 GeV Λ =.5 GeV	μ = 1 GeV Λ =.5 GeV
SLAC	1-30	57%	36%	13%
	5-30	82%	64%	31%
CERN } Fermilab	1-100	60%	38%	13%
	5-100	85%	65%	33%
	10-100	89%	71%	43%
Doubler	1-200	60%	39%	14%
	5-200	86%	68%	35%
	10-200	89%	75%	46%
Future	100-10,000	99.8%	97%	90%

Table 1. Percentage of the total scaling violation in the N=5 moment coming from the leading-twist term.

data below Q^2=5 GeV^2 are used the higher-twist effects are correspondingly larger. In summary, if we take the pessimistic viewpoint that the mass scale appropriate for higher-twist effects is of order 1 GeV, then the majority of the scaling violation which has been seen to date is not coming from the leading-twist term; and even under the more optimistic viewpoint that Λ describes the scale of higher-twist terms, non-leading twist effects are likely to be a significant component in the total scaling violation. The final entry in the table shows that in the Q^2 range of 100-10,000 GeV^2 everyone, optimist and pessimist alike, will agree that almost all of the scaling violation will be described by the leading-twist QCD prediction. Finally, I should stress that the results in table 1 do not critically depend on the fact that I have taken the $1/Q^2$ higher-twist term in Eq. (4). For example, the $1/Q^4$ term although suppressed by an extra factor of $1/Q^2$ is enhanced (according to quark counting arguments)[9,10] by an extra factor of N and leads to numbers very close to those shown in Table 1. Thus, in spite of our simplifying assumptions, Table 1 probably represents a reasonable picture of all higher-twist effects.

Although higher-twist effects may be with us until future accelerators reach giant Q^2 values, one can hope to disentangle the logarithmic Q^2-dependence of the leading-twist term from the power-law dependence of the higher-twist terms by detailed study of very precise data. This may be possible in the future if very accurate data are available over a wide x range and muon scattering experiments[13] may play an important role in this regard. However, using neutrino[14] and electron scattering data,[5] we have found that such a separation is, at present, impossible.[2] One can obtain an excellent fit to all of this data using just the leading-twist, leading-order QCD results. This is shown in Figures 4-6 for F_2^p and F_2^p-F_2^n from electroproduction and for xF_3 from neutrino scattering experiments. The solid lines in these figures show the leading prediction of QCD as fitted to these data. In Fig. 4, we have shown CHIO muon data[13] although this was not used in our fits. The data in Fig. 6 is from the CDHS experiment at CERN.[14] The dashed curves in Figs. 4-6, which can be seen to give an

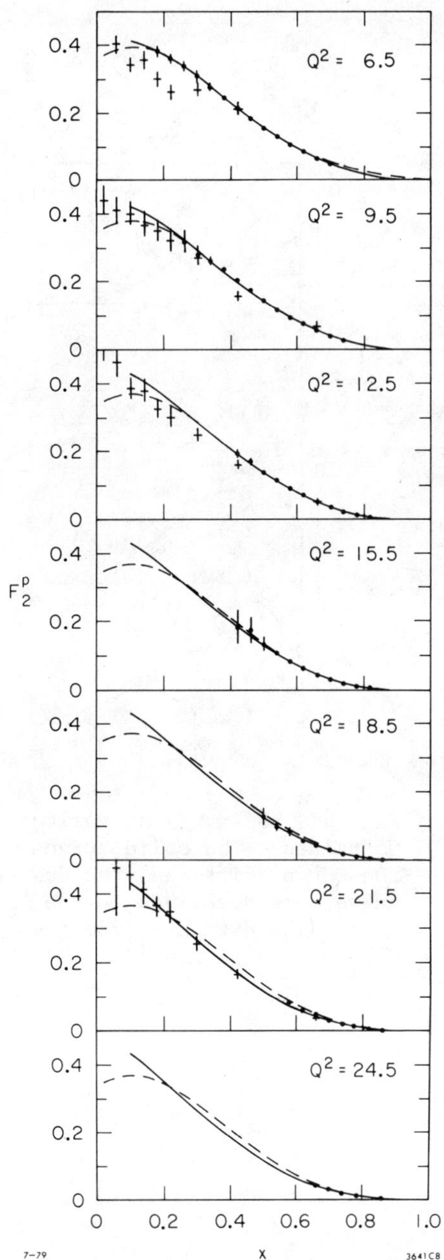

equally good fit to the data, were generated by assuming that <u>all</u> of the scaling violation in the data comes from higher-twist effects. This is an extreme assumption, but it does demonstrate the point that logarithmic and power-law Q^2-dependences cannot be distinguished in the data we have considered. In fact, various combinations of logarithmic Q^2-dependence, parametrized by Λ, and power-law dependence, parametrized by a mass scale μ, can fit the data. The various combinations of μ and Λ which give the best fits are indicated[15] in Fig. 7. The case i=1 corresponds to a form like Eq. (4) where the higher-twist term goes like μ_1^2/Q^2 whereas i=2 represents a higher-twist contribution proportional to $(\mu_2^2/Q^2)^2$.

Let us turn now to the correction terms in Eq. (2) coming from higher orders of QCD perturbation theory. Ignoring for this discussion the higher-twist effects and keeping only the second-order correction, we can write

$$M_N(Q^2) \approx \frac{A_N}{(\ln Q^2/\Lambda^2)^{d_N}}\left\{1 + \frac{C_{10}^{(N)}+C_{11}^{(N)}\ln\ln Q^2/\Lambda^2}{\ln Q^2/\Lambda^2} + \mathcal{O}\left(\frac{1}{\ln^2 Q^2/\Lambda^2}\right)\right\} \quad (5)$$

As I have mentioned the values of d_N, $C_{10}^{(N)}$ and $C_{11}^{(N)}$ are known.[1,7] One might then think it a simple matter to determine the size of the second-order corrections. However, by rescaling Λ we can make these second-order corrections for a given moment as large or as small

Fig. 4. $F_2(x,Q^2)$ on protons at various Q^2 values. The solid curves are the QCD predictions; the dashed curves are described in the text. The fit was done using only SLAC-MIT data[5] (solid dots), but the CHIO data[13] are also shown. The error bars are often smaller than the dots.

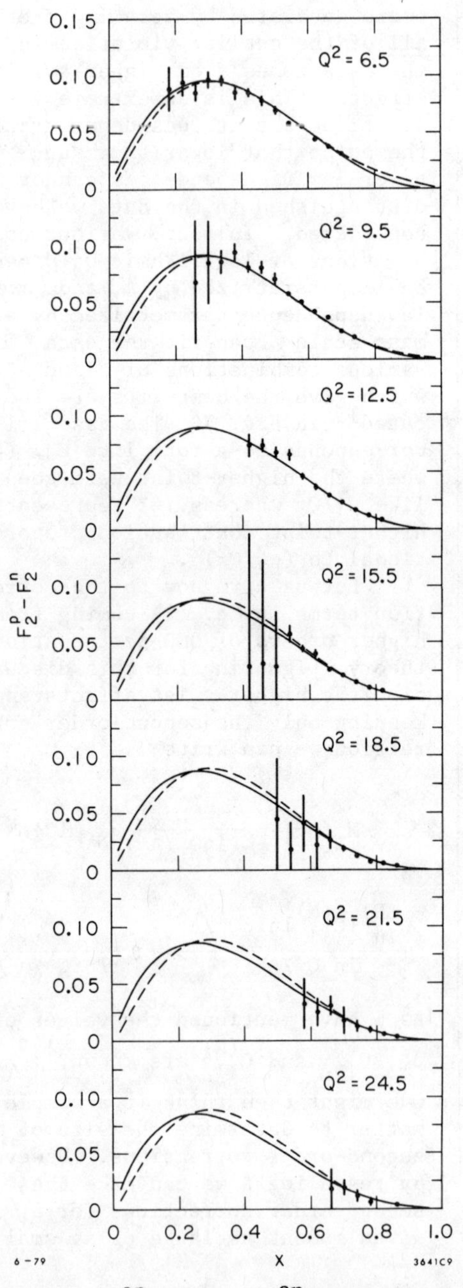

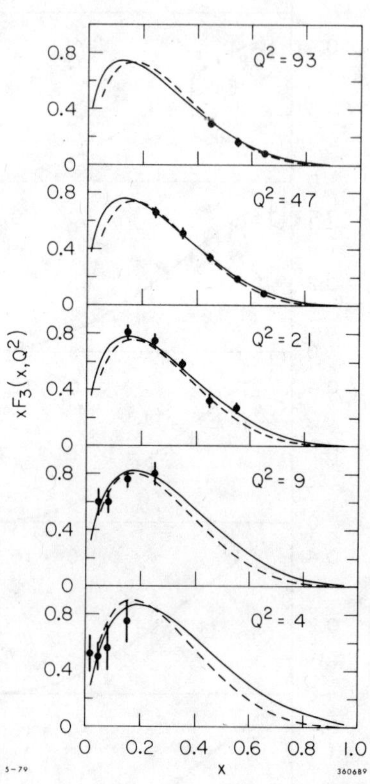

Fig. 6. $xF_3(x,Q^2)$ at various Q^2 values. The solid curves are the QCD predictions; the dashed curves are described in the text. The data are from the CDHS collaboration.[14]

Fig. 5. $F_2^{ep}(x,Q^2) - F_2^{en}(x,Q^2)$ at various Q^2 values. The solid curves are the QCD predictions; the dashed curves are described in the text. The data are from a compilation of SLAC-MIT data.[5]

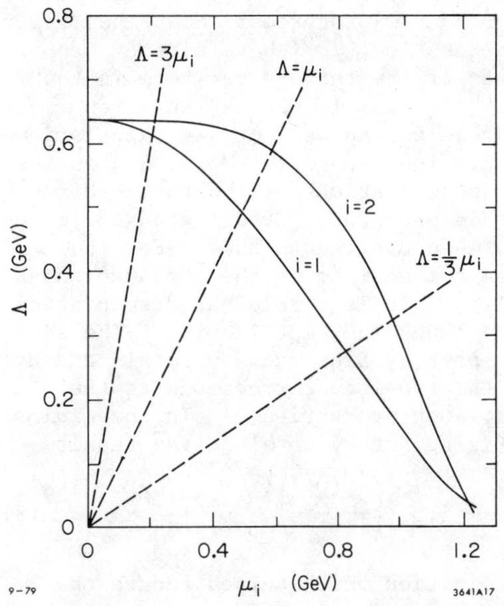

Fig. 7. The value of Λ obtained when higher-twist contributions have been assumed.[15] μ_1 and μ_2 indicate the magnitude of the higher-twist terms where the two forms considered are as discussed in the text.

as we like.[7] For example, define

$$\bar{\Lambda} = \Lambda e^{\frac{1}{2}\rho} \tag{6}$$

where ρ is some number. Expanding Eq. (5) in terms of this Λ we find that

$$M_N(Q^2) = \frac{A_N}{(\ln Q^2/\bar{\Lambda}^2)^{d_N}}\left\{1 + \frac{C_{10}^{(N)}+C_{11}^{(N)}\ln\ln Q^2/\bar{\Lambda}^2 - \rho d_N}{\ln Q^2/\bar{\Lambda}^2} + \mathcal{O}\left(\frac{1}{\ln^2 Q^2/\bar{\Lambda}^2}\right)\right\} \tag{7}$$

For a moderate Q^2 range and a given value of N we can always choose ρ so that

$$C_{10}^{(N)}+C_{11}^{(N)}\ln\ln Q^2/\bar{\Lambda}^2 - \rho d_N \approx 0 \tag{8}$$

For this reason the second-order Q^2 corrections are not very important in phenomenological applications. Also the question of their magnitude becomes much more subtle than one might originally have thought. Nevertheless, the computation of the second-order corrections of QCD for this and other processes probably represents the greatest advance that we have made in this field in the last few years. This is because we can now study how the QCD perturbation series is behaving and we can compare QCD predictions for different processes.

Two questions come to mind when we write down a condition like Eq. (8). First, if we choose a definition of $\bar{\Lambda}$ so that the second-order corrections are small for the moments of interest in phenomenology, do large corrections reappear in the third-order terms? And second, if we choose this $\bar{\Lambda}$ definition are the second-order terms also minimized in the QCD predictions for other processes? Moshe has given a speculative answer to the first question.[16] I have already mentioned that two of the coefficients of the third-order term in Eq. (2) are already known, $C_{21}^{(N)}$ and $C_{22}^{(N)}$. This is because in a renormalizable field theory part of the three-loop answer for any calculation can be determined by knowing the two-loop results. Moshe has estimated the remaining terms[16] to get approximations for the total third-order result. He has shown that his procedure gives a good estimate of the known second-order results when applied to them. For the third-order results he finds large corrections in the $\bar{\Lambda}$ scheme which makes the second-order terms small. This is a somewhat discouraging result, but its speculative nature should be kept in mind. The second question has been answered in part by Dine and Saperstein and by Chetyrkin, et al. who calculated the two-loop corrections to the e^+e^- cross-section ratio for hadron vs. muon production.[17] In the minimal subtraction scheme, without redefining Λ they obtained the result

$$R = \sum_i Q_i^2 \left\{ 1 + \frac{\alpha_s(Q^2)}{\pi} + 5.58 \left(\frac{\alpha_s(Q^2)}{\pi} \right)^2 \right\} \tag{9}$$

However, when they went to a $\bar{\Lambda}$ definition which helped reduce the magnitude of second-order effects in deep-inelastic results they found

$$R = \sum_i Q_i^2 \left\{ 1 + \frac{\alpha_s(Q^2)}{\pi} + 1.52 \left(\frac{\alpha_s(Q^2)}{\pi} \right)^2 \right\} \tag{10}$$

Thus, the same scheme which reduces second-order corrections in deep-inelastic scattering also reduces them in e^+e^- results. This, in principle, allows for the first time a comparison of Λ values obtained from different types of experiments.

Now I would like to consider another R, the ratio of longitudinal to transverse cross sections.[18] This is the only quantity measured in deep-inelastic scattering which does not agree with the leading-twist, leading-order QCD predictions.[2,10,19] The data and the QCD predictions (for various Q^2 values covered by the experiments) are shown in Fig. 8. The solid curves which fall considerably below the data points are the QCD predictions. The errors on these data are largely systematic so the situation is not, in fact, as bad as it looks in

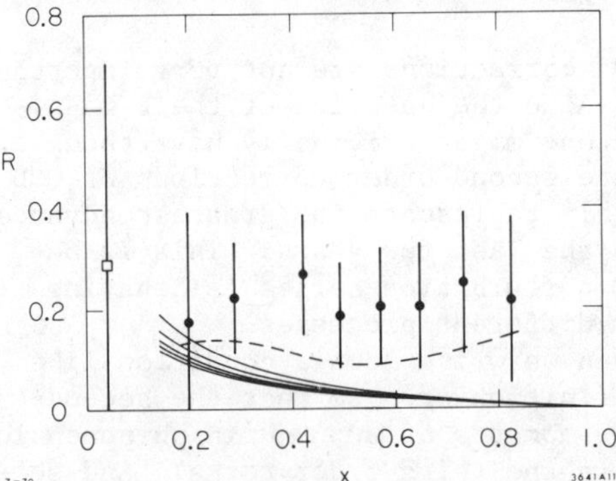

Fig. 8. $R \equiv \sigma_L/\sigma_T$ versus x. Using SLAC-MIT data;[18] the square point is from CHIO.[13] The solid curves show QCD with no higher-twist contributions for the Q^2 values covered by the data. The dashed curve is QCD plus diquark model (Ref. 20) of higher-twist; the curve reflects the average Q^2 of the data points through which it is drawn.

Fig. 8. The QCD prediction is about 1½-2 standard deviations away from the data.

The dashed curve in Fig. 8 comes from assuming, within the context of QCD, that the proton has a dynamical diquark substructure so that diquark scattering makes a significant contribution to the deep-inelastic cross section.[20] Diquark scattering is a higher-twist effect which goes like $1/Q^4$. Because the diquark can be a scalar, diquark scattering leads to a much larger value of the longitudinal cross-section than the leading-twist predictions of QCD would give. This diquark model was developed in collaboration with Ed Berger, Dick Blankenbecler and Gordie Kane.[20] The results of the model are compared with the data in more detail in Fig. 9. The model gives a reasonable though not perfect fit to the data.

In conclusion, the interpretation of present deep-inelastic scattering data in the context of QCD depends critically on assumptions about the magnitude of higher-twist effects. These effects may be showing up in the large ratio of longitudinal to transverse cross sections measured at SLAC. The QCD perturbation series seems to be behaving quite reasonably and due to recent computations, we should for the first time be able to compare Λ values from different types of experiments.

Acknowledgements: I am most grateful to my collaborators W. Atwood, M. Barnett, E. Berger, R. Blankenbecler, and G. Kane. In addition, I thank S. Brodsky, F. Gilman and M. Wise for helpful discussions. This work was supported by the Department of Energy under contracts DE-AC03-76SF00515 and E(11-1)3230.

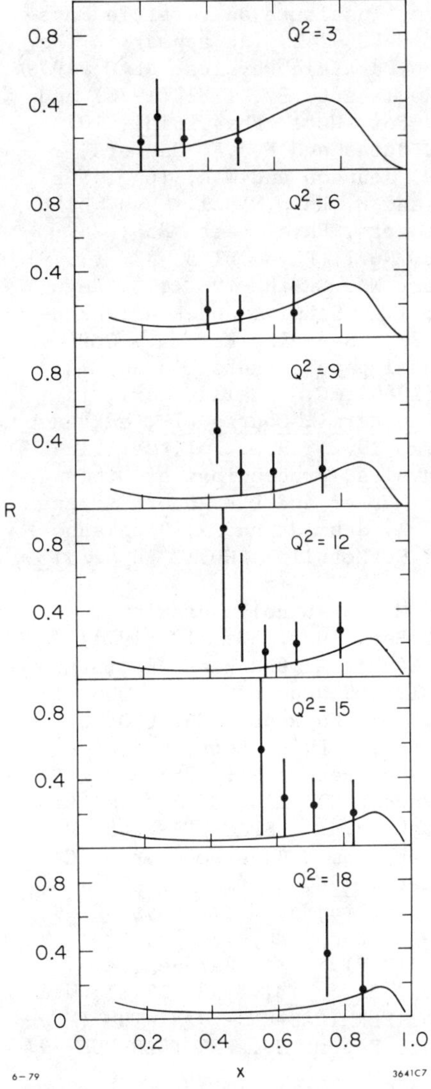

Fig. 9. $R \equiv \sigma_L/\sigma_T$ at various Q^2 values. The data are a compilation of all SLAC-MIT data;[18] the error bars are mostly systematic. The curves show the results of QCD when a higher-twist contribution from diquark scattering (using the model of Ref. 20) is added.

REFERENCES

1. H.D. Politzer, Phys. Rev. Lett. 26, 1346 (1973); D.J. Gross and F. Wilczek, Phys. Rev. Lett. 26, 1343 (1973) and Phys. Rev. D8, 3633 (1973) and D9, 980 (1974); A. Zee, F. Wilczek and S.B. Treiman, Phys. Rev. D10, 2881 (1974); H. Georgi and H.D. Politzer, Phys. Rev. D9, 416 (1974).
2. L.F. Abbott, W.B. Atwood and R.M. Barnett, Report No. SLAC-PUB-2400; L.F. Abbott and R.M. Barnett, Report No. SLAC-PUB-2325 (May 1979), to be published in Annals Phys.; L.F. Abbott, Report No. SLAC-PUB-2296 (to appear in the Proceedings of Orbis-Scientiae, 1979, Coral Gables); R.M. Barnett, Report No. SLAC-PUB-2396 (to appear in the Proceedings of Summer Institute on Particle Physics, SLAC, 1979); W.B. Atwood, SLAC-PUB-2428 (to appear in the Proceedings of Summer Institute on Particle Physics, SLAC, 1979).
3. V. Baluni and E. Eichten, Phys. Rev. Lett. 37, 1181 (1976) and Phys. Rev. D14, 3045 (1976); G.C. Fox, Nucl. Phys. B131, 107 (1977) and B134, 269 (1978); A.J. Buras and K.J.F. Gaemers, Nucl. Phys. B132, 249 (1978); P.W. Johnson and W.K. Tung, Nucl. Phys. B121, 270 (1977); M. Gluck and E. Reya, Nucl. Phys. B156, 456 (1979); D.W. Duke and R.G. Roberts, Phys. Lett. 85B, 289 (1979); L. Baulieu and C. Kounnas, Nucl. Phys. B155, 429 (1979); C. Aviliz et al., Strasbourg Report No. CRN/HE-79-11; E. Reya, Report No. DESY-79/30 (May 1979); I.A. Schmidt and R. Blankenbecler, Phys. Rev. D16, 1318 (1977). See also Ref. 7. QCD analyses also appear in experimental papers, Refs. 13 and 14.
4. K. Wilson, Phys. Rev. 179, 1499 (1969); C.G. Callan, Jr., in Proc. Int. School of Physics Enrico Fermi, Course LIV, edited by R. Gatto (Academic Press, New York, 1972); H.D. Politzer, in Deeper Pathways in High Energy Physics, proceedings of Orbis Scientiae, 1977, edited by A. Perlmutter and L.F. Scott (Plenum Press, New York, 1977), p. 621; I.A. Schmidt and R. Blankenbecler, Phys. Rev. D16, 1318 (1977); S. Gottlieb, Nucl. Phys. B139, 125 (1978).
5. Experiments used from the SLAC and MIT-SLAC collaborations. E49a: J.S. Poucher *et al.*, Phys. Rev. Lett. 32, 118 (1974); J.S. Poucher, PhD Thesis, MIT (1971); E49b: A. Bodek *et al.*, SLAC-PUB-1327; A. Bodek, PhD Thesis, MIT LNS Report No. COO-3069-116 (1972); E.M. Riordan, PhD Thesis, MIT LNS Report No. COO-3069-176 (1972); E89-1: W.B. Atwood *et al.*, Phys. Lett. 64B, 479 (1976); W.B. Atwood, PhD Thesis, SLAC Report No. 185; E89-2: M.D. Mestayer, PhD Thesis, SLAC Report No. 214; E87: E.M. Riordan *et al.*, SLAC-PUB-1634; A. Bodek et al., SLAC-PUB-2248. These data are available on request from W.B. Atwood at SLAC.
6. See for example M. Moshe, Phys. Lett. 79B, 88 (1978).
7. E.G. Floratos, D.A. Ross and C.T. Sachrajda, Nucl. Phys. B129, 66 (1977)(erratum-B139, 545 (1978)), Nucl. Phys. B152, 493 (1979), and Phys. Lett. 80B, 269 (1979). W.A. Bardeen, A.J. Buras, D.W. Duke and T. Muta, Phys. Rev. D18, 3998 (1978); W.A. Bardeen and A.J. Buras, Report No. FERMILAB-PUB-79/31-THY (May 1979); for a review see A.J. Buras, Report No. FERMILAB-PUB-79/17-THY (January 1979).

8. G. Altarelli and G. Parisi, Nucl. Phys. B126, 298 (1977); Yu. L. Dokshitzer, D.I. Dyakonov and S.I. Troyan, Report No. SLAC-TRANS-0183 (June 1978), translated from Proceedings of the 13th Leningrad Winter School on Elementary Particle Physics, 1978, pp. 1-89; J. Kogut and L. Susskind, Phys. Rev. D9, 697 and 3391 (1974); G. Parisi, Phys. Lett. 43B, 207 (1973); D.J. Gross, Phys. Rev. Lett. 32, 1071 (1974); C.H. Llewellyn Smith, Report No. OXFORD-TP 47/78 (February 1978); W.R. Frazer and J.F. Gunion, Phys. Rev. D19, 2447 (1979).
9. S.J. Brodsky and G. Farrar, Phys. Rev. Lett. 31, 1153 (1973) and Phys. Rev. D11, 1309 (1975); R. Blankenbecler and S.J. Brodsky, Phys. Rev. D10, 2973 (1974); I.A. Schmidt and R. Blankenbecler, Phys. Rev. D16, 1318 (1977); H.D. Politzer, in Deeper Pathways in High-Energy Physics, Proceedings of Orbis Scientiae, 1977, edited by A. Perlmutter and L.F. Scott (Plenum Press, New York, 1977), p. 621.
10. A. De Rujula, H. Georgi and H.D. Politzer, Ann. Phys. 103, 315.
11. O. Nachtmann, Nucl. Phys. B63, 237 (1973), H. Georgi and H.D. Politzer, Phys. Rev. D14, 1829 (1976) and Phys. Rev. Lett. 36, 1281 (1976); S. Wandzura, Nucl. Phys. B122, 412 (1977).
12. E. Bloom and F. Gilman, Phys. Rev. Lett. 25, 1140 (1970); A. De Rujula, H. Georgi and H.D. Politzer, Phys. Lett. 64B, 428 (1977) and Ann. Phys. 103, 315 (1977); R. Barbieri, J. Ellis, M.K. Gaillard and G.G. Ross, Phys. Lett. 64B, 171 (1976) and Nucl. Phys. B117, 50 (1976); R.K. Ellis, R. Petronzio and G. Parisi, Phys. Lett. 64B, 97 (1976). D.J. Gross, S.B. Treiman and F.A. Wilczek, Phys. Rev. D15, 2486 (1977).
13. H.L. Anderson, et al. (CHIO), Report No. FERMILAB-PUB-79/30-EXP (May 1979); R.C. Ball, et al., Phys. Rev. Lett. 42, 866 (1979); B.A. Gordon, et al., Phys. Rev. Lett. 41, 615 (1978); C. Chang, et al., Phys. Rev. Lett. 35, 901 (1975).
14. P.C. Bosetti, et al., (BG) Nucl. Phys. B142, 1 (1978); we thank W. Scott for providing us with up-to-date data from BEBC-Gargamelle; J.G.H. DeGroot, et al. (CDHS), Phys. Lett. 82B, 292 and 456 (1979) and Zeit. f. Phys. C1, 142 (1979).
15. We thank H. Georgi for suggesting this figure.
16. M. Moshe, Tel-Aviv Report No. TAUP-769-79 (1979).
17. M. Dine and J. Saperstein, Phys. Rev. Lett. 43, 668 (1979); K.G. Chetyrkin, et al., Institute for Nuclear Research, Moscow, Report No. 126 (1979).
18. M.D. Mestayer, PhD Thesis, SLAC Report No. 214; E.M. Riordan et al., Report No. SLAC-PUB-1634; A. Bodek, et al., Report No. SLAC-PUB-2248.
19. G.C. Fox, Nucl. Phys. B131, 107 (1977); M. Calvo, Phys. Rev. D15, 730 (1977); A. Zee, F. Wilczek and S.B. Treiman, Phys. Rev. D10, 2881 (1974); D. Nanopoulos and G.G. Ross, Phys. Lett. 58B, 105 (1975); I. Hinchliffe and C.H. Llewellyn-Smith, Nucl. Phys. B128, 93 (1977).
20. L.F. Abbott, E.L. Berger, R. Blankenbecler and G.L. Kane, Report No. SLAC-PUB-2327 (May 1979), to be published in Phys. Lett.

HEAVY QUARKS AND NEW PARTICLES*

Jonathan L. Rosner
School of Physics and Astronomy
University of Minnesota, Minneapolis, Minnesota 55455

ABSTRACT

Some aspects of particles containing c, b, and heavier quarks are reviewed. It is shown how these particles can provide tests of the short-range nature of the strong interactions, and how elementary quantum mechanics can reveal the color and charge of the constituents and the number of states below flavor threshold. Some new results are presented on the inverse problem for confining potentials. Properties of flavored hadrons (masses and lifetimes), of gluonic bound states, and of particles containing very heavy objects are also discussed.

INTRODUCTION

New particles containing heavy quarks were first discovered just five years ago.[1] There now appears to be a rich variety of states containing the charmed quark ($m_c \simeq 1\frac{1}{2}$ GeV/c^2),[1,2] and particles containing the b quark ($m_b \simeq 5$ GeV/c^2) are beginning to appear.[2-4] There is very likely at least one heavier quark and its corresponding family of hadrons.[5-7] What can be learned from these new particles?

First, bound systems involving heavy quarks shed light on the strong interactions. The large masses of these quarks suggest simple but powerful nonrelativistic methods,[8] and provide short-distance information not available from the lighter (u,d,s) quarks. Second, the masses, charges and decay properties of the quarks themselves help one understand the weak interactions and unification schemes.

In this review we shall discuss properties of states containing c, b, and heavier quarks from both standpoints: as a window on the strong interactions and as sources of information on the quarks.

In Sec. II, we shall be concerned with deeply bound states of a quark with the corresponding antiquark. The masses and leptonic widths of these quarkonium systems provide information on the short-distance quark-antiquark force, and their decays depend on coupling to (and properties of) the emitted quanta.

Certain properties of quarkonium states are not peculiar to a specific theory of strong interactions but follow only from known data, quantum mechanics, and assumptions about the constituents. This is the subject of Sec. III. There, the charge and color of the quarks in quarkonium states will be learned from observed leptonic widths of these states, and the number of bound states below threshold for decay of these states to flavored pairs will be predicted.

When they are far enough apart, the quark and antiquark in

*Invited talk at the 1979 Meeting of the Division of Particles and Fields of the American Physical Society, October 25-27, Montreal, Quebec.

ISSN:0094-243X/80/590325-27$1.50

quarkonium each can be incorporated into a new (flavored) hadron, by the production of a light quark-antiquark pair. Remarks on this process and on properties of flavored hadrons are contained in Sec. IV. We include there a discussion of gluonic bound states ("glueballs") and particles containing very heavy objects (e.g., stable color sextet quarks).

Section V is devoted to a summary.

II. DEEP BINDING.

The quanta (gluons) of the strong interactions as described by quantum chromodynamics (QCD)[9,10] can manifest themselves in several ways.[2]

Gluon exchange can give rise to a Coulomb-like potential between quarks and antiquarks.[11] It can lead to hyperfine splittings, for example between $^3S_1$ and $^1S_0$ quark-antiquark states.[12] Gluons can be emitted in the annihilation of a quark-antiquark pair.[12] For deeply bound states of heavy quarks, as first encountered in the J/ψ , this is a dominant process, since the quarks cannot separate from one another enough to dissociate into two flavored hadrons. Virtual gluon emission and absorption can significantly alter the structure of all processes involving heavy quarks.[13]

Our starting point for much of this discussion is the Schrodinger equation for a nonrelativistic system of reduced mass μ :

$$-\frac{\nabla^2}{2\mu}\Psi(\underset{\sim}{r}) + V(\underset{\sim}{r})\,\Psi(\underset{\sim}{r}) = E\,\Psi(\underset{\sim}{r}) \quad . \tag{1}$$

The interaction between a quark and antiquark in a color singlet state can be described by the potential[11]

$$V(r) = -\frac{4}{3}\frac{\alpha_s}{r} + ar \tag{2}$$

The first term describes single-gluon exchange, while the second is an approximation to the expected long-range quark-confining force. An effective strong coupling constant $\alpha_s \approx 0.4$ is stipulated by the charmonium and Υ spectra.[11,14,15]

Alternatively, any trend toward short-distance Coulomb-like effects may be extracted directly from the behavior of masses and leptonic widths as functions of quark mass and quantum numbers. This may be done with the help of scaling properties of the Schrödinger equation (subsection A),[8,16] or via inverse scattering methods (subsection B).[17-19] It is possible to compare the value of α_s obtained in Eq. (1) with those obtained from other sources (subsection C).

A. Scaling.

A useful idealization of the quark-antiquark interaction is an effective-power-law potential

$$V(r) \sim r^{\nu} \tag{3}$$

If the true interaction is (1), the effective power ν will approach 1 for large distances (light quarks, highly excited states) and -1 for short distances (heavy quarks, deeply bound states).

By scaling the Schrödinger equation, and applying semiclassical methods to determine the behavior of S wave levels as a function of principal quantum number n, one obtains the dependences shown in Table 1.

Table 1. Scaling of the Schrödinger equation for a potential $V \sim r^{\nu}$.

Quantity \ Variable	Reduced mass μ	Principal quantum number n	
Level spacing ΔE	$\mu^{-\nu/(2+\nu)}$	$(n - \frac{1}{4})^{2\nu/(2+\nu)}$	$\nu > 0$
		$[n-\frac{1+\nu}{2(2+\nu)}]^{2\nu/(2+\nu)}$	$\nu < 0$
Probability density at r = 0 $\|\Psi(0)\|^2$	$\mu^{3/(2+\nu)}$	$(n-\frac{1}{4})^{2(\nu-1)/(2+\nu)}$	$\nu > 0$
		$[n-\frac{1+\nu}{2(2+\nu)}]^{(\nu-2)/(2+\nu)}$	$\nu < 0$

Examples:

	Coulomb (ν = -1)	Linear (ν = 1)
ΔE	$\mu\, n^{-2}$	$\mu^{-1/3}(n-\frac{1}{4})^{2/3}$
$\|\Psi(0)\|^2$	$\mu^3 n^{-3}$	μ

Since[20]

$$|\Psi(0)|^2 = \frac{\mu}{2\pi} \left\langle \frac{dV}{dr} \right\rangle , \tag{4}$$

the probability at the origin does not depend on n for the linear potential, for which $\langle dV/dr \rangle$ = const.

A number of the observables quoted in Table 1 specify ν for the J/ψ and Υ families. Not all of these values need agree with one another, since they probe different parts of the potential.

ΔE vs. $m_Q = 2\mu$. The level spacings[21-23]

$$\frac{\Upsilon'(2S) - \Upsilon(1S)}{\psi'(2S) - \psi(1S)} = \frac{558 \pm 10\ \text{MeV}}{589\ \text{MeV}} = 0.95 \pm 0.02 \tag{5}$$

yield, for $3 \leq m_b/m_c \leq 4$, the range

$$\nu = 0.09 \pm 0.05 \tag{6}$$

ΔE vs. quantum numbers. The position of the spin-averaged P wave charmonium levels, $\bar{\chi}$ = 3522 MeV, with respect to ψ and ψ' corresponds to[24]

$$\nu = 0.15 \tag{7}$$

The ratio (3S-2S)/(2S-1S), which is 0.59 ± 0.02 for $c\bar{c}$ and 0.70±0.07 for $b\bar{b}$,[25] yields[8]

$$\nu(c\bar{c}) = 0.20 \pm 0.06 \tag{8}$$

$$\nu(b\bar{b}) = 0.45 \pm 0.25 \quad . \tag{9}$$

$|\Psi(0)|^2$ vs. m_Q. The leptonic widths of $Q\bar{Q}$ vector mesons V are related to squares of wave functions at the origin by[26]

$$\Gamma(V \to e^+e^-) = \frac{16\pi\ \alpha^2 e_Q^{\ 2} N}{3\ M_V^{\ 2}} \ |\Psi(0)|^2 \ a \tag{10}$$

Here e_Q is the charge of Q in units of the proton's charge, and N is the number of colors (3 for ordinary quarks like c and b). The factor a accounts for strong radiative corrections[27]; a = 1 will be taken for the present nonrelativistic discussion. For $c\bar{c}$ states,[28]

$$1S: \quad \Gamma(\psi \to e^+e^-) = 4.8 \pm 0.6\ \text{keV} \ , \tag{11}$$

$$2S: \quad \Gamma(\psi' \to e^+e^-) = 2.1 \pm 0.3\ \text{keV}, \tag{12}$$

while for $b\bar{b}$ states[21]

$$1S: \quad \Gamma(\Upsilon \to e^+e^-) = 1.32 \pm 0.09\ \text{keV}, \tag{13}$$

$$2S: \quad \Gamma(\Upsilon' \to e^+e^-) = 0.33 \pm 0.10\ \text{keV} \quad . \tag{14}$$

We then find, for $3 \leq m_b/m_c \leq 4$, that Eq. (10) and Table 1 imply[29]

$$-0.66 \leq \nu \leq -0.09 \qquad (1S) \tag{15}$$

or

$$-0.2 \leq \nu \leq 1.7 \qquad (2S) \tag{16}$$

$|\Psi(0)|^2$ vs. n. The leptonic widths of higher S wave $c\bar{c}$ states can be incorporated into a fit for ν, yielding the range[8]

$$-0.09 \leq \nu \leq 0.14 \tag{17}$$

The leptonic widths (13) and (14) of ϒ and ϒ' lead to a lower value

$$\nu = -0.54 \pm 0.24 \quad . \tag{18}$$

The values of ν obtained above are compared with one another in Figure 1. There is not a clear-cut trend toward lower values as the quark mass increases, though one might have expected the heavier b

Fig. 1. Effective power ν from charmonium and upsilon data.

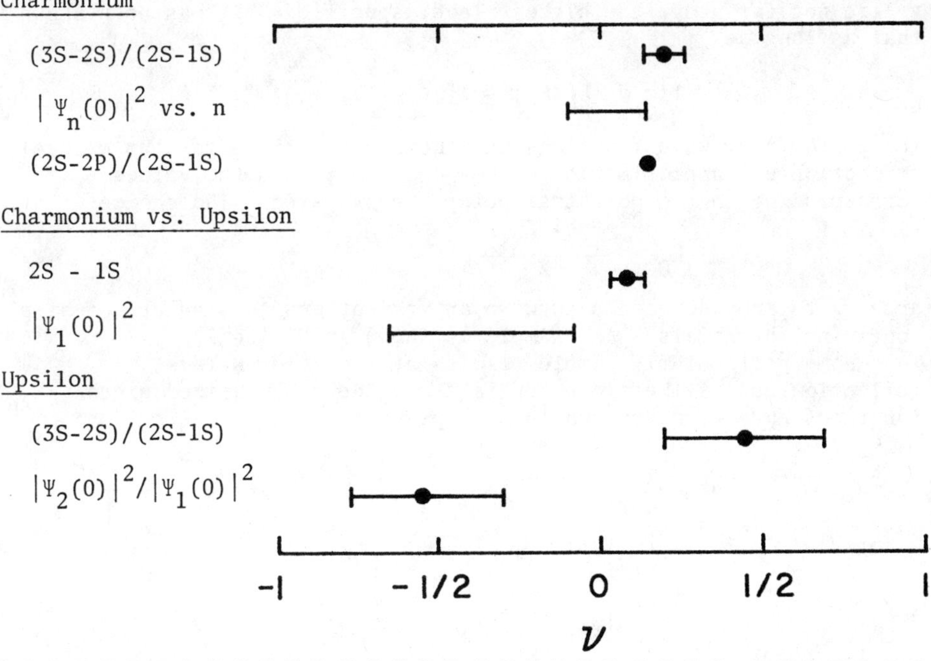

quarks in the ϒ family to "see" shorter distances -- and, hence, more striking Coulomb-like effects -- than those in charmonium. This trend does appear to be visible in the ratios of 2S and 1S leptonic widths. This ratio should decrease further as the quark mass is raised beyond m_b if there is really a short-distance Coulomb-like force between quarks and antiquarks.

B. Inverse scattering results.

Quarkonium data can directly determine the potential between quarks. This "inverse method" enjoys a vast literature,[30,31] which has been adapted and somewhat extended for the present purpose.[17-19,32,33]

In one dimension, we may approximate an infinitely rising symmetric potential by a symmetric, reflectionless one having the same first N levels.[17] The absence of reflection is a strong limitation on the form of the approximate potential, and is assumed only so that the problem reduces to linear algebra rather than inversion of integral equations. (See, however, Ref. 19 for a further discussion.)

One must also choose the lowest continuum energy E_0 of the approximate potential:

$$E_0 \equiv \lim_{|x|\to\infty} V(x) \quad , \tag{19}$$

to lie not far above the highest level specified. It has been shown[19] that as long as

$$\lim_{N\to\infty} (E_0/E_N) = 1 \; , \tag{20}$$

the Schrödinger wave functions and their derivatives for the N-level reflectionless approximation converge at $x = 0$ to the values corresponding to the potential being approximated. The choice[17]

$$E_0 = \frac{E_N + E_{N+1}}{2} \tag{21}$$

appears to provide good pointwise approximations to a number of simple confining potentials. An example is shown in Figure 2.

Some particularly simple results at $x = 0$ for N-level reflectionless, symmetric potentials and their normalized eigenfunctions have been derived.[19] Examples are:

$(K_n^{\;2} \equiv E_0 - E_n)$:

$$V(0) = E_0 + 2 \sum_{i=1}^{N} (-1)^i K_i^{\;2} \quad \text{(any N)} \tag{22}$$

$$= E_0 - 4 \sum_{i=1}^{N} |\psi'_{2i}(0)|^2/K_i \quad \text{(N even)} \tag{23}$$

$$|\psi_i(0)|^2 = \frac{K_i}{2} \; \frac{\prod\limits_{j\ \mathrm{even}}^{N-1} |K_j^{\;2} - K_i^{\;2}|}{\prod\limits_{j\ \mathrm{odd} \neq i}^{N} |K_j^{\;2} - K_i^{\;2}|} \quad \text{(i odd, N odd)} \tag{24}$$

$$|\psi'_i(0)|^2 = \frac{K_i}{2} \; \frac{\prod\limits_{j\ \mathrm{odd}}^{N-1} |K_j^{\;2} - K_i^{\;2}|}{\prod\limits_{j\ \mathrm{even} \neq i}^{N} |K_j^{\;2} - K_i^{\;2}|} \quad \text{(i even, N even)} \tag{25}$$

Figure 2. Eight-level approximation (dashed curve) to the harmonic oscillator potential $V(x) = x^2$ (solid curve). Levels (horizontal solid lines) are at $E_n = 2n = 1$ ($n = 1, \ldots, 8$). Here $E_0 = 16$ (horizontal dashed line).

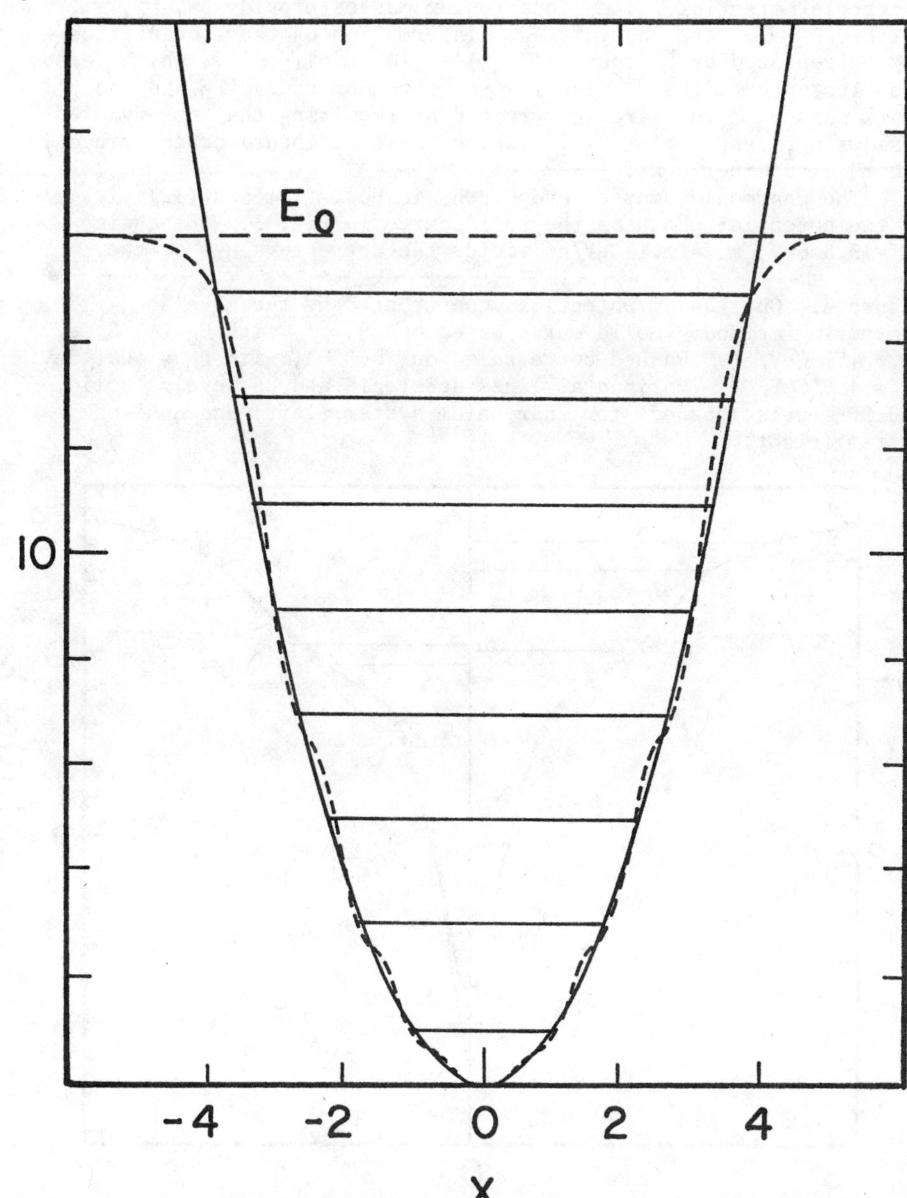

Equation 22, for example, dictates the choice (21) as the correct one for the harmonic oscillator.

The S wave levels of the three dimensional problem correspond to the odd-parity levels of the one-dimensional problem with a symmetric potential. Thus, quarkonium masses provide E_2, E_4, ..., but E_1, E_3, ... are unspecified. Information on these quantities may be replaced by that on $|\psi_2'(0)|^2$, $|\psi_4'(0)|^2$, ..., which comes from leptonic widths.[34] The lowest continuum energy E_0 and the quark mass $m_Q = 2\mu$ are constrained by requiring that observables besides E_{2i} and $|\psi'_{2i}(0)|^2$ (which serve as input) be described in accord with experiment.[17,18]

The charmonium masses and widths in Eqs. (11) and (12) give rise to the potential shown as the solid curve in Fig. 3. The choice $E_0 = 3.8$ GeV, $m_c = 1.1$ GeV/c^2 yields the correct values of the

Figure 3. Quarkonium potentials constructed by the inverse scattering method. Solid curve based on ψ, ψ', with $E_0 = 3.8$ GeV, $m_c = 1.1$ GeV/c^2. Dashed curve based on Υ, Υ', with $E_0 = 10.1$ GeV, $m_b = 4.5$ GeV/c^2. Horizontal lines denote 1S and 2S levels (solid) and 2P levels (dashed) for charmonium system (left) and upsilon system (right).

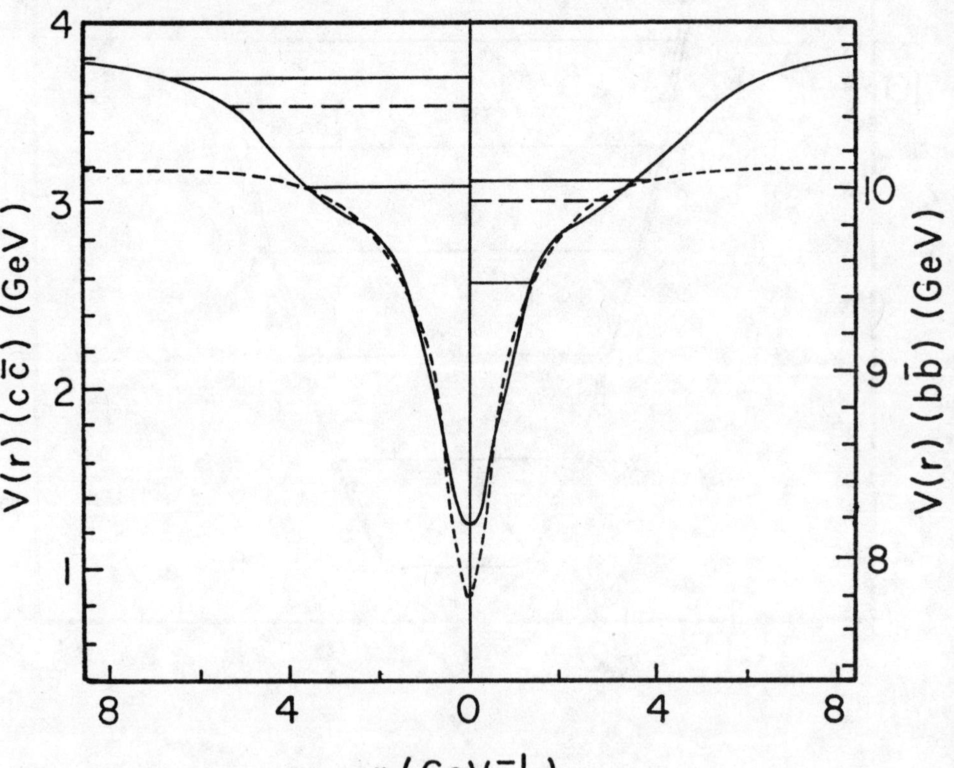

average P wave charmonium mass, the Υ - Υ' spacing, and the Υ, Υ' leptonic widths when the Schrödinger equation is solved for this potential. The 2S and 2P Υ levels lie closer to one another than the corresponding charmonium ones. The predicted ratio $\Gamma_{ee}(2S)/\Gamma_{ee}(1S)$ decreases as one passes from charmonium to upsilons.

The charmonium potential can be compared with one constructed directly from the Υ levels, shown as the dashed curve in Fig. 3.[18] In the region of r probed by the Υ and Υ' levels, the two potentials agree well. Information on Υ'' and Υ''' levels will extend this comparison to larger interquark separations.

The agreement of the two potentials in Fig. 3 provides evidence for flavor independence of the quark-antiquark force. It also supports the conclusion that the quarks in the Υ family have charge -1/3 and are ordinary color triplets. The values of $|\Psi(0)|^2$ inferred from Eq. (10) would be cut in four if one replaced $e_Q = -1/3$ by $e_Q = 2/3$, and in half if $N = 3 \rightarrow N = 6$. This would have drastic effects on the shape of the potentials reconstructed from Υ and Υ';[18] these potentials become much shallower, cease to be monotonic, and look nothing at all like Fig. 3. This is most easily seen from Eq. (23), which says that the depth of the potential is directly proportional to a weighted average of values of $|\Psi(0)|^2$.

A direct transcription from QED[35] to QCD[27] suggests that Eq. (10) receives important radiative corrections:

$$a = 1 - 16\alpha_s/3\pi < 1. \tag{26}$$

These, in fact, seem necessary if one is to describe the observed charmonium and upsilon leptonic widths with a potential of the form (2).[11,14]

To explore this possibility we have constructed potentials from ψ and ψ' as described above,[17] but with a common value of $a < 1$ in Eq. (10) for both charmonium and upsilon systems. The results are shown in Table 2.

As the importance of the radiative corrections increases, and $|\Psi(0)|^2$ is scaled upward from its nonrelativistic value [$a = 1$ in Eq. (10)], several trends are visible. The reconstructed potential becomes deeper at the origin. The lowest quarkonium systems become spatially more compact.[36] The ratios $\Gamma_{ee}(\Upsilon')/\Gamma_{ee}(\Upsilon)$ and $\Gamma_{ee}(\Upsilon'')/\Gamma_{ee}(\Upsilon)$ decrease. The $\Upsilon'(2S)$-$\chi_b(2P)$ splitting tends to zero. All these trends entail an increased role for short-distance Coulomb effects, and confirm the ability of the inverse method to make good approximations to singular potentials.[17-19]

The Υ''-Υ' spacing in Table 2 increases as the potentials become deeper. The fourth Υ level lies at 10.62 ± 0.01 GeV, within 60 MeV of flavor threshold $E_{th}(b\bar{b})$ (estimated in Sec. III). The near constancy of E_{th} reflects the small range[37] in Table 2:

$$3.42 \leq m_b - m_c \leq 3.45 \text{ GeV}/c^2 \tag{27}$$

Table 2. Quarkonium potentials fitting ψ, ψ', χ, Υ'-Υ, $\Gamma_{ee}(\Upsilon)$.

a^{-1}	1	2	3
E_o(GeV)	3.8	3.775	3.75
m_c(GeV/c^2)	1.1	1.6	1.9
m_b(GeV/c^2)	4.554	5.025	5.334
V(0) (GeV)	1.22	0.64	-0.31
$\Gamma_{ee}(\Upsilon')$ (keV)	0.32	0.26	0.25
$\Gamma_{ee}(\Upsilon'')$ (keV)	0.33	0.20	0.15
Υ'' - Υ' (GeV)	0.34	0.40	0.44
Υ' - χ_b (GeV)	0.09	0.05	0.01
Υ''' (GeV)	10.63	10.62	10.61
$E_{th}(b\bar{b})$ (GeV)	10.69	10.62	10.64

C. Gluon emission and exchange.

There are several sources of information about the strong coupling constant α_s from charmonium. Many expressions involving gluon emission or exchange involve $P \equiv |\Psi_\psi(0)|^2/m_c{}^2$, which appears fortuitously to remain nearly constant in Table 2. The results of extracting α_s from charmonium observables are summarized in Table 3. Here we have assumed $|\Psi_\psi(0)^2 \simeq |\Psi_{\eta c}(0)|^2$.

The results of Table 3 may be compared with the lowest-order renormalization-group formula[10,45]

$$\alpha_s(Q^2) = \frac{12\pi}{(33-2n_f)\ \ln(Q^2/\Lambda^2)} \tag{28}$$

by estimating that α_s ([charmonium size]$^{-2}$). $\simeq \alpha_s$ ([$\frac{1}{2}$GeV]^2) $\simeq$ 0.39.[46] Then

$$\Lambda = 84 \text{ MeV}; \tag{29}$$

$$\alpha_s(m_c^2) = 0.24. \tag{30}$$

Only the observed decay width for $J/\psi \to$hadrons appears to conflict with Eqs. (28)-(30) (in the absence of radiative corrections.[38]) It would favor an even smaller value of Λ.[47] The radiative corrections to $\eta_c \to 2g$ appear to be large and positive,[13,48] so that the

Table 3. Information on α_s from charmonium.

Source of information	Distance scale	Corrections to right-hand side	Data	α_s (uncorrected)
$V(r) = -\frac{4}{3}\frac{\alpha_s}{r} + ar$	2 GeV^{-1} (Fig. 3)	Softening at short distances	Fit to charmonium levels (Ref. 11)	0.39
$\Gamma(J/\psi\to 3g) = \frac{40}{81}(\pi^2-9)\alpha_s^3 P$	m_c^{-1}	Unknown[38]	$\Gamma(J/\psi\to$hadrons$)$ = 45 ± 12 keV (Ref. 39)	0.15 ±0.02
$\Gamma(\eta_c\to 2g) = \frac{8\pi}{3}\alpha_s^2 P$	m_c^{-1}	$1 + O(1/v)$ $+18.8\alpha_s/\pi$ (Ref. 13)	$\Gamma(\eta_c\to$hadrons$)$ ≈ 20 ± 10 MeV (Ref. 40)	0.28 ±0.07
$\Gamma(J/\psi\to 2g+\gamma) = \frac{128}{81}(\pi^2-9)\ \alpha_s^2\alpha P$	m_c^{-1}	Unknown	Mark II (Ref.41,42)	Consistent with ≈ 0.2
$M(J/\psi)-M(\eta_c) = \frac{32\pi}{9}\ \alpha_s\ P$	m_c^{-1}	For some effects see Refs. 43-44	≈ 115 MeV (Ref. 40)	0.33 ±0.05

$$P \equiv |\Psi(0)|^2/m_c^2 = 0.031 \pm 0.004 \text{ GeV (Table 2)}$$

- -

uncorrected α_s quoted for this process in Table 3 probably is an upper limit. The observed rate for $J/\psi\to 2g+\gamma$ is compatible with Eqs. (28)-(30), though the observed photon spectrum[42] is distorted with respect to a free-gluon model.[49] Finally, the value of α_s extracted from the J/ψ-η_c hyperfine splitting is consistent with (30), given theoretical uncertainties[43-44] surrounding the prediction.

The gluon-jet events measured at PETRA[50] appear consistent with the prediction of Eqs. (28)-(29):

$$\alpha_s([\alpha_s E]^2) \simeq 0.2 \text{ (E=16 GeV)}. \tag{31}$$

Several more predictions for $c\bar{c}$ radiative decays follow from the near-constancy of $P \equiv |\Psi(0)|^2/m_c^2$. These are summarized in Table 4.

Table 4. Predictions for $c\bar{c}$ radiative decays.

Process	Corrections to right-hand side	Prediction (uncorrected)	Data
$\Gamma(\eta_c \to \gamma\gamma) = \frac{64\pi\alpha^2}{27} P$	$1 + 0\,(1/v)$ $-3.4\,\alpha_s/\pi$ (Ref. 13)	12 keV	See Ref. 51
$\Gamma(J/\psi \to 3\gamma) = \frac{2^{10}}{3^7}(\pi^2-9)\alpha^3 P$	$1-12.7\alpha_s/\pi$ (Ref. 52)	5 eV	$\leq$ 5.1 eV (Ref. 22,53) See Ref. 51
$\Gamma(J/\psi \to \gamma\eta_c) = 4\alpha e_c^2 k^3/m_c^2$	Many (Refs. 54,55)	For m_c=1.1, 1.9: 5, 1.7 keV	~ 0.7 keV (Ref. 56)

The first two differ from standard ones[2] only by kinematic factors which vanish in the zero-binding limit.[57]

The value

$$\alpha_s(m_b^2) = 0.17 \tag{32}$$

obtained from Eqs. (28) and (29) implies $\Gamma(\Upsilon \to 3g) \cong 50$ keV.[58] (For $0.15 \leq \alpha_s \leq 0.2$, this number ranges between 30 and 90 keV.) At present only $\Gamma(\Upsilon \to \text{all}) \geq 25$ keV is known.[21] Estimates like those in Table 3 predict

$$.34 \text{ MeV} \leq \Upsilon - \eta_b \leq J/\psi - \eta_c \approx 115 \text{ MeV}, \tag{33}$$

where the right-hand inequality comes from a scaling estimate of the largest possible value of $|\Psi(0)|^2_{b\bar{b}}$. As a consequence of the wide range in (33), we expect 2 eV $\leq \Gamma(\Upsilon \to \gamma\eta_b) \leq$ 66 eV for m_b = 5 GeV. The smaller charge of the b quarks make the detection of $\Upsilon \to \gamma\eta_b$ harder than $J/\psi \to \gamma\eta_c$. (See, however, Ref. 54).

III. BELOW FLAVOR THRESHOLD

Some aspects of quarkonium physics are not characteristic of short distances and, in fact, have little to do with specific form of the binding. Simple scaling and semiclassical arguments relate one family (e.g., charmonium) to another (upsilons, ...), revealing the charge of the constituents and the number of states below flavor threshold.

A. Quark charge.

The ratio $R = \sigma(e^+e^- \to \text{hadrons})/\sigma(e^+e^-\to\mu^+\mu^-)$ has not always been helpful in determining quark charges. The step in R near charm threshold reflected the additional unit contributed by the charged lepton τ. The small step in R above the Υ levels[50] indeed rules out a charge 2/3 quark, but has not yet been established at the expected ($e_b = -1/3$) value $\Delta R = 1/3$. A step in R of 4/3 (above the highest present PETRA energy) could be the expected t quark ($e_t = 2/3$),[5,6] a degenerate charge -1/3 quark "h" and a heavy lepton τ', or more exotic combinations.

One seeks independent information on quark charges, which comes from leptonic widths of quarkonium states. These also depend via Eq. (10) on $|\Psi(0)|^2$, on which bounds can be placed. The comparison of charmonium and upsilon levels in Fig. 3 shows that, while the lowest Υ level probes a short-range part of the potential not well specified by the lighter charmonium levels, the behavior of $\langle dV/dr\rangle$ (and hence, through Eq. (4), of $|\Psi(0)|^2$) for excited Υ levels is much more constrained by charmonium.[17,59] It is thus the observed Υ' leptonic widths[21,60] that have led most firmly to the conclusion that $e_b=-1/3$ for the quarks in the upsilon family.

If V(r) is monotonic in r, one might expect on semiclassical grounds for heavier quarks to have more spatially compact wave functions.[8,61] If dV/dr is monotonically decreasing in r (as in Eq. (1)), $\langle dV/dr\rangle$ will increase with quark mass. Then, for each nS $Q\bar{Q}$ state,

$$|\Psi_{nS}(0)|^2_{m_2} \geq |\Psi_{nS}(0)|^2_{m_1} \quad \text{for} \quad m_2 \geq m_1 \tag{34}$$

Eq. (34) can be proved rigorously for ground states,[59,62] and for all excited states in power-law potentials with $\nu < 1$ (See Table 1). It can be turned, by means of Eq. (10), into a powerful tool for analyzing the minimum signal due to a vector meson in e^+e^- annihilations.[59,63]

B. Thresholds.

A simple semiclassical argument counts the number of quarkonium levels below flavor threshold.[64]

The zero of energy may be taken at $2m_Q$ both for quarkonium levels and for flavored pairs. The energy of a pair of the two lowest flavored particles $D = Q\bar{q}$ is then

$$\delta \equiv 2m_D - 2m_Q. \tag{35}$$

Semiclassically, this energy corresponds to the $n^{\underline{th}}$ quarkonium level, where n is given by[65]

$$\int_0^{V(r)=\delta} dr \sqrt{m_Q[\delta - V(r)]} = (n - \tfrac{1}{4})\pi. \tag{36}$$

Just as the mass of an atom is governed mainly by that of its constituents, one expects δ to vary slowly as a function of m_Q, approaching a fixed limit $\delta \to \delta_\infty$ as $m_Q \to \infty$.[66] Then, for large m_Q,

$$n \text{ (flavor threshold)} \sim \sqrt{m_Q}\ . \tag{37}$$

Since flavor threshold lies just above the $\psi'(2S)$,

$$n \simeq 2(m_Q/m_c)^{1/2}\ . \tag{38}$$

Since $m_b/m_c = 3$ to 4, one expects three (or, barely, four) narrow $^3S_1$ Υ levels. This agrees with many specific potential models,[66] and with Table 2. For $m_Q > 15$ GeV/c^2, there should be at least six $^3S_1$ quarkonium levels below flavor threshold.

Eq. (36) and Fig. 3 imply that flavor threshold occurs at an interquark separation

$$r = V^{-1}(\delta) = 7\text{-}8 \text{ GeV}^{-1} \simeq 1\tfrac{1}{2} \text{ fm}, \tag{39}$$

which should be just about the distance at which even very heavy quarks dissociate from one another into $Q\bar{q} + \bar{Q}q$.

Estimates of the actual excitation energy for flavor threshold, $E_{th}-E_1$, are more model-dependent. For $b\bar{b}$ production, we use the small variation (27) in $m_b - m_c$. For the lowest (0^-) states,

$$\begin{aligned} m_{D_b} &\simeq m_D + m_b - m_c + (33 \text{ MeV})(1 - \frac{m_c}{m_b}) \\ &= 5.31 - 5.34 \text{ GeV}/c^2 \end{aligned} \tag{40}$$

The correction term on the right-hand side is an estimate of hyperfine and reduced-mass effects.[64,66]

A state decaying to $\psi K\pi$ has been reported[4] in the mass range (40). This decay mode has been suggested for the D_b.[67]

For the illustrative heavy quark mass $m_Q = 16$ GeV one finds

$$1\tfrac{1}{2} \text{ GeV} < E_{th} - E_1 < 2 \text{ GeV}. \tag{41}$$

The lower bound comes from scaling arguments and the assumption $\nu < 0$;[68] the upper bound is based on the potential of Ref. 11.[69,70]

IV. ABOVE FLAVOR THRESHOLD.

Using what is learned from systems involving heavy quarks alone, one can discuss some properties of other particles.

A. Decays to flavored pairs.

One model for $Q\bar{Q} \to (Q\bar{q}) + (\bar{Q}q)$ (Q is heavy, q is light) envisions the light $q\bar{q}$ pair as created with no quantum numbers, i.e., in a $^3P_0$ state.[71] This model implies that an S wave $Q\bar{Q}$ state should decay to $D\bar{D}$, $D\bar{D}^* + D^*\bar{D}$ in the ratios 1:4:7, aside from kinematic corrections.[72] Furthermore, the model specifies that $D^*\bar{D}^*$ should be produced in a combination of states with total spin S = 0 and S = 2 in such a way that every allowed helicity amplitude for $Q\bar{Q} \to (Q\bar{q}) + (\bar{q}Q)$ occurs with equal magnitude.[73] This leads to an intensity ratio $I(S = 2)/I(S = 0) = 20$, and a definite angular distribution for the decay pions in

$$e^+e^- \to (Q\bar{Q})_{L=0} \to D^*\bar{D}^* \quad , \qquad \hookrightarrow \pi^\circ D \tag{42}$$

namely:

$$W(\theta) \sim 1 - \frac{1}{5}\cos^2\theta \quad . \tag{43}$$

Here θ is measured between the outgoing pion and the $e^\pm$ beam axis.[74]

The 1:4:7 prediction is badly violated for $\psi(4.028)$.[11,75] One suggested reason is the different nodes of D and D* radial wave functions.[11] This would not alter the prediction (43), however. If (43) were found not to hold, one would have to conclude that the $\psi(4.028)$ is not a pure $^3S_1$ state[76], or simply that the quark pair creation model is wrong.[77]

B. Masses of flavored hadrons.

Some control over hyperfine splittings is possible for flavored hadrons despite the relativistic nature of their light quarks. Crudely, we expect

$$\Delta E_{hfs} \sim \alpha_s |\Psi(0)|^2 / m_Q m_q \tag{44}$$

for any meson composed of $Q\bar{q}$. Relations among baryon splittings also can be obtained. Here are some examples.

1. F*-F splitting. Since $m_s > m_u \simeq m_d$, both quark masses (in the denominator of (44)) and reduced mass effects in $|\Psi(0)|^2$ affect the

comparison of F*-F and D*-D splittings. Neglecting variations in α_s between F and D, and using a scaling law of Table 1 (for an effective power $\nu \simeq 1$, characteristic of systems containing light quarks), we find

$$F^* - F \lesssim D^* - D \approx 143 \text{ MeV}. \tag{45}$$

2. Charmed baryons. Three baryons can be formed out of a charmed quark and two nonstrange quarks in an S wave.[78] They are the C_0 (cud), C_1, and $C_1{}^*$ (both cdd, cud, or cuu), also called Λ_c, Σ_c, and $\Sigma_c{}^*$.[79] The C_1-C_0 mass difference was predicted to be 160 MeV,[79] a figure supported by experiment.[80] It is worth reviewing[81,82] the prediction, since the report[83] of a candidate for a doubly charged, weakly decaying object makes one ask whether $m(C_1{}^{++})-m(C_0{}^+)$ could be less than m_π, rendering $C_1{}^{++}$ stable with regard to strong decays.

If hyperfine splittings between quarks i and j are proportional to $-\underset{\sim}{S}_i \cdot \underset{\sim}{S}_j / m_i m_j$, the masses of the Λ-Σ-Σ^* and C_0-C_1-$C_1{}^*$ systems may be parametrized by

$$m(\Sigma) - m(\Lambda) = z - 2x \quad , \tag{46}$$

$$m(\Sigma^*) - m(\Lambda) = z + x \quad , \tag{47}$$

$$m(C_1) - m(C_0) = z - 2y \quad , \tag{48}$$

$$m(C_1{}^*) - M(C_0) = z + y \quad , \tag{49}$$

The parameter $z \simeq 206$ MeV is the mass difference between the I = S = 0 and I = S = 1 combinations of nonstrange quarks. The hyperfine interaction of the latter object with the strange or charmed quark is proportional to x ($\approx$ 63 MeV) or y.

Previous estimates[81,82] of y/x = (D*-D)/(K*-K) entail

$$m(C_1) \simeq m(C_0) + 160 \text{ MeV}/c^2 \simeq 2430 \text{ MeV}/c^2 \tag{50}$$

$$m(C_1{}^*) \simeq m(C_0) + 230 \text{ MeV}/c^2 \simeq 2500 \text{ MeV}/c^2. \tag{51}$$

Reduced mass effects in $|\Psi(0)|^2$ (Eq. (44)) can be controlled with the help of Table 1. They increase y/x, but not enough to decrease $m(C_1)-m(C_0)$ below m_π.

The charmed-strange baryons csu, csd (2490-2650 MeV/c^2) and css (2750-2790 MeV/c^2)[81,82] could contribute to inclusive proton production in e^+e^- annihilations[41,84] at higher SPEAR energies. Their favored nonleptonic decays lead to strangeness -2 (csu, csd) or -3 (css) final states. One way to produce them might be via beams of hyperons, which already have a strange quark.

3. D_b*-D_b splitting. The wave functions of the 0^- and 1^- $b\bar{q}$ states should resemble those of $c\bar{q}$, so that the major difference in their hyperfine interactions should arise from the heavy quark mass:

$$D_b{}^*-D_b = \frac{m_c}{m_b}[D^*-D]$$

$$\simeq 35 - 50 \text{ MeV}. \tag{52}$$

In Sec. III we estimated $2m_{D_b} = 10.65 \pm 0.04$ GeV/c.^{2} By $E_{c.m.} = 10.8$ GeV there should then appear some "action" in the e^+e^- cross section, corresponding to the opening of all the thresholds for $D_b\bar{D}_b$, $D_b\bar{D}_b{}^*$ + c.c., and $D_b{}^*\bar{D}_b{}^*$. According to Table 2, the fourth S wave Υ level may be able to decay only to $D_b\bar{D}_b$, if it is above flavor threshold at all. The fifth S wave level may be a copious source of $D_b{}^*\bar{D}_b{}^*$.

C. Lifetimes.

It appears that $\tau(D^+) > \tau(D^o)$.[41,83,85] The $\Delta S = \Delta C$ nonleptonic decays of D^+ lead to an exotic final state, while those of D^o do not.[78] A specific mechanism which utilizes this difference to give selective enhancement of D^o decays has been suggested.[86] It makes use of the weak transition $c\bar{u} \to s\bar{d}$ and invokes a special role for the non-exotic final state.[87] Gluon emission by the initial u quark is assumed to evade the helicity suppression that would normally affect an initial $0^-c\bar{u}$ state and light final quarks.[88]

If nonleptonic decays of $F^+ = c\bar{s} \to u\bar{d}$ are enhanced by a similar mechanism, channels like $F^+ \to 2\pi^+\pi^-$ further dilute the already small signals[89] expected for F decays in any one mode.

Depending on the value of the pseudoscalar decay constant f_F,[90] the decay $F^+ \to \tau^+\nu$[91] could become an important source of τ and ν_τ, the latter through beam-dump experiments.[92] The lifetime of τ itself can serve as a calibration of short-track detectors and as a check of the sequential nature of τ and ν_τ, which entails the prediction[93]

$$\tau(\tau) = 2.9 \times 10^{-13} \text{ sec}. \tag{53}$$

The b quark probably lives somewhat (but not much) less than 10^{-13} sec.[94] The ratio $\Gamma(b\to u)/\Gamma(b\to c)$ can be measured by the spectrum, multiplicity, and charges of leptons in semileptonic b decays.[95]

D. Exotic objects.

1. Glueballs. Several structures have been suggested for bound states of gluons, looking (for example) like bags, bagels, and bacteria.[96-100] Are these objects heavy or light? In a lattice gauge theory,[98] a glueball is a closed flux loop ($\geq$ 4 links), while the lightest $q\bar{q}$ states contain one link joining q and $\bar{q}$. Then

$$M\ (\text{glueball}) \simeq 4\ M(q\bar{q}) = 0\ (2\ \text{GeV}) \tag{54}$$

A toroidal glueball in the bag model (topologically, a bagel) weighs even more: around 8 GeV.[99] A "bacterial" (long, thin) glueball is predicted around 2 1/2 GeV in a string model based on OCD.[100] Other models also permit glueballs in this mass range.[101]

A large mass scale also is characteristic of QCD sum rules for gluons.[102] Despite this, it has been suggested that η' contains a substantial amount of glue.[102] The electromagnetic properties of these states are successfully described without this contribution, however. Moreover, the orthogonal states ought to exist if it really makes sense to talk of constituent gluons.

No evidence for narrow glueballs in the decays

$$J/\psi \rightarrow \gamma + (\text{glueball}) \tag{55}$$

occurs for photon energies above 0.6 E_{max}, i.e., for glueball masses below 2 GeV.[42] The reaction (55) still is of interest for glueballs between 2 and 3 GeV. The photon spectrum is distorted with respect to the free (γ + 2 gluon) one in such a way as to favor this high-mass region.

2. Heavy stable quarks. The lightest meson (π) is 800 MeV lighter than the lightest baryon (p), but this difference decreases rapidly when we go to charmed particles: $m(C_0^+) - m(D) \simeq 400$ MeV. About 100 MeV of this, moreover, is due to the hyperfine depression of the D^o ($^1S_0$) mass; this depression will vanish when the charmed quark is replaced by a very heavy one. Thus as $m_Q \rightarrow \infty$ we expect the lowest baryon Qqq to lie only about 300 MeV above the lightest meson $Q\bar{q}$. If Q is absolutely stable, there are then two stable objects. The situation is even richer for color (anti) sextet quarks,[103] as summarized in Table 5. We have used elementary color-spin arguments,[104,105] such as those which give $m(\Lambda) < m(\Sigma)$ or $m(C_0) < m(C_1)$, to tell which combination of light quarks is lightest.

Table 5 contains states separated by a unit of baryon number, $\leq$ 1 unit of charge, and only $\approx m_p/3$ in mass. This should be a characteristic signal in mass-spectroscopic searches, which already have reached an impressive level for certain mass ranges.[106-108]

3. Charged scalar bosons can occur in theories[109] that seek to explain the origin of W and Z masses. The lightest bosons may be accessible to present e^+e^- experiments: M = O(10 GeV). They will behave in these experiments as pointlike objects with a small asymptotic contribution to R:

Table 5. Lightest states involving heavy quarks Q.

Color of Q	State[a]	Mass	Charge
3	$Q\bar{u}$	M	$e_Q-2/3$
	Q[ud]	M + 300 MeV/c^2	$e_Q+1/3$
6*	Q(ud)	M_1	$e_Q+1/3$
	$Q[\bar{u}\bar{d}]u$	M_1 + 300 MeV/c^2	$e_Q+1/3$
	$Q\{[\bar{u}\bar{d}]\}^2$	M_1 + 600 MeV/c^2	$e_Q-2/3$

a) Square brackets denote color antitriplet with I = S = 0; round brackets denote color sextet with I = 0, S = 1.

$$\Delta R \text{ (spin 0)} = \frac{1}{4}\, \Delta R \text{ (spin } \tfrac{1}{2}) \tag{56}$$

which sets in with a (slow) P-wave threshold factor. Colored scalar mesons could bind to form discrete states with small leptonic widths.

For all these reasons, charged scalar mesons are easy to miss in e^+e^- annihilations, and probably easier to miss elsewhere. No systematic exclusion of them has been performed in present data.[110] Some of the techniques of Sec. II and III may apply to them if they are discovered.

V. CONCLUSIONS

The quarkonium families (J/ψ, Υ) have been of tremendous importance in illustrating how bound states of heavy quarks behave, and in shedding light on properties of the quarks themselves.

Quarkonium levels have provided a laboratory for testing quantum chromodynamics. The tests are necessarily crude for the J/ψ family. The $O(\alpha_s)$ corrections to gluon or photon emission are large for $Q^2 \approx m_c^2$, making the use of first-order perturbation theory unreliable. The effects of Coulomb interactions are not dominant in $c\bar{c}$ systems. Gluon jets are not visible in J/ψ decays. Nonetheless, some qualitative successes of QCD emerge, if one accepts a scale factor Λ below ~ 0.1 GeV.

The chances of testing QCD in the Υ family are markedly better.[111] The decays of Υ contain hints of three gluon jets,[112] kindling some hope for the perturbation expansion. The small ratio of Υ' to Υ leptonic widths suggests an approach to short-distance Coulomb-like behavior.

The next family of heavy vector mesons might be a still better

QCD laboratory, especially if its quarks have charge 2/3 (aiding production in e^+e^- reactions). This family will be heavy enough (if it exists at all) that its lowest states will display a number of features of any short-distance Coulomb interaction between quarks.[113]

Some of the best QCD tests in quarkonium await further work on radiative corrections. These must act to suppress $^3S_1 \to 3g$ decays (e.g., $J/\psi \to$ hadrons) if the picture of Sec. II.C. is to be self-consistent.

Some properties of quarkonium systems, distinct from tests of QCD, follow from ordinary quantum mechanics. Dependences on quark mass and quantum numbers often may be discerned by scaling, semiclassical, and inverse scattering methods. In passing, some new results have been obtained on the inverse problem for confining potentials.

The comparison of charmonium and upsilon families reveals the flavor independence of the quark-antiquark interaction. The conclusion $e_b = -1/3$ also was first obtained from this comparison, and the number of narrow upsilon levels was predicted to be three or (barely) four independent of details of the interaction.

The properties of quarkonium families involving heavier quarks are now well enough fixed by charmonium and upsilon data that these families can be told apart from more exotic objects, like composites of scalars or sextet quarks. When confronted with a step in R, one now knows where to look for narrow resonances in $e^+e^- \to$ hadrons, and how many should be found below flavor threshold, given that the observed ΔR comes from heavy color triplet quarks. The absence of narrow resonances in the predicted range would imply that ΔR is due to objects that do not form bound states (e.g. heavy leptons), or bind more shallowly. More deeply bound narrow resonances might be characteristic of the binding of highly colored (e.g., sextet) quarks.

Charged scalars (and $e_Q = -1/3$ quarks) are not excluded by present e^+e^- data in portions of the currently available energy range. The methods discussed here will not help to fill this *lacuna* but will be useful if such objects turn up in precise measurements of R.

Flavored objects containing the new quarks have been a source of information about both QCD and QFD (quantum flavor dynamics, the province of unified theories of the weak and electromagnetic interactions). The observed lifetimes of charmed particles entail an interplay of both, particularly in nonleptonic decays.

The successful prediction of charmed particle masses[79] has been a major triumph of simple QCD and quark model ideas. These now can be extrapolated with some confidence to heavier systems.

Specific final states in weak decays of heavier quarks (b,$\cdots$) begin to shed light on areas where no detailed theoretical road map[114] exists. Partial information is provided by effects like $D_b^o - \overline{D}_b^o$ mixing, and by CP violation[115] (which welcomes extra quarks[5]).

The elementary methods we have discussed should be able to help determine properties of new quarks from experimental data, and to help tell whether new effects are due to quarks at all.

ACKNOWLEDGMENTS

This work was supported in part by the U. S. Department of Energy under Contract No. EY-76-C-02-1764. I would like to acknowledge the hospitality of the Fermilab theory group during part of the preparation of this report. A particular debt is due to Chris Quigg for enjoyable collaborations on much of this material and to Waikwok Kwong, Jonathan Schonfeld, and Hank Thacker for joint work with the two of us on the inverse problem. I am also grateful to W. A. Bardeen, J. D. Bjorken, R. Cahn, T. P. Cheng, G. Feldman, P. Fishbane, S. Gasiorowicz, D. Geffen, B. Kayser, P. Langacker, G. P. Lepage, S. Meshkov, L. B. Okun', C. Peck, H. Sadrozinski, H. Suura, M. B. Voloshin, W. Wilson, and V. I. Zakharov, for many useful discussions.

REFERENCES

1. J. J. Aubert, et al., Phys. Rev. Lett. 33, 1404 (1974); J.-E. Augustin, et al., Ibid., 1406 (1974).
2. For recent reviews, see V. A. Novikov, et al., Phys. Reports 41C, 1 (1978); T. Appelquist, R. M. Barnett, and K. Lane, Ann. Rev. Nucl. Part. Sci. 28, 387 (1978); E. Eichten, et al., Phys. Rev. D17, 3090 (1978) and Cornell University report CLNS-425, 1979, to be published; J. D. Jackson, C. Quigg, and J. L. Rosner, in Proc. XIX Int. Conf. on High Energy Physics, Tokyo, 1978, edited by S. Homma, M. Kawaguchi, and H. Miyazawa (Phys. Soc. Japan, Tokyo, 1978), p. 391; M. Krammer and H. Krasemann, DESY 78/66, (Lectures at Advanced Summer Institute, Karlsruhe, Setp. 1-15, 1978, unpublished) and DESY 79/20 (Lectures at 18th Int. Universitatswochen fur Kernphysik, Schladming, Austria, Feb. 28-Mar. 10, 1979, unpublished), and C. Quigg, FERMILAB-CONF-79/74-THY, Sept., 1979, to be published in Proc. Int. Symp. on Lepton and Photon Interactions at High Energies, Batavia, Ill., Aug. 23-27, 1979.
3. S. W. Herb, et al., Phys. Rev. Lett. 39, 252 (1977); W. R. Innes, et al., Phys. Rev. Lett. 39, 1240, 1640 (E) (1977).
4. R. Barate, et al., presented at 9th Int. Symp. on Lepton and Photon Interactions at High Energies, Batavia, Ill., Aug. 23-27, 1979, paper no. 184.
5. M. Kobayashi and T. Maskawa, Prog. Theor. Phys. 49, 652 (1973), suggest a sixth quark "t" with $e_t=2/3$. Models based on exceptional groups (e.g., E_6) suggest the existence of a sixth quark "h" with $e_h=-1/3$. For a review, see F. Gürsey, Yale Univ. report COO-3075-178, Mar., 1977, to be published in Proc. of the Conf. on Non-Associative Algebras, Charlottesville, Va., Mar. 18, 1977.
6. For a review of six-quark models, see H. Harari, Phys. Reports 42C, 235 (1978).
7. For a collection of references to literature on heavier quarks and other new particles, see J. Rosner, "Resource Letter NP-1: New Particles," 1979, to be published in Am. J. Phys.

8. These methods are reviewed by C. Quigg and Jonathan L. Rosner, FERMILAB-PUB-79/22-THY, to be published in Phys. Reports.
9. D. Gross and F. Wilczek, Phys. Rev. Lett. 30, 1343 (1973); H. D. Politzer, Ibid., 1346 (1973).
10. Reviews of QCD include those by H. D. Politzer, Phys. Rep. 14C, 129 (1974); W. Marciano and H. Pagels, Phys. Rep. 36C, 137 (1978); A. Peterman, Phys. Rep. 53, 157 (1979); and A. J. Buras, FERMILAB-PUB-79/17-THY, to be published in Rev. Mod. Phys.
11. Eichten, et al., Ref.2. These papers also contain extensive references to earlier work.
12. T. Appelquist and H. D. Politzer, Phys. Rev. Lett. 34, 43 (1975); A. De Rujula and S. L. Glashow, Ibid., 34, 46 (1975).
13. R. Barbieri, et al., Nucl. Phys. B154, 535 (1979).
14. C. Quigg and Jonathan L. Rosner, Phys. Lett. 71B, 153 (1977).
15. A. B. Henriques, B. H. Kellett, and R. G. Moorhouse, Phys. Lett. 64B, 85 (1976).
16. C. Quigg and J. L. Rosner, Comments Nucl. Part. Phys. 8, 11 (1978).
17. H. B. Thacker, C. Quigg, and Jonathan L. Rosner, Phys. Rev. D18, 274, 287 (1978).
18. C. Quigg, H. B. Thacker, and Jonathan L. Rosner, FERMILAB-PUB-79/52-THY, to be published in Phys. Rev. D.
19. Jonathan F. Schonfeld, Waikwok Kwong, Jonathan L. Rosner, C. Quigg, and H. B. Thacker, FERMILAB-PUB-79/ 77 -THY, to be submitted to Ann. Phys. (N.Y.).
20. J. Schwinger, Harvard lecture notes (unpublished); E. Eichten, et al., Phys. Rev. Lett. 34, 369 (1975).
21. H. Spitzer, DESY Internal Report PLUTO-79/03, lectures given at VII Int. Winter Meeting on Fundamental Physics, Segovia (Spain), Feb. 5-10, 1979 (unpublished).
22. Particle Data Group, Phys. Lett. 75B, 1 (1978).
23. We shall use the symbol for a particle to denote its mass.
24. See, in particular, Fig. 13 of Ref. 8.
25. We have used 3S-2S = 0.35 ± 0.01 GeV/c^2 ($c\bar{c}$) and 0.39 ± 0.04 GeV/c^2 ($b\bar{b}$). See Ref. 22; B. Wiik and G. Wolf, DESY Report No. 78/23; and K. Ueno, et al., Phys. Rev. Lett. 42, 486 (1979). A more precise measurement of the Υ'' mass will be possible soon, e.g., at CESR (R. Siemann, this conference).
26. R. Van Royen and V. F. Weisskopf, Nuovo Cim. 50, 617 (1967); 51, 583 (1967).
27. R. Barbieri, et al., Phys. Lett. 57B, 455 (1975), Nucl. Phys. B105, 125 (1976); W. Celmaster, Phys. Rev. D19, 1517 (1979); Enrico Poggio and Howard J. Schnitzer, Phys. Rev. D20, 1175 (1979); L. Bergström, H. Snellmann, and G. Tengstrand, Phys. Lett. 80B, 242 (1979); Ibid., 82B, 419 (1979); Royal Inst. of Technology (Stockholm) preprint TRITA-TFY-79-10, 1979 (unpublished).
28. P. A. Rapidis, et al., Phys. Rev. Lett. 39, 526, 974 (E) (1977).
29. The ψ, ψ', Υ, Υ' masses are 3.095, 3.684, 9.46, 10.02 GeV/c^2.
30. I. M. Gel'fand and B. M. Levitan, Am. Math. Soc. Trans. 1, 253 (1955); I. Kay and H. E. Moses, J. Appl. Phys. 27, 1503 (1956). Further references are given in the first of Refs. 17.

31. A review of the inverse scattering method is given by A. C. Scott, F. Y. F. Chu, and D. W. McLaughlin, Proc. IEEE 61, 1443 (1973).
32. H. Grosse and A. Martin, Nucl. Phys. B148, 413 (1979).
33. Bruce McWilliams, Phys. Rev. D20, 1221 (1979).
34. We use Eq. (10) and the connection $2\pi \mid \Psi_i(0) \mid^2 = |\psi'_{2i}(0)|^2$ between the three-dimensional and one-dimensional Schrödinger wave functions.
35. R. Karplus and A. Klein, Phys. Rev. 87, 848 (1952).
36. As a result, predicted rates for E1 transitions $\Gamma(\psi' \to \gamma\chi)$ decrease. These rates are predicted to be about a factor of 3 too large if a = 1. [See Refs. 11 and 17, and the discussion by C. Quigg, Ref. 2].
37. See also S. Nussinov, Saclay preprint DPh-T/79/100, July, 1979, submitted to Zeit. Phys. C., A. Martin, CERN report TH. 2741, Sept. 1979 (unpublished), and R. Bertlmann and A. Martin, CERN report TH. 2772, to be published in Nucl. Phys. B.
38. G. P. Lepage informs me that a calculation of these corrections by authors of Ref. 13 is contemplated.
39. This value is obtained by subtracting $4\frac{1}{2}\ \Gamma(J/\psi \to e^+e^-)$ (corresponding to $J/\psi \to \gamma^* \to$ all) from the total width $\Gamma(J/\psi \to$ all) = 67 ± 12 keV quoted in Ref. 22.
40. C. Peck (this conference), report on Crystal Ball data. The errors on $\Gamma(\eta_c \to$ hadrons) in Table 3 are my own estimate.
41. J. Dorfan (this conference) report on Mark II data.
42. G. S. Abrams, et al., SLAC-PUB-2415, Oct., 1979 (unpublished).
43. H. J. Schnitzer, Phys. Rev. D19, 1566 (1979).
44. C. Callan, et al., Phys. Rev. D18, 4684 (1978); E. Eichten and F. Feinberg, Phys. Rev. Lett. 43, 1205 (1979).
45. We take the number of light flavors $n_f = 3$ for charmonium. See T. Appelquist and J. Carazzone, Phys. Rev. D11, 2856 (1975).
46. The estimate of charmonium size is read off Fig. 3.
47. Small values of Λ have been advocated by the I.T.E.P. group. See Novikov, et al., Ref. 2; L. B. Okun', presented at EPS High Energy Physics Conf., Geneva, Switz., Jun. 27-Jul. 4, 1979; M. B. Voloshin, preprints ITEP-176-1978 and ITEP-54-1979 (unpublished) and Yad. Fiz. 29, 1368 (1979). Larger values usually are found (L. F. Abbott, this conference) from scaling violations in deep inelastic scattering. One specific potential model also has a larger Λ. See John L. Richardson, Phys. Lett. 82B, 272 (1979).
48. The O(1/v) term in Table 2 may be absorbed into the definition of $|\Psi(0)|^2$. I thank G. Peter Lepage for a discussion of this point.
49. T. Appelquist, A. DeRújula, H. D. Politzer, and S. L. Glashow, Phys. Rev. Lett. 34, 365 (1975); M. Chanowitz, Phys. Rev. D12, 918 (1975); Novikov, et al., Ref. 2; K. Koller and T. F. Walsh, Nucl. Phys. B140, 449 (1978); S. J. Brodsky, D. G. Coyne, T. A. DeGrand, and R. R. Horgan, Phys. Lett. 73B, 203 (1978).
50. B. Wiik, this conference.
51. New, improved bounds on these processes are now possible as a result of the Crystal Ball data as reported, for example, in Ref. 40.

52. This number is calculated from the work of William E. Caswell, G. Peter Lepage, and Jonathan Sapirstein, Phys. Rev. Lett. 38, 488 (1977), by dropping the fermion loop term (their Fig. 1f) and replacing α by $(4/3)\alpha_s$ in the remaining terms.
53. W. Braunschweig, et al., Phys. Lett. 67B, 243 (1977).
54. D. Geffen and W. Wilson, "Magnetic Properties of the Low-Lying Hadrons," Univ. of Minnesota preprint, 1979, submitted for publication. These authors suggest that QCD may give rise to a diminution of the c quark magnetic moment and an increased b quark moment with respect to those for pointlike quarks.
55. J. Sucher, Rep. Prog. Phys. 41, 1781 (1978) has reviewed some effects which could decrease M1 transition rates.
56. The value $B(J/\psi \rightarrow \gamma\, \eta_c) \simeq 1\%$, quoted in Ref. 40 for $\Gamma(\eta_c) = 20$ MeV, is strongly dependent on the latter quantity.
57. Thus, one often sees the predictions $\Gamma(\eta_c \rightarrow \gamma\gamma) = 4/3$ $\Gamma(J/\psi \rightarrow e^+e^-) = 6.4$ keV, and $\Gamma(J/\psi \rightarrow 3\gamma) = [2^6(\pi^2-9)/3^5\,\pi]$ $\Gamma(J/\psi \rightarrow e^+e^-) = 2.6$ eV. Such predictions are discussed extensively by Novikov, et al., Ref. 2.
58. Similar estimates were obtained by J. Ellis, M. K. Gaillard, D. V. Nanopoulos, and S. Rudaz, Nucl. Phys. B131, 285 (1977).
59. J. Rosner, C. Quigg, and H. Thacker, Phys. Lett. 74B, 350 (1978).
60. J. K. Bienlein, et al., Phys. Lett. 78B, 360 (1978); C. W. Darden, et. al., Ibid., 364 (1978).
61. A. Martin has kindly supplied a proof. (See Ref. 8).
62. C. N. Leung and J. L. Rosner, J. Math. Phys. 20, 1435 (1979); A. Martin, unpublished.
63. For a specific numerical illustration relevant to the PEP and PETRA energy range, see C. Quigg, Ref. 2.
64. C. Quigg and J. Rosner, Phys. Lett. 72B, 462 (1978).
65. The $-1/4$ in Eq. (36) is appropriate for non-singular potentials; it is replaced by $-(1+\nu)/[2(2+\nu)]$ for potentials behaving at the origin as r^ν, $\nu < 0$. (See Table 1 and Ref. 8.)
66. E. Eichten and K. Gottfried, Phys. Lett. 66B, 286 (1977).
67. H. Fritzsch, Phys. Lett. 86B, 343, 164 (1979); Mark B. Wise, SLAC-PUB-2399, Sept., 1979, submitted to Phys. Lett.
68. This should be a good approximation for heavy quarks. The case of $\nu = 0$ ($V(r) \sim \ell n\, r$), examined in Ref. 64, leads to the specific prediction $(E_{th}-E_1)_{m'_Q} = (E_{th}-E_1)_{m_Q} - (0.37\text{ GeV})\ \ell n\,(m'_Q/m_Q)$, giving the lower bound in Eq. (41).
69. Asymptotic freedom corrections, which would weaken the Coulomb singularity at short distances and decrease the estimate of $E_{th}-E_1$, are not included in Ref. 11.
70. A similar range is encountered in the potentials of G. Bhanot and S. Rudaz, Phys. Lett. 78B, 199 (1979), and H. Krasemann and S. Ono, Nucl. Phys. B154, 283 (1979).
71. J. C. Carter and M. E. M. Head, Phys. Rev. 176, 1808 (1968); L. Micu, Nucl. Phys. B10, 521 (1969); D. Horn and Y. Ne'eman, Phys. Rev. D1, 2710 (1970); R. Carlitz and M. Kislinger, Ibid., D2, 336 (1970); E. W. Colglazier and J. Rosner, Nucl. Phys. B27, 349 (1971); W. P. Petersen and J. Rosner, Phys. Rev.

D6, 820 (1972), Ibid., D7, 747 (1973); A. Le Yaouanc, L. Oliver, O. Pène, and J.-C. Raynal, Ibid., D8, 2223 (1973).

72. J. Kogut and L. Susskind, Phys. Rev. D12, 1742 (1975); E. Eichten, et. al. Cornell Univ. Report CLNS-316, 1975 (unpublished), and Ref. 2; A. De Rújula, H. Georgi, and S. L. Glashow, Phys. Rev. Lett. 37, 398 (1976); E. Eichten and K. Lane, Ibid., 477 (1976); Novikov, et al., Ref. 2. An early calculation with different results is given by R. Barbieri, R. Kögerler, Z. Kunszt, and R. Gatto, Phys. Lett. 56B, 477 (1975).
73. F. E. Close, Phys. Lett. 65B, 55 (1976).
74. B. Kayser and R. Cahn, private communication. For a similar discussion of $e^+e^- \to F^*\overline{F}^* \to F_1\gamma\ \overline{F}_2\gamma$ see R. Cahn, Y. Eylon, and S. Nussinov, Univ. of Calif. (Davis) report, 1979 (unpublished).
75. G. Goldhaber, et al., Phys. Lett. 69B, 503 (1977).
76. It could have substantial $^3D_1$ or gluonic admixtures, or could be a "molecular charmonium" state in which the light quarks cannot be ignored. For further discussion and references see Novikov, et al., Ref. 2; A. De Rújula, H. Georgi, and S. L. Glashow, Phys. Rev. Lett. 38, 317 (1977); D. Horn and D. Novoseller, Phys. Rev. D18, 4035 (1978); and A. Arneodo, J. L. Femenias, and F. Guerin, Univ. of Nice reports N TH 79/3 and N TH 79/6, 1979.
77. This model does have some experimental support from light-quark physics. For a discussion and further references see J. Rosner, Phys. Reports 11, 189 (1974).
78. M. K. Gaillard, B. W. Lee, and J. L. Rosner, Rev. Mod. Phys. 47, 277 (1975).
79. A. De Rújula, H. Georgi, and S. L. Glashow, Phys. Rev. D12, 147 (1975).
80. C. Baltay, this conference; C. Baltay, et al., Phys. Rev. Lett. 42, 1721 (1979).
81. A. De Rujula, H. Georgi, and S. L. Glashow, Phys. Rev. Letters 37, 785 (1976).
82. B. W. Lee, C. Quigg and J. Rosner, Phys. Rev. D15, 157 (1977).
83. N. W. Reay, this conference.
84. G. S. Abrams, et al., SLAC-PUB-2406/LBL-9855, Sept., 1979, submitted to Phys. Rev. Letters.
85. J. Kirkby, SLAC-PUB-2419, to be published in Proc. Int. Symp. on Lepton and Photon Interactions at High Energies, Batavia, Ill., Aug. 23-27, 1979.
86. M. Bander, D. Silverman, and A. Soni, Univ. Calif. (Irvine) preprint, 1979 (unpublished).
87. M. Einhorn and C. Quigg, Phys. Rev. D12, 2015 (1975); S. P. Rosen, Los Alamos preprints LA-UR-79-2619 and LA-UR-79-2702, 1979, unpublished.
88. J. Ellis, M. K. Gaillard, and D. V. Nanopoulos, Nucl. Phys. B100, 313 (1975).
89. C. Quigg and J. Rosner, Phys. Rev. D17, 239 (1978).
90. S. S. Gershtein and M. Yu Khlopov, Sov. Phys. JETP Lett. 23, 338 (1976); Novikov, et al., Ref. 2 and Phys. Rev. Lett. 38, 626 (1977); E. G. Floratos, S. Narison, E. de Rafael, Nucl. Phys. B155, 115 (1979); A. Ali, et al., Zeit. Phys. C1, 269 (1979).
91. I. Karliner, Phys. Rev. Lett. 36, 759 (C) (1976).

92. C. Albright and R. Shrock, Princeton Univ. Preprint PU-COO-3072-96, May, 1979 (unpublished).
93. J. Rosner, in Cosmic Rays and Particle Physics-1978 (Bartol Conference), edited by T. K. Gaisser (American Institute of Physics, New York, 1979), p. 297.
94. Recent estimates are given by S. Pakvasa, Univ. of Hawaii report UH-511-346-79, to be published in proc. of Neutrino '79 Conf. Bergen, Norway and V. Barger, W. F. Long, and S. Pakvasa, Univ. of Hawaii report UH-511-339-79, Apr., 1979 (unpublished). A somewhat more conservative estimate is given by R. E. Shrock, S. B. Treiman, and L. L. Wang, Phys. Rev. Lett. 42, 1589 (1979): $6 \times 10^{-15}s \leq \tau_b \leq 5 \times 10^{-13}s$.
95. A. Ali, Zeit. Phys. C1, 25 (1979); C. Quigg and J. Rosner, Phys. Rev. D19, 1532 (1979).
96. D. Robson, Nucl. Phys. B130, 328 (1977).
97. J. D. Bjorken, SLAC-PUB-2366, Aug. 1979, to be published in Proc. of the EPS High Energy Physics Conf., Geneva, Switz., Jun. 27-Jul. 4, 1979.
98. J. Kogut, D. K. Sinclair, and L. Susskind, Nucl. Phys. B114, 199 (1976).
99. K. Johnson, private communication.
100. H. Suura, Phys. Rev. D20, 1412 (1979), and "Relativistic Wave Equation and Mass Spectrum of Gluonium", Univ. of Minn. preprint, 1979, submitted to Phys. Rev. Letters.
101. K. Ishikawa, Phys. Rev. D20, 731 (1979).
102. V. A. Novikov, M. A. Shifman, A. I. Vainshtein, and V. I. Zakharov, Phys. Lett. 86B, 347 (1979); preprint ITEP-73-1979 (unpublished).
103. This was noticed by H. Georgi and S. L. Glashow, Harvard Univ. preprint, HUTP-79/A027, May, 1979 (unpublished). We differ with their suggestion that the Υ contains color sextet quarks: See Refs. 18 and 4. Constituent gluons could alter the stability of the states in the second part of Table 5, since states $Q\ g\ \overline{u}$ then become possible. See P. Freund and C. Hill, Nature 276, 250 (1978) and Phys. Rev. D19, 2755 (1979) and Jackson, et al., Ref. 2 for references and further discussion of color sextets.
104. T. DeGrand, et al., Phys. Rev. D12, 2060 (1975); R. Jaffe, Phys. Rev. Lett. 38, 195, 617 (E) (1977).
105. C. Dover, T. Gaisser, and G. Steigman, Phys. Rev. Lett. 42, 1117 (1979).
106. R. Muller, L. Alvarez, W. Holley, and E. Stephenson, Science 196, 521 (1977).
107. R. Middleton, R. W. Zurmühle, J. Klein, and R. V. Kollarits, Phys. Rev. Lett. 43, 429 (1979).
108. R. N. Boyd, et al., Phys. Rev. Lett. 43, 1288 (1979).
109. L. Susskind, these proceedings, E. Farhi and L. Susskind, SLAC-PUB-2361, July, 1979 (unpublished); M. A. B. Bég, H. D. Politzer, and P. Ramond, Rockefeller Institute report COO-2232B-189, 1979 (unpublished).
110. M. K. Gaillard, in Proc. Int. Symp. on Lepton and Photon Interactions at High Energies, Batavia, Ill., Aug. 23-27, 1979, to be published; and C. Albright, J. Smith, and S.H.H.Tye,

FERMILAB-PUB-79/69-THY, Sept., 1979 (unpublished).
111. In this context, see M. B. Voloshin, preprints ITEP-176-1978 and ITEP-54-1979 (unpublished).
112. F. H. Heimlich, et al., Phys. Lett. 86B, 399 (1979).
113. A. Duncan, Phys. Rev. D13, 2866 (1975); M. B. Voloshin and V. I. Zakharov, private communication.
114. S. L. Glashow, J. Iliopoulos, and L. Maiani, Phys. Rev. D2, 1285 (1970).
115. L. Wolfenstein, these proceedings.

A SIMPLE MODEL OF THE GROUND STATE OF QUANTUM CHROMODYNAMICS *

K. Johnson †
Stanford Linear Accelerator Center
Stanford University, Stanford, California 94305

ABSTRACT

A proposal for the form of the ground state wave function of quantum chromodynamics is made. It is shown to lead to the phenomenology of the MIT Bag Model. The parameters of this model are related to the fundamental scale parameter of QCD.

I. INTRODUCTION

There have been many ideas about the ground state wave function of quantum chromodynamics. One common feature has been a state which involves some sort of condensed phase with color magnetic properties.[1] Here I shall suggest a version of this which is extremely simple, but which leads to a quantitative model, in this case, the "static" bag model with some additional features, not the least of which is a relationship between the ad-hoc bag constant "B" and the scale of the running coupling constant of QCD.

The discussion will proceed in the following order. First, I will review the main features of the static bag model with particular reference to those aspects which are universal for all hadrons. Next, I will make a suggestion of how the strong coupling regime of QCD may be handled quantitatively, and show how the MIT Bag Model evolves from it. I will then discuss how the large N (equals number of quark colors) limit appears. I will briefly allude to the inclusion of light quarks. Finally, I will, also briefly, discuss how the model works phenomenologically with a few examples, and make the conclusions.

II. STATIC BAG

The bag[2,3] has provided a reasonably simple and successful model of hadron structure. Colored quark constituents are confined; all hadrons are composite, color singlet states. These results are obtained in a natural way by the enforcement of simple boundary conditions on the constituent wave functions at the surface of the bag. In addition, the confinement is associated with a term in the energy of the hadron of the form BV, where V is the volume occupied by the valence quark wave functions and $B \sim 55\ \mathrm{MeV/f}^3$ or $B^{1/4} = .145$ GeV. B is the same for all particles. Finally, the interior of the bag is assumed to be described by the perturbative QCD vacuum, that is, quarks interact within a bag by ordinary, perturbative QCD.

---

*This work was supported by the Department of Energy under contract number DE-AC03-76SF00515.

†Permanent address: Center for Theoretical Physics, Massachusetts Institute of Technology, Cambridge Massachusetts 02139.

ISSN:0094-243X/80/590353-11$1.50

In the more elaborate version of the bag model[3] used to compute the spectrum of light quark hadrons, an additional term of the form $-Z/R$ was added to the energy on an ad hod basis (R is the radius of the bag). As for the volume term, this term was taken to be the same for all hadrons, and the parameter Z determined by a fit to the spectrum, which gave $Z \sim 1.8$. It was subsequently realized that a contribution to the mass of this form is associated with a center of mass effect.[4] Since the bag is a localized hadron, it is not a momentum eigenstate. If it is assumed that the cost of localization is associated with the total momentum of the valence quarks, a correction to the mass equal to $-Z/R$ with $Z \sim .75$ and independent of the number of quarks is obtained. Thus, if this is removed from the Z/R term a universal contribution to the energy of all hadrons which is equal to

$$E^{o} = BV - Z'/R \tag{2.1}$$

with $Z' \sim 1$, and $B^{1/4} = .145$ GeV, has been found to yield the best phenomenology. The energy E^{o} can be thought of as an "inside" vacuum energy. One may note that the principal effect of the Z'/R term has been to lower the energy difference between the vacuum inside a bag, and the true vacuum outside. At the same time, the confinement "pressure"

$$p^{o} = \partial E^{o}/\partial V = B + Z'/4\pi R^{4} \tag{2.2}$$

has been left about the same. In practice, for a typical state where $R = 5\ \text{GeV}^{-1}$, we find a substantial cancellation in the energy; $(4\pi/3)BR^3 = .23$ GeV, whereas, $-Z'/R = -.20$ GeV. At the same time the integrated pressures, $4\pi R^2 p$, are $4\pi BR^2 = .14\ \text{GeV}^2$, and $Z'/R^2 = .04\ \text{GeV}^2$, so Z'/R^2 is a small part of the total.

III. A MODEL OF THE QCD VACUUM

Here I shall try to relate some of these bag ideas to a proposal for the form of the ground state wave function of QCD. I shall show that the ad hoc "inside" vacuum energy is related to a simple ansatz for the form of this wave function. I will derive an improved form for this energy, and obtain an expression for the constant B.

QCD involves no parameters, only a scale Λ, which gives the variation of the running coupling constant in arbitrary units. Hence $B^{1/4}$ should come out proportional to Λ, with a specified numerical coefficient.

Several authors have suggested ideas about the ground state of QCD.[1] The proposal most closely related to the one which will be developed here is that of a Bose condensate with color magnetic properties. I will suggest a model for such a condensate which permits one to make quantitative and manifestly gauge invariant calculations.

In an asymptotically free theory, for momentum changes which are large on the scale Λ, the interactions are weak. For changes small in comparison to Λ, the interactions are strong. It is reasonable that these separate regions be treated by distinct approximations. It is obvious that the weak coupling domain should be treated perturbatively. The strong interaction is presumably associated with confinement. Confinement should mean that it costs a lot of energy to separate colors over distances long in comparison to $1/\Lambda$.

One would then believe that in QCD ground state wave functional colors are not so separated, that is, the magnitude of the wave functional should be negligible if it corresponds to a configuration of separated colors. This suggests the use of boundary conditions, viewed as a strong coupling approximation, to define a trial wave functional in which color separation is absolutely forbidden. The use of gauge invariant boundary conditions to handle the strong coupling features, seems particularly useful in a non-Abelian gauge theory. At the same time, perturbation theory should be used in the domain of its validity. Thus, start with the fields in a box with volume V, which is subdivided into N smaller equal boxes with volume $V_0 = V/N$. One may take the small boxes, small enough to permit the use of perturbative QCD.[5] That is, each small box contains a perturbative vacuum. On the walls of the small boxes, color confining boundary conditions are imposed;

$$\hat{n} \times \vec{E}_a = 0, \quad \hat{n} \cdot \vec{B}_a = 0 \quad . \tag{3.1}$$

These are gauge invariant boundary conditions which specify a complete set of operators for each box. That is, together with the use of perturbation theory, we have defined a vacuum wave function for each small box, and hence also for the large box. I shall briefly discuss the addition of quarks to the boxes in section V. Their introduction can be made in a straightforward way. The imposition of color confining boundary conditions means that color does not flow between boxes. Color separation in the large box is ruled out over distances longer than the scale defined by the size of the small boxes. This would be expressed by the absence of correlations in color density fluctuations in distinct boxes. The boundary conditions (3.1) imply that a thin layer of induced color magnetic poles (and currents) lie "between" the boxes (that is, on the surfaces). The effect of this on the energy of the boxes is expressed in the dependence on the size of the box of the perturbative energy associated with each box. Let E_0 and V_0 be the energy and volume, respectively, of the small boxes. The total energy density is then,

$$E/V = E_0 / V_0 \quad . \tag{3.2}$$

The field energy associated with each box may be calculated using the following method.

Consider the small boxes. Each is surrounded by several neighbors. If one takes quasi-spherical Wigner-Seitz boxes, each would have fourteen neighbors. No matter how the large box is subdivided, each small box will be surrounded by many neighbors each containing a perturbative vacuum. As the walls are moved, the total number of modes in the large box is left unchanged. We assume that one may approximate the energy of each small box by calculating the effect on the zero point energy stored in the perturbative vacuum of a shell enclosing a sphere with radius R in empty space. One now has the classic, Casimir[6] stress problem first estimated for a sphere by

Boyer.[7] The term in the energy which depends on the radius R was obtained recently[8,9] with great accuracy with the result,

$$E_{QED} = a_{QED}/R , \quad a_{QED} = .04618 .$$

Although E was calculated in QED for a conducting sphere ($n \cdot E = 0$, $n \times B = 0$), the symmetry of the free Maxwell equations under the interchange $E \to B$, $B \to -E$, means that the energy with the boundary conditions (3.1) is the same. Since there are eight vector fields in QCD, the magnitude will be eight times larger, that is

$$E_{QCD} = a/R , \quad a = .3694 . \qquad (3.3)$$

It is important to discuss briefly the physical basis of this result. In the case of the Casimir stress the computation is made by introducing a cut off on the frequency of the fields. This cut off may be imagined to be of the order of the plasma frequency ω_p associated with the material of the metal boundaries. The metal excludes fields with lower frequencies, that is, the boundary condition is maintained by the dynamics of the metal. Since the dependence of the energy of the shell on its radius is insensitive to the cut off, as long as $1/R << \omega_p$, the stress on the shell can be computed with the aid of the boundary condition. The boundary condition expresses the effect of the strong coupling between the field fluctuations and the induced currents in the metal. It is important to stress that the finite result (3.3) depends upon a cancellation between cut off dependent terms which came from <u>short</u> wave length fluctuations localized near the inside <u>and outside</u> of the spherical shell. That is, the energy (3.3) should be regarded as being closely associated with the surface carrying the induced charges and currents.

In the QCD problem one may have a similar picture. The difference is that in the QCD case, the boundary conditions correspond to a screening by induced color magnetic poles and currents; these sources are present in the non-Abelian gauge fields. They will be produced automatically if the ground state can lower its energy with them present. The boundary dependent terms represent the field energy associated with these sources. However, one may see that if there is only the term indicated in (3.3), then

$$E/V = E^o/V^o = a/(4\pi/3)R^4 . \qquad (3.4)$$

The lowest energy comes with one large box, i.e., $R = \infty$.

It is in the first order that the effects of the non-Abelian gauge theory within each box become apparent. It has been emphasized by several people that the Feynman diagrams indicated in Fig. 1 lead to an attractive interaction with gluons in a color singlet state. This suggests a pairing instability in the QCD vacuum and the

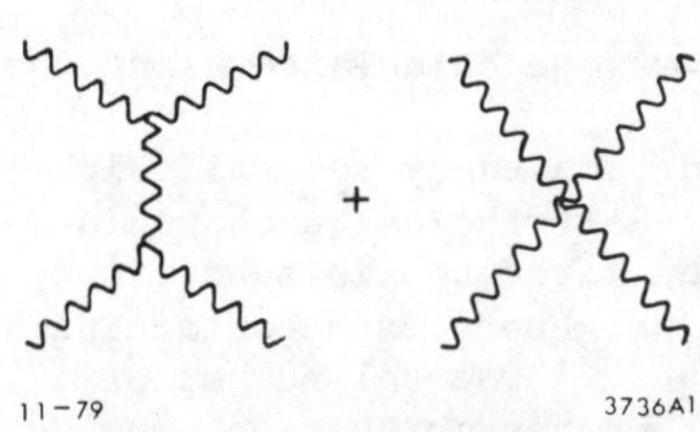

Fig. 1. Diagrams which lead to an attractive interaction between colorless pairs of gluons.

formation of a condensate with color screening properties. In the present case, all color singlet states in the vacuum will be paired in each box. (In lowest order we have a "pairing" effect, in higher orders we get multiple gluon effects). Since the gluons attract each other, when they are confined in a box the attraction should be enhanced, that is, increase as $R \to 0$. Thus, one expects a finite term of the form

$$-\ (b/R)\alpha_s \quad , \tag{3.5}$$

with $b > 0$. This energy corresponds to the sum of Feynman diagrams indicated in Fig. 2. On including the higher order effects which transform the perturbative coupling constant into a running coupling constant (3.5) becomes

$$-\ (b/R)\ \alpha_s\ (\Lambda R) \tag{3.6}$$

where Λ is the basic scale of QCD. Unfortunately, the diagrams in Fig. 2b are difficult to evaluate, and so at present the numerical value of b has not been obtained. Thus, the energy per unit volume, taken to second order is,

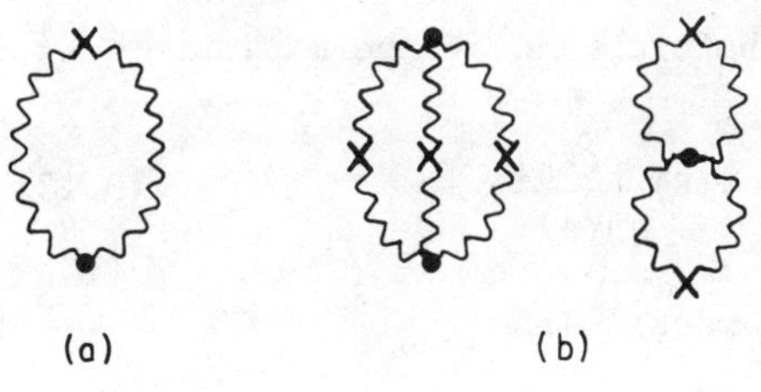

Fig. 2. (a) Diagram corresponding to the term (3.4) in the energy. The x indicates the vector field propagator for all space, with the boundary condition (3.1) imposed on the surface of a sphere. (b) Diagrams corresponding to the energy in the next order.

$$\frac{E}{V} = \frac{1}{\frac{4\pi}{3}R^3}\left(\frac{a}{R} - \frac{b}{R}\alpha_s(R\Lambda)\right) + \rho_0 \tag{3.7}$$

with

$$\begin{aligned} a &= .3694 \\ b &= ?\ (>0) \end{aligned} \tag{3.8}$$

and where ρ_0 is the (divergent) boundary independent vacuum energy. As a consequence of asymptotic freedom, as $R \to 0$, $E/V \sim +1/R^4$. However, as R increases, $\alpha_s(\Lambda R)$ also increases and when R is such that $\alpha_s > a/b$, the vacuum subdivided into boxes has an energy density below ρ_0. As $R \to \infty$, the boundary dependent term vanishes, so (3.7) has a minimum. Thus, the subdivided vacuum has its lowest energy for "boxes" with a size given by the solution of

$$\frac{\partial}{\partial R}\left[\frac{1}{R^4}\left(a - b\,\alpha_s\,(R)\right)\right]_{R=R_0} = 0 \quad . \tag{3.9}$$

Boxes with this size shall be called "empty bags".

It is now possible to consider the excited states of this system, that is, the hadron spectrum. For convenience let

$$e(R) = \frac{1}{R}\left(a - \alpha_s(R\Lambda)b\right) \quad . \tag{3.10}$$

One may now imagine an excitation of this ground state which corresponds to locally exciting the system in one of the boxes. It will be convenient to call the valence energy $e_V(R)$. The box with the excited modes is taken to have a size R. For simplicity, the excited box is assumed to have the same shape as the empty bags. At the same time, assume that the remaining empty bags change their size to R_0'. With V as the total volume, the number of empty bags will be $\left(V - V(R)\right) / \left(V(R_0')\right)$, and the total energy is therefore

$$E = [\rho_0 V(R) + e(R) + e_V(R)] + \frac{V - V(R)}{V(R_0')} [\rho_0 V(R_0') + e(R_0')]$$

$$= \rho_0 V + e(R) + e_V(R) + \frac{V - V(R)}{V(R_0')} e(R_0') \qquad (3.11)$$

The minimum occurs with $R_0' = R_0$, as before, and at the minimum with respect to R of

$$E_{BAG}(R) = e(R) + e_V(R) - V(R) \frac{e(R_0)}{V(R_0)} \quad . \qquad (3.12)$$

This is recognized as the static bag model with

$$B = \frac{-e(R_0)}{V(R_0)} = -\frac{1}{\frac{4\pi}{3} R_0^4} \left(a - \alpha_s(R_0) b\right) \quad . \qquad (3.13)$$

Since $e(R_0)$ at the minimum given by (3.9) is negative, B, is of course positive.

Using this simple model for the vacuum, an effective "inside" vacuum energy equal to

$$B \frac{4\pi}{3} R^3 + \frac{1}{R} \left(a - \alpha_s(R) b\right) = E(R) \qquad (3.14)$$

has been obtained. The empty bags with fixed sizes outside provide the pressure B, and (3.13) together with (3.9) determines B in terms of the scale parameter of QCD. The model has also provided an extra term which has roughly the same phenomenological effect as the ad hoc $-Z/R$ term. This is because for $R > R_0$, the second term in (3.14) is negative, and this acts to reduce the cost of "drilling" the hole in the vacuum occupied by the valence particles. Indeed, not only does (3.14) vanish when $R = R_0$, but its derivative is also zero at $R = R_0$, that is, the "inside" vacuum energy E(R) takes the form

$$\frac{1}{2} E''(R_0) \, (R - R_0)^2 \quad .$$

near R_0. Thus for low energy excitations the cost of enlarging the bag is small. The enlarged bag associated with a given hadron is close to the size of the empty bags found in the vacuum. The "inside" and "outside" vacua are not very different. This explains qualitatively why large renormalization effects on quark operators are absent. The fact that the vacuum is filled with such bags also explains why it was consistent to assume that the only cost for localizing a bag is that associated with the total momentum of the valence particles. These features of earlier work were the principal motivation for the proposal made here.

In summary, I have shown that a vacuum densely filled with bubbles of perturbative vacua, with the pairing effect of the many colored gluons in each computed in lowest non-trivial order, has a minimum energy at a finite size R_0 of the order $1/\Lambda$. In such a vacuum, color density correlations between neighboring bubbles are absent. Neighboring bubbles are screened by a surface layer of induced color magnetic poles and currents.

IV. LARGE N

Because the behavior of non-Abelian gauge theories for large N (order of color group) has been of continuing interest,[10] it is appropriate to study the model of the ground state which has been proposed here in this limit. Vacuum energy is of order N^2. Thus, the constants a and b are of order N^2. Consequently the radius of the empty vacuum bags given by the solution of (3.9) is independent of N, for large N. Since a color singlet gluon field (or meson) exitation with just two valence gluons (or a quark and antiquark) can be made, the "valence" term is of order zero in N, so the energy (3.12) is dominated by the "inside" vacuum term which is of order N^2 and has its minimum (= 0) at R_0. The "glueball" (or meson) mass is thus independent of N. For baryons which are color singlets, the valence term is of order N, so the "inside" vacuum term still dominates. The minimum radius is asymptotically equal to R_0 and the mass is of order N. These results are consistent with what is known about the spectrum for large N. It should be remarked that as N increases a ground state in which the condensed empty bags organize themselves to spontaneously break translation invariance (and also Lorentz invariance) might not be unexpected since the repulsion between empty bags increases as N^2. We are, of course, optimistically assuming that for $N = 3$, when translation invariance is restored to our ansatz, the symmetry will not be spontaneously broken.

V. THE ADDITION OF LIGHT QUARKS TO THE VACUUM

Quarks whose "bare" (QCD independent) masses are small in comparison to the scale Λ, also can produce long range separation of color in the ground state. In practice only the up down and, perhaps marginally, the strange quarks have such bare masses. One would thus also expect a surface boundary condition of the color confining form,

$$\hat{n} \cdot \bar{q}_a \vec{\gamma} q_b = 0 \tag{5.1}$$

on the light quark wave functions. That is, the trial ground state for the light quarks will consist of filled Dirac seas of light quarks in each small box. The complete sets of wave functions associated with each box are defined by solutions of the free Dirac equation with a linear boundary condition which implies (5.1) on the surface of each small box.

Although for massless quarks, the boundary condition (5.1) and free Dirac equation are chirally symmetric, the realization of (5.1) in a linear form requires the breaking of chiral symmetry. The most general linear boundary condition which implies (5.1) is

$$-i\,\gamma\cdot\hat{n}\,q_a \;=\; e^{\frac{i}{2}\omega_\alpha \lambda_\alpha \gamma_5}\,q_a \tag{5.2}$$

where λ_α are the flavor generators together with the singlet generator and ω_α is arbitrary. Since there is no particular reason also to break flavor symmetry, one may take $\omega_\alpha = 0$ (also for the U(1) component). Chiral symmetry is now broken in order to prevent the high energy cost of color separation in the ground state. Since the energy of the light quarks is independent of which of the boundary conditions is chosen, we have the signal of a spontaneously[11] broken chiral symmetry.

One can now include in the vacuum energy, the energy of the light quarks. The Feynman diagrams are pictured in Fig. 3.

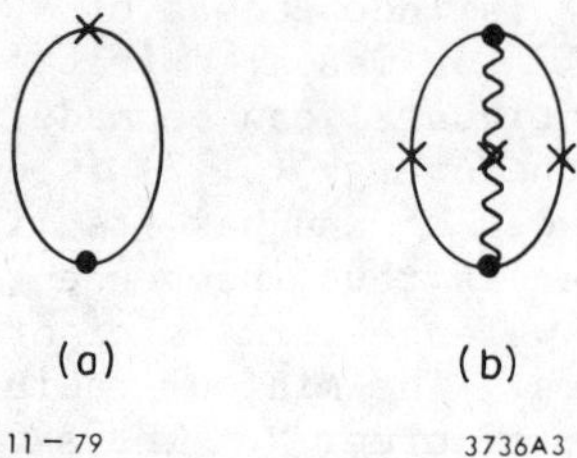

Fig. 3. Diagrams for the quark contributions to a and b. The x indicates a quark propagator for all space with (5.2) imposed on the surface of a sphere.

The additional contribution to "a" of massless up, down and strange quarks is

$$a_{\text{Quarks}} \cong \frac{2}{64} \times \underset{\text{color}}{3} \times \underset{\text{flavor}}{3} = .28 \tag{5.3}$$

The number 2/64 is based upon an approximation[12] equivalent to one made for the vector fields,[8] which gives $a_{QED} = 3/64$. The total value of "a" from both gluons and light quarks is then

$$a_{\text{Total}} = .37 + .28 = .65 \quad .$$

As for the gluons, one expects that the second order term will be attractive, so the quark contribution to "b" will increase it.

VI. PHENOMENOLOGICAL APPLICATION - THE SPECTRUM OF THE UP AND DOWN QUARK STATES, π, ρ, N, Δ

Because the constant "b" has not been calculated, a detailed quantitative comparison of this vacuum model with reality is not possible. However, a comparison of the predictions of the model with the observed mass sprectrum can be made for values of "b" which are such that $a - \alpha_s b$ becomes negative in the region where $\alpha_s \sim 1$. Since "a" is approximately .65, there are no free parameters other than "b". The energy of the valence quarks has not been computed beyond order α_s so to be consistent one should use a "lowest order" form for α_s,

$$\alpha_s = \frac{1}{\frac{9}{2\pi} \ln\left(\frac{1}{R\Lambda} + 1\right)} . \tag{6.1}$$

In (6.1) the liberty has been taken to fix a form which is not singular at $R \sim 1/\Lambda$, since if propagators associated with the boxes are used there is no infrared singularity in the energy for finite values of R. The nature of the singularity at $R = \infty$ is irrelevant in this application since $R\Lambda$ will be of order 1.

With (6.1) chosen for $\alpha_s(R\Lambda)$, in Table I the parameters $B^{1/4}/\Lambda$ and $(R_0\Lambda)$, as determined by (3.13) and (3.9), are given for several values of "b". The value of "a" is taken to be .65. α_s for the corresponding values of $R_0\Lambda$ is also shown.

TABLE I. "Empty Bag" Parameters for Various Values of "b"

b/a	b	$R_0\Lambda$	$B^{1/4}/\Lambda$	$R_0 B^{1/4}$	$\alpha_s(R_0\Lambda)$
.6	.39	2.56	.177	.453	2.12
.8	.52	1.78	.250	.445	1.57
1	.65	1.32	.332	.438	1.24
1.2	.78	1.01	.424	.428	1.01
1.4	.91	.80	.528	.422	.86
1.6	1.04	.65	.645	.419	.75

The scale R corresponding to the bag size, and the momentum transfer q used in perturbative QCD are not necessarily related in the simple form $q = 1/R$, so Λ cannot be directly related to the corresponding QCD scale. They presumably are related within a factor of 2 or 3. To judge which of the values in Table I corresponds most closely to previously obtained results using the bag model, one may determine Λ by the requirement that $\alpha_s = 2$ when $R \sim 5\ \text{GeV}^{-1}$ which is what was obtained previously.[3] In this case, $\Lambda \sim .5$ GeV. Since, in Ref. (3) $B^{1/4} = .145$ GeV, one may see that the case of $b = .65$ agrees well ($B^{1/4} = .165$ GeV) with the earlier work. One should also note that with $\Lambda \sim .5$ GeV, the empty bag size R_0 is such that $1/R_0 \approx .38$ GeV. $1/R_0$ should be compared with the "primordial" transverse momentum observed in strong interactions, since the quark constituents of hadrons are made in "empty bags" with the size R_0.

To make a crude but more accurate assessment using previous bag model results which also includes a "center of mass" correction one may use the formula

$$E_{BAG}^2 = M^2 + \langle P_{cm}^2 \rangle \tag{6.2}$$

with

$$\langle P_{cm}^2 \rangle = n(\bar{x}/R)^2 \tag{6.3}$$

and with

$$E_{BAG} = n\,\frac{2.04}{R} + \left(\frac{4\pi}{3}\,BR^3 + \left(a - \alpha_s(R\Lambda)b\right)\frac{1}{R}\right) + \mu\,\frac{\alpha_s(R\Lambda)}{R}\,. \tag{6.4}$$

Here, n = number of quarks. To roughly estimate the center of mass effect we have included the term $n(\bar{x}/R)^2 = \langle P_{cm}^2 \rangle$. We fit $\bar{x}$ so that $m_\pi \sim 0$. For consistency, $\bar{x}/R$ should be close to the momentum of a valence quark, that is, 2.04/R. We then use the same value for all other states. μ is determined by the free quark wave functions[3] and the color-spin matrix elements in the various states, $\mu_\pi = -.70$, $\mu_\rho = .70/3$, $\mu_N = -.70/2$, $\mu_\Delta = .70/2$. We determine the minimum M^2 as a function of R. The results for two different values of "b" are given in Table II. We again see that $b \sim .65$ corresponds well with the previous bag model calculations (and the observed masses).

TABLE II. Hadron Mass Spectrum for Two Values of "b"

CASE I	Particle	π	ρ	N	Δ
b = .65 $\bar{x}$ = 2.21 Λ = .436 GeV $B^{1/4}$ = .145 GeV = .332 Λ	M/Λ M(GeV) R(GeV^{-1}) α_s	<u>0</u> <u>0</u> 3.9 1.5	1.84 .80 4.7 1.7	2.16 <u>.94</u> 5.2 1.9	2.86 1.25 5.5 2.0
CASE II	Particle	π	ρ	N	Δ
b = 1.04 $\bar{x}$ = 2.49 Λ = .223 GeV $B^{1/4}$ = .144 GeV = .645 Λ	M/Λ M(GeV) R(GeV^{-1}) α_s	<u>0</u> <u>0</u> 3.4 .83	3.01 .67 4.3 .97	4.21 <u>.94</u> 5.0 1.1	5.09 1.14 5.3 1.1

Note: Ground state masses of hadrons composed of "bare" massless up and down quarks, for two values of "b". The momentum spread of the valence quarks is parameterized by $\bar{x}$ and this is adjusted to make $m_\pi = 0$. Λ is taken to fit the nucleon mass, .94 GeV.

VII. CONCLUSIONS

I have indicated how the attractive interaction between colored particles can lead to a vacuum densely filled with bubbles of perturbative vacua consisting of paired quanta. The bubbles are separated by walls of induced color magnetic poles and currents which screen long range color density correlations. Naturally, only the long wave length fluctuations will actually experience this pairing, but the energy of the ground state is insensitive to the short scale fluctuations. We have shown how this vacuum state leads directly to the static bag model phenomenology. It should be needless to remark on all the difficiencies which exist in this treatment. Although the vacuum wave function is manifestly gauge invariant, it is not translation invariant. It is also not Lorentz invariant. We have indicated how chiral symmetry may be spontaneously broken, but we have not given a complete treatment of this symmetry breaking. Since α_s must be of order unity to accommodate the large spin-spin interaction observed in the mass spectrum, the calculation must be extended to higher orders of α_s.

In spite of these problems, I believe that this picture provides a simple, intuitive, and quantitative basis for the development of more detailed phenomenologies for the theory of hadron structure.

ACKNOWLEDGMENT

I would like to thank many colleagues for discussions about these ideas. In particular, I should like to thank R. P. Feynman, H. B. Nielsen, L. Susskind and C. Thorn for very helpful insights.

REFERENCES

1. S. Mandelstam, Phys. Reports 23C, 245 (1976); R. Fukuda, T. Kugo, Prog. Theor. Phys. 60, 565 (1978); C. G. Callen, Jr., R. A. Dashen, D. J. Gross, Phys. Rev. D19, 1826 (1979); H. B. Nielsen, P. Olesen, NBI-HE-79-17 (preprint) (1979); H. B. Nielsen, M. Ninomiya, Nucl. Phys. (to be published); C. B. Thorn, Phys. Rev. D19, 639 (1979).
2. A. Chodos et al., Phys. Rev. D9, 3471 (1974); K. Johnson, Acta Phys. Pol. B6, 865 (1975); R. Friedberg, T. D. Lee, Phys. Rev. D18, 2623 (1978); T. D. Lee, Phys. Rev. D19, 1802 (1979); C. Detar, UUHEP-79-6 (University of Utah) (1979); R. L. Jaffe, MIT-CTP-814 (1979).
3. T. A. deGrand et al., Phys. Rev. D12, 2060 (1975).
4. J. F. Donoghue, K. Johnson, MIT-CTP-802 (1979).
5. J. D. Bjorken, SLAC-PUB-2372 (1979).
6. H. B. G. Casimir, Proc. Kon. Ned. Akad. Wetensch. 51, 793 (1948).
7. T. H. Boyer, Phys. Rev. 174, 1764 (1968).
8. K. A. Milton, L. L. DeRaad, J. Schwinger, Ann. Phys. [N.Y.] 115, 388 (1978).
9. R. Balian , B. Duplantier, Ann. Phys. [N.Y.] 112, 165 (1978). This reference gives methods which will allow one to calculate the vacuum energy for boxes with arbitrary shapes.
10. G. 't Hooft, Nucl. Phys. B72, 461 (1974); R. Brower, R. Giles, C. B. Thorn, Phys. Rev. 18, 484 (1978).
11. Y. Nambu, G. Jona-Lasinio, Phys. Rev. 122, 345 (1961).
12. Details will be published later.

GAUGE MODELS OF CP VIOLATION

L. Wolfenstein
Carnegie-Mellon University, Pittsburgh, PA 15213

Abstract

Experimental consequences of milliweak gauge models of CP violation that may distinguish them from superweak are discussed. Included are η_{oo} in $K^o \to 2\pi$ decay, η_{+-o} in $K^o \to 3\pi$ decay, and the electric dipole moment of the neutron. Emphasis is on the six-quark SU(2) × U(1) model (Kobayashi-Maskawa CP violation).

The Weinberg-Salam unified gauge theory of weak and electromagnetic interaction as it appeared in the early seventies involved four leptons, four quarks, three heavy vector bosons, and a single Higgs doublet. That theory predicted the now-established phenomenology of neutral currents and required the existence of charmed particles. In spite of its great success, that model could not accommodate the phenonemon of CP violation discovered in K^o decay in 1964. Three general ways of extending the model to incorporate CP violation have been suggested: (1) increase the number of quarks from four to six or more,[1] (2) increase the number of Higgs doublets from one to two or more,[2,3] and (3) increase the number of intermediate vector bosons by going beyond the group SU(2) × U(1).[4]

In this talk I want to discuss experimental consequences of such gauge models.[5] I will limit myself to models which are milliweak; that is, the CP-violating interaction effective for K^o decays is of the order of 10^{-3} times the ordinary weak interaction. In the case of models with extra Higgs or vector bosons, this factor of 10^{-3} may arise from the heavy mass of these extra particles, while in the model with extra quarks it may arise in part from the smallness of the mixing angles connecting the new quarks with the old. The experimental effects we are interested in are those that distinguish a milliweak model from a superweak model in which the effective CP-violating interaction is of the order 10^{-9} times the ordinary weak interaction. It is possible also to design gauge models which are superweak[6] but usually only by recourse to superheavy vector bosons or Higgs bosons.

I shall focus here on three experiments now being initiated or proposed that should be able to detect such milliweak effects:

(1) Determination of $|\eta_{oo}/\eta_{+-}|$. The study of CP-violation in K^o decays can determine the quantity

$$\left| \frac{\eta_{oo}}{\eta_{+-}} \right|^2 = \frac{\Gamma(K_L \to \pi^o\pi^o)/\Gamma(K_S \to \pi^o\pi^o)}{\Gamma(K_L \to \pi^+\pi^-)/\Gamma(K_S \to \pi^+\pi^-)}$$

In the superweak model $\eta_{oo} = \eta_{+-}$ since the CP-violating K_L decay occurs only because of the mixture of the CP-even K_1 in K_L. Past

ISSN:0094-243X/80/590365-7$1.50

experiments yielded $\left|\frac{\eta_{oo}}{\eta_{+-}}\right| = 1.00\pm.06$; proposed experiments[7] aim at an accuracy of 1%.

(2) Measurement of the CP-violating parameter in the 3π decay of K^0

$$\eta_{+-o} = \frac{A(K_S \to \pi^+\pi^-\pi^o)}{A(K_L \to \pi^+\pi^-\pi^o)}$$

While there exists a CP-conserving $K_S \to 3\pi$ decay (to the I=2 final state) the CP-violating amplitude can be isolated by looking for an interference effect between K_L and K_S. In the superweak model $\eta_{+-o} = \eta_{+-}$ since the same mixing effect is responsible for both. At present there is no interesting limit on η_{+-o}, but a proposal[8] recently submitted to FNAL aims to measure the difference $(\eta_{+-o}-\eta_{+-})/\eta_{+-}$ to an accuracy of the order of $\pm.25$.

(3) Electric dipole moment of the neutron. In many milliweak models one may expect D_n to be of the order 10^{-23} to 10^{-24} e-cm, in contrast to a value of less than 10^{-29} for the superweak model.[9] The present upper limit $D_n \leq 10^{-24}$ e-cm may be improved by two orders of magnitude by experiments[10] now starting at Grenoble.

Let me first discuss the six-quark (KM) model, which has been the center of interest since the evidence in favor of a fifth quark. It was first emphasized by Kobayashi and Maskawa[1] that with three doublets of quarks the single Cabibbo angle becomes replaced by three mixing angles (θ_1, θ_2, θ_3) and one significant CP-violating phase δ. In this case the charged-current Hamiltonian mediated by the usual $W^\pm$ is CP-violating; in particular, the effective $\Delta S=1$ piece can be written

$$\begin{aligned} H_{\Delta S=1} &= (G_F/\sqrt{2}) \sin\theta_c \cos\theta_c \\ &\quad \bar{d}\,\gamma^\lambda[(u\bar{u} - c\bar{c}) + (K+iK')(c\bar{c}-t\bar{t})]\gamma_\lambda s + \text{h.c.} \\ K &= s_2^2 + s_2c_2t_3 \cos\delta/\cos\theta_c \\ K' &= s_2c_2t_3 \sin\delta/\cos\theta_c \end{aligned} \tag{1}$$

where u, d, etc. are left-handed quark fields; $s_i = \sin\theta_i$, $c_i = \cos\theta_i$, $t_i = \tan\theta_i$; and θ_c is essentially the Cabibbo angle. The mass matrix also depends only on K and K' so that the effects of going from four to six quarks on $\Delta S=1$ physics depends only on these two combinations of θ_2, θ_3, and δ. The form of Eq.(1) involves a definite phase convention, for which the CP-violating term proportional to K' satisfies the $\Delta I=1/2$ rule. If we define Im A_I as the CP violating amplitude to the final $\pi\pi$ state with isospin I, then Im A_2=0 but Im $A_o \neq 0$.[11] If the $\Delta I=1/2$ rule were also exact for the CP-conserving interaction, then η_{oo}/η_{+-} would have to equal unity because both K_L and K_S would have a ratio of 2 to 1 for $\pi^+\pi^-/\pi^o\pi^o$. Thus the deviation from unity is proportional to the small $\Delta I=1/2$ rule violation (Re A_2/Re A_o) in the K_S CP-conserving decay. An analysis yields[12]

$$\left|\frac{1 - \eta_{oo}/\eta_{+-}}{3}\right| = \left|\frac{\varepsilon'}{\varepsilon}\right| \approx .05 \left[\frac{m'/\Delta m}{\text{Im } A_o/\text{Re } A_o} + 1\right]^{-1} \tag{2}$$

where m' is the CP-violating piece of the mass matrix and $.05 = (\text{Re } A_2/\text{Re } A_o)$.

The estimate of ε'/ε now depends on the relative value of CP violation in the mass matrix measured by $m'/\Delta m$ to CP violation in the decay amplitude measured by $\text{Im } A_o/\text{Re } A_o$. The larger the relative value of $m'/\Delta m$ the closer one gets to the superweak result. In an early discussion of this question, Ellis et al[13] suggested that $\text{Im } A_o/\text{Re } A_o$ was relatively small because the "Zweig rule" inhibits contributions from terms in H involving c and t quarks, which is the case for the K' term in Eq.(1). They concluded ε'/ε might be of the order .002. Recently Gilman and Wise[14] suggested that this argument fails if penguin graphs dominate the $K \to 2\pi$ decay as has been suggested[15] in order to explain the $\Delta I=1/2$ rule. In the penguin diagrams the basic weak transition is $s \to d$ + gluon with the u, c, or t quark involved in a virtual loop from which the gluon emerges; the gluon then ties on to the $\bar{d}$ of the K^o so that the decay is proportional to the matrix element of the form

$$M = \langle 2\pi | \bar{d}_R d_R \bar{d}\ s | K^o \rangle$$

independently of which quark u, c, or t was involved.

In this case we can write schematically

$$\frac{\text{Im } A_o}{\text{Re } A_o} = \frac{f\ K'\ M\ P\ (tc)}{M[P(uc) + K\ P(tc)]} \tag{3}$$

where f is the fraction of $\text{Re } A_o$ due to penguin graphs and P(ac) is the penguin graph amplitude arising from the term in Eq.(1) involving a and c. Unfortunately, the calculations of both P(uc) and M depend on our understanding of almost-confined quarks and so are quite uncertain. P(tc) involving only heavy quarks can be calculated more reliably. Gilman and Wise[14] calculate P(uc)/P(tc) so that M cancels out from their result. Some of their results are shown in Table 1; they depend on a low-energy cutoff μ parameterized by $\alpha(\mu^2)$. Guberina and Peccei[16] on the other hand do not trust the evaluation of P(uc) so that they use a theoretical value[15] for the matrix element M to evaluate the numerator of Eq.(3) and then use the empirical value of $\text{Re } A_o$ for the denominator. An example of their considerably lower results are also shown in Table 1. The difference reflects the uncertainty in our understanding of nonleptonic decays.

There is still another problem in these evaluations of ε'/ε, in my opinion, which arises from the evaluation of $m'/\Delta m$. The assumption is made that Δm is accurately given by the second-order box diagram involving quarks originally calculated by Gaillard and Lee.[17] Since m'is given by a similar box diagram, the ratio $m'/\Delta m$ can be calculated without evaluating the matrix element B of the $\Delta S=2$ quark operator between K^o and $\overline{K^o}$

$$B = \langle \overline{K^o} | \bar{s}\ d\ \bar{s}\ d | K^o \rangle$$

I believe that there is a significant contribution to Δm (of the order of Δm itself) that cannot be calculated using quark diagrams; this comes from low mass intermediate states such as π, η, and 2π. If I call this contribution ($D\ \Delta m$) then I find for a fixed value of K that Eq.(2) is correct if the box diagram value is used for $(m'/\Delta m)$ and at the same time the factor .05 is replaced by $.05(1-D)^{-1}$. By including the uncertainty in the value of D, I get a range of values of ε'/ε as shown in Table 1.[18] I have constrained the value of 1-D by two conditions, (a) $\Delta m - D\ \Delta m = m_{box}$, (b) a reasonable limit $|D| < 2$. Since m_{box} is proportional to B, the first condition gives a range of values of (1-D) equal to the range of reasonable values of B; because I consider a factor of 5 variation in B, I obtain in Table 1 for K = .07 a factor 5 variation in ε'/ε.

The results discussed so far are given for an arbitrary value of K equal to .07, which can be realized by setting $\theta_2 = 15^o$ with a small value for θ_3. To find a minimum value of $|\varepsilon'/\varepsilon|$ I set K equal to the upper limit allowed by the condition $|D| < 2$; this corresponds to $m_{box} = 3\ \Delta m$ and so depends on B. These minimum values for the same range of B are shown in the last line of Table 1.

Unfortunately, the conclusion of this discussion must be rather weak. I would say that the KM model predicts that $\eta_{+-}/\eta_{oo} > 1$ by a small amount, perhaps of the order 1% to 3%, probably between 1/2% to 5%.

We now turn to the parameter η_{+-o} in $K^o \rightarrow 3\pi$ decay. Whereas it is well-known that $\eta_{+-} - \eta_{oo}$ is expected to be small because of the approximate $\Delta I=1/2$ rule, one might expect on an a priori basis that η_{+-o} and η_{+-} might differ by as much as 100%. A recent analysis by Li and myself[19] shows that $\eta_{+-o} - \eta_{+-}$ is small in the KM model. The analysis is based on standard current algebra and soft pion techniques, which can be used in the KM model because only left-handed currents enter the weak interaction. If we write

$$K_L = K_2 + \rho\ K_1$$

$$K_S = K_1 + \rho\ K_2$$

we have

$$\eta_{+-} = \rho + i\ \mathrm{Im}\ A_o/\mathrm{Re}\ A_o$$

$$\eta_{+-o} = \rho + \mathrm{Im}\ A_{+-o}/\mathrm{Re}\ A_{+-o}$$

where Re(Im) A_{+-o} is the decay amplitude of $K_L(K_S)$ to the predominant I=1 final state of the 3π's. The soft-pion results relate A_{+-o} to A_o and if the small ΔI=3/2 amplitudes are neglected yield

$$\text{Im } A_{+-o}/\text{Re } A_{+-o} = \text{Im } A_o/\text{Re } A_o$$

so that $\eta_{+-o} = \eta_{+-}$ as in the superweak model. The deviations from this equality are then of the order of violations of the ΔI=1/2 rule and so are small. Including the ΔI=1/2 rule violations, we obtain the approximate result

$$\eta_{+-o} - \eta_{+-} \approx \eta_{oo} - \eta_{+-}$$

indicating that the difference is at most a few percent. While this result may have significant corrections due to the limitations of the soft-pion method, it indicates that as far as the KM model is concerned the proposed 25% measurement of η_{+-o} cannot compete with the proposed 1% measurement of $|\eta_{oo}/\eta_{+-}|$.

Calculations of the neutron electric dipole moment D_n are usually carried out by first calculating the electric dipole moments d_q of the quarks. In the KM model it is obvious that d_q vanishes in lowest order, and it has been shown[20] that d_q also vanishes in second order. Recent calculations[21,22] have shown that D_n need not vanish in second order if diagrams involving two of the neutron's quarks are used. These calculations yield values of D_n between 10^{-30} and 10^{-32} e-cm. It has also been argued[22] that d_q does not vanish in second-order when gluon corrections are included. In any case, it is clear that the value of D_n caused by the KM mechanism is too low to be found in prospective experiments.

We now turn briefly to some other models the experimental consequences of which distinguish them from the KM model. In the Weinberg model[3] CP violation is associated with the exchange of charged Higgs bosons. Weinberg calculated the quark electric dipole moments d_u and d_d and from them a value of D_n of about 2×10^{-24} e-cm just about equal to the present experimental limit. Anselm and D'Yakonov using the same model but somewhat different dynamical assumptions find values of d_u and d_d and thus of D_n a factor of 10 lower. However, with their assumptions the electric dipole moment d_s of the s quark and thus of the Λ is 10^{-22} e-cm. Since the well-known strong interactions couple the proton to Λ+K it seems certain that if $D_\Lambda \sim 10^{-22}$ then D_n must be at least 10^{-24} e-cm. Thus, if this model is the basic cause of CP violation, the forthcoming experiments measuring D_n should find a significant non-zero result.

In the $SU(2)_L \times SU(2)_R \times U(1)$ model of Mohapatra and Pati[4] CP violation is associated with the exchange of the heavy W_R gauge bosons. The CP violation can be shown to show up as a relative phase φ between the P-even and P-odd parts of the Hamiltonian. Thus, for observables that involve one of these parts there will be no CP-violating effect. It follows that in lowest order $\eta_{oo} = \eta_{+-}$ and D_n vanishes. An observable that is particularly sensitive to this relative phase is $|\eta_{+-o} - \eta_{+-}|$, which just equals $|\tan\phi|$. Thus this is a model for which the measurement of η_{+-o} provides the most sensitive test.

In discussing different gauge models it is important to note that the various models are not mutually exclusive. One point of view is to imagine that we start with a grand model that contains many vector bosons, many Higgs bosons, and at least six quarks. Furthermore, this model violates CP invariance everywhere it can. The different models we have discussed would then represent different reductions of the grand model. It may well be that one of the models, or, better, one of the mechanisms, may provide nearly all the CP violation in $K^o \to 2\pi$ decay. However, the same mechanism might not dominate all CP-violating effects. For example, if the KM mechamism is the correct explanation of the $K^o \to 2\pi$ CP violation it does not necessarily mean that D_n will be as small as 10^{-30} e-cm, since if there is also CP violation in the Higgs sector this might dominate D_n.[23]

The problem of CP violation has been with us for fifteen years. The unified gauge models have provided no fundamental understanding of CP violation, but they have led to new insights on the phenomenology of CP violation. As a result there is reason to believe that prospective experiments may finally prove the superweak model incorrect. That would be a step forward in solving the CP problem, perhaps.

I am grateful for discussions with F. Gilman, R.D. Peccei, D. V. Manopoulos, V. I. Zakharov, and M. B. Voloshin. This work was supported in part by the U.S. Department of Energy.

Table 1

Calculations of ε'/ε in the KM Model[(a)]

	K	f	Cutoff[(b)]	ε'/ε
Gilman-Wise (Ref. 14b)	.07	3/4	$\alpha(\mu^2) = 0.75$	.02
			$\alpha(\mu^2) = 1.0$	.0125
			$\alpha(\mu^2) = 1.25$	.01
Guberina-Peccei (Ref. 16)	.07	(c)	$\mu = 0.5$ Gev	.003
Wolfenstein	.07	3/4	$\alpha(\mu^2) = 1.0$	.006-.030
	max. value	3/4	$\alpha(\mu^2) = 1.0$	.004-.007

(a) $m_t = 15$ Gev, $m_c = 1.5$ Gev are used for all results.

(b) The cutoff procedures in Refs. 14b and 16 are not directly comparable; however, μ=0.5 Gev in Ref.16 corresponds closely to $\alpha(\mu^2)$=1.25 Gev in Ref. 14b. The results of Ref.16 are much less sensitive to the cutoff than those of Ref. 14b.

(c) f is not used in Ref. 16. However, the calculations of P(uc) in Ref. 16 taken literally together with the value used for M would yield $f \approx 1/4$ for this case.

REFERENCES

1. M. Kobayashi and T. Maskawa, Progr. Theor. Phys. 49, 652 (1973).
2. T. D. Lee, Phys. Reports 9C, 143 (1974).
3. S. Weinberg, Phys. Rev. Letters 37, 657 (1976).
4. An example of particular interest we discuss briefly is the gauge group $SU(2)_L \times SU(2)_R \times U(1)$, R. Mohapatra and J.C. Pati, Phys. Rev. D 11, 566 (1975).
5. This talk closely parallels my talk given at the Neutrino 79 Conference in Bergen, to be published in the Proceedings. Some more recent work (Refs. 14b, 16, and 19) is included here leading to a broader range of values for ε'/ε in the KM model and a more definite prediction for η_{+-o}. For a still earlier version see Proceedings of the International Neutrino Conference, Aachen, 1976 (Vieweg, 1977), p. 530.
6. See, for example, Mohapatra, Pati, and Wolfenstein, Phys. Rev. D 11, 3319 (1975).
7. B. Winstein *et al*., Fermilab proposal;
   R. K. Adair *et al*., Brookhaven AGS proposal.
8. G. Thomson *et al*., Fermilab proposal.
9. L. Wolfenstein, Nucl. Phys. B77, 375 (1974).
10. N. Ramsey, private communication.
11. Note this is different from the Wu-Yang phase convention. Equations for ε and ε' for a general phase convention are given in Ref. 12.
12. L. Wolfenstein, Nucl. Phys. B, to be published (Carnegie-Mellon preprint COO-3066-124).
13. Ellis, Gaillard, and Nanopoulos, Nucl. Phys. B109, 213 (1976).
14. a) F. Gilman and M.B. Wise, Phys. Lett. 83B, 83 (1979).
    b) F. Gilman and M.B. Wise, Phys. Rev. D, to be published, (SLAC preprint SLAC-PUB-2341).
15. M. A. Shifman *et al*., JETP Lett. 22, 55 (1975); Nucl. Phys. B120, 316 (1977); Sov. Phys. JETP 45, 670 (1977).
16. Guberina and Peccei, Nucl. Phys. B, to be published. Similar results have been given by Prokhorov, JETP Letters. to be published.
17. M. K. Gaillard and B. W. Lee, Phys. Rev. D 10, 897 (1974).
18. I have followed the approach of Ref. 14b in evaluating Eq.(3). The results in Ref. 12 are different because there I followed Ref. 14a.
19. L.F. Li and L. Wolfenstein, Phys. Rev. D. to be published, LBL preprint LBL-9579, Sept., 1979.
20. E. P. Shabalin; ITEP preprint ITEP-31 (1978).
21. B. Morel, Harvard preprint HUTP-79/A009.
22. D. V. Nanopoulos *et al*., Harvard preprints HUTP-79/A024 and A048.
23. Still another possible contribution to D_n might come from the θ term in the QCD Hamiltonian; see, for example, V. Baluni, Phys. Rev. D 19, 2227 (1979).

$SIN^2\theta_W$, GRAND UNIFIED GAUGE THEORIES AND PROTON DECAY

William J. Marciano
The Rockefeller University, New York, N.Y. 10021

ABSTRACT

The role of $\sin^2\theta_W$ in the Weinberg-Salam SU(2) x U(1) model is reviewed. Effects of radiative corrections on neutral current determinations of this parameter are described. It is pointed out that such effects can potentially shift the real value of $\sin^2\theta_W$ away from the present experimental world average of 0.23±0.015. Predictions by grand unified gauge theories for the effective low energy value of $\sin^2\theta_W$ are also reviewed. Implications of those results for the Georgi-Glashow SU(5) model and the lifetime of the proton are outlined. Combining that analysis with the Goldman-Ross estimate for the super- heavy boson mass $M_S \simeq 3\times10^{14}$GeV., the SU(5) model is found to predict $\sin^2\theta_W \simeq .21-.22$ and a proton lifetime $\tau_p \simeq 10^{29}-10^{30}$ yrs.

TABLE OF CONTENTS

INTRODUCTION

Strong, weak and electromagnetic interactions all seem to be very well described by the standard $SU(3)_c$ x SU(2) x U(1) model, i. e. quantum chromodynamics (QCD)[1] x Weinberg-Salam (W-S) model[2]. Some features of the standard model are: It contains three independent gauge couplings g_3, g_2 and g_1 corresponding to the distinct gauge groups $SU(3)_c$, SU(2) and U(1) respectively. They separately determine the different interaction strengths. There are eight massless color gluons g^a, a=1,2 ... 8, responsible for strong interactions, three massive intermediate vector bosons $W^\pm$ and Z^o which

ISSN:0094-243X/80/590373-24$150

mediate weak charged and neutral current interactions and the massless photon γ responsible for electromagnetic interactions. In addition there must be a neutral massive Higgs scalar ϕ^o, which is a remnant of spontaneous symmetry breaking. (I leave open the possibility that there may be additional neutral and charged Higgs scalars.) Fermions can be grouped into sequential generations of leptons and quarks of which three are presently known 1) ν_e,e,u,d 2) ν_μ,μ,c,s 3) ν_τ,τ,t,b (t flavor is as yet undiscovered) with each quark flavor coming in three distinct colors as specified by QCD. Left-handed components of these fermion fields transform as doublets (T=1/2) under weak SU(2) isospin while right-handed components are singlets (T=0). The electric charge of each field is given by

$$Q=T_3+Y/2 \tag{1}$$

where T_3 is the third component of weak isospin and Y is the U(1) hypercharge.

In the W-S model, one can trade off the coupling parameters g_1 and g_2 for e the unit of electric charge and θ_W the weak mixing angle via the relationships

$$e=g_1g_2/\sqrt{g_1^2+g_2^2} \quad , \quad \theta_W=\tan^{-1}g_1/g_2 \tag{2}$$

The value of e is well determined, $e^2/4\pi=\alpha\simeq1/137$ and θ_W is beginning to be very precisely measured in weak neutral current experiments. At present the world average is[3,4]

$$\sin^2\theta_W=0.23\pm0.015 \text{ (experimental average);} \tag{3}$$

the value should eventually become even more precisely known. Because of the already small errors in (3), it now seems appropriate and timely to scrutinize the quantity $\sin^2\theta_W$ more closely. In that regard, one can ask: What is the relationship between the theoretical and experimental definitions of $\sin^2\theta_W$? Or equivalently, what are the effects of radiative corrections on experimental neutral current determinations of $\sin^2\theta_W$? In section II of this talk, I will describe some ongoing work[5] by Alberto Sirlin and myself which should provide answers to these questions.

When the standard model is embedded in a simple gauge group G: $G\supset SU(3)_c \times SU(2) \times U(1)$, strong, weak and electromagnetic interactions become truly unified. That is because G possesses only one bare gauge coupling g_{G_0} [subscript or superscript zero is used to denote bare (unrenormalized) quantities]; so the three originally independent bare couplings g_{3_0}, g_{2_0} and g_{1_0} must all be equal to g_{G_0} (up to group-theoretic weighting factors). In such a scheme the observed unequal strengths of strong, weak, and electromagnetic interactions is a consequence of performing present-day experiments at relatively low energies, so that higher-order radiative corrections effectively enhance or diminish the various interaction strengths; they become equal only at superhigh energies above all mass scales in the theory (i.e. at very very short distances). Such simple gauge group schemes, the best known of which is G=SU(5), the Georgi-Glashow model[6], are collectively called grand unified theories (GUTS)[7,8].

With regard to the weak mixing angle, an appealing property of grand unification is that $\sin^2\theta_W^o$ is predicted to be some definite number, rather than an (infinite) adjustable counterterm parameter (its role in the W-S model when considered alone). This must be the case because $\sin^2\theta_W^o = g_{1_0}^2/(g_{1_0}^2+g_{2_0}^2)$ and $g_{1_0}^2 = C g_{2_0}^2$ with C a constant determined simply by the group structure of G and the definition of the electric charge. For those of us who are less agile with group theory, there exists a nice formula relating $\sin^2\theta_W^o$ to the values of T_{3_i} (third component of weak isospin) and Q_i (electric charge) of the members of any representation R of G (whether it be fermions, gauge fields or scalars)[9]

$$\sin^2\theta_W^o = \frac{g_{1_0}^2}{g_{1_0}^2+g_{2_0}^2} = \frac{\sum_{i\varepsilon R} T_{3_i}^2}{\sum_{i\varepsilon R} Q_i^2} \qquad (4)$$

So, for example, in a theory such as SU(5) which requires only sequential fermions (left-handed doublets and right-handed singlets), one finds from summing over the three known generations of quarks and leptons

$$\sin^2\theta_W^o = \frac{12(1/2)^2 + 12(-1/2)^2}{6(-1)^2+18(2/3)^2+18(-1/3)^2} = 3/8 \qquad (5)$$

(a factor of 3 for color has been included, left and right-handed components counted separately). For consistency, a single generation or any other representation of SU(5) (scalars or gauge fields) must also give 3/8 when summed over in (4) and they of course do. A lesson that we learn from (5) is that any grand unified model will predict $\sin^2\theta_W^o=3/8$ unless it contains exotic fermions, i.e. non-sequantial additions with Q and/or T_3 not equal to zero. We note, however, that just as effective interaction strengths depend on the experimental energy scale examined, so does the effective value of $\sin^2\theta_W$, since it is a ratio of couplings. Therefore, within the framework of GUTS, the effective value of $\sin^2\theta_W$ measured in present-day experiments (see eq. (3)) may differ significantly from its bare asymptotic value, $\sin^2\theta_W^o$, because of large radiative corrections (finite renormalization effects); a feature initially pointed out and explored by Georgi, Quinn and Weinberg (G-Q-W)[10]. I will outline their results shortly and then in section III discuss further these radiative corrections, present a simple expression for the effective quantity $\sin^2\theta_W$ which can be compared with experiment and describe theoretical sources of uncertainty in my analysis.

In addition to predicting $\sin^2\theta_W^o$, GUTS exhibit other very nice properties such as natural charge quantization, mass ratio predictions[11,12] and massless (or very small mass) neutrinos. I won't have time to discuss these features; they have already been well reviewed.[13]

A very interesting prediction which is difficult (although not impossible[8]) to avoid in realistic grand unification schemes is the existence of baryon number violating interactions. This feature when combined with CP violation offers a nice explanation for the baryon asymmetry of the universe (the observed predominance of matter over antimatter).[14,15] It also implies that the proton can decay, albeit with a very long half-life. The present experimental bounds on the lifetime of the proton are already quite good[16]

$$\tau_p \gtrsim 10^{29}\text{-}10^{30}\,\text{yrs. (Exp. bound)} \tag{6}$$

but these numbers, as we shall see, are very close to the present best estimates for τ_p in the SU(5) model. So planned experiments[17] designed to search for proton decay up to $\tau_p \approx 10^{33}$ yrs may have an exciting discovery awaiting them.

As I already stated, although the bare couplings g_{3_0}, g_{2_0} and g_{1_0} are all equal in GUTS (up to group theory factors), the coupling strengths measured in present day experiments are very different because of radiative corrections. So, for example, in the SU(5) model the QED tree level amplitude in fig. 1. is 3/8 (i.e. $\sin^2\theta_W^o$) times the magnitude of the corresponding QCD amplitude. But we know experimentally that at low q^2, strong interactions are much stronger than 8/3 x electromagnetic interactions. What happens is that quantum loop effects are very different for the two cases (see fig. 1). The photon couples to loops of charged massive intermediate vector bosons $W^\pm$, $X^{\pm 4/3}$ and $Y^{\pm 1/3}$ while the gluon couples to loops of colored vector bosons which include not only X and Y but also other massless colored gluons. The latter because of their masslessness cause a sizeable logarithmic variation in the effective strong coupling (asymptotic freedom) as q^2 changes (for all values of q^2) whereas the effective electromagnetic coupling $\alpha(q^2)$ is relatively insensitive to variations in q^2. The 3/8 constant of proportionality will only be realized at q^2 values much larger than any $(\text{mass})^2$ in the theory. (In the SU(5) model, unification is realized for $q^2 \gg M_S^2 \simeq 10^{29}\,\text{GeV}^2$.)

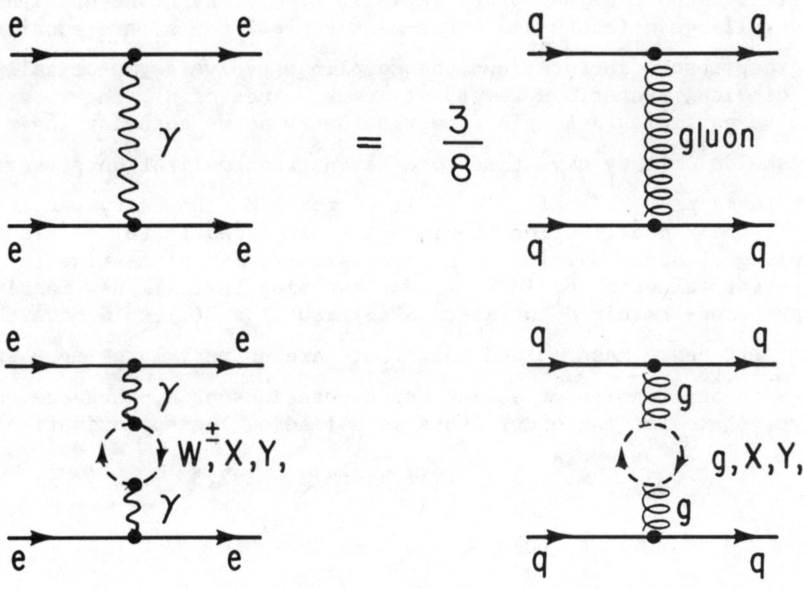

Fig. 1. QED versus QCD amplitudes in the SU(5) model. Asymptotically, at very large q^2, the constant 3/8 is realized as the ratio of these interaction strengths; however, at lower q^2 the strong QCD interaction is enhanced by gauge boson vacuum polarization effects such as those illustrated above. (For gauge invariance, catastrophic graphs must also be included.)

Since radiative corrections are so important in GUTS, one needs a systematic procedure for handling such effects. Fortunately, the most important parts (logarithmic contributions) can be summed to all orders by renormalization group techniques. This method was first applied to GUTS in the pioneering work of G-Q-W[10]; I will briefly outline some of their results so as to have a reference for comparison when I describe attempts to extend and refine that anslysis.

The effective couplings $g_i(\mu)$, i=1,2,3 appropriate for describing physical processes at $q^2=-\mu^2$ are governed by their respective beta functions

$$\mu \frac{\partial}{\partial\mu} g_i(\mu)=\beta_i(g_i)=b_i g_i^3+ \ldots \tag{7}$$

where ... denote higher order terms in g. In any grand unified model, at very large μ (above all other mass scales) the b_i are equal (up to group theory factors) and the couplings evolve asymptotically in an identical manner. However, at lower values of μ, the very massive particles decouple from the theory as we go below their thresholds[18] (they give rise to constant renormalization effects plus terms proportional to q^2/M^2); so for $\mu<M$ (the heavy particle mass), their contribution should not be included in (6). This decoupling of heavy particles in domains where $\mu<M$ gives rise to different values of b_i in (6). Now assuming that all new massive gauge bosons required to extend $SU(3)_c \times SU(2) \times U(1)$ to G have the same very heavy mass M_S and that there are no very large mass splittings in any fermion or scalar representations of G, then one can solve (6) to leading order (this is a leading log approximation)

$$\begin{aligned} g_i^{-2}(\mu) &\simeq g_G^{-2}(M)+2b_i \ln M/\mu, \quad i=2,3 \\ g_1^{-2}(\mu) &\simeq \cot^2\theta_W^o g_G^{-2}(M)+2b_1 \ln M/\mu \end{aligned} \tag{8}$$

where M is the approximate unification point ($M \approx M_S$) and the effective values of b_i are given by[10,19]

$$b_3=b_2-\frac{11}{48\pi^2}=b_1\tan^2\theta_W^o-\frac{11}{16\pi^2} \tag{9}$$

(I should note that my definition of g_{1_0} is $\tan\theta_W^o$ times the one originally defined in ref. 10.) Manipulating the three equations in (8), one obtains for $\sin^2\theta_W^o=3/8$,

$$\ln M/\mu \simeq \frac{8\pi}{33}\left(\frac{3}{8\alpha}-\frac{1}{\alpha_s}\right) \tag{10a}$$

$$\sin^2\theta_W \simeq \frac{1}{6}\left(1+\frac{10}{3}\frac{\alpha}{\alpha_s}\right) \qquad (10b)$$

where $\alpha=e^2/4\pi=g_2^2\sin^2\theta_W/4\pi$ and $\alpha_s=g_3^2/4\pi$ (the effective values at μ). These are the results obtained by G-Q-W.[10] If as a first approximation, one takes $\alpha\simeq1/137$, $\alpha_s\simeq.15$ for $\mu\approx M_W\approx85$GeV., then (10a and b) imply $M\simeq5\times10^{16}$GeV. and $\sin^2\theta_W\simeq.194$. Furthermore, in the SU(5) model the proton lifetime τ_p is estimated to be[20]

$$\tau_p\simeq2\times10^{-29}(M_S \text{ in GeV.})^4 \text{ yrs.} \qquad (11)$$

where M_S is the mass of the intermediate vector bosons that mediate proton decay. So, if we approximate $M_S \simeq M$ determined above, then we find $\tau_p\simeq10^{38}$yrs. This value is 8 or 9 orders of magnitude larger than the present bound in (5) and out of reach of conceivable experimental searches. If we were satisfied with this first look at renormalization effects, we would conclude (within the SU(5) framework) that the planned experiments[17] will fail to observe proton decay and the predicted value of $\sin^2\theta_W$ is a little more than two standard deviations below the world average in (3).

Of course, the results in (10) are only a first approximation to the true properties of the SU(5) model and they are liable to considerable change when a more precise comparison of theory and experiment is undertaken. In that regard, one would like to answer the following questions: What is the relationship between M (the unification mass scale) and M_S? What values of α and α_s should be compared and at what μ? What is the relationship between theoretical predictions for $\sin^2\theta_W$ and present-day determinations of $\sin^2\theta_W$ (see eq. (3))? What about non-leading log effects? I will try to answer or at least address some of these questions in the remainder of this talk; but before going into details let me state some results of recent work on these questions.

The work of Buras et al.[12], Ross[21], and Goldman and Ross[22] has been directed at comparing α and α_s to determine M_S, $\sin^2\theta_W$ and τ_p (as was done in ref. 10); but they have refined this approach by including threshold effects, 2 terms in the β functions, Higgs contributions, and fermion loop effects which give rise to an effective $\alpha>1/137$ that is more appropriate at $\mu\approx M_W$ the comparison point. This approach has culminated in the results of Goldman and Ross[22]. They find

$$M_S\simeq3\times10^{14}\text{GeV.} \qquad (12)$$

which strictly implies from (11), $\tau_p \simeq 10^{29}$ yrs.; however because of the uncertainty in (11)[20] I will say that their estimate of M_S implies

$$\tau_p \simeq 10^{29}-10^{30}\text{yrs.} \tag{13}$$

This prediction is right at the present experimental bound in (6). (I note, however, that even a small shift in M_S can severely affect the prediction in (13).) There are some questions about the procedures employed. In particular, the practice of including threshold effects in β function considerations, because that may lead to gauge dependent results. I would prefer to use β functions to sum logs and then in terms of the effective coupling defined in that way to calculate other corrections order by order for the experimental process used to determine α_s. I also feel that there is considerable uncertainty in attempts to extract α_s from present day experiments because the values of q^2 probed are too small to reliably trust QCD perturbation theory. In any case, having noted these sources of uncertainty, I will accept for the purpose of this talk $M_S \simeq 3\text{x}10^{14}$GeV. as at present the best SU(5) estimate.

My own approach[23] has been to ignore the QCD sector of GUTS (and its uncertainty) as much as possible and to concentrate instead on the W-S sector where I trust perturbation theory. In that way I obtain directly[23]

$$\sin^2\theta_W = \frac{3}{8}[1-\frac{\alpha(M_W)}{\pi}\frac{109}{18}\ln M_S/M_W + O(\alpha)] \tag{14}$$

for the SU(5) model, where M_S is the actual super-heavy bosons' mass, $M_W = 38.5/\sin\theta_W$ (the W's mass) and $\alpha(M_W) \simeq 1/128.5$ is the effective QED coupling appropriate for this anaylsis. I presently estimate the uncertainty in this result from ordinary $O(\alpha)$ contributions and higher orders to be $\lesssim 2\%$ (the uncertainty will be discussed in section III). Combining (14) with (11), one finds

$$\tau_p \simeq 1\text{x}10^{33}\left(\frac{.200}{\sin^2\theta_W}\right)^2 \exp[-711(\sin^2\theta_W - .200)]\ \text{yrs.}; \tag{15}$$

so if $\sin^2\theta_W$ were very precisely known, one could use (15) to pinpoint the SU(5) model's prediction for τ_p. A criticism of this procedure has been its sensitivity to small variations in $\sin^2\theta_W$. In any case, on the basis of the results in (3) for $\sin^2\theta_W$, I would say that (15) suggests $\tau_p \simeq 10^{29}-10^{33}$ yrs. with lower values favored. This is consistent with the results of Goldman and Ross.[22]

If instead of ignoring QCD, I combine the Goldman-Ross value $M_S \simeq 3\times10^{14}$GeV. with my calculation in (14), then I estimate

$$\sin^2\theta_W \simeq .21-.22 \text{ (for } M_S \simeq 3\times10^{14}\text{GeV.)}. \tag{16}$$

(The range comes from possible theoretical uncertainty in this estimate.) The result in (16) is very close to the present world average for $\sin^2\theta_W$ in (3) (it differs from the prediction .19-.20 often associated with SU(5)) and suggests that the time is right to reexamine $\sin^2\theta_W$ and its experimental determination very carefully. The next section describes work in that direction.

II. THE WEINBERG-SALAM SU(2)xU(1) MODEL

1. Definition of $\sin^2\theta_W$. In the W-S model[2] the bare weak mixing angle θ_W^o is defined by

$$\theta_W^o \equiv \tan^{-1} g_{1_0}/g_{2_0} \tag{17}$$

and the bare electric charge is

$$e_0 = g_{2_0} \sin\theta_W^o \tag{18}$$

Furthermore, if only Higgs isodoublets are used to break the gauge symmetry, then θ_W^o is involved in another relationship

$$\cos\theta_W^o = M_W^o/M_Z^o \tag{19}$$

the ratio of the $W^{\pm}$ and Z^o's bare masses. In that case, these equations can be combined to give the natural relationship

$$\sin^2\theta_W^o = e_0^2/g_{2_0}^2 = 1 - M_W^{o2}/M_Z^{o2} \tag{20}$$

between bare (unrenormalized) quantities. Because it is natural, (20) requires that the renormalized quantities must satisfy

$$\sin^2\theta_W = (1-M_W^2/M_Z^2)(1+\text{finite } O(\alpha)) = e^2/g_2^2\ (1+\text{finite } O(\alpha)) \tag{21}$$

where the finite $O(\alpha)$ corrections depend on the specific definitions of $\sin^2\theta_W$, e and g_2 employed in (21) (M_W and M_Z are physical on mass shell values). In the case of $\sin^2\theta_W$, we would like to choose a definition which makes $\sin^2\theta_W$ easily accessible to an accurate measurement and minimizes somewhat the higher order radiative corrections to neutral current processes. With these criteria in mind, I will employ throughout this talk the following definition[5]

$$\cos\theta_W \equiv M_W/M_Z \tag{22}$$

or

$$\sin^2\theta_W \equiv 1-M_W^2/M_Z^2$$

[This definition differs by very small $O(\alpha)$ corrections from the quantity $\sin^2\theta_W(M_W)$ defined in ref. 23; presently, the difference is quantitatively unimportant.] When M_W and M_Z are accurately measured (after their discovery), θ_W can be very precisely determined via (22).

The next step is to calculate the relationship between bare and renormalized couplings in the W-S model. The results of one loop calculations give[23]

$$e(0)=e_0[1-\frac{e_0^2}{16\pi^2}(\frac{-11}{3}\frac{1}{n-4}+\sum_f \frac{4}{3}Q_f^2 \ln M_W/m_f - 1/3)+O(e_0^4)] \tag{23a}$$

$$g_{2_R}=g_{2_0}[1-\frac{g_{2_0}^2}{16\pi^2}(\frac{19}{6}\frac{1}{n-4}+\text{cons})+O(g_{2_0}^4)] \tag{23b}$$

where dimensional regularization (n=dim. of space-time) has been used to regulate ultraviolet divergences and the unit of mass M_W has been introduced to keep couplings dimensionless.[24] The quantity e(0) is the electric charge renormalized at zero momentum transfer that appears in the usual fine structure constant

$$\frac{e^2(0)}{4\pi}=\frac{1}{137.035987} \tag{24}$$

Because this is a long distance coupling, vacuum polarization effects give rise to fermion mass singularities ($\ln M_W/m_f$ terms) in (23a) which are quite large. (All fermions, quarks and leptons are included in the summation.) The renormalized coupling g_{2_R} is presently best defined using the muon total decay rate[23]

$$\Gamma(\mu\to e\nu\bar{\nu})\equiv \frac{g_{2_R}^4 m_\mu^5}{3\cdot 2^{11}\pi^3 M_W^4}=4.551384\text{x}10^5\text{sec}^{-1} \tag{25}$$

i.e. g_{2_R} is defined so as to absorb all higher order corrections to this process, making (25) exact (up to m_e^2/m_μ^2 terms) to all orders in perturbation theory. Using this definition, the constant in (23b) is guaranteed to be free of fermion mass singularities by the Kinoshita-Sirlin theorem[25] and hence should not be very large. Similarly, one finds by explicit calculation of the $W^\pm$ and Z^o self-energies

$$1-M_W^2/M_Z^2=\sin^2\theta_W=\sin^2\theta_W^o[1+\frac{e_0^2}{8\pi^2}\{(\frac{11}{3}+\frac{19}{6\sin^2\theta_W^o})\frac{1}{n-4}+\text{cons.}\}+O(e_0^4)] \tag{26}$$

where the constant in (26) can be determined from the results in ref. 26. (It is free of fermion mass singularities and fairly insignificant.)

From (23) and (26) one finds (summing leading logs to all orders)

$$\sin^2\theta_W\equiv 1-M_W^2/M_Z^2=\frac{e^2(0)}{g_{2_R}^2}\{\frac{1}{1-\frac{2\alpha}{3\pi}\sum_f Q_f^2 \ln M_W/m_f}+O(\alpha)\} \tag{27}$$

I previously[23] approximated the bracketed terms to be about 1.066 with an uncertainty of about ±.01; so (27) becomes

$$\sin^2\theta_W\simeq 1.066e^2(0)/g_{2_R}^2 \tag{28}$$

Many people (often unknowingly) define $\sin^2\theta_W$ by $e^2(0)/g_{2_R}^2$ (the quantity called $\sin^2\theta_W(0)$ in ref. 23). The estimate in (28) shows that quantity differs by about 6.6% from the definition advocated here.

Some implications regarding the W-S model that follow from (28) are: The masses M_W and M_Z are about 3.2% larger than lowest order predictions, i.e. from (24), (25) and (28) one finds

$$M_W=\frac{38.53}{\sin\theta_W}\text{GeV.},\qquad M_Z=\frac{77.06}{\sin 2\theta_W}\text{GeV.} \tag{29}$$

with θ_W defined in (22). I estimate the uncertainty in (29) to be about .5%; so if $\sin\theta_W$ were very precisely known, (29) would provide a testable prediction of the W-S model at the loop level. Furthermore, since the decay rates of the $W^\pm$ and Z^o are proportional to $G_F M^3$, this 3.2% mass increase becomes a 10% increase in their predicted decay rates.[27]

What are the prospects for a very precise experimental determination of $\sin^2\theta_W$? Eventually, it should be quite accurately known, after the $W^\pm$ and Z^o are discovered and their masses and decay rates are very precisely measured. In the meantime, neutral current experiments provide the best available information about this parameter. The present situation as reviewed by Baltay[4] and summarized in (3) is already quite good, with deep-inelastic neutrino scattering providing the best data. Indeed, the errors in (3) are becoming small enough for the following question to become relevant. Is $\sin^2\theta_W$ in (3) the same as $\sin^2\theta_W \equiv 1-M_W^2/M_Z^2$? To quantitatively answer this question, Alberto Sirlin and I are calculating the one loop corrections to neutral and charged current neutrino scattering[5]. Although that work is still in progress, I will now report some preliminary results that we have found.

2. Radiative Corrections and Neutral Currents. In deep-inelastic neutrino scattering experiments, the ratio of neutral to charged current cross-sections (R_ν and $R_{\bar{\nu}}$) is measured. To lowest order, these two processes are described in the W-S model by the following amplitudes (for $|q^2| \ll M_W^2$)

$$\mathcal{M}_{N.C.} = -i\frac{G_F}{\sqrt{2}}\bar{\nu}\gamma_\alpha(1-\gamma_5)\nu\bar{q}\gamma^\alpha(T_3^q(1-\gamma_5)-2Q_q\sin^2\theta_W)q \quad (30)$$

$$\mathcal{M}_{C.C.} = -i\frac{G_F}{\sqrt{2}}\bar{\mu}\gamma_\alpha(1-\gamma_5)\nu\bar{q}'\gamma^\alpha(1-\gamma_5)q \quad (31)$$

corresponding to the tree diagrams in fig. 2.

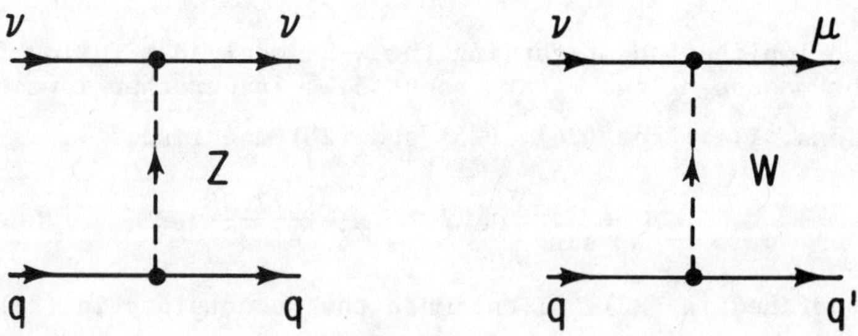

Fig. 2. Tree level diagrams which give rise to the lowest order neutrino induced neutral and charged current amplitudes in (30) and (31).

In these expressions q stands for a generic quark in the target, Q_q is its electric charge and $T_3^q = \pm 1/2$ is its third component of weak isospin. The simplicity of (30) and (31) allow one parameter ($\sin^2\theta_W$) fits to R_ν and $R_{\bar{\nu}}$ data. However, for really precise determinations of $\sin^2\theta_W$, one should modify these lowest order amplitudes by including order α corrections to both neutral and charged current processes. Some of the diagrams which contribute at that order are illustrated in fig. 3.

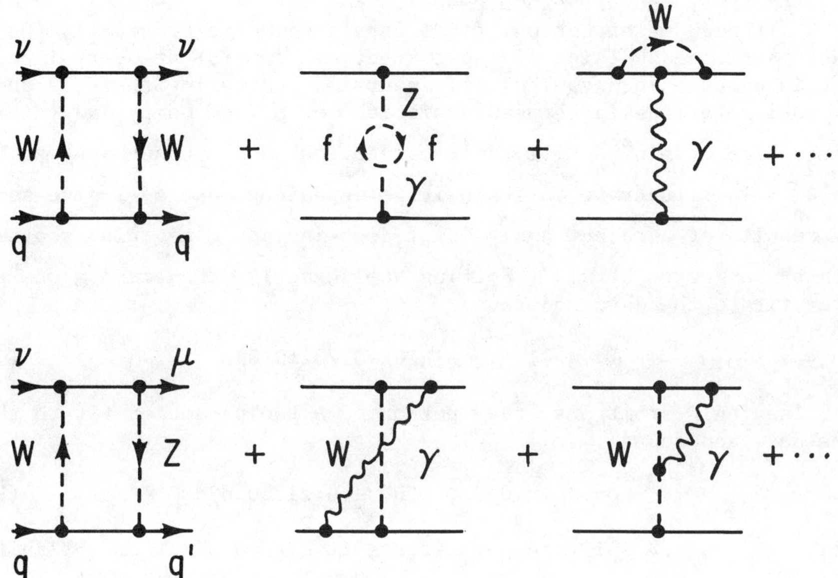

Fig. 3. Some one loop radiative corrections to deep-inelastic neutral and charged current interactions.

There are many such diagrams; fortunately, however, for inclusive processes the sum of all such radiative corrections (to both neutral and charged currents) can be simply accounted for by replacing (30) with the effective amplitude

$$\mathcal{M}^{\text{eff.}}_{\text{N.C.}}=-i\,\frac{G_F}{\sqrt{2}}\,\rho\bar{\nu}\gamma_\alpha(1-\gamma_5)\nu\bar{q}\gamma^\alpha(T_3^q(1-\gamma_5)-2Q_q\,\kappa\sin^2\theta_W)q \qquad (32)$$

where

$$\rho=1+0(\alpha) \quad \text{and} \quad \kappa=1+0(\alpha) \qquad (33)$$

(The quantity ρ includes contributions from charged current corrections.)

Although I cannot give the final results for ρ and κ, let me comment on some of their properties. (Actually the ρ and κ depend to some extent on the q considered, i.e. they are flavor dependent.) In the case of κ the radiative corrections are quite small <2% because of our definition $\sin^2\theta_W\equiv 1-M_W^2/M_Z^2$; if we had chosen $e^2(0)/g_{2_R}^2$ as our definition, they would have been increased by an additional 6.6%. This finding shows that the definition of $\sin^2\theta_W$ in

(22) satisfies our second criterion; it naturally leads to rather small radiative corrections. In the case of ρ , some of the individual diagrams contribute $\approx\pm1\%$ corrections; however cancellations occur. The final sum of all corrections to ρ may turn out to be fairly small $\lesssim 2\%$ (for truly inclusive processes).

Although we expect our final result for ρ to be near 1, the exact magnitude and sign of the corrections in (33) are very important. That is because in deep-inelastic neutrino scattering there is about a 4 to 1 relationship between deviations of ρ from unity and the implied value of $\sin^2\theta_W$, i.e. a 1% correction to $\rho \rightarrow$ about a 4% shift in $\sin^2\theta_W$. To illustrate this sensitive dependence on ρ, I give some of the results of a recent analysis of deep-inelastic neutrino scattering data by Langacker et al.[3] Setting $\rho=1$ exactly, they made a one parameter fit to the data and found

$$\rho=1 \quad , \quad \sin^2\theta_W=0.232\pm0.012 \tag{34}$$

Then leaving ρ arbitrary, they performed a two parameter fit to the same data and found

$$\rho=.981\pm.037 \;,\; \sin^2\theta_W=0.213\pm0.038 \tag{35}$$

Notice that a -2% shift in ρ manifests itself in about an -8% shift in the central value of $\sin^2\theta_W$ as expected. So although their analysis[3] indicates that ρ is very near 1, any small deviation must be accounted for before a precise determination of $\sin^2\theta_W$ can be claimed.

Because of the sensitive dependence of $\sin^2\theta_W$'s determination on ρ, it is important that we complete our calculation of the radiative corrections to this parameter. The algebraic sign of the corrections will be particularly interesting; a correction of $\pm1\%$ to ρ would imply a shift of about $\pm.01$ in (3) (since deep-inelastic scattering dominates the world average). Such a shift would have important implications for GUTS.

3. Higgs, Multiplets with $T\neq1/2$. A possibility which could also lead to a shift in ρ (in addition to radiative corrections) is the existence of Higgs multiplets with $T\neq1/2$ which break the relationship $\cos\theta_W^o=M_W^o/M_Z^o$ in (19). (In that case $\sin^2\theta_W(M_W)$ defined in ref. 23 would be a better renormalized parameter.) Usually, such a possibility is not seriously considered because ρ is known experimentally to be very close to 1 (see (35)); however, a very small vacuum expectation value for such exotic scalar multiplets is not ruled out. For that reason, I present a brief discussion of this possibility and its consequences.

In general the effect of any Higgs multiplet on the values of M_W^o and M_Z^o is specified by giving its weak isospin T, third component of weak isospin T_3 of its neutral member and the vacuum expectation value v_o of that field. In terms of those quantities, the resultant contributions to vector bosons' masses are given by

$$M_W^{o2} = \frac{1}{2} v_o^2 g_{2_0}^2 (T(T+1)-T_3^2) \tag{36}$$

$$M_Z^{o2} = v_o^2 (g_{1_0}^2 + g_{2_0}^2) T_3^2 \tag{37}$$

So summing over all possible multiplets, one finds the following general (lowest order) result

$$\rho = \frac{M_W^{o2}}{M_Z^{o2} \cos^2\theta_W^o} = \frac{\sum_i v_o^{i2} [T^i(T^i+1)-T_3^{i2}]}{2\sum_i v_o^{i2} [T_3^{i2}]} \tag{38}$$

If all Higgs multiplets have T=1/2, T_3=±1/2, then ρ=1 in lowest order; the situation usually assumed. (Note that there are other multiplets that satisfy $3T_3^2=T(T+1)$, e.g. T=3, T_3=±2, which also leave ρ=1.[28]) The next least exotic possibility is an additional Higgs triplet with T=1, T_3=0 (this actually occurs in the SU(5) model). Such an admixture would imply ρ>1. All other possibilities require multiply charged physical Higgs scalars, an exotic possibility. In general, however, if no constraint is placed on Higgs multiplets, ρ may be >1, < 1 or equal to 1 in lowest order.

It would be nice if two parameter fits such as the one in (35) could pinpoint the value of ρ more precisely. Then after radiative corrections are accounted for, one might be able to say something more definite about the existence of non-doublet Higgs representations.

III. SIN$^2\theta_W$ AND GRAND UNIFIED GAUGE THEORIES

In the W-S model, $\sin^2\theta_W^o$ is an infinite adjustable counterterm parameter which is related to the renormalized quantity $\sin^2\theta_W$ by eq. (26)

$$\sin^2\theta_W = \sin^2\theta_W^o\left[1+ \frac{e_0^2}{8\pi^2}\left(\frac{11}{3} + \frac{19}{6\sin^2\theta_W^o}\right)\left(\frac{1}{n-4} + \ln M_W + \text{cons.}\right) + 0(e_0^4)\right] \tag{39}$$

where $\ln M_W$ now appears because I have not assumed a unit of mass in the dimensional regularization procedure. In any GUT, $\sin^2\theta_W^o$ is a rational number (eg. 3/8); so it can no longer be used to cancel the divergence in (39). What must happen (and does) is that the new massive particles introduced in extending the standard model to G must contribute an equal but opposite divergence which cancels that in (39). If all such new particles have the same super-heavy mass M_S, they give rise to a contribution

$$\frac{-e_0^2}{8\pi^2}\left(\frac{11}{3}+\frac{19}{6\sin^2\theta_W^o}\right)\left(\frac{1}{n-4}+\ln M_S\right)+O(e_0^4) \tag{40}$$

in the brackets of (39) which leads to a finite relationship between $\sin^2\theta_W$ (as defined in (22)) and $\sin^2\theta_W^o$

$$\sin^2\theta_W=\sin^2\theta_W^o\left[1-\frac{\alpha}{2\pi}\left(\left(\frac{11}{3}+\frac{19}{6\sin^2\theta_W^o}\right)\ln M_S/M_W+\text{ cons.}\right)+O(\alpha^2)\right] \tag{41}$$

Although the divergences cancel, a large finite logarithmic correction is left (finite renormalization effect) which relates $\sin^2\theta_W$ and $\sin^2\theta_W^o$. The source of this large correction is nicely illustrated for neutrino scattering in fig. 4. The lowest order tree diagram depends on $\sin^2\theta_W^o$ while the γ-Z mixing diagrams at low q^2 give rise to the large logarithmic shift in (41). Together they effectively produce a tree diagram amplitude with $\sin^2\theta_W^o$ replaced by $\sin^2\theta_W$. So, within the GUT framework, one would say that experimentalists who measure $\sin^2\theta_W$ are observing to a large extent γ-Z mixing due to super-heavy boson loop effects. (They have discovered super-heavy bosons!)

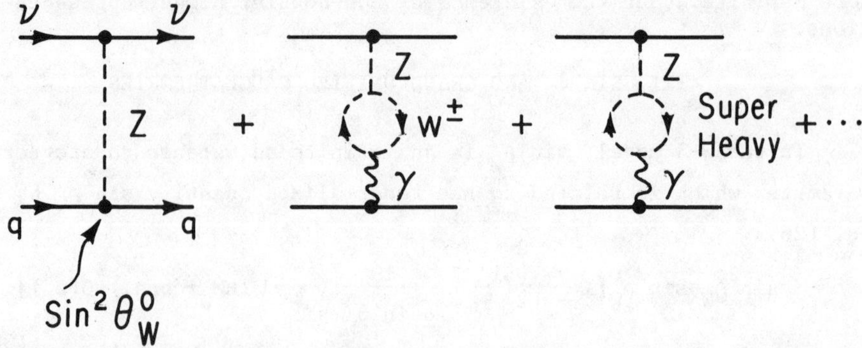

Fig. 4. An illustration of the source of the large logarithmic shift in (41) in the unitary gauge. (Catastrophic graphs must also be included.) The sum of these amplitudes gives rise to the effective $\sin^2\theta_W$ measured in present day experiments.

The following observations regarding the result in (41) are in order. Because of the definition of θ_W employed (see (22)), the constant term in (41) can be obtained from ref. 26; it is small, <1% effect. Using the renormalization group, one can show[23] that the effects of higher order logarithms $(\alpha/\pi)^2 \ln^2 M_S/M_W$ etc. can be included by merely using $\alpha(M_W)$, the effective QED coupling at $q^2=-M_W^2$, in (41) so that it becomes

$$\sin^2\theta_W = \sin^2\theta_W^o[1-\frac{\alpha(M_W)}{2\pi}((\frac{11}{3}+\frac{19}{6\sin^2\theta_W^o})\ln M_S/M_W + \text{cons.}) + O(\alpha^2(M_W))] \tag{42}$$

I previosuly estimated $\alpha(M_W)$ to be[23]

$$\alpha(M_W) \simeq 1/128.5 \tag{43}$$

There is of course some ambiguity in the definition of $\alpha(M_W)$, as in any effective coupling. So, for example, if threshold effects are included (in addition to leading logs) one would find $\alpha(M_W) \simeq 1/130.$[29] Which of these is the better expansion parameter? Clearly that one that makes the $O(\alpha^2(M_W))$ terms in (42) smaller; however, that cannot be discerned without computing those higher-order effects. Combining this uncertainty in $\alpha(M_W)$ with the constant term in (42), I estimate an uncertainty of about 2% in (42) when the logarithmic shift is used alone.

If there are actually N_H light Higgs isodoublets and N_F flavors of quarks (and N_F lepton flavors), then (42) generalizes to[23]

$$\sin^2\theta_W = \sin^2\theta_W^o[1-\frac{\alpha(M_W)}{2\pi}\{\frac{22}{3}\cot^2\theta_W^o - \frac{1}{6}N_H(\frac{1}{\sin^2\theta_W^o}-2) - \frac{2}{3}N_F(\frac{1}{\sin^2\theta_W^o} - \frac{8}{3})\}\ln M_S/M_W + \text{small corrections}] \tag{44}$$

This is my general result for the prediction of $\sin^2\theta_W$ in any GUT which has the standard model embedded in it and only one heavy mass scale M_S.[30] Some features of (44) are: 1) Because $\alpha(M_W) \simeq 1/128.5$ appears in this expression, all leading logs (or all logs for an appropriate $\alpha(M_W)$) are included: 2) The non-logarithmic $O(\alpha)$ corrections to (44) are small. 3) The actual masses M_S and M_W appear in the log, not unification and low energy mass scales. (This follows from explicit one loop calculations.) 4) Adding more Higgs isodoublets ($N_H>1$) tends to increase $\sin^2\theta_W$ (if $\sin^2\theta^o_W<1/2$). 5) The estimated uncertainty in (44) is about 2%.

IV. THE GEORGI-GLASHOW SU(5) MODEL

1. Basic Features. The SU(5) model of Georgi and Glashow is the most economical of all GUTS. It is minimal in that it requires only the usual sequential fermions which make up three generations (6 flavors), each of which is composed of a $\underline{5}+\overline{10}$ representation of SU(5). This implies $\sin^2\theta_W^o = 3/8$ as demonstrated in (5).

There are 24 gauge bosons in the SU(5) model. Twelve of these are the ordinary standard model's gauge fields (8 gluons, $W^\pm$, Z^o and γ) while the other twelve are very exotic in that they carry color and have fractional electric charges ±1/3 and ±4/3. The latter form an SU(2) doublet and hence are degenerate up to very small SU(2) breaking. These bosons and the pattern of symmetry breaking in the SU(5) model are illustrated in fig. 5.

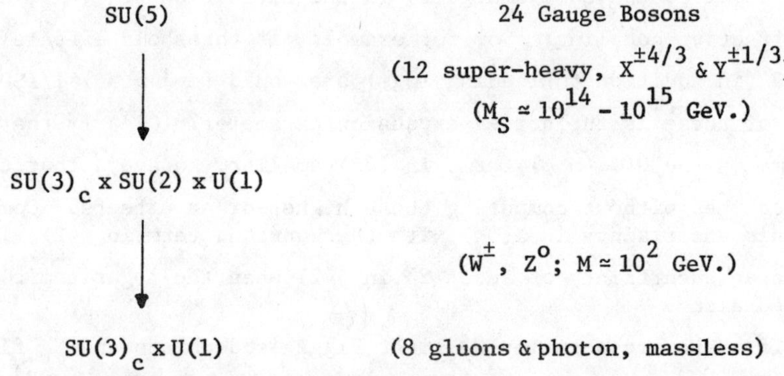

Fig. 5. Pattern of symmetry breaking and resulting mass scales of the gauge bosons in the Georgi-Glashow SU(5) model.

The minimum Higgs scheme required to break the symmetry and give fermion masses is a real 24-plet and a complex 5-plet (there may in general be many such 5-plets). In addition, to get fermion mass relationships right, a 45-plet of Higgs scalars may be necessary.[6] The masses of the physical scalar particles in this model are somewhat arbitrary. I will assume that all physical scalars originating from the 24-plet and all fractionally charged scalars coming from the 5-plets (and 45) have the same super-heavy mass M_S, while the N_H iso-doublets under SU(2) weak isospin have mass $\simeq M_W$.

With these assumptions, the general result in (44) becomes for the SU(5) model[23]

$$\sin^2\theta_W \simeq \frac{3}{8}[1 - \frac{\alpha(M_W)}{2\pi}(\frac{110-N_H}{9})\ln M_S/M_W]. \qquad (45)$$

Using $\alpha(M_W) = 1/128.5$ and assuming $N_H = 1$, this becomes

$$\sin^2\theta_W \simeq \frac{3}{8}[1 - 0.015 \ln M_S/M_W] \qquad (46)$$

The estimated uncertainty in this result is about 2% (for N_H=1).

2. Proton Decay. The exotic super-heavy gauge bosons $X^{\pm 4/3}$ and $Y^{\pm 1/3}$ in fig. 5 mediate baryon number violating processes such as proton decay. The predicted rate for this (as yet unobserved) process is extremely small because M_S is very heavy. Some of the proton decay modes are schematically illustrated in fig. 6.

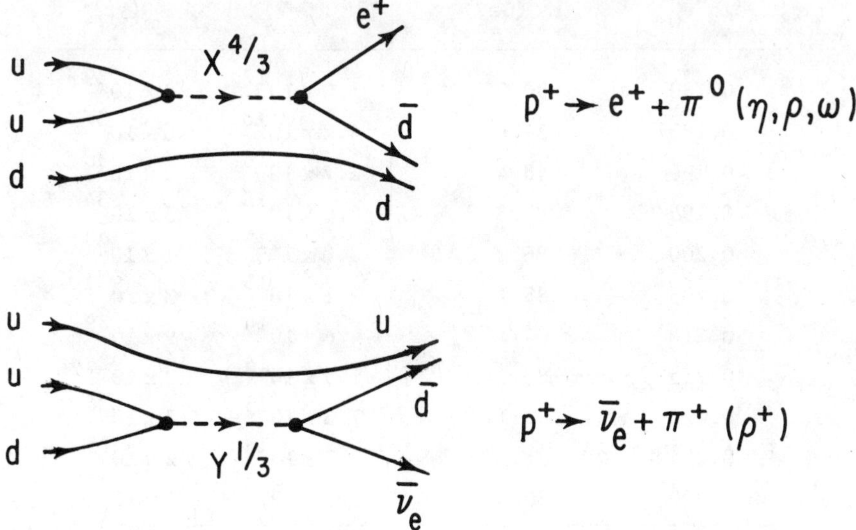

Fig. 6. Some of the proton decay channels in the SU(5) model.

Combining all the possible modes, Jarlskog and Yndurain[20] have estimated

$$\tau_p \simeq 2 \times 10^{-29} (M_S \text{ in GeV.})^4 \text{ yrs.} \qquad (47)$$

for the proton's lifetime. Although there is considerable uncertainty in (47), (an overall factor of 10 or more), I will use it in the following analysis. Combining the estimate in (47) with (46) and (29), one finds

$$M_S \simeq \frac{38.53}{\sin\theta_W} \exp\left[\frac{1-\frac{8}{3}\sin^2\theta_W}{.015}\right] \mathrm{GeV}. \qquad (48)$$

$$\tau_p \simeq 1\times 10^{33}\left(\frac{.200}{\sin^2\theta_W}\right)^2 \exp[-711(\sin^2\theta_W - .200)]\ \mathrm{yrs}. \qquad (49)$$

To illustrate these formulas, I have given in table I, values of M_W, M_S and τ_p corresponding to $\sin^2\theta_W$ in the range 0.180 to 0.230. Notice the sensitive dependence of τ_p on $\sin^2\theta_W$.

$\sin^2\theta_W$	M_W (GeV)	M_S (GeV)	τ_p (yr)
0.180	90.8	1.0×10^{17}	2×10^{39}
0.185	89.6	4.2×10^{16}	6×10^{37}
0.190	88.4	1.7×10^{16}	2×10^{36}
0.195	87.3	6.9×10^{15}	5×10^{34}
0.200	86.2	2.8×10^{15}	1×10^{33}
0.205	85.1	1.1×10^{15}	3×10^{31}
0.210	84.1	4.6×10^{14}	9×10^{29}
0.215	83.1	1.9×10^{14}	3×10^{28}
0.220	82.1	7.6×10^{13}	7×10^{26}
0.225	81.2	3.1×10^{13}	2×10^{25}
0.230	80.3	1.3×10^{13}	6×10^{23}

TABLE I. Values of M_W (W-boson mass), M_S (super-heavy mass), and τ_p (proton lifetime) for a given $\sin^2\theta_W$ in the SU(5) model.

If $\sin^2\theta_W$ were very precisely determined, we could read off the prediction for τ_p from this table. Given the present world average in (3), it seems reasonable to assume $\sin^2\theta_W > .20$ (unless drastically modified by radiative corrections). In that case, (49) implies 10^{29} yrs. $< \tau_p < 10^{33}$ yrs. in the SU(5) model. If $\sin^2\theta_W \simeq .215$, then τ_p should be close to the present experimental bound.

It would be very nice to know $\sin^2\theta_W$ more precisely, so that a more definite conclusion regarding the validity of the SU(5) model and its prediction for τ_p could be reached. In that respect, the calculation of the radiative corrections to deep-inelastic neutrino scattering will be helpful and more experimental data would be well appreciated. (Of course, the existence of additional Higgs multiplets[23], $N_H>1$, or the extension to unified theories with more than one mass scale[30] can change the $\sin^2\theta_W$ prediction.)

V. DISCUSSION

Combining my analysis of $\sin^2\theta_W$ in the SU(5) model (see eq. (46)) with the Goldman-Ross estimate[22] $M_S\simeq 3\times 10^{14}$ GeV., I estimate that the SU(5) model predicts

$$\sin^2\theta_W\simeq .21-.22 \tag{50}$$

$$\tau_p\simeq 10^{29}-10^{30} \text{ yrs.} \tag{51}$$

These values are consistent with the experimental values in (3) and (6). (Although the τ_p estimate in (51) should perhaps be extended by another factor of 10 due to uncertainty.) They suggest that the coming series of proton decay experiments may quickly observe such a decay. That would be an incredible discovery.

During the coming months we can expect the following developments to occur. The radiative corrections to ρ and κ should be completed, and can then be used to accurately determine $\sin^2\theta_W$ experimentally. The estimate of M_S obtained by comparing α and α_s will be closely scrutinized. Calculations of τ_p as a function of M_S will be carried out using several different approaches; thus allowing us to judge the uncertainty in those calculations. Most importantly, the proposed proton decay experiments will begin their search. They will provide the real test of these ideas.

ACKNOWLEDGEMENT

I would like to thank A. Sirlin for his collaboration in the work of section II and for many enlightening discussions on the subject of radiative corrections.

REFERENCES

1. For a review of QCD, see W. Marciano and H. Pagels, Phys. Rep. 36C, 137 (1978).
2. S. Weinberg, Phys. Rev. Lett. 19, 1264 (1967); A. Salam in Elementary Particle Physics: Relativistic Groups and Analyticity (Nobel Symposium No. 8), edited by N. Svartholm (Almqvist and Wiksells, Stockholm, 1968), p.367.
3. P. Langacker, J. Kim, M. Levine, H. Williams and D. Sidhu, Univ. of Penn. preprint COO-3071-243 (1979).
4. C. Baltay, proceedings of this conference.
5. W. Marciano and A. Sirlin, to be published.
6. H. Georgi and S. Glashow, Phys. Rev. Lett. 32, 438 (1974).
7. Other grand unification schemes are discussed in H. Fritzsch and P. Minkowski, Ann. Phys. (N.Y.) 93, 193 (1975).
8. For a review of grand unified gauge theories and the question of proton stability, see M. Gell-Mann, P. Ramond, and R. Slansky, Rev. Mod. Phys. 50, 721 (1978).
9. J.D. Bjorken, in Weak Interactions at High Energy and the Production of New Particles, proceedings of the SLAC Summer Institute on Particle Physics, 1976, edited by M. Zipf (SLAC, Stanford, 1977), p.1.
10. H. Georgi, H. Quinn, and S. Weinberg, Phys. Rev. Lett. 33, 451 (1974).
11. M. Chanowitz, J. Ellis, and M. Gaillard, Nucl. Phys. B128, 506 (1977).
12. A. Buras, J. Ellis, M. Gaillard, and D. Nanopoulos, Nucl. Phys. B135, 66 (1978).
13. D. Nanopoulos, Protons are not forever, Harvard preprint HUTP-78/A062 (1978); J. Ellis, SU(5), CERN preprint (1979).
14. The discussion of baryon asymmetry within the framework of GUTS was initiated by M. Yoshimura, Phys. Rev. Lett. 41, 281 (1978); 42, 746(E) 1979. For an up to date list of publications on this subject, see ref. 15.
15. General discussions of baryon nonconserving processes and implications regarding GUTS that can be deduced from the observation of proton decay have been given by S. Weinberg, Phys. Rev. Lett. 43, 1566 (1979); F. Wilczek and A. Zee, Phys. Rev. Lett. 43, 1571 (1979). See also, F. Wilczek, proceedings of this conference.
16. F. Reines and M. Crouch, Phys. Rev. Lett. 32, 493 (1974); J. Learned, F. Reines and A. Soni, Phys. Rev. Lett. 43, 907 (1979).
17. Proposals to search for proton decay up to $\tau_p \simeq 10^{33}$ yrs. have been put forward by two groups, M. Goldhaber et al., Irvine-Michigan-Brookhaven collaboration; J. Blandino et al., Harvard-Minnesota-Purdue - Wisconsin collaboration. For a review, see L. Sulak, Proceedings of the International Conference on Neutrino Physics, Bergen 1979.
18. T. Appelquist and J. Carazzone, Phys. Rev. D11, 2856 (1975).
19. D. Gross and F. Wilczek, Phys. Rev. D8, 3633 (1973).

20. C. Jarlskog and F. Yndurain, Nucl. Phys. B149, 29 (1979). I use the estimate of these authors throughout this talk. Several groups are refining this calculation and preliminary results (private communication from J. Donohue) indicate that τ_p may be an order of magnitude larger than the estimate in (10). It is also possible that due to fermion mixing angle effects the decay modes of the proton are further suppressed.
21. D. Ross, Nucl. Phys. B140, 1 (1978).
22. T. Goldman and D. Ross, Phys. Lett. 84B, 208 (1979).
23. W. Marciano, Phys. Rev. D20, 274 (1979).
24. W. Marciano, Phys. Rev. D12, 3861 (1975).
25. T. Kinoshita and A. Sirlin, Phys. Rev. 113, 1652 (1959).
26. W. Marciano, Nucl. Phys. B84, 132 (1975). This paper contains errors in eqs. (19) and (21). In the expression for A, -54R+38 should be -6R-10 and in C, 227/2 should be 418/3. Also, those results must be modified to include additional fermions.
27. D. Albert, W. Marciano, D. Wyler and Z. Parsa, Rockefeller University preprint COO-2232B-190 (1979).
28. Such possibilities were pointed out to me by A. Carter (unpublished).
29. K. Mahanthappa and M. Sher, Univ. of Colorado preprint (1979).
30. The situation in which more than one super-heavy mass scale is present has been analyzed by S. Dawson and H. Georgi, Phys. Rev. Lett. 43, 821 (1979).

QUESTIONS

J. Moffat (University of Toronto). You mentioned that almost all grand unified gauge models predict proton decay. Actually, it is easy to construct gauge models in which the proton doesn't decay and these models appear to be in agreement with the data (e.g. SU(n) with an extra U(1) symmetry and others). Of course if the proton is seen to decay, then this would clearly be important. If it doesn't decay up to $\sim 10^{35}$ yrs., then not much, if anything, is lost. Baryon number conservation is one of the most fundamental laws of nature (stability of matter), and one could argue with conviction that it is most natural to propose theories in which the proton is absolutely stable. At present there are no compelling theoretical or experimental reasons to choose gauge models with proton decay.

W. Marciano. Although grand unified models with exact baryon number conservation can be constructed (C.f. ref. 8), they lack elegance and often naturalness (in the technical sense). Also, if one imposes exact baryon number conservation, then the nice explanation for the baryon number asymmetry of the universe that models such as SU(5) offer is lost. Indeed, if one accepts big bang cosmology, then some violation of baryon number is necessary to explain the observed predominance of matter over antimatter in the universe. Unfortunately the predicted lifetime of the proton, τ_p, is very sensitive to small changes in the model; so even if proton decay up to $\tau_p \simeq 10^{33}$ yrs. is not seen, models such as SU(5) can be easily modified to accommodate such a finding. All we can say is that in the simplest most natural grand unified schemes our best estimates indicate that τ_p is near the present bound of $10^{29} - 10^{30}$ yrs. and therefore may be seen by the proposed experiments. There are no guarantees.

A. Bodek (University of Rochester). You have emphasized the importance of radiative corrections in the extraction of $\sin^2\theta_W$ from deep inelastic ν scattering data. You have only included virtual photon processes. However. the emission of real photons introduces a real difference between the charged and neutral current case due to the charge of the final state muon. This may be as important as the virtual photon corrections because experimentally, the photon is included in the hadron energy cuts for R_ν and $R_{\bar{\nu}}$ (e.g. CDHS and CITF $E_H > 12$ GeV cuts).

W. Marciano. Although my illustrations only show virtual photons, we do include real photon emission in our calculation. For totally inclusive processes (which we are assuming), virtual and real photon effects cancel to a large extent and the remaining radiative corrections are controllable by general theorems. Unfortunately, experimental cuts may render the experimental cross-sections not totally inclusive. We will have to face up to this problem when we compare our calculations with the actual experiments (it shouldn't be a severe obstacle).

HYPERWEAK INTERACTIONS*

Frank Wilczek
Joseph Henry Laboratories
Princeton University
Princeton, New Jersey 08544

ABSTRACT

The form of the effective Hamiltonian for $\Delta B \neq 0$ processes in superunified theories and its phenomenological consequences are reviewed.

*Talk given at the APS "Particles and Fields-1979" Meeting at McGill University

ISSN:0094-243X/80/590397-10$1.50

This is a review talk about a subject which does not yet exist. Not a single $\Delta B \neq 0$ decay has been observed. If such decays do occur they are surely mediated by interactions far weaker than the usual weak interactions, which we will therefore refer to as hyperweak interactions. I shall carry through the exercise of supposing that after heroic experimental efforts many hyperweak decays are collected. What will we learn from such data?

Of course this exercise is motivated by recent theories incorporating the very successful SU(3) gauge theory of strong interactions and SU(2) x U(1) gauge theory of electroweak interactions into a single gauge theory based on a larger group. These theories, of which the prototype is the SU(5) model of Georgi and Glashow,[1] inevitably involve baryon number violating processes. Quantitative treatments, following the lead of Georgi, Quinn, and Weinberg,[2] have made it plausible that if this whole framework of ideas makes sense then hyperweak interactions have barely escaped detection in previous searches[3] and should be seen in profusion in a new generation of more sensitive experiments now being prepared.

One thing we obviously learn from observations of hyperweak interactions is the lifetime of the nucleus or nucleon under study. However, as Marciano[4] has reviewed for us, it is difficult accurately to relate the raw lifetimes to the microscopic theory. Not only is the calculated decay rate exponentially related to imperfectly measured low-energy parameters, but also it is sensitive to any intermediate structures between 10^3 and 10^{15} GeV — and most of all it involves hadronic matrix elements which can only be estimated crudely.

For these reasons (and also to have something to say different from Marciano) I shall consider some quantities which may not be so immediately accessible to experiment—within reason, I hope— but which probe the underlying microscopic theory in a more meaningful way.

1. OPERATOR ANALYSIS— CONSEQUENCES OF SU(3) x SU(2) x U(1)[4]

Hyperweak interactions are expected to be mediated by very heavy particles. There is a standard method for analyzing the effects of heavy particles at low energies. These effects can be summarized in an effective Hamiltonian involving only light particles. The most important operators in this Hamiltonian will be those of minimal dimension for the appropriate quantum numbers, which respect the full symmetry of the light particle theory with the light particles regarded as massless.

The application of these methods to nucleon decay is straightforward and leads to significant results. Operators changing baryon number necessarily have dimension ≥ 6 (at least three quark fields, and hence four fermion fields to make a Lorentz scalar). Thus we are lead to the mathematical problem of classifying operators of dimension 6, formed from quark and lepton fields, which change baryon number and are singlets under SU(3) x SU(2) x U(1). The following is an exhaustive list (up to Hermitean conjugates):

<u>Funniness = 0</u>

$$O_1 = (\overline{Cu})_L \gamma_\mu u_L (\overline{Cd})_L \gamma_\mu e_L - (\overline{Cu})_L \gamma_\mu d_L (\overline{Cd})_L \gamma_\mu \nu_L$$

$$O_2 = (\overline{Cd})_R \gamma_\mu u_R (\overline{Cu})_R \gamma_\mu e_R$$

$$O_3 = (\overline{Cu})_R d_L (\overline{Cu})_R e_L - (\overline{Cu})_R d_L (\overline{Cd})_R \nu_L$$

$$O_4 = (\overline{Cu})_L d_R (\overline{Cu})_L e_R$$

<u>Funniness = 1</u>

$$O_1' = (\overline{Cu})_L \gamma_\mu u_L (\overline{Cs})_L \gamma_\mu e_L - (\overline{Cu})_L \gamma_\mu d_L (\overline{Cs})_L \gamma_\mu \nu_L$$

$$O_1'' = (\overline{Cu})_L \gamma_\mu c_L (\overline{Cd})_L \gamma_\mu e_L - (\overline{Cu})_L \gamma_\mu s_L (\overline{Cd})_L \gamma_\mu \nu_L$$

$O'_2 = (\overline{Cd})_R \gamma_\mu u_R (\overline{Cc})_R \gamma_\mu e_R - (\overline{Cu})_R \gamma_\mu u_R (\overline{Cs})_R \gamma_\mu e_R$

$O'_3 = (\overline{Cu})_R d_L (\overline{Cc})_R e_L - (\overline{Cu})_R d_L (\overline{Cs})_R \nu_L$

$O''_3 = (\overline{Cc})_R d_L (\overline{Cu})_R e_L + (\overline{Cu})_R c_L (\overline{Cd})_R e_L$

$+ (\overline{Cd})_R s_L (\overline{Cu})_R \nu_L + (\overline{Cs})_R u_L (\overline{Cd})_R \nu_L$

$O'_4 = (\overline{Cs})_L u_R (\overline{Cu})_R e_R$

Conventions: C = change conjugation, color indices suppressed. The electr and its neutrino can be replaced by the muon and its neutrino. "Funniness" is the number of different families (SU(2) multiplets) of quarks involved the operator minus one. Properly the charge -1/3 quarks should be Cabibbo rotated, and there are also funniness = 2 operators, but I will not belabor these fine points.

From the general classification some surprisingly strong conclusions emerge:

a) $\Delta(B-L) = 0$. Baryon minus lepton number should be conserved — e.g. $p \to e^+$ but $p \not\to e^- \pi^+ \pi^+$.

b) $\Delta S \geq 0$. For instance, $n \not\to \mu^+ K^-$.

c) Isospin and helicity structure. The operators with $\Delta S = 0$ all change strong isospin by ½ unit. Moreover the structure of the Hamiltonian is very restricted. There are two types of operators: O_1 and O_2, which represent terms that could arise from exchange of a single vector boson, and O_3 and O_4, which represent terms which could arise from exchange of a single scalar boson. If either pair is dominant — e. if nucleon decay is mediated primarily by exchange of vector gauge field then the effective Hamiltonian takes the form

$$\mathcal{H} \propto M^+ e_L - M^- \nu_{eL} + rM^{+'} e_R + K(M^+ \mu_L - M^- \nu_{\mu L} + r'M^{+'} \mu_R) \quad (1)$$

where (M^+, M^-) forms a strong isodoublet, $M^{+'}$ is the parity transform of M^+, and r is a parameter coming from the microscopic theory.

In the S = 1 sector the operator associated with decays into charged leptons have strong isospin I = 1, $I_3 = -1$. If the interaction is dominated by either vector or scalar exchange (but not both) then the hadronic operators associated with charged leptons of opposite helicities are parity conjugates.

2. TESTS OF ISOSPIN AND HELICITY STRUCTURE[5]

We have seen that under very general hypotheses $\Delta I = \frac{1}{2}$ for $\Delta S = 0$ for B-violating processes. This can be tested using some relations it implies among total decay rates:

$$\Gamma(p \to \pi^o e^+) = \tfrac{1}{2}\,\Gamma(n \to \pi^- e^+)$$

$$\Gamma(p \to e^+\pi^+ X_{s=o}) = \Gamma(p \to e^+ {}^- X_{s=o})$$

$$\Gamma(0^{16} \to e^+\pi^+ X_{s=o}) + \Gamma(0^{16} \to e^+\pi^- X_{s=o}) = 2\ (0^{16} \to e^+\pi^o X_{s=o})$$

The last of these relations may be especislly useful in proposed experiments using H_2O.

If the helicity of the final lepton is measured there are additional relations such as

$$\Gamma(0^{16} \to \mu_R^+ + X_{s=o}) = \Gamma(0^{16} \to \bar{\nu}_\mu + X_{s=o})$$

In practice neither the neutrino type nor the electron polarization is accessible and the testable result is the inequality

$$\Gamma(0^{16} \to \mu_R^+ + X_{s=o}) + \Gamma(0^{16} \to e^+ + X_{s=o}) \geq \Gamma(0^{16} \to \nu + X_{s=o})$$

— even this is assuming that $\bar{\nu}_\tau$ plays no part.

In the pure scalar or pure vector case the form of the Hamiltonian simplifies as discussed above and some more specific predictions are possible:

i) Universal polarization: in any exclusive or inclusive $\Delta S=0$ channel the ratio of μ_R^+ to μ_L^+ emission is the same.

ii) Isospin: from the form (1) of $\mathcal{H}$ many isospin predictions follow, e.g.

$$\Gamma(p \rightarrow e^+ X_{s=o}) = (1+r^2)\ \Gamma(n \rightarrow \bar{\nu}_e X_{s=o})$$

$$\Gamma(p \rightarrow \mu^+ X_{s=o}) = (1+r'^2)\ \ (n \rightarrow \bar{\nu}_\mu X_{s=o})$$

$$\Gamma(O^{16} \rightarrow e^+ X_{s=o}) = (1+r^2)\ \Gamma(O^{16} \rightarrow \bar{\nu}_e X_{s=o})$$

$$\Gamma(O^{16} \rightarrow \mu^+ X_{s=o}) = (1+r'^2)\ \Gamma(O^{16} \rightarrow \bar{\nu}_\mu X_{s=o})$$

and many others, all described by the parameters r, r'. To test these one must again hope $\bar{\nu}_\tau$ does not appear (see KH, below) and add e^+ and μ^+ events. Since r' is separately measurable from the muon polarization, the resulting confusion of r and r' can be disentangled. The practical result is then

$$\frac{1}{1+r^2} = \frac{\Gamma(n \rightarrow \bar{\nu} X) - 1/1+r'^2\ \Gamma(n \rightarrow \mu^+ X)}{\Gamma(p \rightarrow e^+ X)}$$

$$= \frac{\Gamma(O^{16} \rightarrow \bar{\nu} X) - 1/1+r'^2\ \Gamma(O^{16} \rightarrow \mu^+ X)}{\Gamma(p \rightarrow O^{16}\ e^+ X)}$$

and similarly for the other relations discussed in Ref. 4,5.

3. KINSHIP AND RENORMALIZATION

We have seen that, at least if vector exchange is dominant, the effective Hamiltonian for hyperweak decays is of a very restricted form. Different possibilities for the underlying microscopic theory can be distinguished crudely y the overall rate and more precisely by the values of the parameters r,r' s discussed above.

In formulating the predictions for r,r' there is the possible omplication of new Cabibbo-like angles occurring.[6] In the absence of data t seems reasonable to make the simplest consistent assumption about these ngles, which A. Zee and I have called the kinship hypothesis (KH). This ypothesis says that in hyperweak interactions the charge 2/3 quarks are otated just as in the weak interactions relative to the charge -1/3 quarks, hile the charge -1 leptons are not rotated at all relative to the charge -1/3 uarks. This structure necessarily follows if the fermions acquire mass from he simplest Higgs representation ($\underline{5}$) in SU(5) — an assumption which is ualitatively successful in relating the τ and b-quark masses[7] but also gives he poor relation $m_e/m_\mu = m_d/m_s$. For this reason I'm afraid KH is most probably rong, but sufficiently interesting to be worth discussing.

Assuming KH then to a first approximation (neglecting $\sin\theta_c$ effects) = 2 and r' = 1 in SU(5). Two other striking qualitative consequences are hat nucleon decays satisfy $\Delta(S+L_\mu) = 0$ (L_μ = muon number) to a good approximaion and that while in the $\Delta S = 1$ decays the emitted μ^+ will be nearly unpolarized n the rare $\Delta S = 0$ decays with an emitted μ^+ the μ^+ will be left-handed.

These predictions are subject to renormalization effects which however ave been computed to be small (but see §5 below).

. POSSIBLE T-VIOLATION[8]

In ordinary weak interactions T-violation is extraordinarily well hidden. n the theory of Kobayashi and Maskawa[9] this is understood in terms of the mallness of weak mixing among the three families.

Hyperweak interactions suggest the possibility of much larger effects, since in these interactions T-violation could occur even without interfamily mixing.

Recall our general classification led to an effective Hamiltonian of the form

$$H = \alpha(\overline{Cu})_L \gamma_\mu u_L \; (\overline{Cd})_L \gamma_\mu \mu_L + \beta(\overline{Cu})_R \gamma_\mu u_R \; (\overline{Cd})_R \gamma_\mu \mu_R$$

$$+ \gamma(\overline{Cu})_R d_L \; (\overline{Cu})_R \mu_L + \delta(\overline{Cu})_L d_R \; (\overline{Cu})_L \mu_R \qquad (2)$$

$$\equiv \left(\frac{\alpha+\beta}{2}\right) O_v^+ + \left(\frac{\alpha-\beta}{2}\right) O_v^- + \left(\frac{\gamma+\delta}{2}\right) O_s^+ + \left(\frac{\gamma-\delta}{2}\right) O_s^-$$

for $\Delta S = 0$ decays into μ^+. The superscripts +,- indicate operators even or odd under parity respectively and the subscripts v,s indicate operators which arise in tree graphs from vector or scalar boson exchange respectively. Now if the coefficients $\frac{\alpha+\beta}{2}$, $\frac{\alpha-\beta}{2}$, $\frac{\gamma+\delta}{2}$, $\frac{\gamma-\delta}{2}$ are relatively complex they cannot be made real by redefining the fields without spoiling the reality of masses, since the fields in these operators differ only in handedness. The presence of phases signals T-violation.

Experimentally, the interesting signature for T-violation is a dependence of the decay rate for $p \to \pi^+ \pi^- \mu^+$ on the variable

$$q = (E_{\pi+} - E_{\pi-}) \;\; k\cdot(\vec{p} \times \vec{s})$$

where $\vec{k}$ is the relative momentum of the pions and $\vec{p}, \vec{s}$ are the μ momentum and spin. As discussed in Ref. 8, the interpretation of such an asymmetry would be relatively free of final-state interaction problems.

There is one important point to notice. If in the effective Hamiltonian (2) either the vector-type or scalar-type interactions are dominant then T-violation from the interference of say O_v^+ and O_v^- is inevitably accompanied by P-violation. Since q is even under P no dependence of the decay on q would be possible in this case. We have not succeeded in finding physically reasonably probes for P odd, T odd interactions so one will probably have to rely on interference of vector and scalar type interactions for an observable T-violating effect.

The T-violation in nucleon decay is especially interesting because it ties up with the proposed explanation of the cosmological asymmetry between matter and antimatter in terms of T-violating hyperweak interactions in the early universe.[10]

5. OTHER SCENARIOS

A. Zee and I[11] constructed a special model in which the dominant $\Delta B \neq 0$ interactions are mediated by Higgs bosons and inevitably associated with SU(2) x U(1) breaking and thus rules like $\Delta(B-L) = 0$ would be violated in nucleon decay. (In fact in this model $\Delta(B+L) = 0$!) This effect is proportional to the "small" SU(2) x U(1) breaking parameter $\langle\phi\rangle \approx 300$ GeV and thus suppressed by additional powers of ν/μ where μ is the mass of the exhanged boson. It could only be experimentally interesting if μ is intermediate between the weak (10^3 GeV) and unification (10^{15} GeV) scales, say $\mu \approx 10^9$ GeV. This model is too arbitrary to be regarded very seriously, but it does illustrate that no argument in physics is ever "completely general".

Georgi and collaborators have analyzed in some detail implications of gauge theories extending SU(5) for the prediction of θ_w and nucleon lifetimes. Such theories deserve much attention, especially in connection with tidying up the unsatisfactory 3 x $\overline{10}$ + 3 x 5 representation of fermions in SU(5). Their main conclusion is that the prediction for the nucleon lifetime is remarkably insensitive to the details of the theory if the experimental value of θ_w is used as input. I think it is important to make these calculations more precise, in particular by including threshold effects and renormalizations in the effective Hamiltonian more accurately. It is interesting that the renormalization effects described above, in principle accessible to experiment, do give a measure of the effective coupling at the unification scale and are sensitive to departures from simple SU(5).

References

1. H. Georgi, S. Glashow, Phys. Rev. Lett. 32, 438 (1974).
2. H. Georgi, H. Quinn, S. Weinberg, Phys. Rev. Lett. 33, 451 (1974).
3. F. Reines, M. Crouch, Phys. Rev. Lett. 32, 493 (1974).
4. S. Weinberg, Phys. Rev. Lett. 43, 1566 (1979);
   F. Wilczek, A. Zee, Phys. Rev. Lett. 43, 1571 (1979).
5. This section incorporates results from Ref. 4 and from a paper by A. Hurlbert, F. Wilczek (in preparation).
6. C. Jarlskog, Phys. Lett. 82B, 401 (1979).
7. A. Buras, J. Ellis, M. Gaillard, D. Nanopoulos, Nucl. Phys. B135, 66 (1978
8. A. Hurlbert, F. Wilczek, "Possibility and Consequences of T-Violation in Nucleon Decay" (Princeton preprint).
9. S. Kobayashi, K. Maskawa, Prog. Theor. Phys. 49, 652 (1973).
10. M. Yoshimura, Phys. Rev. Lett. 41, 281 (1978);
    S. Dimopoulos, L. Susskind, Phys. Rev. D18, 4500 (1978);
    D. Toussaint, S. Treiman, F. Wilczek, A. Zee, Phys. Rev. D19, 1036 (1979);
    S. Weinberg, Phys. Rev. Lett. 42, 850 (1979);
    J. Ellis, M. Gaillard, D. Nanopoulos, Phys. Lett. 80B, 360 (1979).

TECHNICOLOUR

S. Dimopoulos, L. Susskind and S. Raby
Institute of Theoretical Physics
Stanford University, Stanford CA 94305

I. INTRODUCTION

a) Brief Motivation

The Extended Technicolour[1,2] (ETC) models are the only known candidates for theories with the following desirable features:

(1) They do not have elementary scalar fields.

(2) They can generate masses for quarks, leptons, and vector bosons.

(3) They offer the possibility of introducing a small CP-violation without the appearance of the notorious Strong CP problem[3].

We should remind the reader that scalars are unwanted because: (a) They carry several couplings (Yukawa couplings, quartic couplings, and vacuum expectation values) about which gauge theories have nothing intelligent to say. Thus they spoil the predictive power of gauge theories. Even Grand Unified 3 family models need $\sim$ 20 undetermined parameters[4,5]. Often these couplings have unnaturally small values ($g_{Yukawa} \sim 10^{-6}$, e.g.). These facts suggest that elementary scalars are nothing but a good way to parametrize low energy ($E \ll 300$ GeV) phenomenology. (b) They cannot naturally account for the huge difference between the electroweak scale ($\sim$ 300 GeV) and the Grand Unified Scale[6,7] $\gtrsim 10^{15}$ GeV[5,8]. There exist extremely strong acausal correlations between the small distance ($r \lesssim 10^{-28}$ cm) and large distance physics ($r \gtrsim 10^{-16}$ cm). Thus, for example, a 1% change of a small distance quantity can change the values of quantities of the large distance physics by 25 orders of magnitude![7]

ISSN:0094-243X/80/590407-46$1.50

b) Quick Review

Historically the ETC models came into existence[1] in an effort to fix up a big drawback of the old Technicoloured[9,7] (TC) models: the fact that TC models could not generate masses for quarks and leptons. They did generate masses only for the electroweak vector bosons and naturally accounted for the heavy Isospin symmetry relation $M_W = M_Z \cos \Theta_W$[7]. The old TC models consisted of (a) a new Strong interaction, called Technicolour, which becomes strong at a mass scale of $\sim$ 1 TeV; and (b) one doublet of massless "Techniquarks," denoted by U and D. The Techniquarks bind via the technicolour forces and form Technihadrons with masses of order of $\gtrsim$ 1 TeV. Some bosons, for symmetry reasons, escape from receiving any mass. These are the 3 Technipions. They are the Nambu-Goldstone bosons of the spontaneously broken chiral $SU_{2L} \times SU_{2R}$ associated with the two massless Techniquarks U and D. The massless Technipions have exactly the right quantum numbers to play the role of the usual Nambu-Goldstone bosons of the Higgs mechanism of the standard $SU_{2L} \times U_1$ electroweak model[10]. Thus they can be eaten and give masses to the $W^{\pm}$ and Z^o.

The most important new feature of the ETC models is that they allow transitions between ordinary fermions and Technifermions. These transitions are mediated by a new set of heavy (Mass $\sim$ 1-100 TeV) ETC vector bosons. It is these transitions that allow the quarks and leptons to acquire their masses.

c) Some Consequences

An extremely important feature of the ETC models[1,2] that is not shared by the old TC models is the apparent necessity for having more than one electroweak doublet of Technifermions, denoted by U_1, D_2^-; U_2, D_2; U_3, D_3, etc. This appears to be a necessary feature of all known quasirealistic models(11,12) and it is almost certain that it should be true for realistic models This feature has the amusing consequence that there exist several Technipion-like almost massless bosons with masses ranging from about ∿ 1 GeV to ∿ 300 GeV. They are called Pseudo-Nambu-Goldstone bosons or, for short, Pseudos. Many quasirealistic ETC models have either 60 or 116 or 132 such bosons. Thus it appears difficult to miss such an overpopulated Pseudo-world living less than an order of magnitude of energy away from us. In fact, the upcoming generation of accelerators, especially the FNAL TEVATRON ($p\bar{p}$ 1000 + 1000 GeV; 1984?), should find these particles.

Since more than 90% of these Pseudo-Nambu-Goldstone bosons carry ordinary SU_3(colour) (see §III) we expect that the $p\bar{p}$ and pp machines will be very useful for finding them and studying their properties.

II. SCALEONTOLOGY

In this section we discuss the masses of the new types of hadrons that exist in ETC models. There are at least two new strong gauge forces in ETC models(1,2). (A) The first one becomes strong at a mass scale of order of ∿ 10 TeV. Thus we expect to find several hadrons at this mass range. This sector of the theory is the least understood so we shall not discuss it any further. We should however emphasize that this sector may produce some light Pseudos or true Nambu-Goldstone particles with exotic properties. (B) The second strong gauge force is Technicolour[9,7]. It becomes strong at a scale of order of 1 TeV and gives mass to the $W^{\pm}$ and Z^{o}. The exact scale at which the Technicolour forces become strong is very important, for it is this scale that determines the masses of the Technihadrons and the numerous Pseudos mentioned before. The value of this scale μ_{TC} depends crucially on the number of Technifermion electroweak doublets. μ_{TC} decreases as the number of Technifermion doublets goes up. An important feature of the ETC models is that the number of Technifermion doublets almost inevitably has to go up. This comes about, in all quasirealistic models,[11,12] because the Technifermions have to give masses to several quarks and leptons.

For the sake of definiteness, from now on we shall assume that we have one family of Technifermions consisting of 16 chiral members[11,12,13]. This comes about naturally when we attempt to construct Grand Unified models which contain Technicolour[11,12,13]. Let us denote these Technifermions by U_c^L, D_c^L, E^L, N^L; U_c^R, D_c^R, E^R, N^R, where c = 1,2,3 is a colour index. Let us for the moment turn off all the orfinary SU_3 (colour) x SU_{2L} x U_1 and the forces mediated by the heavy ETC vector bosons[1,2] and keep only the Technicoloured forces. Let us furthermore assume that all the technifermions belong to the same complex representation of the Technicolour group. (The case of real representation of the TC group will be considered later.) Then the theory has a global SU_{8L} x SU_{8R} chiral symmetry. At the scale of $\mu_{TC} \sim$ TeV the Technicoloured forces become strong and form the following 8 condensates[1,2] (c=1,2,3; no sum on c):

$$\langle \bar{U}_c^R U_c^L \rangle = \langle \bar{D}_c^R D_c^L \rangle = \langle \bar{N}^R N^L \rangle = \langle \bar{E}^R E^L \rangle \qquad \text{(II.1)}$$

The point that we wish to emphasize here is that there exist four electroweak isodoublets of Technifermions which form condensates and therefore effectively four Technipions. Let $F_{T\pi}$ be the T-πion decay constant associated with each electroweak doublet. Then this corresponds to a Higgs-type model with four Higgs fields ϕ_1, ϕ_2, ϕ_3, ϕ_4 and with vaccuum expectation values $\langle\phi_1\rangle = \langle\phi_2\rangle = \langle\phi_3\rangle = \langle\phi_4\rangle = F_{T\pi}$[7]. Thus the W mass is given by $M_W^2 = 1/4\ g_2^2\ 4\ F_{T\pi}^2$ and therefore $2\ F_{T\pi} \simeq 250$ GeV or $F_{T\pi} \simeq 125$ GeV. In general, if we have N electroweak doublets of Technifermions then the corresponding $F_{T\pi}$ will be $F_{T\pi} \simeq N^{-1/2}\ 250$ GeV. This is a useful fact because it means that as the Technihadron world becomes more populated it becomes lighter and therefore more accessible to the future accelerators. The same holds true for Pseudos. The Pseudos, as we shall discuss, owe their existence as well as their relatively light masses to the existence of several electroweak doublets of Fermions which is necessary in ETC models.[13]

To estimate the masses of the Technihadrons, let us assume that they scale naively, i.e., in proportion to $F_{T\pi}/f_\pi$. This yields a mass for the Techniρho of order

$$M_{T\rho} \sim \frac{F_{T\pi}}{f_\pi} m_\rho \sim \frac{125 \text{ GeV}}{93 \text{ MeV}} m_\rho \sim 1039 \text{ GeV} \qquad \text{(II.2)}$$

These estimates are too naive since they assume that the ratio m_ρ/f_π is group independent. Note that the mass of the Techniρho would be twice as large if we had only one electroweak doublet of T-fermions. Having four electroweak doublets also increased the number of Techniρhos. Now instead of four Technivector mesons (Techniρho and Techniωmega) we shall have 64 of them all with (naive masses of order of $\sim$ 1040 GeV. They will have the same colour and electroweak quantum numbers as the Pseudos shown in Table Ia.

III. PSEUDOTYPES

a) The 16 Colour Pseudomultiplets That Occur When The Technifermions' Representation Is Complex

In this section we wish to discuss the Pseudo-Nambu-Goldstones in these theories. As in the previous section, we assume for definiteness that we have one family of Technifermions. As mentioned, this occurs naturally in GUT models which include Technicolour[11,12]. All the groundwork for discussing the Pseudos has been done in the part of the previous section preceding equation II.1.

Let us begin our discussion of Pseudos from equation II.1 where the form of the condensates was explicitly displayed. These condensates break the chiral $SU_{8L} \times SU_{8R}$ symmetry down to the $SU_{8(L+R)}$ symmetry. The broken 63 axial generators give rise to 63 massless Nambu-Goldstone pseudoscalars. They correspond to all but one (the U_{1A} piece) of the ground state pseudoscalars that can be formed by combining a T-fermion and an anti-T-fermion. They are all shown in Table Ia, except for charge conjugates, together with their colour, electric charge, and masses (to be estimated in §IV). Three of these are eaten by the $W^{\pm}$ and the Z^{o}. The remaining 60 receive masses when the $SU_3 \times SU_2 \times U_1$ forces and the forces mediated by the heavy ETC generators are turned on. They can be divided into the following types:

(1) Coloured Pseudos: All but four of the pseudos are coloured. There exist four octets, four triplets, and four antitriplets.

(1a) The four colour octets: They consist of an Isotriplet and an Isosinglet. They can be written

$$\Pi_f^a = \bar{Q}\gamma_5\lambda^a\tau_f Q \qquad \begin{pmatrix} a=1,2,\ldots 8 \\ f=0,1,2,3 \end{pmatrix} \qquad \text{III.1}$$

Here a=1,2,...8 labels a colour SU_3 Gell-Mann matrix, f=1,2,3 labels a Pauli matrix, and f=0 labels the identity matrix, $Q=\binom{U}{D}$. Π_f^a f=1,2,3 are just like coloured and heavy (mass $\lesssim$ 300 GeV) versions of the pion. We call them Coloured T-pions. Π_o^a is called Coloured T-eta. They are quite amusing for several reasons. For one thing, they, together with the gluons (and colour pseudosextets), would be the strongest carriers of ordinary colour. Thus they should be abundantly produced at very high energy $p\bar{p}$ collisions ($E_{cm} \gtrsim$ 10 TeV). In fact, each colour component proliferates as much as the $W_L^\pm$ and Z_L^o (7) (which proliferate because they too are like πions). Just like all other Pseudos they can be singly produced. More on these later.

(1b) The four colour triplets: They also consist of an Isotriplet and an Isosinglet. They can be written:

$$K_f^c = \bar{L}\gamma_5\tau_f Q^c \qquad \text{III.2}$$

Here $L = \binom{N}{E}$, c = 1,2,3 = colour index. We shall call them Technileptoquarks. They are expected to have masses of order of $\lesssim$ 200 GeV. They are expected to be quite narrow ($\Gamma \sim$ 1 MeV) in general. Details later.

(2) Colourless Pseudos: There are four colourless Pseudos again consisting of an isotriplet and an isosinglet. They can be written:

$$\Pi_f^{TL} = \bar{L}\gamma_5\tau_f L \qquad (f=0,1,2,3) \qquad \text{III.3}$$

Let us introduce also

$$\Pi_f^{TQ} = \bar{Q}^c\gamma_5\tau_f Q^c \qquad (f=0,1,2,3) \quad (c=1,2,3) \qquad \text{III.4}$$

Then it is clear that:

$\Pi_i^{TQ} + \Pi_i^{TL}$ (i=1,2,3) are eaten by $W^{\pm}$ and Z^o.

$\Pi_o^{TQ} + \Pi_o^{TL}$ is eaten by the Technicoloured Anomaly.

$\Pi_f^{TF} \equiv \Pi_f^{TQ} - \Pi_f^{TL}$ (f=0,1,2,3) are Pseudos. They consist of an isosinglet and an isotriplet. We shall refer to them as the Axion family.

$\Pi_1^{TF} \pm i\Pi_2^{TF} \equiv \Pi_{\pm}^{TF}$ is an electrically charged Pseudo and therefore it receives mass from the electroweak interactions.

The neutral Pseudos Π_o^{TF} and Π_3^{TF} work as follows. The linear combination of Π_o^{TF} and Π_3^{TF} which is the Nambu-Goldstone boson of the difference $Y_\mu^{TQ} - Y_\mu^{TL}$ of the techniquark and technilepton hypercharges does <u>not</u> receive any mass from the electroweak interactions since $Y_\mu^{TQ} - Y_\mu^{TL}$ commutes with $SU_{2L} \times U_{1Y}$. This Nambu-Goldstone boson is the Axion[14] of these dynamical models[3,15,2]. This axion is strongly analogous to the neutral Π^o of QCD which is also massless in the chiral limit.

The only place where the axion can receive mass is the ∿ 10 TeV heavy Extended Technicolour generators which are also responsible for quark and lepton mass generation[1,2]. This is an important constraint for ETC model building. It means that, for example, the quarks and the leptons should <u>not</u> get their masses from two <u>disjoint</u> sets of Technifermions (called T-quarks and T-leptons respectively). For in this case we could define an exactly conserved ETC singlet current, namely $Y_\mu^{TQ} + Y_\mu^{quarks} - Y_\mu^{TL} - Y_\mu^{leptons}$, which would be spontaneously broken and therefore would yield a massless Nambu-Goldstone boson. In general, this constraint means that we should <u>not</u> have several reducible ETC multiplets[15,2]. We will assume that these constraints are satisfied and that the axion receives a mass from the ∿ 10 TeV ETC generators[1,2]. Similar discussion applies to the Nambu-Goldstone boson orthogonal to the axion, which we call paraxion. The paraxion is essentially the Nambu-Goldstone boson coupling to the difference of the T-Quarks' and T-Leptons' U_1(Axial) currents [3,15] in the same way that the axion essentially corresponds to the difference of the I_{3R}'s of T-Quarks and T-Leptons.

b) There Exist 18 Additional Colour Pseudomultiplets When The Technifermions Belong To A Real Representation With An Antisymmetric Metric

When the Technicolour group is real and we turn off all other interactions (i.e., $SU_{3c} \times SU_{2L} \times U_1 \times \frac{ETC}{TC}$) then handedness has no dynamical meaning. All particles can be considered left-handed. Thus, instead of an $SU_{8L} \times SU_{8R}$ chiral symmetry we have an SU_{16L} symmetry. If the metric of the technifermions' representation is antisymmetric then the condensates of equation II.1 break this SU_{16L} down to Sp(16) leaving 119 massless Nambu-Goldstone bosons (in the absence of all other forces). Of these, three are the Technipions that will be eaten by $W^{\pm}$ and Z^{o}, 60 are those that also arise in models with complex Technicolour groups and have been discussed in section IIIa. The remaining 56 exist only because of the reality of the Technicolour group and are shown in Table Ib. We call them Pseudodifermions. From this table we see that there are three colour antitriplets, one sextet, four triplets, one singlet, and their antiparticles, i.e., 18 colour multiplets of Pseudos.

c) There Exist 22 Additional Colour Multiplets When The Technifermions Belong To A Real Representation With A Symmetric Metric

As we mentioned in §III.c, when the Technicolour group is real the global chiral symmetry associated with on Technifamily is not $SU_{8L} \times SU_{8R}$ but $SU_L(16)$. If the Technirepresentation is characterized by a symmetric invariant tensor, δ_{ab}, then the condensates of equation II.1 break the symmetry down to SO(16), leaving us with 135 Nambu-Goldstone bosons. Three of these are the Technipions that are eaten by $W^{\pm}$ and Z°. Sixty of them are identical to the ones that appear in models with complex technirepresentation and are shown in Table Ia. The remaining 72 Pseudos signal the fact that the Technirepresentation is real and has a symmetric metric. The 72 Pseudos form 22 colour multiplets and are shown in Table Ic (except for charge conjugates). The differences with the 18 additional Pseudotypes that occur in the case of an antisymmetric metric (see Table Ib) are the following. In Table Ic we have six (instead of two) colour sextet Pseudos. In particular the isotriplet of Pseudos, consisting of UU, UD, DD are now colour sextets instead of colour $\bar{3}$'s.

TABLE I

A. THE 16 MULTIPLETS OF PSEUDOS THAT OCCUR WHEN THE TECHNIFERMION REPRESENTATION IS COMPLEX

Pseudotype	Colour	Charge	~Mass(GeV)	Name
$\bar{U}_c U_c, \bar{D}_c D_c \cdots$	1	0,±1	0	Techniπions
$\bar{U}_c D_c', \cdots$	8	0,±1,0	300	Coloured T-πions and T-eta
$\bar{E}U$	3	5/3	240	T-leptoquark
$\bar{E}D$	3	2/3	205	"
$\bar{N}U$	3	2/3	205	"
$\bar{N}D$	3	-1/3	200	"
$\bar{N}E - \bar{U}D$	1	1	few-100	Charged axions
$\bar{E}E - \bar{N}N + \bar{U}U - \bar{D}D$	1	0	few-100	Axion
$\bar{N}N + \bar{E}E - \bar{U}U - \bar{D}D$	1	0	few-100	Paraxion

B. THE 18 ADDITIONAL COLOUR MULTIPLETS THAT OCCUR WHEN THE TECHNIFERMIONS BELONG TO A REAL REPRESENTATION WITH AN ANTISYMMETRIC INVARIANT SYMBOL

Pseudotype	Colour	Charge	~Mass(GeV)	Name
UU	$\bar{3}$	4/3	225	Ditechniquarks
[UD]	$\bar{3}$	1/3	200	"
DD	$\bar{3}$	-2/3	205	"
{UD}	6	1/3	320	"
UN	3	2/3	205	
UE	3	-1/3	200	
DN	3	-1/3	200	
DE	3	-4/3	225	
EN	1	-1	few-100	Ditechnilepton

TABLE I

C. THE 22 ADDITIONAL COLOUR MULTIPLETS THAT OCCUR WHEN THE TECHNIFERMIONS BELONG TO A REAL REPRESENTATION WITH A SYMMETRIC METRIC

Pseudotype	Colour	Charge	∼Mass(GeV)	Name
UU	6	4/3	340	
DD	6	-2/3	330	
{UD}	6	1/3	320	
[UD]	$\bar{3}$	1/3	200	
UN	3	2/3	205	
UE	3	-1/3	200	
DN	3	-1/3	200	
DE	3	-4/3	225	
EN	1	-1	few-100	Ditechnilepton Triplet
EE	1	-2	few-100	Ditechnilepton Triplet
NN	1	0	few-100	Ditechnilepton Triplet

On Tables I.A, I.B, and I.C we indicate the Pseudos that appear in the various types of models with one complete family of Technifermions. The masses shown are only rough approximate estimates. The reason for the great ambiguity in the masses of the colourless Pseudos is the fact that the contribution of the ETC generators is quite model dependent (see §IV). In models with more families of Technifermions we shall have more replicas of these Pseudos. Their masses will be smaller in proportion to the square root of the number of families.

Very important new additions in Table Ic are the four Colourless ditechnileptonic Pseudos: EE, NN, $\overline{E}\overline{E}$, $\overline{N}\overline{N}$. EE, EN, and NN form an electroweak triplet of colour singlet Pseudos. These Pseudos can in fact be quite light in a large class of models with possible masses between a few and 100 GeV . The most amusing one is the Doubly Charged EE. It contributes one unit to $R(e^+e^- \rightarrow \gamma \rightarrow$ anyth.). Another interesting feature of these possibly light, colourless, dileptopseudos is the fact that in a large class of models they are very narrow and can decay only via emission of four quarks and/or leptons(11). This comes about because they can easily be protected by lepton and baryon number type conservation laws against decays into two ordinary fermions(11) (see §VII.f).

IV. PSEUDOMASSES

In this section we discuss some approximate estimates for the masses of the various Pseudos. The Pseudos receive their masses because the symmetries to which they correspond are explicitly broken by the ordinary SU_3 (colour) interactions, electroweak interactions, and the heavy $\sim$ 10 TeV ETC generators. We begin with a discussion of the masses of the Coloured Pseudos.

The Coloured Pseudos receive their mass via the graphs of Figure 1. The computation of these graphs is isomorphic to that of the electromagnetic mass difference of the πions . All we need to do is scale up all masses by $F_{T\pi}/f_\pi$ and replace α by $\alpha_c(F_{T\pi})C_2(R)$, where $C_2(R)$ is the quadratic Casimir operator of the Pseudo's representation and $\alpha_c(F_{T\pi}) \sim 0.1$ the SU_3 (colour) coupling constant at $F_{T\pi}$. To do this note that $m^2_{\pi^+} - m^2_{\pi^\circ}$ does not change much in the chiral limit where

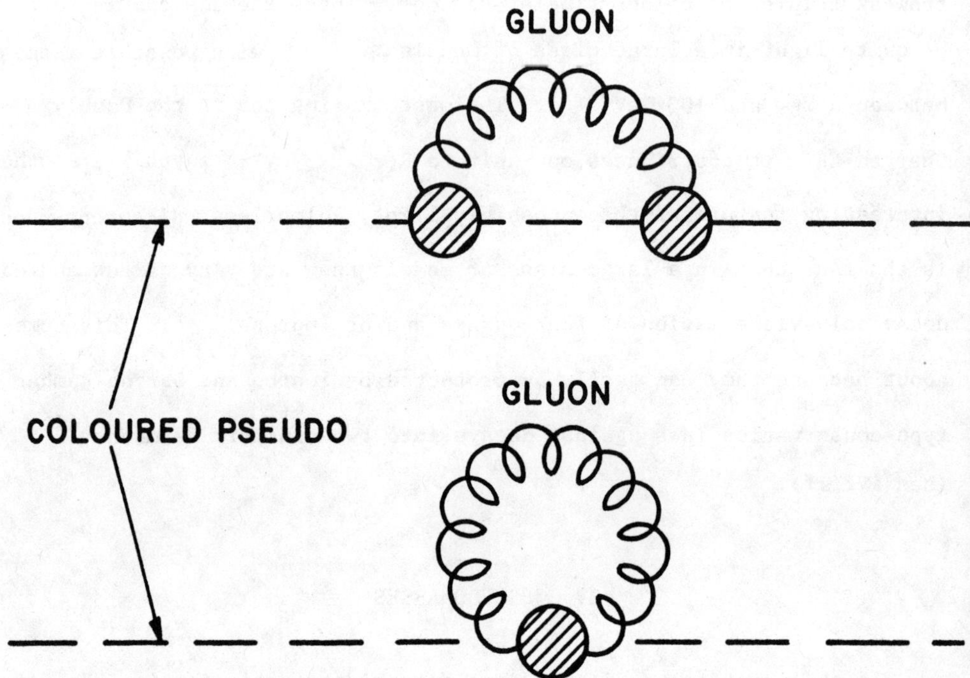

Fig. 1 Graphs that contribute mass to the Coloured Pseudos

$m_{quark} = 0$ and $m_{\pi^\circ} = 0$. Thus the mass of π in the chiral limit is purely electromagnetic and equal to $\simeq (m^2_{\pi^+} - m^2_{\pi^\circ})^{1/2} \simeq 35.5$ MeV. Thus the SU_3 (colour) contribution to the mass of a coloured Pseudo is of order

$$M_c \sim \left(\frac{C_2(R)\alpha_c(F_{T\pi})}{\alpha} \right)^{1/2} \frac{F_{T\pi}}{f_\pi} 35.5 \text{ MeV} \simeq 170\sqrt{C_2(R)} \text{ GeV} \quad \text{(IV.1)}$$

Here $f_\pi \simeq 95$ MeV.

Next we discuss the electroweak contribution to the mass of a Pseudo. The electromagnetic contribution to the mass of a Pseudo is

$$M_{EM} \sim e_{ps} \frac{F_{T\pi}}{f_\pi} 35.5 \text{ MeV} \simeq e_{ps} 47 \text{ GeV} \quad \text{(IV.2)}$$

Here e_{ps} is the electric charge of the Pseudo in unite of the πion's electric charge. The contribution of the complete Electroweak gauge group to the mass of a Pseudo (which is more relevant at these scales $F_{T\pi}$), barring possible cancellations[(2)], can be of order of 100 GeV f the Pseudo carries SU_{2L} quantum numbers. If a Pseudo carries both U_3 (colour) and Electroweak charges then its mass is given by $\sim \sqrt{m_c^2 + m_{EW}^2}$.

The third source from which Pseudos receive a contribution to their ass is the heavy $\sim$ 10 TeV ETC generators. This mass contribution called m_{ETC}) is quite model dependent and Pseudo dependent. It can e anywhere from 0 GeV to order $\sim$ 100 GeV, depending on the model and n the Pseudo. For example, in the Fahri-Susskind SU(7) model there s a massless neutral axion[(11)]. This is a general feature of models ith several decoupled representations[(2,15)].

If the group is enlarged to O(14), then the representations become coupled at ∿ 100 TeV and the axion receives a mass which is roughly of order of 1 GeV. In general, the ETC models have to have ETC vector bosons with masses as small as 1 TeV if there exists a quark or lepton with mass of 30 GeV (see §VII.a). If these relatively light ETC vector bosons explicitly break the symmetry associated with a given Pseudo then it is easily seen that this Pseudo receives a contribution of $M^2_{ETC} \sim (100\text{ GeV})^2$ to its mass squared . The reason for this relatively large contribution is that the scale of Technicoloured spontaneous symmetry breaking of order of $\langle\bar{E}E\rangle^{1/3} \sim 300$ GeV (see §VII.a) and the masses of the lightest ∿ 1 TeV ETC vector bosons which explicitly break the chiral symmetry corresponding to the Pseudo are not very far apart. Therefore the chiral symmetry is broken relatively strongly and the corresponding mass² contribution to the Pseudo is not small $\sim (100\text{ GeV})^2$. This contribution is not in general very important for the Pseudos that already have masses² contribution from SU_3 (colour) or the electroweak SU_{2L} interactions. However, this contribution is very important for the axionic objects and for the charged partners $\Pi^{TF}_{\pm}$ of the axion that were shown , by Eichten and Lane[2], to receive only a small mass from the electroweak interactions. The only way for axions or their charged partners to avoid getting large masses ∿ 100 GeV from the 1 TeV ETC generators is if they correspond to currents which are singlets of a subgroup of the ETC group which is connected to the remainder of the ETC group by generators that have masses of order of 3 TeV or heavier.

V. DECAYS OF PSEUDOS

In this section we wish to make some comments on the dominant decays of Pseudos. We have two types of dominant possibilities. The Pseudo can decay into $SU_3 \times SU_{2L} \times U_1$ gauge vector bosons or into ordinary fermions. The dominant decays of the Coloured Techniπions and techniηta are into pairs of gauge bosons. The dominant decays of the majority of Pseudos are into pairs of ordinary fermions. Occasionally global conservation laws forbid such a decay[11]. In this case a Pseudo will have to decay into four ordinary fermions and it will be quite narrow.

Let us begin our discussion with the Coloured T-πions and T-eta. The dominant decay modes of these pseudos are (see figures 2):

$$\Pi^{c\pm} \longrightarrow W^{\pm} + \text{gluon}$$

$$\Pi^{c3} \longrightarrow Z \ \underline{\text{or}}\ \gamma + \text{gluon}$$

$$\Pi^{co} \longrightarrow 2 \text{ gluons} \qquad (V.1)$$

To give a very rough estimate of the orders of magnitude of the widths of these particles based on scaling arguments we simply note that the relevant mass scales in the problem are $\langle\bar{E}E\rangle^{1/3} \sim 300$ GeV, Pseudo-masses ~ 300 GeV, and $F_{T\pi} \sim 125$. They are all of the same order. Thus we expect an overestimate of all these widths to be of order

$$\Gamma(\Pi\text{'s}) \sim \alpha_s^2 \, 300 \text{ GeV} \sim \frac{\alpha_s \alpha}{\sin^2 \theta_W} \, 300 \text{ GeV} \sim 3 \text{ GeV} \qquad (V.2)$$

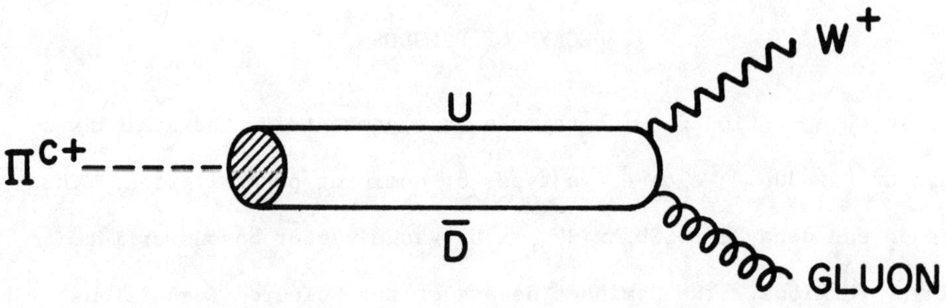

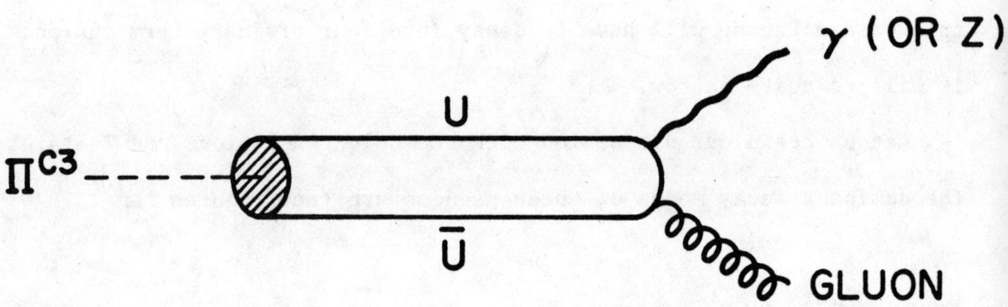

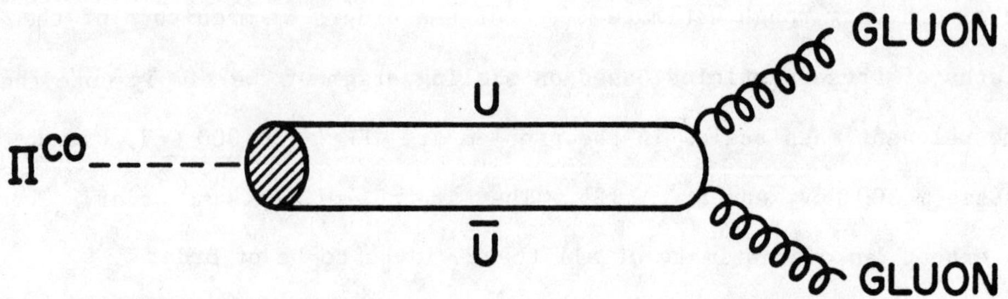

Fig. 2 Important decay modes of the Coloured Technipions ($\Pi^{c\pm}$, Π^{c3}) and the Coloured Technieta

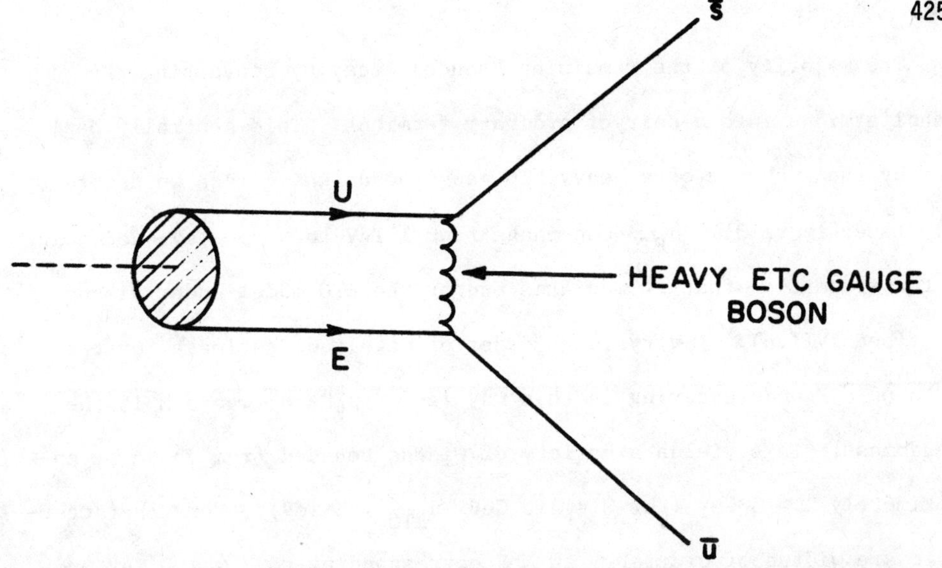

Fig. 3 A usual decay mode of a common Pseudo into two ordinary fermions

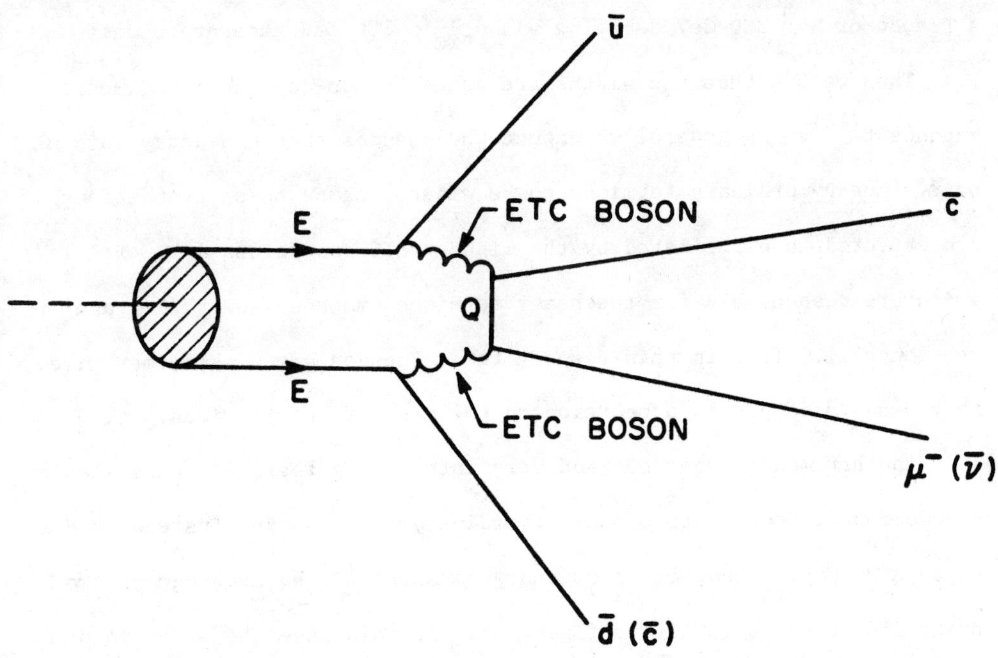

Fig. 4 A (dileptonic) Pseudo decays into four ordinary fermions. This is the dominant decay mode of Pseudos which are forbidden to decay into two fermions by some conservation laws.

The majority of the remaining Pseudos decay by converting the Technifermions into a pair of ordinary fermions. This generally happens by the exchange of a heavy ETC gauge boson whose mass we denote by μ_{ETC} (see figure 3). μ_{ETC} can range from 1 TeV to $\sim$ 250 TeV, depending on the transition that it mediates and/or the ETC model under discussion (see §VII.a). A very rough order of magnitude estimate for the width of a Pseudo decaying in this way is $\Gamma \sim \mu_{ETC}^{-4} M^5$ where M is the Pseudomass. This yields a variety of widths bounded from above by approximately $\Gamma \sim 3$ GeV (for $M \simeq 310$ GeV, $\mu_{ETC} \sim 1$ TeV). More characteristic are widths of order $\Gamma \sim 20$ MeV corresponding to $M \sim 200$ GeV and $\mu_{ETC} \sim 2$ TeV). The width can be as small as a few electronvolts for a Pseudo of $M \sim 200$ GeV decaying via $\mu_{ETC} \sim 100$ TeV generators only.

Thus we see that the widths are quite Pseudo-dependent and model-dependent[(16)]. In general we expect the Pseudos that can decay into a pair of _heavy_ ordinary fermions to be wider because these transitions are expected to be mediated by the lighter ETC generators ($\mu_{ETC} \sim 1$ TeV) which are responsible for the heavy fermions' masses (see §VII.a and reference 1). Thus in this respect these Pseudos are Higgsomimes, i.e., they tend to couple in proportion to the mass of the fermions.

Another model-dependent and very interesting feature occurs when a Pseudo cannot decay into a pair of ordinary fermions and instead it decays into, say, a quartet of ordinary fermions[(11)] by exchange of two heavy ETC vector bosons (see figure 4). In this case the width is of order of $\Gamma \sim \frac{M^9}{\mu_{ETC}^8}$. In the Fahri-Susskind model this occurs for the di-technilepton EN($\rightarrow$ cud $\bar{\nu}_\mu$). This Pseudo is protected against 2-body decays by an exact fermion number conservation. In general, such conservation law are quite natural and necessary in ETC models. They are nothing but the statements of baryon and lepton number conservations generalized to include T-coloured and ETC-coloured particles.

In the case of the doubly charged detechnielectron EE (see Table I.c) it is easy to find sufficient conditions under which it is forbidden to decay into a pair of ordinary fermions. The pair of ordinary fermions into which EE decays would have to consist of two negatively charged leptons because these are the only two fermion states with charge -2. Thus if the theory does not have ETC vertices (see figure 5) in which a technielectron E becomes a negatively charged lepton by the emission of an ETC gauge boson then EE cannot decay to any two-fermion state. It can decay into a four-fermion state via graphs like the one shown in figures 4. This is exactly what happens in theories where the left-handed Anti-technifermions form electroweak doublets[11,12]. In fact, in such theories none of the dileptonic Pseudos can decay into two ordinary fermions. To see this explicitly note that in such theories the only allowed ETC vertices that change a Technifermion into an ordinary fermion are:

$$\begin{aligned} &\bar{L} \rightarrow \ell + g \\ &Q \rightarrow \bar{q} + \bar{b} \\ &\bar{Q} \rightarrow \ell + \bar{b} \\ &\bar{L} \rightarrow q + \bar{b} \end{aligned} \qquad (V.3)$$

Here $\bar{b}$ and g denote ETC gauge bosons. The reason that no other such processes are allowed follows from the fact that the chirality and the SU_2 (electroweak) quantum numbers of the fermions are conserved in the processes V.3. The latter follows because the ETC gauge bosons do not

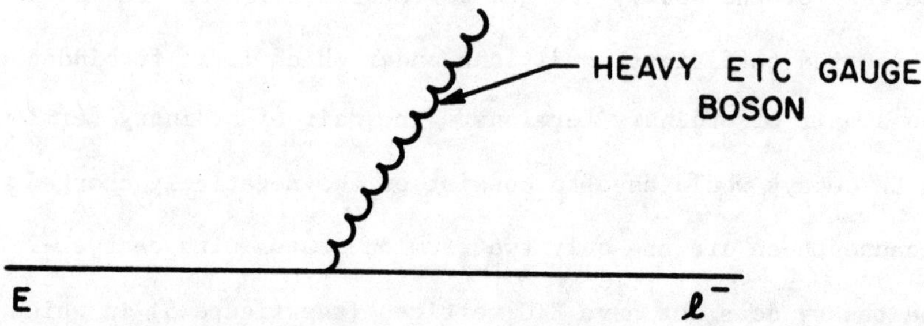

Fig. 5 A Technielectron E turns into a negative lepton ℓ^- by the emission of an ETC gauge boson. The absence of such graphs is sufficient to ensure that the ditechnielectron EE decays in higher order( ) via graphs like figure 4. These graphs are absent in theories where the anti-Techniquarks form left-handed electroweak doublets.

carry SU_2 (electroweak) quantum numbers. (The SU_2(EW) group breaks off at the GUT scale $\gtrsim 10^{14}$ GeV.)

The processes V.3 conserve the following numbers:

$$N_1 = \frac{3}{2} N_g + N_q - \frac{3}{2} N_L - \frac{1}{2} N_Q - \frac{1}{2} N_b$$

$$N_2 = -\frac{3}{2} N_g + N_\ell + \frac{1}{2} N_L - \frac{1}{2} N_Q + \frac{1}{2} N_b \qquad \text{(V.4)}$$

These conservation laws forbid the decay of any ditechnilepton ($N_1 = -3$) into two ordinary fermions ($N_1 \geq -2$).

The four members of the axionic family Π_f^{TF} (f=0,1,2,3) decay as follows: Π_o^{TF} likes to decay into two gluons, $\Pi_o^{TF} \rightarrow$ 2 gluons, with a width of very rough order $\sim \alpha_s^2 M(\Pi_o^{TF})$. Another possible important decay mode is $\Pi_o^{TF} \rightarrow \bar{q}_H + q_H$ where the subscript H denotes the heaviest quark for which this decay can go. Similarly Π_3^{TF} decays into $\Pi_3^{TF} \rightarrow \bar{q}_H + q_H$, also $\Pi_3^{T\pi} \rightarrow \gamma + \gamma$ and $\Pi_+^{TF} \rightarrow \bar{d}_H + u_H$. If Π_3^{TF} and Π_+^{TF} are sufficiently heavy then the modes $\Pi_3^{TF} \rightarrow \gamma + Z$ and $\Pi_+^{TF} \rightarrow W^+ + \gamma$ will be important.

Careful estimates of the decays of the various Pseudos will be presented in a future paper.

VI. PRODUCTION OF PSEUDOS

a) Pseudos Can Be Singly Produced

One of the interesting and useful features of Pseudos is that they can be singly produced. This may lead to their discovery in the nearer future. However, only a limited number of Pseudos have significant couplings to ordinary light quarks and leptons and have a good chance of being singly produced. These include the Coloured Techniπions, the Coloured Technieta, and the colour singlets Π_f^{TF} (f=0,1,2,3) which are the paraxion, axion, and the two charged partners of the axion. In figures 6 we show some graphs, most of which are Technivector meson dominance type graphs, by which these Pseudos can be singly produced.

The rest of the Pseudos, which includes all the Pseudos of non-vanishing triality and the dileptonics, can also be singly produced but with a cross-section which is quite small. The reason is again the fact that these Pseudos couple to fermions in proportion to the fermions' masses and therefore their couplings to ordinary light leptons are very small (see §V).

The computation of the single production cross sections for the various Pseudos is not free of subtleties due to the pseudo nature of Pseudos[(17)]. The order of magnitude of the cross sections shown in figures 6 is in the nanobarn range. Naive estimates can be obtained either by direct application of Technivector meson dominance (see figures 6) or by direct scaling up of the isomorphic process $e^+e^- \to \Pi^\circ\gamma$ cross section (figure 7) at corresponding energies and coupling constants.

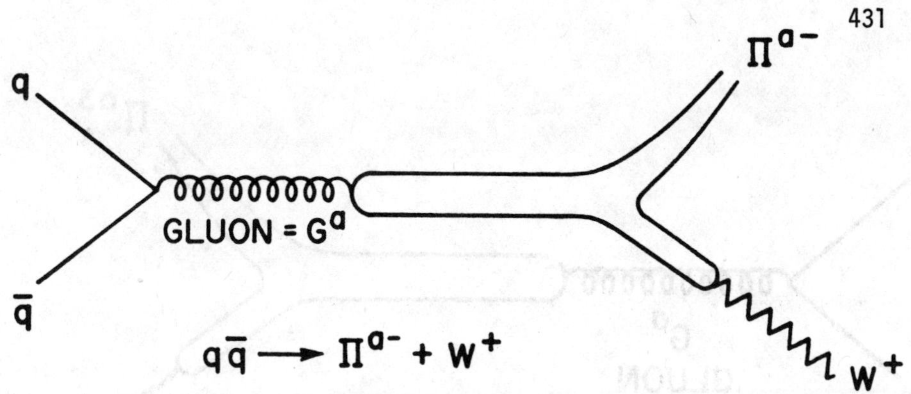
q
q̄
GLUON = G^a
Π^{a-}
W^+
$q\bar{q} \longrightarrow \Pi^{a-} + W^+$

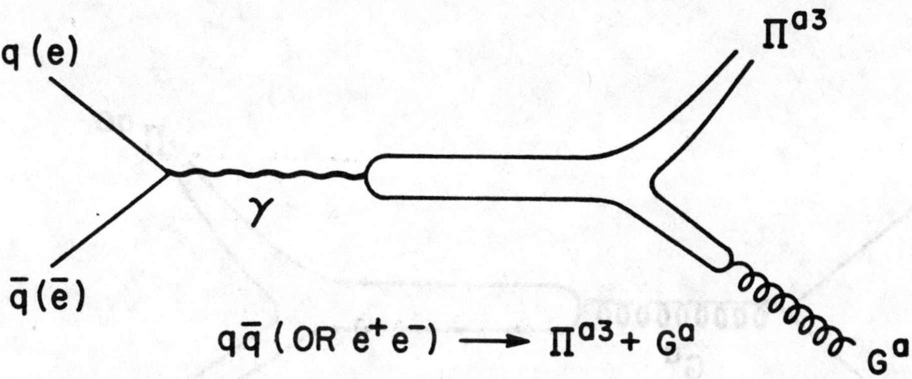
q (e)
q̄ (ē)
γ
Π^{a3}
G^a
$q\bar{q}$ (OR e^+e^-) $\longrightarrow \Pi^{a3} + G^a$

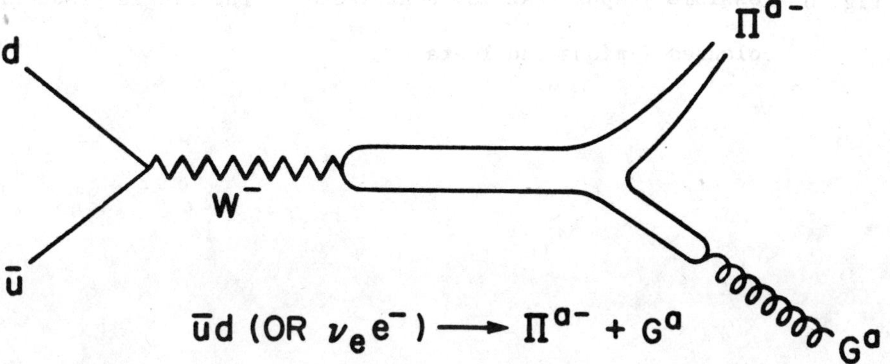
d
ū
W^-
Π^{a-}
G^a
$\bar{u}d$ (OR $\nu_e e^-$) $\longrightarrow \Pi^{a-} + G^a$

Figures 6

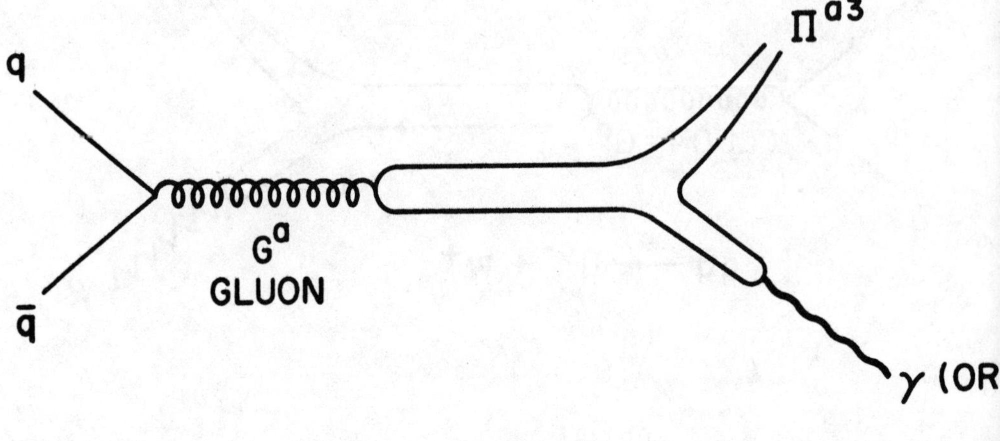

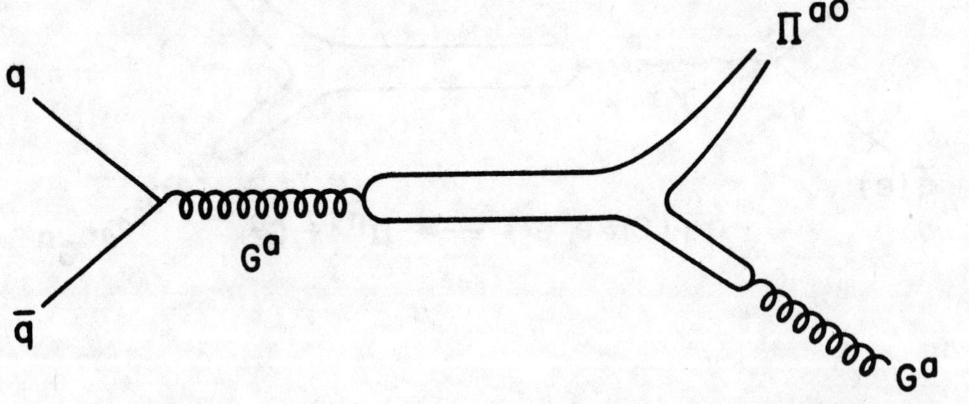

Fig. 6 Possible graphs that may contribute to the single production o: coloured T-πions and T-eta

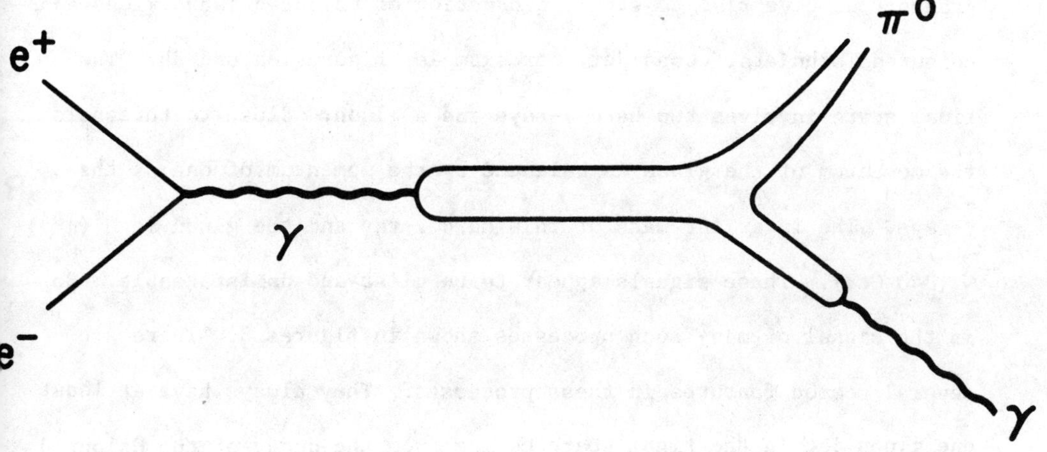

Fig. 7 A low energy vector meson dominance type graph which contributes to the single production of Π°. It is a scaled down version of the processes shown in figure 6.

When these singly produced Pseudos decay they give rise to several interesting final states involving jets (see figures 8 and 9). Many of these are quite exotic and should be clearly differentiable from ordinary types of QCD processes involving jets. In figures 8 we show some graphs that give rise to single production of coloured Techniπions and coloured Technieta. Consider, for example, figures 8a and 8b. The final state involves two hard γ-rays and a gluon. Close to threshold the momentum of the gluon is balanced by the momentum of one of the γ-rays. The invariant mass of this hard γ-ray and the gluon is $M^2(\Pi^{a3})$ ∿ (340 GeV). These signals appear to be clear and unmistakeable. So is the signal of many such processes shown in figures 8. There are several common features in these processes. They always have at least one gluon jet in the final state coming from the decay of the Coloured T-πions or T-eta. The momentum of this hadronic gluon jet is very often balanced by the momentum of leptons or a photon. For example, in figure 8c the momentum of this gluon is balanced by a lepton pair momentum, while in figures 8a,b,e it is balanced by a photon. The gluon jet combines with these leptons or photon to a system with invariant mass $M^2(\Pi) \sim (340\ \text{GeV})^2$. All such processes will contribute a significant number of events at large angles with respect to the original beam direction since they are associated with the opening of a new threshold.

Another feature of the single T-πion and T-eta production is the fact that these processes do not yield more than two electroweak vector bosons. Thus the final state never contains more than four quark jets or four leptons. If we see one photon or one lepton we cannot have more than two quark jets. This should be contrasted with what

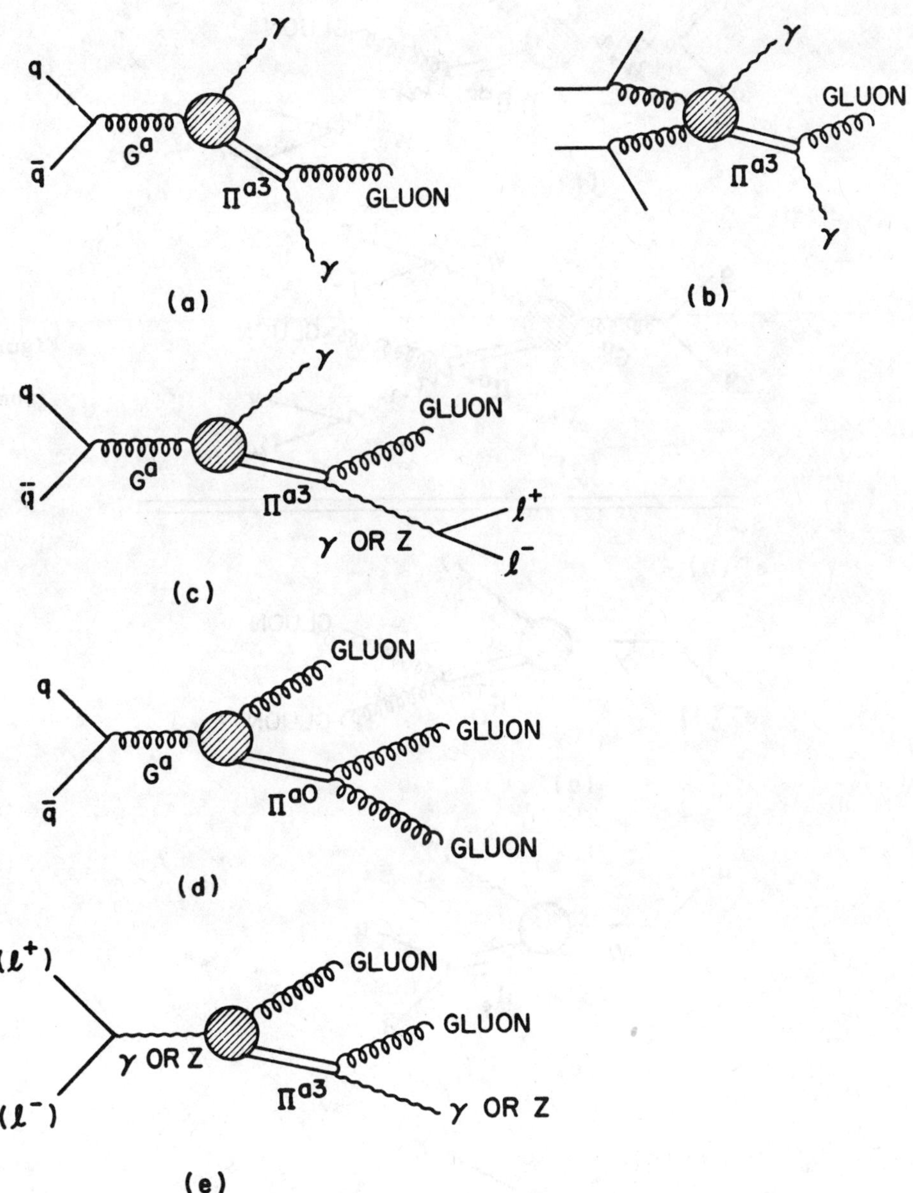

Fig. 8 Processes in which Coloured T-πions and T-eta are singly produced result in some interesting events containing jets. Many of them have clear and unmistakeable signatures. For example, figure 8a has a single gluon jet and two hard photons.

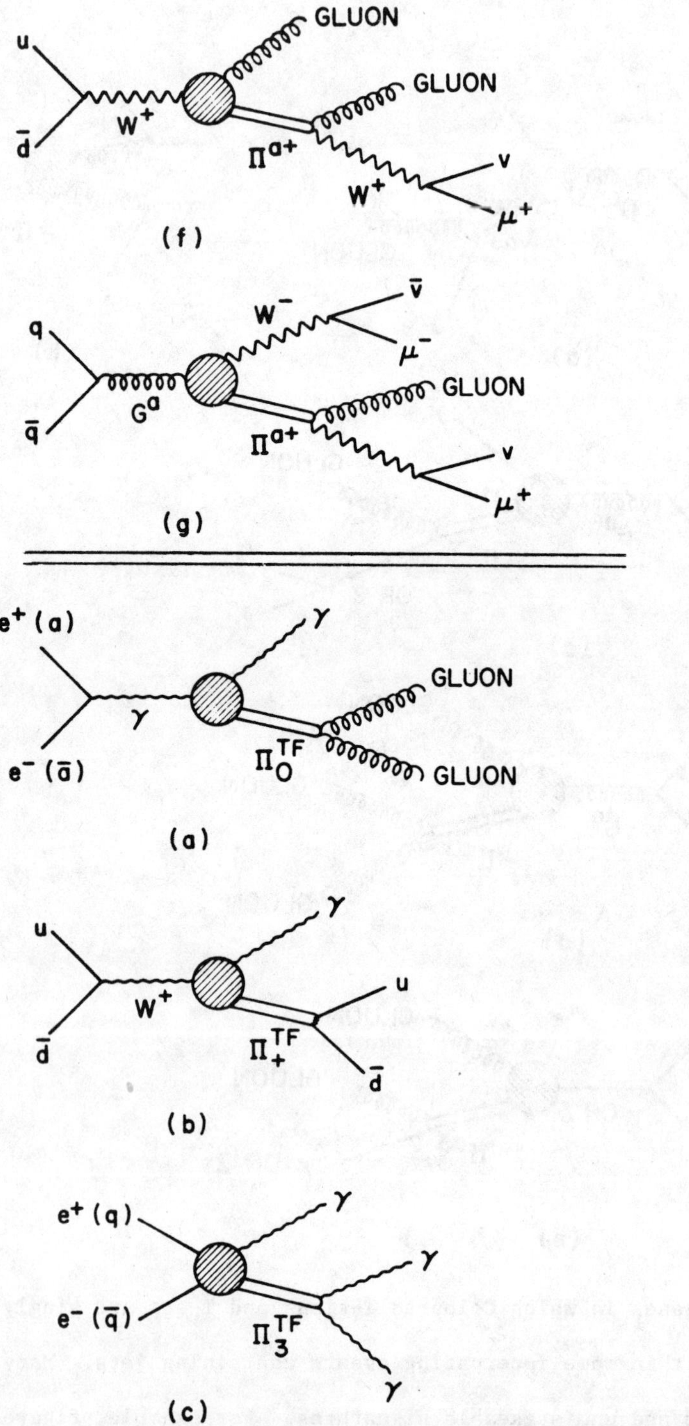

Figures 8

(cont.)

Fig. 9 Possible events arising from the single production of members of the axionic family.

happens when we reach a heavy quark threshold. The heavy quark and antiquark (with masses $\sim$ 150 GeV) decay into light quarks via the emission of several (at least six) $W^{\pm}$ bosons. Thus the final state is quite complicated and contains several ($\sim$ 14) quark jets and/or leptons.

Finally, note figures 9 where some possible graphs contributing to the single production of the Π_f^{TF} (f=0,1,2,3) are shown. These types of graphs are of great interest (if pseudo-unsuppressed)[17] the Π_f^{TF}'s can be relatively light (m $\sim$ tens of GeV) in a large class of models[2]. They are analogous to the single production of Higgs H, e.g., in $e^+e^- \to H + \gamma$. In fact, it will not be very easy to differerentiate this single Higgs production from the single axion Π_3^{TF} production $e^+e^- \to \Pi_3^{TF} + \gamma$ except , e.g., through details arising from the different parities of Π_3^{TF} and H.

b) Pair Production Of Pseudos And Emerging Jets

The possibility of pair producing Pseudos is to become viable in the not-so-distant future. For example, the FERMILAB TEVATRON machine ($\sim$ 1984) will be a $\bar{p}p$ machine running at a c.m. energy $\sim$ 1 TeV + 1 TeV or more. Since the probability for a quark to carry half of the proton momentum is not small we shall often have $q\bar{q}$ collisions at an energy $\sim$ 1 TeV. We shall similarly have often gluon-gluon collisions at an energy of several hundreds of GeV. Thus we expect to have the possibility of pair producing even the heaviest of the Pseudos shown in Tables Ia, b, and c (mass $\sim$ 300 GeV). In figures 10 we show a few processes that result in pair production of Pseudos. Again the decays of these Pseudos are going to give rise to

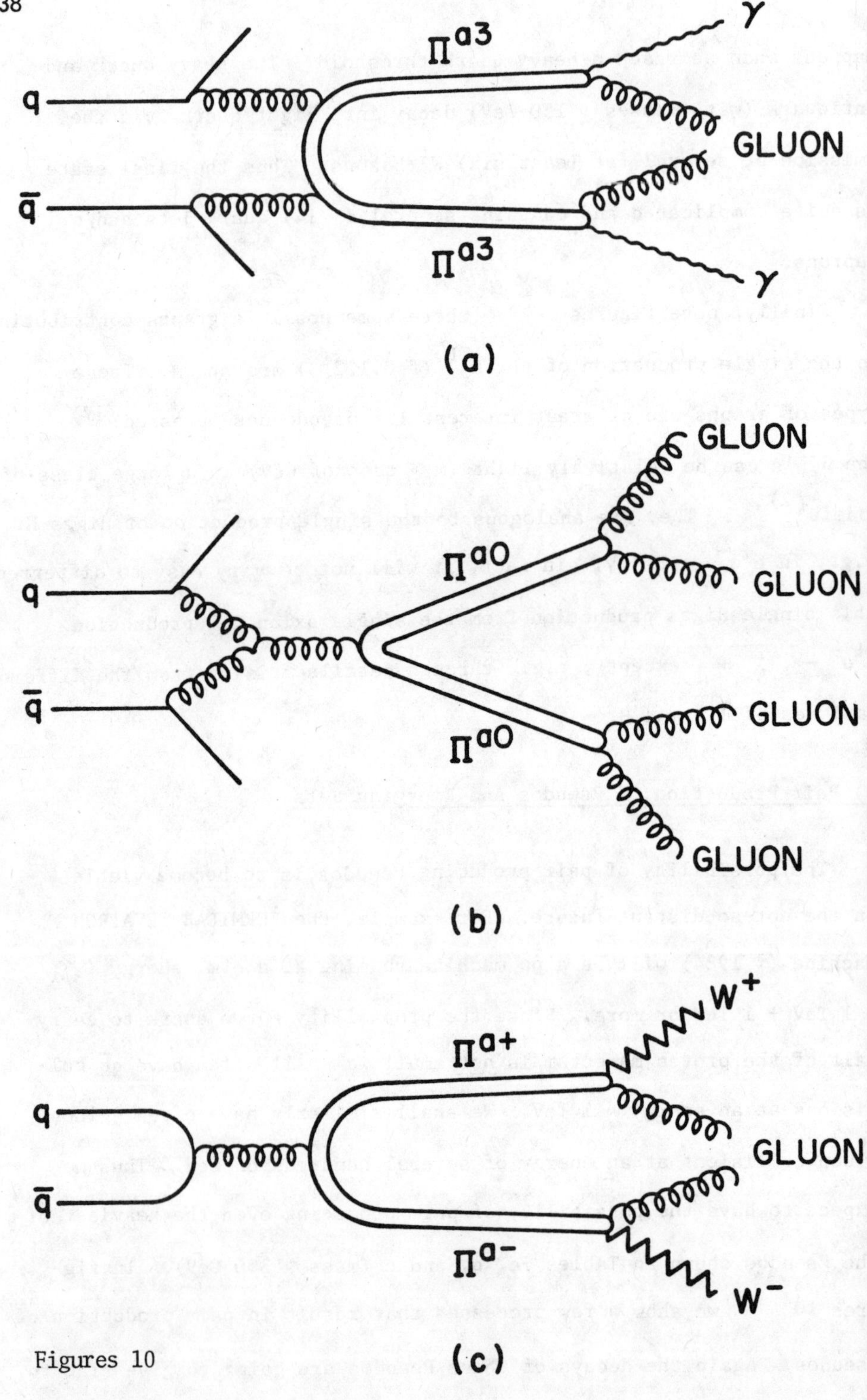
q
q̄
Π^{a3}
Π^{a3}
γ
GLUON
γ
(a)
GLUON
Π^{a0}
GLUON
GLUON
Π^{a0}
GLUON
(b)
W^+
Π^{a+}
GLUON
Π^{a-}
W^-
(c)

Figures 10

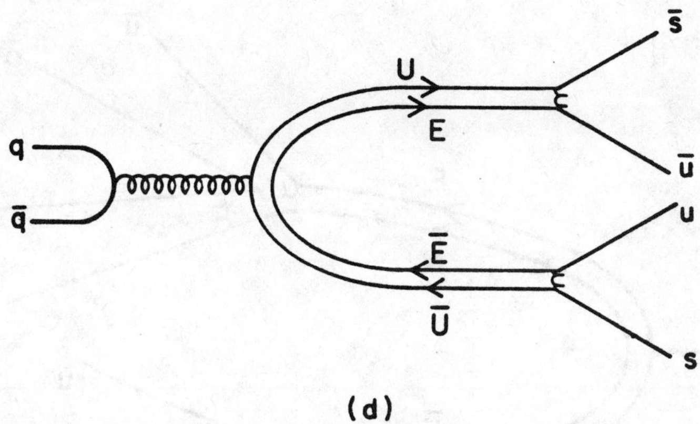

(d)

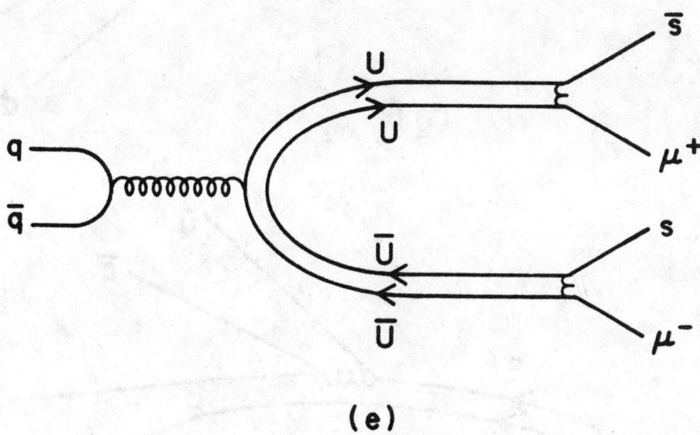

(e)

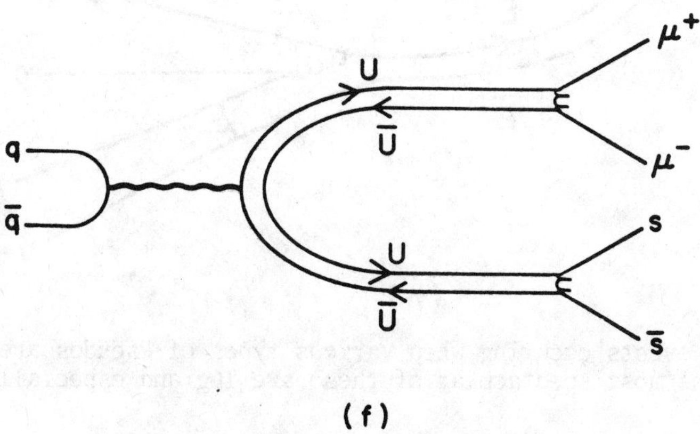

(f)

Figures 10

(cont.)

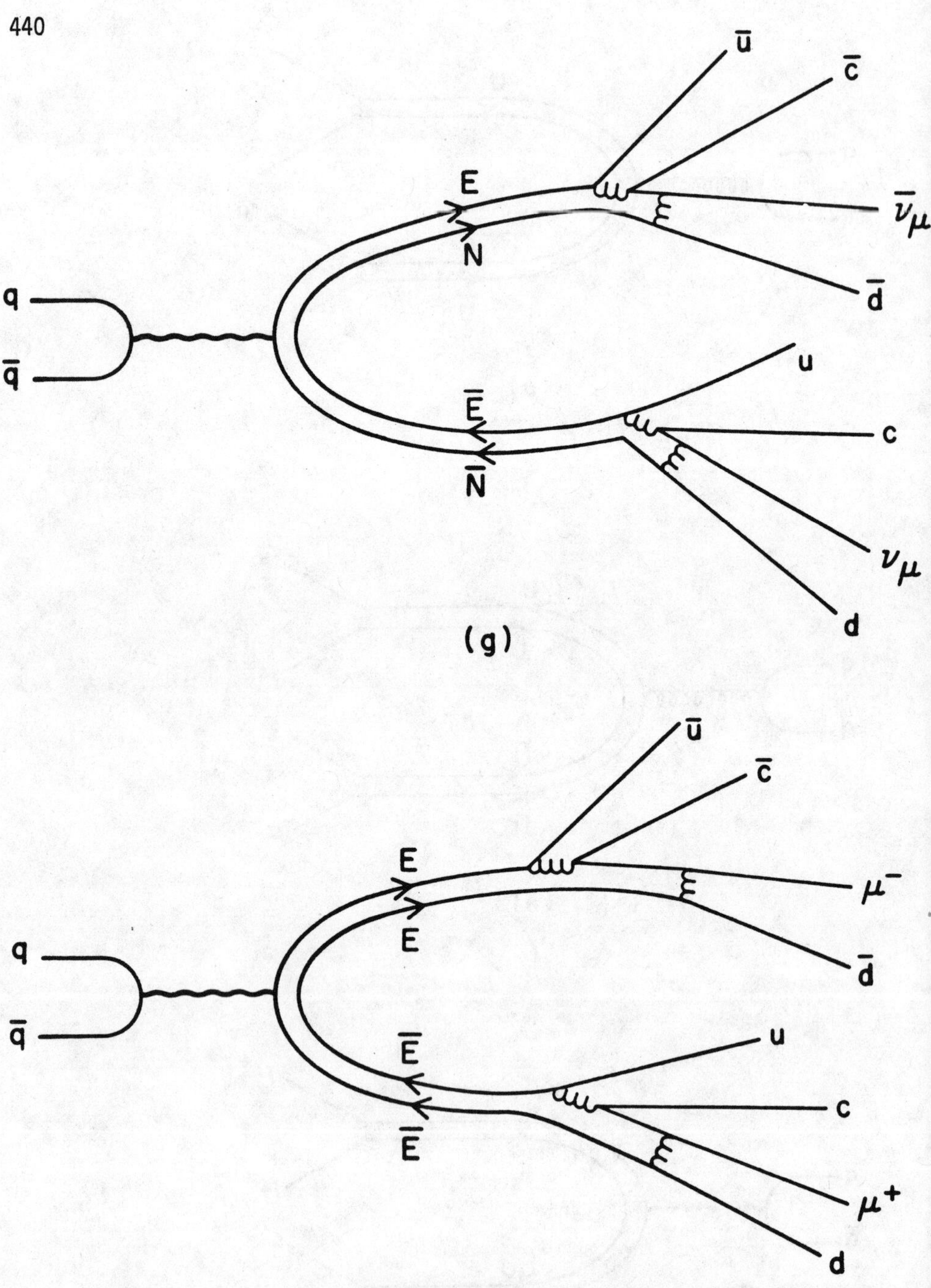

Fig. 10 Some jetful events occuring when various types of Pseudos are pair produced. The most spectacular of these are 10g and especially 10h.

several interesting final states containing jets, some with amusing and unmistakeable signatures. For example, in figure 10a we have an event with two gluon jets and two hard photons. Again, not far from threshold, each gluon's momentum is balanced by that of a hard γ-ray. The invariant masses of this γ-ray and the gluon combine to $m(\Pi)^2 \sim (300\ \text{GeV})^2$. In figure 10b we have a four gluon jet event. Many Pseudo pair productions lead to four-quark jets or two-quark jet and two lepton events shown in figures 10d,e,f. More amusing are the frequently occuring events of the type shown in figure 10e.

These events have two quark jets and two leptons. However the momentum of the quark jet is balanced by a lepton's momentum. Most amusing of all is the decay of the objects which, just like the EN of the Fahri-Susskind model, can only decay into a four-fermion final state because of fermion number conservation. These particles, when they are pair produced and decay, can yield spectacular jet events (see figures 10g,h). Note that because of their long lifetime these Pseudos may travel relatively long distances $\gtrsim$ mm before they decay.

VII. MISCELLANEOUS COMMENTS

This section is a compendium of remarks which did not naturally fit in the previous discussions.

a) The Spectrum Of ETC Scales

The mass m_f of an ordinary quark or lepton is given by graphs of the type shown in figure 11[1,2]. The magnitude of this mass is[1,2]

$$m_F \simeq G_{ETC} \langle \bar{Q}Q \rangle \tag{VII.1}$$

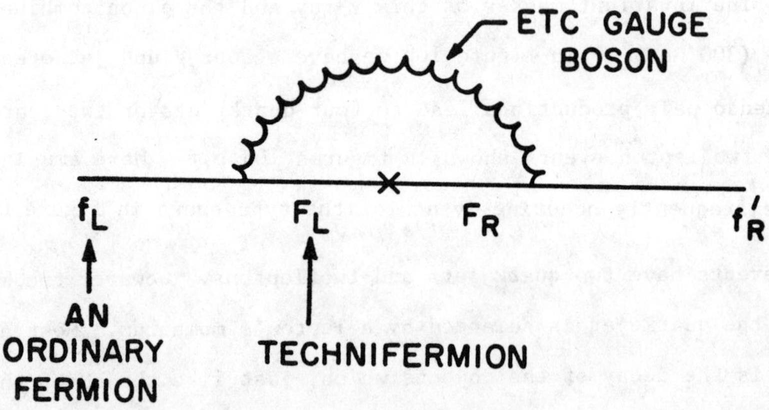

Fig. 11 The general type of graphs that contribute to the ordinary fermions' mass matrix (to lowest order). The cross x denotes the action of the techniconstituent mass ∿ 300 GeV.

Here G_{ETC} is an ETC effective Fermi coupling and $<\bar{Q}Q>$ is any one of the eight condensates of equation II.1 (i.e., no colour summation is implied). An approximate value for $<\bar{Q}Q>$ is given by assuming that the dimensionless ration $<\bar{Q}Q>/F^3_{T\pi}$ is the same as the corresponding ratio of ordinary SU_3 (colour) strong interactions. This does not take into account the fact that this ratio is probably group dependent and yields $<\bar{Q}Q>^{1/3} = (F_{T\pi}/f_\pi) <\bar{q}q>^{1/3} \simeq 317$ GeV. This relation together with VII.1 implies that for a quark or lepton of a mass of order of ~ 30 GeV[18] the corresponding ETC Fermi coupling satisfies

$$G_{ETC}^{-1/2} \simeq 1 \text{ TeV} \qquad \text{(VII.2)}$$

The mass of the ETC gauge boson mediating such a transition is expected to be of order of 1 TeV. This is because $\frac{g^2_{ETC}}{M^2_{ETC}} = G_{ETC}$ [2,19] and $g_{ETC} \sim 1$ at this scale. Thus these would be 1 TeV vector boson carrying the same Technicolour quantum numbers as Technifermions[20]. They bind with each other and with the Technifermions via the Technicolour forces and form new types of Technihadrons with masses of order of 1 TeV. Thus, e.g., there will exist ~ 1 TeV Technibaryon made out of such an ETC gauge boson and a Technifermion

On the other extreme in theories where the lightest quarks and leptons receive their mass by the lowest order graphs of figure 11 there must exist ETC gauge bosons with masses $G_{ETC}^{-1/2} \sim 250$ TeV.

b) The SU_3 (Colour) β-function In The Presence Of Several Coloured Pseudos

The large proliferation of coloured Pseudos (see Table I) has some interesting effects on the β-function of ordinary colour forces. These effects mainly arise at an energy region centered around 600 GeV which corresponds to twice the mass of the Pseudos that carry large colour (sextets and octets). Consider, for example, the b-function for energies between 600 GeV and a mass of order of the Techniρho mass $\gtrsim$ 1000 GeV. In this region the value of the b-function is expected to be dictated by the ordinary quarks, gluons, and the Coloured Pseudos. The formula for SU_3 (colour) is:

$$-b = 11 - \frac{2}{3} N_F(3) - \frac{1}{6} \sum_s T(R_s)$$

$$= 11 - \frac{2}{3} N_F(3) - \frac{1}{12} N_s(3) - \frac{1}{2} N_s(8) - \frac{5}{12} N_s(6) \qquad \text{(VII.3)}$$

Here $N_F(3)$ is the number of 4-component Dirac fermion triplets, and $N_s(3)$, $N_s(8)$, and $N_s(6)$ are the numbers of Real Scalar Triplets, Octets, and Sextets, respectively. Consider now a model with a real Technicolour group with an antisymmetric invariant symbol. The Pseudo content of such a model is shown in Tables Ia and Ib. If we assume three families of ordinary quarks, then the b becomes equal to $b \sim 2\frac{1}{3}$ instead of the $b \sim 7$ which we would have in the absence of Pseudos. For four quark families $b \sim 1$ instead of $b \sim 5\frac{2}{3}$.

In models where the Techniquarks belong to a real representation with a symmetric invariant symbol the colour contents of Pseudos are slightly different (see Table I.c). In particular the number of sextets increases from two to six. This changes the value of b from $2\frac{1}{3}$ to 1 for three quark families. For four quark families b becomes negative $b \simeq -1/3$. At energies quite a bit higher than the Techniρho mass > 1 TeV the β-function receives contributions only from the coloured

T-quarks, the ordinary quarks and gluons, and also from the lightest coloured ETC vector bosons whose masses are $\sim$ 1 TeV (see §VII.a).

c) Stability Of Technihadrons

The question of whether there exists a Technifermion number which is conserved (except at the superunification[4,8,5] scale $\gtrsim 10^{14}$ GeV) is an interesting one. The answer to this question is model dependent and no simple criterion has yet been formulated. It does not, for example, have to do with the technicolour group *per se*. To prove this, consider an SU_5 (ETC) which breaks in the pattern SU_5(ETC) $\longrightarrow$ SU_2(TC) x SU_3(C). The fermion content, say, consists of SU_5 quintets which break down to T-quark doublets and quark colour triplets. Then the T-Baryons consist of two T-Quarks and are stable because they carry baryon number equal to 2/3 and no colour singlet state light quark state carries fermion number = 2/3. In contrast, if we enlarge this model in straight-forward ways[11,12,15] to include leptons, always keeping the T-Colour group SU_2, then the T-Baryon can in general decay into fermion pairs[11,12,15]. Thus the stability of T-baryons does not have to do only with the TC group *per se* but also with how the TC group fits with the rest of the symmetries.

In general we expect the theories with stable T-baryon to be problematic in view of the fact that they would naively be expected to predict comparable numbers of Technibaryons and baryons in the universe[21]. This is because quarks and T-Quarks become unified at energies of order of 10-100 TeV which is much earlier than the energies of $\gtrsim 10^{14}$ GeV[4,5,8] at which the fermion numbers of the universe were created[21].

d) Horizontal Transitions

An interesting feature of the ETC models is the necessity of unifying all of the quark and lepton flavours at a scale of $\lesssim$ 100 TeV. This has to be done in order that quarks and leptons receive their masses without the appearence of extremely light Pseudos.

This early unification has several important consequences. It implies the existence of new horizontal currents which cause transitions between quark flavours, leptons, and also between quarks and leptons. Many of these horizontal currents can be dangerous and screen out several models because they mediate rare transitions at unacceptable levels. these will be discussed in a future paper(22). A possibly interesting and novel feature of ETC models which we want to just mention is the occurence of direct quark-lepton transitions. These can happen via the emission of an ETC boson (or a coloured Pseudo). They will contribute several interesting effects in the future $p\bar{p}$, e^+e^-, and especially in the possible ep machines(22). These direct $q \leftarrow\rightarrow \ell$ transitions at sufficiently large energies are going to be more important than the direct electroweak annihilation graphs which contribute to the Drell-Yan type processes $q\bar{q} \leftarrow\rightarrow \ell\bar{\ell}$. This is due to the obvious fact that the direct annihilation cross sections die out as $\sim \frac{\alpha^2}{s}$ while the t-exchange direct q-ℓ transitions do not. Also, the angular dependences will be different in the two cases. In the high energy ep machines the direct q-ℓ transition will yield a quark jet in the original e-direction and vice versa.

e) R

A quantity of very futuristic interest is $R(\ell^+\ell^- \to \text{anything})$ (ℓ = lepton) at the TeV range. Let us for simplicity discuss $R(\ell^+\ell^- \to \gamma \to \text{any})$. The contribution ΔR_{ps} to this R coming from the Pseudos of Tables Ia and Ib is $\Delta R_{ps} \simeq 9\frac{1}{3}$. This contribution is effective at energies between 600 GeV and ∿ TeV. At higher energies the contribution ΔR_{TQ} to R comes mainly from Techniquarks and also the lighter ETC vector bosons. The contribution of the T-quarks is $\Delta R_{TQ} \simeq \frac{8}{3} N$ where N is the size of the T-colour representation to which the T-fermions belong. Thus for $N \geq 4$ the contribution of the T-Quarks is bigger than that of the Pseudos.

f) The Lightest Pseudos

It is important and urgent to study in detail the properties of the colourless Pseudos in various types of large classes of models. For they are expected to have masses between few GeV and $\lesssim$ 100 GeV (see § IV) which are smaller than the smallest expected masses of coloured Pseudos ∿ 200 GeV. They should be seen in the $\bar{p}p$ CERN machine (1986?) and possibly in the upcoming lepton machines. Unfortunately, their dominant properties are much more model dependent than those of the heavier Coloured Pseudos (especially the colour octets). This is because their dominant properties strongly depend on the heavy ETC sector which is both model dependent and poorly understood.

The colourless Pseudos are the four axionic Pseudos Π_f^{TF} (f=0,1,2,3) and, more amusingly, the three possible ditechnileptons EE, EN, and

NN. The latter Pseudos are likely to be very fascinating. It is, e.g., quite possible that baryon and lepton conservation laws allow them to decay only into final states involving at least four ordinary fermions. In fact, as discussed before (§V), the doubly charged dielectron EE cannot decay into a pair of ordinary fermions if the ETC vertices of figure 11 are not present in a theory. In this case it would decay into four fermions and would be long enough lived to leave visible tracks. such a particle would yield spectacular jetful events.

VIII. SUMMARY

In this paper we made some general observations on some likely features of the scalarless models with Extended Technicolour. We mainly discussed the spectrum and gross properties of the Pseudos that occur in these models if we assume the existence of one family of Technifermions. This assumption was motivated by Superunified GUT type models. Obviously none of the features that we mentioned are inescapabl Several of the possible consequences mentioned would constitute a relatively low energy evidence of the ETC ideas. In particular, finding Pseudos with the right combinations of strong-electroweak quantum numbers in roughly the right mass range and with the right mass pattern would be a suggestive evidence for ETC. It is unnatural and unlikely to manufacture such a pseudospectrum in models with elementary scalars.

An important feature is the possible existence of several coloured Pseudos. In the future we are going to have available high energy $\bar{p}p$ machines of hopefully decent luminocities. These machines will also be relatively luminous gluon-gluon machines. This will greatly facilitate the production and study of new coloured states. The dominant process at high energies will be 2-gluon collisions a la the 2-photon collisions of Brodsky, Takahashi, and Terazawa[(23)]. At still higher energies (say, the possible 20 + 20 TeV machines) we will have a scaled up replica of ordinary strong interaction physics, e.g., almost constant cross sections, multiperipheral processes, de-coupling of the triple technipomeron vertex, etc.

REFERENCES

1) S. Dimopoulos and L. Susskind, Nucl. Phys. B 155 (1979) 237

2) E. Eichten and K. Lane, "Dynamical Breaking of Weak Interaction Symmetries," Harvard Preprint HUTP 79/A002 (1979)

3) S. Dimopoulos and L. Susskind, "A Technicoloured Solution to the Strong CP Problem," Columbia-Stanford Preprint (February 1979); W. Fishler, "The CP Problem in Two Dimensions," Los Alamos Preprint 79-0263

4) H. Georgi and S.L. Glashow, Phys. Rev. Lett. 32 (1974) 438; J. Patti and A. Salam, Phys. Rev. D8 (1973) 1240; D10 (1974) 275; H. Fritzch and P. Minkowski, Ann. Phys. 93 (1975) 193; F. Hadjioannou, Univ. of Athens preprint (1978); H. Georgi and D.V. Nanopoulos, Harvard preprints HUTP-78/A039 and 79/A001

5) A.J. Buras, J. Ellis, M.K. Gaillard, D.V. Nanopoulos, Nucl. Phys. B 135 (1978) 66

6) K. Wilson, unpublished

7) L. Susskind, "Dynamics of Spontaneous Symmetry breaking in the Weinberg-Salam Theory," SLAC preprint (May, 1978); S. Weinberg, D 19 (1978) 1277

8) H. Georgi, H.K. Quinn, S. Weinberg, Phys. Rev. Lett. 33, 451

9) S. Weinberg, Phys. Rev. D 13 (1976) 974

10) S. Weinberg, Phys. Rev. Lett. 19 (1967) 1264; A. Salam in Elementary Particle Physics, ed. N. Svartholm (Almquist and Wiksells, Stockholm, 1968) p. 367

11) E. Fahri and L. Susskind, "A Technicoloured GUT" SLAC Pub. 2361 (1979) and unpublished

12) S. Raby, unpublished

13) In some quasirealistic models (see ref. 11 and 12) with three or four families of ordinary fermions one gets more than one family of Technifermions. This has the effect of decreasing the masses and increasing the number of Pseudos.

14) R. Peccei and H. Quinn, Phys. Rev. Lett. 38 (1977) 1440; Stanford University Report No. ITP 572, 1977; S. Weinberg, Phys. Rev. Lett. 40 (1978) 223; F. Wilczek, Phys. Rev. Lett. 40 (1978) 279

15) S. Dimopoulos and L. Susskind, unpublished

16) Much of the model dependence arises becuase ordinary fermions may either all get their masses from Technifermions via ETC guage bosons of varying masses (see figure 11) or the light fermions get their masses from heavy ordinary fermions via not so heavy Technicolourless ETC gauge bosons.

17) We refer to the various soft decoupling theorems as well as the fact that some decay and production graphs can only go through anomalies (a la $\Pi^\circ \rightarrow \gamma\gamma$).

18) J. Bjorken, "Speculations on the Pattern of Quark and Lepton Masses," SLAC Pub. 2195 (1978)

19) This can be seen directly by Fierzing the graph of figure 11 and equation VII.1.

20) The existence of $\sim$ 1 TeV Technicolourless (but possibly colourful and charged) vector bosons is also possible. They can cause transitions among the heaviest ordinary fermions or among technifermions.

21) M. Yoshimura, Phys. Rev. Lett. **41** (1978) 381; S. Dimopoulos and L. Susskind, Phys. Rev. **D18** (1978) 4500; D. Toussaint, S. Treiman, F. Wilzcek and A. Zee, Phys. Rev. **D19** (1979) 1036; S. Weinberg, Cosmological production of baryons, Harvard preprint HUTP-78/A040; S. Dimopoulos and L. Susskind, Baryon Asymmetry in the very early universe, Stanford University Report No. ITP 616; J. Ellis, M.K. Gaillard and D.V. Nanopoulos, CERN preprint (November 1978)

22) S. Dimopoulos, S. Raby and L. Susskind, to be published.

23) For a review see H. Terezawa, Rev. Mod. Phys. **45** (1973) 615.

AIP Conference Proceedings

No.	Title	L.C. Number	ISBN
No.1	Feedback and Dynamic Control of Plasmas	70-141596	0-88318-100-2
No.2	Particles and Fields - 1971 (Rochester)	71-184662	0-88318-101-0
No.3	Thermal Expansion - 1971 (Corning)	72-76970	0-88318-102-9
No.4	Superconductivity in d-and f-Band Metals (Rochester, 1971)	74-18879	0-88318-103-7
No.5	Magnetism and Magnetic Materials - 1971 (2 parts) (Chicago)	59-2468	0-88318-104-5
No.6	Particle Physics (Irvine, 1971)	72-81239	0-88318-105-3
No.7	Exploring the History of Nuclear Physics	72-81883	0-88318-106-1
No.8	Experimental Meson Spectroscopy - 1972	72-88226	0-88318-107-X
No.9	Cyclotrons - 1972 (Vancouver)	72-92798	0-88318-108-8
No.10	Magnetism and Magnetic Materials - 1972	72-623469	0-88318-109-6
No.11	Transport Phenomena - 1973 (Brown University Conference)	73-80682	0-88318-110-X
No.12	Experiments on High Energy Particle Collisions - 1973 (Vanderbilt Conference)	73-81705	0-88318-111-8
No.13	π-π Scattering - 1973 (Tallahassee Conference)	73-81704	0-88318-112-6
No.14	Particles and Fields - 1973 (APS/DPF Berkeley)	73-91923	0-88318-113-4
No.15	High Energy Collisions - 1973 (Stony Brook)	73-92324	0-88318-114-2
No.16	Causality and Physical Theories (Wayne State University, 1973)	73-93420	0-88318-115-0
No.17	Thermal Expansion - 1973 (lake of the Ozarks)	73-94415	0-88318-116-9
No.18	Magnetism and Magnetic Materials - 1973 (2 parts) (Boston)	59-2468	0-88318-117-7
No.19	Physics and the Energy Problem - 1974 (APS Chicago)	73-94416	0-88318-118-5
No.20	Tetrahedrally Bonded Amorphous Semiconductors (Yorktown Heights, 1974)	74-80145	0-88318-119-3
No.21	Experimental Meson Spectroscopy - 1974 (Boston)	74-82628	0-88318-120-7
No.22	Neutrinos - 1974 (Philadelphia)	74-82413	0-88318-121-5
No.23	Particles and Fields - 1974 (APS/DPF Williamsburg)	74-27575	0-88318-122-3

No.24	Magnetism and Magnetic Materials - 1974 (20th Annual Conference, San Francisco)	75-2647	0-88318-123-1
No.25	Efficient Use of Energy (The APS Studies on the Technical Aspects of the More Efficient Use of Energy)	75-18227	0-88318-124-X
No.26	High-Energy Physics and Nuclear Structure - 1975 (Santa Fe and Los Alamos)	75-26411	0-88318-125-8
No.27	Topics in Statistical Mechanics and Biophysics: A Memorial to Julius L. Jackson (Wayne State University, 1975)	75-36309	0-88318-126-6
No.28	Physics and Our World: A Symposium in Honor of Victor F. Weisskopf (M.I.T., 1974)	76-7207	0-88318-127-4
No.29	Magnetism and Magnetic Materials - 1975 (21st Annual Conference, Philadelphia)	76-10931	0-88318-128-2
No.30	Particle Searches and Discoveries - 1976 (Vanderbilt Conference)	76-19949	0-88318-129-0
No.31	Structure and Excitations of Amorphous Solids (Williamsburg, VA., 1976)	76-22279	0-88318-130-4
No.32	Materials Technology - 1975 (APS New York Meeting)	76-27967	0-88318-131-2
No.33	Meson-Nuclear Physics - 1976 (Carnegie-Mellon Conference)	76-26811	0-88318-132-0
No.34	Magnetism and Magnetic Materials - 1976 (Joint MMM-Intermag Conference, Pittsburgh)	76-47106	0-88318-133-9
No.35	High Energy Physics with Polarized Beams and Targets (Argonne, 1976)	76-50181	0-88318-134-7
No.36	Momentum Wave Functions - 1976 (Indiana University)	77-82145	0-88318-135-5
No.37	Weak Interaction Physics - 1977 (Indiana University)	77-83344	0-88318-136-3
No.38	Workshop on New Directions in Mossbauer Spectroscopy (Argonne, 1977)	77-90635	0-88318-137-1
No.39	Physics Careers, Employment and Education (Penn State, 1977)	77-94053	0-88318-138-X
No.40	Electrical Transport and Optical Properties of Inhomogeneous Media (Ohio State University, 1977)	78-54319	0-88318-139-8
No.41	Nucleon-Nucleon Interactions - 1977 (Vancouver)	78-54249	0-88318-140-1
No.42	Higher Energy Polarized Proton Beams (Ann Arbor, 1977)	78-55682	0-88318-141-X
No.43	Particles and Fields - 1977 (APS/DPF, Argonne)	78-55683	0-88318-142-8
No.44	Future Trends in Superconductive Electronics (Charlottesville, 1978)	77-9240	0-88318-143-6

No.45	New Results in High Energy Physics - 1978 (Vanderbilt Conference)	78-67196	0-88318-144-4
No.46	Topics in Nonlinear Dynamics (La Jolla Institute)	78-057870	0-88318-145-2
No.47	Clustering Aspects of Nuclear Structure and Nuclear Reactions (Winnepeg, 1978)	78-64942	0-88318-146-0
No.48	Current Trends in the Theory of Fields (Tallahassee, 1978)	78-72948	0-88318-147-9
No.49	Cosmic Rays and Particle Physics - 1978 (Bartol Conference)	79-50489	0-88318-148-7
No.50	Laser-Solid Interactions and Laser Processing - 1978 (Boston)	79-51564	0-88318-149-5
No.51	High Energy Physics with Polarized Beams and Polarized Targets (Argonne, 1978)	79-64565	0-88318-150-9
No.52	Long-Distance Neutrino Detection - 1978 (C.L. Cowan Memorial Symposium)	79-52078	0-88318-151-7
No.53	Modulated Structures - 1979 (Kailua Kona, Hawaii)	79-53846	0-88318-152-5
No.54	Meson-Nuclear Physics - 1979 (Houston)	79-53978	0-88318-153-3
No.55	Quantum Chromodynamics (La Jolla, 1978)	79-54969	0-88318-154-1
No.56	Particle Acceleration Mechanisms in Astrophysics (La Jolla, 1979)	79-55844	0-88318-155-X
No. 57	Nonlinear Dynamics and the Beam-Beam Interaction (Brookhaven, 1979)	79-57341	0-88318-156-8
No. 58	Inhomogeneous Superconductors - 1979 (Berkeley Springs, W.V.)	79-57620	0-88318-157-6
No. 59	Particles and Fields - 1979 (APS/DPF Montreal)	80-66631	0-88318-158-4